Lecture Notes in Physics

The Lecture Notes in Physics

The series Lecture Notes in Physics (LNP), founded in 1969, reports new developments in physics research and teaching – quickly and informally, but with a high quality and the explicit aim to summarize and communicate current knowledge in an accessible way. Books published in this series are conceived as bridging material between advanced graduate textbooks and the forefront of research to serve the following purposes:

- to be a compact and modern up-to-date source of reference on a well-defined topic;
- to serve as an accessible introduction to the field to postgraduate students and nonspecialist researchers from related areas;
- to be a source of advanced teaching material for specialized seminars, courses and schools.

Both monographs and multi-author volumes will be considered for publication. Edited volumes should, however, consist of a very limited number of contributions only. Proceedings will not be considered for LNP.

Volumes published in LNP are disseminated both in print and in electronic formats, the electronic archive is available at springerlink.com. The series content is indexed, abstracted and referenced by many abstracting and information services, bibliographic networks, subscription agencies, library networks, and consortia.

Proposals should be sent to a member of the Editorial Board, or directly to the managing editor at Springer:

Dr. Christian Caron
Springer Heidelberg
Physics Editorial Department I
Tiergartenstrasse 17
69121 Heidelberg/Germany
christian.caron@springer.com

Jan Dereziński   Heinz Siedentop   (Eds.)

Large Coulomb Systems

Lecture Notes on Mathematical Aspects of QED

Springer

Editors

Professor Dr. Jan Dereziński
Department of Mathematical
Methods in Physics
Warsaw University
ul. Hoża 74
00-682 Warsaw, Poland
E-mail: jan.derezinski@fuw.edu.pl

Professor Dr. Heinz Siedentop
Mathematisches Institut
Ludwig-Maximilians Universität München
Theresienstraße 39
80333 München, Germany
E-mail: h.s@lmu.de

J. Dereziński and H. Siedentop, *Large Coulomb Systems*,
Lect. Notes Phys. 695 (Springer, Berlin Heidelberg 2006), DOI 10.1007/b11607427

ISSN 0075-8450
ISBN 978-3-642-06901-7 e-ISBN 978-3-540-32579-6

Springer is a part of Springer Science+Business Media
springer.com

Softcover reprint of the hardcover 1st edition 2006

Cover design: *design & production* GmbH, Heidelberg

Preface

Since the discovery of its fundamentals by Heisenberg, Born, Jordan, Schrödinger, and Dirac, quantum mechanics turned out to be one of the most successsful theories of modern science. One of its cornerstones, the many-body Schrödinger equation with Coulomb interactions, was applied successfully to explain various properties of matter, such as the properties of light atoms.

In addition to its triumph as a physical theory, quantum mechanics has been mathematically formulated and many of its most basic statements have been proven, e.g., the self-adjointness of atomic and molecular Hamiltonians, asymptotic completeness, and the stability of matter. Various effective one-particle models – such as the Thomas-Fermi and the Hartree-Fock theory – were shown to be limiting models of fully interacting Coulomb systems.

Soon after its discovery, however, physicists realized that it is difficult to formulate a consistent and relativistic theory of matter interacting with the electromagnetic field. These difficulties have been skillfully treated by the great masters of physics belonging to the middle of the last century, e.g., Feynman, Dyson, Schwinger and Tomonaga. In fact, quantum electrodynamics (QED), the theory developed by them, has an admirable quantitative predictive power.

Nevertheless, this contrasts with the mathematical status of QED: its consistent mathematical formulation is unknown. In fact, not even a generally acknowledged multiparticle theory of relativistic electrons is available.

In recent years a renewed interest in these fundamental questions arose. Even if a fully satisfactory mathematical formulation of QED is so far unavailable, various effective models based on QED are within the reach of mathematical methods and generate numerous new and interesting results.

One of the activities in the field is the IHP-network "Analysis and Quantum" financed by the European Union. To introduce young postdocs into the field, it organized the summer school "Large Quantum Systems – QED" in Nordfjordeid, Norway, August 11–18, 2003. The present collection of reviews is based on the lecture notes of this school. In addition, one of the reviews is based on lectures from the summer school "Quantum Field Theory – from a Hamiltonian point of view" held in Sandbjerg Manor, Denmark, 2–9, August 2000, which was organized by the EU Network "PDE and Applications Quantum Mechanics".

We hope that this collection of reviews will be well received by all those interested in these developments.

Warsaw and Munich
November 2005

Jan Dereziński
Heinz Siedentop

Contents

List of Contributors

V. Bach
FB Mathematik und Informatik
Johannes Gutenberg-Universität
D-55099 Mainz
vbach@mathematik.uni-mainz.de

S. De Bièvre
UFR de Mathématiques
et UMR P. Painlevé
Université des Sciences
et Technologies de Lille
59655 Villeneuve d'Ascq
Cedex France
Stephan.De-Bievre
@math.univ-lille1.fr

J. Dereziński
Department of Mathematical
Methods in Physics
Warsaw University
Hoża 74, 00-682, Warszawa, Poland
jan.derezinski@fuw.edu.pl

M. Griesemer
Department of Mathematics
University of Alabama at
Birmingham
Birmingham AL 35294, USA
marcel@math.uab.edu

V. Jakšić
Department of Mathematics and
Statistics
McGill University
805 Sherbrooke Street West
Montreal, QC, H3A 2K6, Canada

E. Kritchevski
Department of Mathematics and
Statistics
McGill University
805 Sherbrooke Street West
Montreal, QC, H3A 2K6, Canada

C.-A. Pillet
CPT-CNRS, UMR 6207
Université de Toulon, B.P. 20132
F-83957 La Garde Cedex, France

R. Seiringer
Department of Physics
Jadwin Hall Princeton University
P.O. Box 708
Princeton NJ 08544, USA
rseiring@math.princeton.edu

V.M. Shabaev
Department of Physics
St. Petersburg State University
Oulianovskaya 1, Petrodvorets
198504, St.Petersburg, Russia
shabaev@pcqnt1.phys.spbu.ru

H. Siedentop
Mathematisches Institut
Ludwig-Maximilians Universität
München
Theresienstraße 39
80333 München, Germany
h.s@lmu.de

A Tutorial Approach to the Renormalization Group and the Smooth Feshbach Map

V. Bach

FB Mathematik und Informatik; Johannes Gutenberg-Universität; 55099 Mainz
vbach@mathematik.uni-mainz.de

1 Introduction

A new method of spectral analysis of Hamiltonians deriving from quantum field theoretic models has been defined by the *Feshbach Map* in [2–5] and, more recently, by the *Smooth Feshbach Map* in [1]. The main goals of those papers were the following:

- The introduction of an appropriate *Banach space* $\mathcal{W}$ *of operators* (in physics often referred to as the "space of theories"),
- which contains a small ball $\mathcal{D} \subset \mathcal{W}$ that is the domain of definition of
- the *renormalization group map* $\mathcal{R}_\rho : \mathcal{D} \to \mathcal{D}$ that assigns to each Hamiltonian operator $H \in \mathcal{D}$ a new, renormalized operator $\mathcal{R}_\rho(H)$ in $\mathcal{D}$.
- The first key property of $\mathcal{R}_\rho$ is its *isospectrality* it inherits from the Smooth Feshbach Map. It guarantees that the spectral properties of H are preserved under the application of $\mathcal{R}_\rho$.
- The second key property of $\mathcal{R}_\rho$ is its *codimension-1 contractivity* on $\mathcal{D}$. This contractivity insures that the iterative application of $\mathcal{R}_\rho$ generates a (time-discrete) dynamical system on $\mathcal{D}$ with a fixed point manifold of dimension 1.
- It turns out that the limiting operator $H_\infty(H) := \lim_{n\to\infty} \mathcal{R}_\rho^n(H)$ can be explicitly diagonalized. By the isospectrality, its spectral properties are the same as those of H, and the goal of the spectral analysis of H is achieved.

V. Bach: *A Tutorial Approach to the Renormalization Group and the Smooth Feshbach Map*, Lect. Notes Phys. **695**, 1–14 (2006)
www.springerlink.com

The purpose of the present notes is to illustrate the character of the renormalization group map $\mathcal{R}_\rho$. More specifically, We present a concrete computation of some terms (in an infinite series of terms) contributing to the image $\mathcal{R}_\rho(H)$ of H under the renormalization group map $\mathcal{R}_\rho$. Although not a mathematically rigorous proof, this sample computation should clarify two important points:

- Three decades after the introduction of the renormalization group by Wilson there exists a diversity of concrete implementations. Yet, they are all based on the same *renormalization group idea.*
- The verification that the image $\mathcal{R}_\rho(H)$ of H under this map belongs again to $\mathcal{D}$ is technically rather involved. Yet, the idea implemented by the renormalization group map $\mathcal{R}_\rho$ is *conceptually simple.*

2 Spin-Boson Model

We start by introducing the model whose spectral properties are to be analyzed. The simplest, nontrivial model is the spin-boson model. It describes the dynamics of a two-level atom which is coupled to a quantized boson field. The Hilbert space containing the states of this system is the tensor product

$$\mathcal{H} := \mathcal{H}_{el} \otimes \mathcal{F} = \mathbb{C}^2 \otimes \mathcal{F}[L^2(\mathbb{R}^3)] , \tag{1}$$

of the state space $\mathcal{H}_{el} = \mathbb{C}^2$ of the two-level atom and $\mathcal{F} := \mathcal{F}[L^2(\mathbb{R}^3)]$ is the boson Fock space over the one-boson Hilbert space $L^2(\mathbb{R}^3)]$ of a quantized scalar field. Note that $\binom{0}{1} \in \mathbb{C}^2$ is the wave function of the atomic ground state, with zero groundstate energy, and $\binom{1}{0} \in \mathbb{C}^2$ represents the wave function of the first (and only) excited atomic level at energy $= 2$. The Fock space $\mathcal{F}$ is given by $\mathcal{F} = \bigoplus_{n=0}^{\infty} \mathcal{F}^{(n)}$, where the so-called n-boson sectors $\mathcal{F}^{(n)}$ are defined by

$$\mathcal{F}^{(n)} = \{\psi_n \in \otimes^n L^2 \,|\, \forall \pi : \psi_n(k_1, \ldots, k_n) = \psi_n(k_{\pi(1)}, \ldots, k_{\pi(n)})\} , \tag{2}$$

for $n \geq 1$, and $\mathcal{F}^{(0)} := \mathbb{C}\,\Omega$ is called the vacuum sector with vacuum vector Ω. On $\mathcal{F}$, we have creation and annihilation operators obeying the usual canonical commutation relations (CCR)

$$[a(k), a(k')] = [a^*(k), a^*(k')] = 0 , \; [a(k), a^*(k')] = \delta(k-k') , \; a(k)\Omega = 0 , \tag{3}$$

for all $k, k' \in \mathbb{R}^3$.

The dynamics is generated by the Hamiltonian

$$H_g := H_0 + gW \tag{4}$$
$$H_0 := H_{el} \otimes \mathbf{1} + \mathbf{1} \otimes H_f , \tag{5}$$
$$H_{el} := \begin{pmatrix} 2 & 0 \\ 0 & 0 \end{pmatrix} , \tag{6}$$
$$H_f := \int d^3k \, a^*(k) \, \omega(k) \, a(k) , \tag{7}$$

and to be as simple as possible, the interaction is assumed to be of the form

$$W := \begin{pmatrix} 0 & 1 \\ 1 & 0 \end{pmatrix} \otimes \left[a^*(G) + a(G) \right] , \tag{8}$$
$$a^*(G) := \int d^3k \, G(k) \, a^*(k) , \tag{9}$$
$$a(G) := \int d^3k \, \overline{G(k)} \, a(k) , \tag{10}$$

where we assume the *coupling functions* $G : \mathbb{R}^3 \to \mathcal{B}[\mathcal{H}_{el}]$ to be bounded operators $G(k)$ on $\mathcal{H}_{el}$, pointwise in $k \in \mathbb{R}^3$, obeying

Hypothesis 1.

$$\Lambda_1 := \left(\int d^3k \, \left(1 + \omega(k)^{-1}\right) \|G(k)\|^2 \right)^{1/2} < \infty , \tag{11}$$
$$\Lambda_5 := \sup_{k \in \mathbb{R}^3} \left\{ \omega(k)^{1/2 - \mu/2} \|G(k)\| \right\} < \infty , \tag{12}$$

for some $\mu > 0$, where $\|G(k)\|$ denotes the operator norm.

We remark that often $\mu = 0$ in (12) in physical applications, which is not covered by Hypothesis 1.

2.1 Relative Bounds on the Interaction

Our first task is to show that W is a relative H_0 form bounded perturbation. This justifies viewing H_g as a small deviation from H_0. Here and henceforth, we leave out trivial tensor factors and simply write A and B instead of $\mathbf{1} \otimes A$ and $B \otimes \mathbf{1}$, respectively.

Lemma 1.

$$\left\| (H_f + 1)^{-1/2} \, W \, (H_f + 1)^{-1/2} \right\| \leq 2\Lambda_1 . \tag{13}$$

Proof: Pick $\varphi, \psi \in \mathcal{H}$. Then, by Schwarz' inequality,

$$
\begin{aligned}
&\left|\left\langle \varphi \;\middle|\; (H_f+1)^{-1/2}\, a(G)\, (H_f+1)^{-1/2}\, \psi \right\rangle\right| \\
&= \left| \int dk \left\langle \varphi \;\middle|\; (H_f+1)^{-1/2}\, G^*(k) \otimes a(k)\, (H_f+1)^{-1/2}\, \psi \right\rangle \right| \\
&\leq \|\varphi\| \int dk \|G(k)\| \, \|a(k))\, (H_f+1)^{-1/2}\, \psi\| \\
&\leq \|\varphi\| \Big(\int dk \, \frac{\|G(k)\|^2}{\omega(k)} \Big)^{1/2} \Big(\int dk\, \omega(k)\, \|a(k))\, (H_f+1)^{-1/2}\, \psi\|^2 \Big)^{1/2} \\
&\leq \Lambda_1 \, \|\varphi\| \, \|{H_f}^{1/2}\, (H_f+1)^{-1/2}\, \psi\| \\
&\leq \Lambda_1 \, \|\varphi\| \, \|\psi\| \, . \qquad \square
\end{aligned}
\tag{14}
$$

2.2 The Feshbach Map and Pull-Through Formula

In this section we describe the *Smooth Feshbach Map* introduced in [1] in a particular situation which allows us to keep the exposition simple.

Let $\mathcal{H}$ be a Hilbert space, $T \geq \varepsilon \cdot \mathbf{1}$ a positive selfadjoint operator on $\mathcal{H}$, with $\varepsilon > 0$, and W a relatively form-bounded perturbation such that $\|T^{-1/2} W T^{-1/2}\|_{\mathrm{op}} < \infty$. Suppose that $\theta : \mathbb{R}_0^+ \to [0,1]$ is Borel measurable and set $\bar{\theta} := \sqrt{1-\chi^2}$. By functional calculus, define $\chi := \theta(T)$ and $\bar{\chi} := \bar{\theta}(T)$, and note that χ, $\bar{\chi}$, and T commute. We set

$$
H := T + W \, , \quad W_\chi := \chi \, W \, \chi \, , \quad W_{\bar{\chi}} := \bar{\chi} \, W \, \bar{\chi} \, , \tag{15}
$$

$$
H_\chi := T + \chi \, W \, \chi \, , \quad H_{\bar{\chi}} := T + \bar{\chi} \, W \, \bar{\chi} \, . \tag{16}
$$

Denoting by P and $\overline{P}$ the projections onto $\operatorname{Ran} \chi$ and $\operatorname{Ran} \bar{\chi}$, respectively, we assume that, for some $\varepsilon > 0$,

$$
\overline{P} \, H_\chi \, \overline{P} \geq \varepsilon \cdot \overline{P} \, , \tag{17}
$$

so that the restriction of H_χ onto $\operatorname{Ran} \bar{\chi}$ is invertible. This is the key assumption for the following definition of the smooth Feshbach map $\mathcal{F}_\chi$,

$$
F_\chi(H,T) := H_\chi \; - \; \chi \, W \, \bar{\chi} \, H_{\bar{\chi}}^{-1} \, \bar{\chi} \, W \, \chi \, , \tag{18}
$$

$$
Q_\chi(H,T) := \chi \; - \; \bar{\chi} H_{\bar{\chi}}^{-1} \, \bar{\chi} \, W \, \chi \, . \tag{19}
$$

Note that if θ is a characteristic function of some Borel set, then $\chi = P$ is the spectral projection of T onto that set, and

$$
F_P(H,T) \;=\; P \, H \, P \;-\; P \, H \, \overline{P} \big(\overline{P} \, H \, \overline{P} \big)^{-1} \overline{P} \, H \, P \tag{20}
$$

reduces to the usual effective Hamiltonian on $\operatorname{Ran} P$ known from the Feshbach projection method.

Lemma 2 (Smooth Feshbach Map). *Assume that $\|T^{-1/2} W T^{-1/2}\|_{\mathrm{op}} < \infty$ and (17), and denote $F := F_\chi(H,T)$ and $Q := Q_\chi(H,T)$. Then*

$$H \text{ is invertible on } \mathcal{H} \quad \Leftrightarrow \quad F \text{ is invertible on } P\mathcal{H}\,; \tag{21}$$
$$H\psi = 0 \text{ and } \psi \neq 0 \quad \Rightarrow \quad \chi\psi \neq 0 \text{ and } F(\chi\psi) = 0\,; \tag{22}$$
$$F\varphi = 0 \text{ and } \varphi = P\varphi \neq 0 \quad \Rightarrow \quad H\,(Q\varphi) = 0 \text{ and } Q\varphi \neq 0\,. \tag{23}$$

We refer the reader to [1] for the proof of Lemma 2 and further details. Another important ingredient of our analysis is the following Pull-Through formula.

Lemma 3 (Pull-Through). *Let* $F : \mathbb{R}_0^+ \to \mathbb{C}$, $F(r) = \mathcal{O}(r = 1)$. *Then* $F[H_f]$ *is defined on* $\mathcal{D}(H_f)$ *and*

$$a(k)F[H_f] = F[H_f + \omega(k)]a(k)\,, \tag{24}$$
$$F[H_f]a^*(k) = a^*(k)F[H_f + \omega(k)]\,. \tag{25}$$

Proof: Let $\psi = \prod_{j=1}^N a^*(k_j)\Omega$. Then

$$F[H_f]\,a^*(k)\,\psi = F\Big[\omega(k) + \sum_{j=1}^N \omega(k_j)\Big]\,a^*(k)\,\psi \tag{26}$$
$$= a^*(k)F\Big[\omega(k) + \sum_{j=1}^N \omega(k_j)\Big]\,\psi \;=\; a^*(k)F[\omega(k) + H_f]\,\psi$$

in the sense of operator-valued distributions. Integrating this identity against suitable test functions finishes the proof. □

2.3 Elimination of High-Energy Degrees of Freedom

As a first application of the Smooth Feshbach map, we use it to eliminate the high-energy degrees of freedom of our spectral problem. To this end, we choose $T := H_0$ and *theta* to be the characteristic function onto $[0,1)$, such that $\chi = \theta(T)$ equals the projection

$$P^{(0)} \;:=\; \chi[H_0 < 1] \;=\; P_{el} \otimes \chi[H_f < 1]\,, \tag{27}$$

where

$$P_{el} \;:=\; \begin{pmatrix} 0\,0 \\ 0\,1 \end{pmatrix} \;=\; |\varphi_{el}\rangle\langle\varphi_{el}|\,, \quad \text{with} \quad \varphi_{el} \;:=\; \begin{pmatrix} 0 \\ 1 \end{pmatrix} \tag{28}$$

being the atomic ground state vector.

Now, we check the positivity of $\overline{P}^{(0)} H_g \overline{P}^{(0)} - \lambda$ on Ran $\overline{P}^{(0)}$, noting that $\overline{P}^{(0)} = \chi[H_0 \geq 1]$. Observe that $\langle \varphi_{el} \otimes \Omega \,|\, H_g(\varphi_{el} \otimes \Omega)\rangle = 0$, thus $E_0(g) \leq 0$, and we may assume $\lambda \leq \frac{1}{2}$ henceforth. Therefore,

$$\overline{P}^{(0)}\,(H_0 + gW - \lambda)\,\overline{P}^{(0)} \;\geq\; (1 - \lambda - 2g\Lambda_1)\,\overline{P}^{(0)} \;\geq\; \frac{1}{4}\,\overline{P}^{(0)}\,, \tag{29}$$

for $|g| \ll 1$, employing Lemma 1. Moreover,

$$0 \leq \frac{(H_0+1)\overline{P}^{(0)}}{H_0-\lambda} \leq \frac{2\,\overline{P}^{(0)}}{1-\lambda} \leq 4\,\overline{P}^{(0)} . \tag{30}$$

We construct the inverse of $\overline{P}^{(0)} H_g \overline{P}^{(0)} - \lambda$ on Ran $\overline{P}^{(0)}$ by a Neumann series expansion,

$$\begin{aligned}
&(H_0+1)^{1/2}\,\overline{P}^{(0)}\,\big(\overline{P}^{(0)} H_g \overline{P}^{(0)} - \lambda\big)^{-1}\,\overline{P}^{(0)}\,(H_0+1)^{1/2} \\
&= \sum_{L=0}^{\infty} \left(\frac{(H_0+1)^{1/2}\,\overline{P}^{(0)}}{H_0-\lambda}\right) \left\{ -\,gW \left(\frac{\overline{P}^{(0)}}{H_0-\lambda}\right)\right\}^{L} \\
&= \sum_{L=0}^{\infty} \underbrace{\left(\frac{(H_0+1)\overline{P}^{(0)}}{H_0-\lambda}\right)}_{\|\cdot\|\leq 4} \\
&\quad \cdot \left\{ \underbrace{\left[(H_0+1)^{-1/2}(-gW)(H_0+1)^{-1/2}\right]}_{\|\cdot\|\leq 2\,g\,\Lambda_1} \underbrace{\left(\frac{(H_0+1)\overline{P}^{(0)}}{H_0-\lambda}\right)}_{\|\cdot\|\leq 4} \right\}^{L} .
\end{aligned} \tag{31}$$

Thus the series is norm-convergent and

$$\begin{aligned}
\big\|(H_0+1)^{1/2}\overline{P}^{(0)}\big(\overline{P}^{(0)} H_g \overline{P}^{(0)} - \lambda\big)^{-1}\overline{P}^{(0)}(H_0+1)^{1/2}\big\| &\leq 4 \sum_{L=0}^{\infty} (8g\Lambda_1)^L \\
&\leq 4 + \mathcal{O}(g) .
\end{aligned} \tag{32}$$

As a result, Lemma 2 implies

Lemma 4. *Let $P_\rho := \mathbf{1}[H_f < \rho]$, for $\rho > 0$, and denote*

$$\mathcal{H}_{red} \;:=\; \mathrm{Ran}(P_1) \;=\; \mathbf{1}[H_f < 1]\mathcal{F} . \tag{33}$$

Then $H_g - \lambda$ is invertible iff $H_{(0)}[\lambda]$ is, where

$$P_{el} \otimes \big(H_{(0)}[\lambda] - \lambda\big) \;:=\; \mathcal{F}_{P^{(0)}}(H_g - \lambda) , \tag{34}$$

so $H_{(0)}[\lambda]$ is the following operator $\in \mathcal{B}[\mathcal{H}_{red}]$,

$$\begin{aligned}
\big(H_{(0)}[\lambda] - \lambda\big) &= P_1(H_f - \lambda + g\langle W\rangle_{el})P_1 \\
&\quad - P_1 \Big\langle gW\,\overline{P}^{(0)}\,(\overline{P}^{(0)} H_g \overline{P}^{(0)} - \lambda)^{-1}\,\overline{P}^{(0)}\,gW \Big\rangle_{el} P_1 ,
\end{aligned} \tag{35}$$

where $\langle\,\cdot\,\rangle_{el} := \langle \varphi_{el}|\,\cdot\,\varphi_{el}\rangle$. Moreover, $H_{(0)}[\lambda]$ can be expanded in a norm-convergent series,

$$H_{(0)}[\lambda] = P_1(H_f - \lambda) + gP_1 \langle W\rangle_{el} P_1 - g^2 P_1 \left\langle W \left(\frac{\overline{P}^{(0)}}{H_0 - \lambda} \right) W \right\rangle_{el} P_1$$
$$+ g^3 P_1 \left\langle W \left(\frac{\overline{P}^{(0)}}{H_0 - \lambda} \right) W \left(\frac{\overline{P}^{(0)}}{H_0 - \lambda} \right) W \right\rangle_{el} P_1 - \dots . \quad (36)$$

2.4 Normal form of Hamiltonians

Our next goal is to write (recall $P_1 \equiv P_1(H_f) = [H_f < 1]$)

$$H_{(0)}[\lambda] = P_1 \left(E_{(0)}[\lambda] + T_{(0)}[\lambda; H_f] + W_{(0)}[\lambda] \right) P_1 , \quad (37)$$

where $E_{(0)}[\lambda] \in \mathbb{C}$ is an energy shift, $T_{(0)}[\lambda; H_f]$ is the new, renormalized, free (=unperturbed) Hamiltonian, and

$$W_{(0)}[\lambda] := \sum_{M+N \geq 1} W_{M,N}^{(0)}[\lambda] , \quad (38)$$

$$W_{M,N}^{(0)} := \int dk^{(\mu)} d\tilde{k}^{(N)} \underbrace{a^*(k^{(\mu)})}_{\prod_{j=1}^{\mu} a^*(k_j)} w_{M,N}^{(0)}[\lambda; H_f; k^{(\mu)}; \tilde{k}^{(N)}] \, a(\tilde{k}^{(N)}) \quad (39)$$

where the operators $T_{(0)}[\lambda; H_f]$, $w_{M,N}^{(0)}[\lambda; H_f; k^{(\mu)}; \tilde{k}^{(N)}]$ are defined pointwise, for each λ, $k^{(\mu)}$, and $\tilde{k}^{(N)}$, by the spectral theorem (functional calculus) for H_f.

Looking at $H_0 + gW - \lambda$, we see that

$$E_{(0)}[\lambda] = \Delta E_{(0)}[\lambda] , \quad (40)$$
$$T_{(0)}[\lambda; H_f] = H_f + \Delta T_{(0)}[\lambda; H_f] , \quad (41)$$
$$W_{(0)}[\lambda] = gW + \Delta W_{(0)}[\lambda] , \quad (42)$$

where $\Delta E_{(0)}[\lambda]$, $\Delta T_{(0)}[\lambda; H_f]$, and $\Delta W_{(0)}[\lambda] = \sum_{M+N\geq 1} \Delta W_{(0)}[\lambda]$ are generated from $H_{(0)}[\lambda]$ as given in the Neumann series (36).

To illustrate the method of deriving $\Delta E_{(0)}, \Delta T_{(0)}$, and $\Delta W_{(0)}$, we do a sample computation. In order g^2, we have

$$\left\langle W \left(\frac{\chi[H_0 \geq 1]}{H_0 - \lambda} \right) W \right\rangle_{el} \quad (43)$$
$$= \underbrace{\left\langle a^*(G) \left(\frac{\chi[H_0 \geq 1]}{H_0 - \lambda} \right) a^*(G) \right\rangle_{el}}_{=:X_1} + \left\langle a^*(G) \left(\frac{\chi[H_0 \geq 1]}{H_0 - \lambda} \right) a(G) \right\rangle_{el}$$
$$+ \left\langle a(G) \left(\frac{\chi[H_0 \geq 1]}{H_0 - \lambda} \right) a(G) \right\rangle_{el} + \underbrace{\left\langle a(G) \left(\frac{\chi[H_0 \geq 1]}{H_0 - \lambda} \right) a^*(G) \right\rangle_{el}}_{=:X_4} .$$

and we restrict our attention to X_1 and X_4. Due to Lemma 3, $F[H_0]a^*(k) = a^*(k)F[H_0 + \omega(k)]$. Thus

$$\begin{aligned} X_1 &= \int dk\, dk' \left\langle G(k) \otimes a^*(k) \left(\frac{\chi[H_0 \geq 1]}{H_0 - \lambda} \right) G(k') \otimes a^*(k') \right\rangle_{el} \\ &= \int dk\, dk' \left\langle G(k) \otimes a^*(k) a^*(k') \left(\frac{\chi[H_0 + \omega(k') \geq 1]}{H_0 + \omega(k') - \lambda} \right) G(k') \otimes 1 \right\rangle_{el} \\ &= \int dk\, dk'\, a^*(k) a^*(k')\, \widetilde{w}_{2,0}[H_f; \lambda, k, k'] , \end{aligned} \tag{44}$$

where the integral kernel $\widetilde{w}_{2,0}[r; \lambda, k, k']$ is given by

$$\widetilde{w}_{2,0}[r; \lambda, k, k'] \;:=\; \left\langle G(k) \left(\frac{\chi[H_{el} + r + \omega(k') \geq 1]}{H_{el} + r + \omega(k') - \lambda} \right) G(k') \right\rangle_{el} \tag{45}$$

Thus, we recognize X_1 as an additive contribution to $W_{2,0}^{(0)}[\lambda]$.

Similarly, additionally using the CCR, we convert X_4,

$$\begin{aligned} X_4 &= \int dk\, dk' \left\langle G(k) \otimes a(k) \left(\frac{\chi[H_0 \geq 1]}{H_0 - \lambda} \right) G(k') \otimes a^*(k') \right\rangle_{el} \\ &= \int dk\, dk' \left\langle G(k) \otimes a(k) a^*(k') \left(\frac{\chi[H_0 + \omega(k') \geq 1]}{H_0 + \omega(k') - \lambda} \right) G(k') \otimes 1 \right\rangle_{el} \\ &= \int dk\, dk' \Big\langle G(k) \otimes \Big(\delta(k - k') + a^*(k') a(k) \Big) \\ &\qquad \times \left(\frac{\chi[H_0 + \omega(k') \geq 1]}{H_0 + \omega(k') - \lambda} \right) G(k') \otimes 1 \Big\rangle_{el} \end{aligned}$$

$$\begin{aligned} &= \int dk \left\langle G(k) \otimes 1 \left(\frac{\chi[H_0 + \omega(k) \geq 1]}{H_0 + \omega(k) - \lambda} \right) G(k) \otimes 1 \right\rangle_{el} + \\ &\quad \int dk\, dk' \left\langle G(k) \otimes a^*(k') \left(\frac{\chi[H_0 + \omega(k) + \omega(k') \geq 1]}{H_0 + \omega(k) + \omega(k') - \lambda} \right) G(k') \otimes a(k) \right\rangle_{el} \\ &= \Delta e_{(0)}[\lambda] \,+\, \Delta t_{(0)}[H_f; \lambda] \,+\, \int dk\, dk'\, a^*(k')\, \widetilde{w}_{1,1}[H_f; \lambda, k, k']\, a(k) , \end{aligned} \tag{46}$$

where

$$\Delta e_{(0)}[\lambda] \;:=\; \int dk \left\langle G(k) \left(\frac{\chi[H_{el} + \omega(k) \geq 1]}{H_{el} + \omega(k) - \lambda} \right) G(k) \right\rangle_{el} , \tag{47}$$

is a number that contributes additively to $\Delta E_{(0)}[\lambda]$,

$$\begin{aligned} &\Delta t_{(0)}[r; \lambda] \;:= \\ &\int dk \left\langle G(k) \left(\frac{\chi[H_{el} + r + \omega(k) \geq 1]}{H_{el} + r + \omega(k) - \lambda} - \frac{\chi[H_{el} + \omega(k) \geq 1]}{H_{el} + \omega(k) - \lambda} \right) G(k) \right\rangle_{el} , \end{aligned} \tag{48}$$

is a function contributing to $\Delta T_{(0)}[\lambda]$, having the property that

$$\Delta t_{(0)}[0;\lambda] \;=\; 0 \quad \text{and} \quad \|\partial_r \Delta t_{(0)}[r;\lambda]\|_\infty \;<\; \infty \,. \tag{49}$$

Finally

$$\widetilde{w}_{1,1}[r;\lambda,k,k'] \;:=\; \left\langle G(k) \left(\frac{\chi[H_{el}+r+\omega(k)+\omega(k') \geq 1]}{H_{el}+r+\omega(k)+\omega(k')-\lambda} \right) G(k') \right\rangle_{el} \tag{50}$$

is an additive contribution to $\Delta W_{1,1}^{(0)}[\lambda]$. We now appeal to the reader's imagination, that this normal-ordering – demonstrated on the examples of X_1 and X_4 – can be systematically carried out for all terms in the Neumann series expansion (36).

2.5 Banach Space of Operators

We systematize the normal-ordering by introducing a Banach space of Hamiltonians as follows. Let

$$\mathcal{H}_{\mathrm{red}} \;:=\; \mathrm{Ran}\,\mathbf{1}[H_f<1] \;\equiv\; \mathbf{1}[H_f<1]\mathcal{F} \;\subseteq\; \mathcal{F} \tag{51}$$

be the spectral subspace corresponding to photon energies less than 1. Our goal is to study Hamiltonians $H \in \mathcal{B}[\mathcal{H}_{\mathrm{red}}]$ of the form (37), i.e.,

$$H \;=\; T[H_f] \;+\; W \;-\; E\cdot\mathbf{1}\,, \tag{52}$$

where $E \in \mathbb{C}$ is a complex number, $T \in C^1(I)$ is a continuously differentiable function on $I := [0,1]$ with $T[0] = 0$, and $T[H_f] \in \mathcal{B}[\mathcal{H}_{\mathrm{red}}]$ is defined by functional calculus. Moreover, $W = \sum_{m+n\geq 1} W_{m,n}$, with

$$\begin{aligned} W_{m,n} \;&\equiv\; W_{m,n}[w_{m,n}] \\ &:= P_{\mathrm{red}} \int_{B_1^{m+n}} \frac{dK^{(m,n)}}{|K^{(m,n)}|^{1/2}}\, a^*(k^{(m)})\, w_{m,n}\big[H_f;K^{(m,n)}\big]\, a(\tilde{k}^{(n)})\, P_{\mathrm{red}}\,, \end{aligned} \tag{53}$$

where B_1 denotes the unit ball in $\mathbb{R}^d$, and P_{red} is a shorthand for $\mathbf{1}[H_f<1]$. We use the notation

$$k^{(m)} \;:=\; (k_1,\ldots,k_m) \,\in\, \mathbb{R}^{dm}\,, \qquad \tilde{k}^{(n)} \;:=\; (\tilde{k}_1,\ldots,\tilde{k}_n) \,\in\, \mathbb{R}^{dn}\,, \tag{54}$$

$$K^{(m,n)} \;:=\; (k^{(m)},\tilde{k}^{(n)})\,, \qquad dK^{(m,n)} \;:=\; \textstyle\prod_{i=1}^m d^dk_i \,\prod_{i=1}^n d^d\tilde{k}_i\,, \tag{55}$$

$$a^*(k^{(m)}) \;:=\; \textstyle\prod_{i=1}^m a^*(k_i)\,, \qquad a(\tilde{k}^{(n)}) \;:=\; \prod_{i=1}^n a(\tilde{k}_i)\,, \tag{56}$$

$$|K^{(m,n)}| \;:=\; |k^{(m)}|\cdot|\tilde{k}^{(n)}|\,, \qquad |k^{(m)}| \;:=\; |k_1|\cdots|k_m|\,, \tag{57}$$

$$\Sigma[k^{(m)}] \;:=\; |k_1|+\ldots+|k_m|\,. \tag{58}$$

The function $w_{m,n}[\,\cdot\,,K^{(m,n)}] \in C^1(I)$ is continuously differentiable on I, pointwise, for almost every $K^{(m,n)} \in B_1^{(m+n)}$, its derivative is denoted by

$\partial_r w_{m,n}$. As a function of $K^{(m,n)}$, it is totally symmetric w. r. t. the variables $k^{(m)} = (k_1, \ldots, k_m)$ and $\tilde{k}^{(n)} = (\tilde{k}_1, \ldots, \tilde{k}_n)$ and obeys the L^2-norm bound

$$\|w_{m,n}\|_\mu^\# \;:=\; \|w_{m,n}\|_\mu \;+\; \|\partial_r w_{m,n}\|_\mu \;<\; \infty\,, \tag{59}$$

where

$$\|w_{m,n}\|_\mu \;:=\; \left(\int_{B_1^{m+n}} \frac{dK^{(m,n)}}{\left|K^{(m,n)}\right|^{3+2\mu}} \sup_{r\in I} \left|w_{m,n}[r; K^{(m,n)}]\right|^2 \right)^{1/2}, \tag{60}$$

for some $\mu > 0$. The Banach space of these functions is denoted by $\mathcal{W}_{m,n}^\#$. Moreover, $W_{m,n}[w_{m,n}]$ stresses the dependence of $W_{m,n}$ on $w_{m,n}$. For a family $\{w_{m,n}\}_{m+n\geq 1}$ we write $\underline{w} := \{w_{m,n}\}_{m+n\geq 1}$.

Further introducing

$$\mathcal{W}_\mu^\# \;:=\; \mathbb{C} \oplus \mathcal{T} \oplus \left(\bigoplus_{M+N\geq 1} \mathcal{W}_\mu^\#(M,N) \right), \tag{61}$$

with

$$\mathcal{T} \;:=\; \left\{ T : [0,1] \to \mathbb{C} \;\middle|\; \|\partial_r T\|_\infty < \infty, \;\; T(0) = 0 \right\}, \tag{62}$$

and, with $B_\rho := \{|k| < \rho\}$,

$$\mathcal{W}_\mu^\#(M,N) \;:=\; \left\{ w_{M,N} : [0,1] \times B_1^M \times B_1^N \to \mathbb{C} \;\middle|\; \|w_{M,N}\|_\mu^\# < \infty \right\}, \tag{63}$$

we observe that $\mathcal{W}_\mu^\#$ is a Banach space with norm

$$\left\|(E, T, \underline{w})\right\|_\mu^\# \;:=\; \max\left\{ |E|\,, \;\; \|\partial_r T\|_\infty\,, \;\; \sup_{M+N\geq 1} \|w_{M,N}\|_\mu^\# \right\}. \tag{64}$$

To every element $(E, T, \underline{w}) \in \mathcal{W}_\mu^\#$ corresponds an operator $H \in \mathcal{B}(\mathcal{H}_{red})$ in *normal form*,

$$H \;=\; P_1 \left(E \,+\, T[H_f] \,+\, W\right) P_1\,, \quad W \;=\; \sum_{M+N\geq 1} W_{M,N}[w_{M,N}]\,, \tag{65}$$

$$W_{M,N} \;=\; \int dK^{(M,N)}\; a^*(k^{(\mu)})\; w_{M,N}[H_f; k^{(\mu)}; \tilde{k}^{(N)}]\; a(\tilde{k}^{(N)})\,. \tag{66}$$

Conversely, the effective Hamiltonian $H^{(0)}[\lambda]$ uniquely determines an element $(E_{(0)}[\lambda], T_{(0)}[\lambda], \underline{w}^{(0)}[\lambda]) \in \mathcal{W}_\mu^\#$ such that

$$H_{(0)}[\lambda] \;=\; P_1 \left(E_{(0)}[\lambda] \,+\, T_{(0)}[\lambda; H_f] \,+\, W_{(0)}[\lambda] \right) P_1\,, \tag{67}$$

as desired in (37). (Of course, it remains to be checked that the norms of $T_{(0)}[\lambda]$ and $\underline{w}^{(0)}[\lambda]$) are all uniformly bounded – we take this for granted, here.) We identify $H \equiv (E, T, \underline{w})$.

Let $D_{1/2} := \{z \in \mathbb{C} \mid |z| < 1/2\}$.

$$\begin{aligned} \mathcal{W}_\mu := \big\{ H[\cdot] : D_{1/2} \to \mathcal{W}_\mu^\# \;\big|\; z \mapsto H[z] \text{ is analytic}, \\ \|H[\cdot]\|_\mu := \sup_{z \in D_{1/2}} \|H[z]\|_\mu^\# < \infty \big\} . \end{aligned} \tag{68}$$

From (40)–(42) we conclude that

$$H_{(0)}[\cdot] \;\in\; \mathcal{W}_\mu \tag{69}$$

and that

$$\|H_{(0)}[z] - H_f\|_\mu^\# \;=\; \big\|(E_{(0)}[z],\, T_{(0)}[z,r] - r,\, \underline{w}^{(0)})\big\| \;\leq\; \mathcal{O}(g) \,. \tag{70}$$

Defining, for $\frac{1}{16} > \delta, \varepsilon > 0$, a polydisc in $\mathcal{W}_\mu$, by

$$\begin{aligned} \mathcal{B}(\delta,\varepsilon) := \big\{ (E,T,\underline{w})[\cdot] \in \mathcal{W}_\mu \;\big| \\ \|\partial_r T[z] - 1\|_\infty \leq \delta,\; |E[z]| \leq \varepsilon,\; \|\underline{w}[z]\|_\mu^\# \leq \varepsilon \big\} , \end{aligned} \tag{71}$$

Equation (70) may be expressed as

$$H_{(0)}[z] \;\in\; \mathcal{B}(cg, cg) \,. \tag{72}$$

2.6 The Renormalization Map $\mathcal{R}_\rho$

Let $\delta > 0$, $0 < \varepsilon < 1$, $\rho \leq \frac{1}{16}$, and $(E,T,\underline{w}) \in \mathcal{B}(\delta,\varepsilon)$. Let furthermore $\theta \in C_0^\infty([0,\frac{3}{4});[0,1])$ be such that $\theta' \leq 0$, that $\theta \equiv 1$ on $[0,\frac{1}{2}]$, and that $\bar\theta = \sqrt{1-\theta^2} \in C^\infty$. For $\tau > 0$, denote

$$\chi_\tau \;:=\; \theta[H_f/\tau] \,, \quad \bar\chi_\tau \;=\; \sqrt{1-\chi_\tau^2} \,, \tag{73}$$

and assume that $|z - E[z]| \leq \rho/2$.

Lemma 5.

$$\left\| \frac{(H_f+\rho)}{T[H_f;z] + E[z] - z} \right\| \leq 8 \,. \tag{74}$$

Proof. Since $|\partial_r T - 1| \leq \delta \leq \frac{1}{16}$, $T[0;z] = 0$

$$\Leftrightarrow |T[z;r] + E[z] - z| \geq (1-\delta)\cdot r - \frac{\rho}{2} \geq \left(1 - \delta - \frac{1}{2}\right)\left(\frac{r+\rho}{2}\right) \,. \quad \square \tag{75}$$

Lemma 6. *If* $\|\underline{w}\|_\mu^\# \leq \varepsilon$ *then*

$$\begin{aligned} &\left\| (H_f+\rho)^{-1/2} P_1[H_f] W_{M,N} P_1[H_f] (H_f+\rho)^{-1/2} \right\| \\ &\leq \varepsilon \cdot \varepsilon^{M+N} \cdot \big(C_d \Gamma(1+\mu)\big)^{M+N} \cdot \rho^{-\delta_{M,0}} \cdot \rho^{-\delta_{N,0}} (M!)^{-1/2} (N!)^{-1/2} \,. \end{aligned} \tag{76}$$

Proof: Similar to Lemma 1.

Lemma 7. $\overline{\chi}_\rho H[z]\overline{\chi}_\rho - z$ *is invertible on Ran* $\overline{\chi}_\rho$.

Proof: By Lemma 6,

$$\left\|(H_f+\rho)^{-1/2}P_1 W P_1 (H_f+\rho)^{-1/2}\right\| \leq \left(1+\mathcal{O}(\xi)\right)\left(\frac{\varepsilon}{\rho^{1/2}}\right) . \tag{77}$$

Hence, Lemma 5 implies the norm-convergence of the Neumann series,

$$\begin{aligned} &\|(\overline{\chi}_\rho H \overline{\chi}_\rho - z)^{-1}\overline{\chi}_\rho\| \\ &= \sum_{L=0}^{\infty}\left\|\left(\frac{\bar{\chi}_\rho}{(H_f+\rho)}\right)\right. \\ &\times\left[\left(\frac{\bar{\chi}_\rho\cdot(H_f+\rho)^{1/2}}{T[H_f;z]+E[z]-z}\right)\cdot\left((H_f+\rho)^{1/2}P_1\,(-W)\,P_1\,(H_f+\rho)^{1/2}\right)\right]^L \\ &\times\left.\left(\frac{\bar{\chi}_\rho\cdot(H_f+\rho)^{1/2}}{T[H_f;z]+E[z]-z}\right)\right\| \\ &\leq \sum_{L=0}^{\infty}\frac{8}{2\rho}\cdot\left[\frac{8\varepsilon(1+\mathcal{O}(\xi)}{\rho^{1/2}}\right]^L \leq \frac{4}{\rho}\left(1+\mathcal{O}(\varepsilon\rho^{-1/2})\right) . \ \square \end{aligned} \tag{78}$$

Lemma 7 tells us that we may apply $\mathcal{F}_{\chi_\rho}$ to $H[z]-z$:

$$\widetilde{H}[z]-z := \mathcal{F}_{\chi_\rho}[H[z]-z] . \tag{79}$$

We then apply a normal-ordering procedure as in Sect. 2.4 to obtain

$$\widetilde{H}[z]-z := \chi_\rho\left(\widetilde{E}[z]-z+\widetilde{T}(z,H_f)+\widetilde{W}[z]\right)\chi_\rho , \tag{80}$$

$$\widetilde{W}[z] = \sum_{M+N\geq 1}\widetilde{W}_{M,N}[z] , \tag{81}$$

$$\widetilde{W}_{M,N}[z] = \int dk^{(\mu)}d\tilde{k}^{(N)}\,a^*\left(k^{(\mu)}\right)\widetilde{w}_{M,N}\left[z;H_F;k^{(\mu)};\tilde{k}^{(N)}\right]a\left(\tilde{k}^{(N)}\right) , \tag{82}$$

with

$$\begin{aligned} &\left|\widetilde{W}_{M,N}\left[z;r;k^{(\mu)};\tilde{k}^{(N)}\right]\right| \\ &\leq \varepsilon\left(1+\mathcal{O}(\varepsilon/\rho^{1/2})\right)\cdot g^{M+N}\cdot\textstyle\prod_{j=1}^{M}\omega(k_j)^{-1/2+\mu/2}\prod_{j=1}^{N}\omega(\tilde{k}_j)^{-1/2+\mu/2} , \end{aligned} \tag{83}$$

using $\|w_{M,N}\|_\Delta^{(\infty)} \leq \varepsilon$. To get back to $\mathcal{H}_{red} = \text{Ran } P_1$ from Ran χ_ρ, we rescale the photon momenta, $k \to \rho k$, by means of a unitary Γ_ρ, so

$$\frac{1}{\rho}\,\Gamma_\rho\,H_f\,\Gamma_\rho^* \;=\; H_f \tag{84}$$
$$\Rightarrow \Gamma_\rho\,\chi_\rho\,\Gamma_\rho^* \;=\; \Gamma_\rho\,\theta\big[\frac{H_f}{\rho}\big]\,\Gamma_\rho^* \;=\; \theta[H_f] \;=\; \chi_1\,.$$

One easily checks that

$$W_{(1)} := \frac{1}{\rho}\Gamma_\rho\,\widetilde{W}\,\Gamma_\rho^* \;=\; \sum_{M+N\geq 1} W_{M,N}^{(1)}\,,$$

with coupling functions

$$w_{M+N}^{(1)}\left[z; r, k^{(\mu)}, \tilde{k}^{(N)}\right] := \rho^{3/2(M+N)-1}\widetilde{w}_{M+N}\left[z; \rho H_f, \rho k^{(\mu)}, \rho\tilde{k}^{(N)}\right]\,. \tag{85}$$

The key point about assuming $\mu > 0$ is that

$$\begin{aligned}
&\left|w_{M+N}^{(1)}[z; r; k^{(\mu)}; \tilde{k}^{(N)}]\right| \qquad (86)\\
&\leq \varepsilon\left[1+\mathcal{O}(\varepsilon\rho^{-1/2})\right] g^{M+N}\cdot\rho^{3/2(M+N)-1}\,.\\
&\quad\prod_{i=1}^{M}\omega(\rho k_i)^{-1/2+\mu/2}\prod_{j=1}^{N}\omega(\rho\tilde{k}_j)^{-1/2+\mu/2}\\
&= \varepsilon\left[1+\mathcal{O}(\varepsilon\rho^{-1/2})\right] g^{M+N}\quad \underbrace{\rho^{(1+\frac{\mu}{2})(M+N)-1}}_{\leq\rho^{\mu/2},\ \text{since}\ \underline{M+N\geq 1}}\\
&= \prod_{i=1}^{M}\omega(k_i)^{-1/2+\mu/2}\prod_{j=1}^{N}\omega(\tilde{k}_j)^{-1/2+\mu/2}\,,
\end{aligned}$$

implying that

$$\|w_{M+N}^{(1)}[z]\|_\mu^\# \;\leq\; \varepsilon\cdot\rho^{\mu/2}\cdot\underbrace{\left[1+\mathcal{O}\left(\varepsilon\rho^{-1/2}\right)\right]}_{\leq 2}. \tag{87}$$

We now define the renormalization map

$$\mathcal{R}_\rho\,:\,\mathcal{W}_\mu\to\mathcal{W}_\mu\,, \tag{88}$$

by

$$H[z] \;\mapsto\; \mathcal{R}_\rho(H)[z] \;:=\; \frac{1}{\rho}\Gamma_\rho\,\mathcal{F}_{\chi_\rho}\left(H\left[Z^{-1}(z)\right]-Z^{-1}(z)\right)\,\Gamma_\rho^* \;-\; z\,, \tag{89}$$

where $Z : \{|z - E[z]| < \rho/2\} \to D_{1/2}$, $Z(\zeta) := \frac{1}{\rho}(\rho - E[\zeta])$ is a suitable rescaling of the spectral parameter, z, to project out an unstable direction of the flow generated by $\mathcal{R}$. Observe that $\mathcal{R}_\rho$ has the isospectral property:

$$H\left[Z^{-1}(z)\right]-Z^{-1}(z)\ \text{invertible on}\ \mathcal{H}_{red} \;\Leftrightarrow\; \mathcal{R}_\rho(H)[z]-z\ \text{invertible on}\ \mathcal{H}_{red}\,. \tag{90}$$

The main fact about $\mathcal{R}_\rho$ is its contraction property:

Theorem 1. *$\mathcal{R}_\rho$ is defined on $\mathcal{B}(\delta,\varepsilon)$, for $0<\delta,\varepsilon\leq 1/16$, and, for $\rho^\mu<1/2$,*

$$\mathcal{R}\rho \,:\, \mathcal{B}(\delta,\varepsilon)\to\mathcal{B}\left(\delta+\frac{\varepsilon}{2},\,\frac{\varepsilon}{2}\right)\,. \tag{91}$$

To apply $\mathcal{R}_\rho$ iteratively, we define $H_{(n)}[z]:=\mathcal{R}^n(H_{(0)})[z]$. According to Theorem 1, we have

$$\begin{aligned} H_{(0)}\in\mathcal{B}(cg,cg) &\Leftrightarrow H_{(1)}\in\mathcal{B}\big(cg+\frac{cg}{2},\,\frac{cg}{2}\big)\\ &\Leftrightarrow H_{(2)}\in\mathcal{B}\big(cg+\frac{cg}{2}+\frac{cg}{4},\,\frac{cg}{4}\big)\\ &\Leftrightarrow\cdots\quad\Leftrightarrow\quad H_{(n)}\in\mathcal{B}\Big(cg\sum_{k=0}^{n}2^{-k},\;cg2^{-n}\Big)\,. \end{aligned} \tag{92}$$

Since we have chosen an appropriate topology, the intersection

$$\{z_*\}\;:=\;\bigcap_{n=1}^{\infty}(Z_1^{-1}\circ\ldots\circ Z_n^{-1})[D_{1/2}] \tag{93}$$

defines a number $z_*\in\mathbb{R}$ which turns out to be the perturbed eigenvalue sought for. Indeed, $H_{(n)}$ converges in the sense of (92), and the limit contains no interaction term anymore,

$$H_{(n)}\;\overset{\|\cdot\|_\Delta}{\to}\;H_{(\infty)}[z_*]\;=\;E_{(\infty)}[z_*]\;+\;T_{(\infty)}(H_f;[z_*])\,. \tag{94}$$

This implies, in particular, that the vacuum vector is an eigenvector of $H_{(\infty)}[z_*]$,

$$H_{(\infty)}[z_*]\,\Omega\;=\;E_{(\infty)}[z_*]\Omega\,. \tag{95}$$

We can now recover the original eigenvector of H_g-z_* by successively applying the operator S_{χ_ρ} from Lemma 2(d).

References

1. V. Bach, T. Chen, J. Fröhlich, and I. M. Sigal. Smooth Feshbach map and operator-theoretic renormalization group methods. *J. Funct. Anal.*, 203(1):44–92, 2003.
2. V. Bach, J. Fröhlich, and I. M. Sigal. Quantum electrodynamics of confined non-relativistic particles. *Adv. in Math.*, 137:299–395, 1998.
3. V. Bach, J. Fröhlich, and I. M. Sigal. Renormalization group analysis of spectral problems in quantum field theory. *Adv. in Math.*, 137:205–298, 1998.
4. V. Bach, J. Fröhlich, and I. M. Sigal. Spectral analysis for systems of atoms and molecules coupled to the quantized radiation field. *Commun. Math. Phys.*, 207(2):249–290, 1999.
5. V. Bach, J. Fröhlich, and I. M. Sigal. Return to equilibrium. *J. Math. Phys.*, 41(6):3985–4060, June 2000.

Local States of Free Bose Fields

S. De Bièvre

UFR de Mathématiques et UMR P. Painlevé, Université des Sciences et Technologies de Lille, 59655 Villeneuve d'Ascq Cedex France
Stephan.De-Bievre@math.univ-lille1.fr

S. De Bièvre: *Local States of Free Bose Fields*, Lect. Notes Phys. **695**, 15–61 (2006)
www.springerlink.com

1 Introduction

These notes contain an extended version of lectures given at the "Summer School on Large Coulomb Systems" in Nordfjordeid, Norway, in august 2003. They furnish a short introduction to some of the most basic aspects of the theory of quantum systems that have a dynamics generated by an equation of the form

$$\ddot{q} = -\Omega^2 q$$

where Ω is a self-adjoint, positive, invertible operator on a dense domain $\mathcal{D}(\Omega)$ in a real Hilbert space $\mathcal{K}$.

Such systems are usually referred to as free bose fields. They are really just harmonic systems and I will occasionally use the term oscillator fields since I will also discuss their *classical* counterparts and because I want to stress the instructive analogy with finite systems of coupled oscillators, which is very helpful when one tries to understand the underlying physical interpretation of the theory.

Many of the simplest systems of classical and quantum mechanics obey an equation of this form. Examples include (see Sect. 2.2):

(i) Finite dimensional systems of coupled oscillators, where $\mathcal{K} = \mathbb{R}^n$ and Ω is a positive definite matrix.
(ii) Lattices or chains of coupled oscillators, where $\mathcal{K} = \ell^2(\mathbb{Z}^d, \mathbb{R})$ and Ω is usually a bounded operator with a possibly unbounded inverse. Those are used to model lattice vibrations in solid state physics.
(iii) The wave equation, where $\mathcal{K} = L^2(K, \mathbb{R})$, $K \subset \mathbb{R}^d$ and $\Omega^2 = -\Delta$ with suitable boundary conditions.
(iv) The massive or massless Klein-Gordon equation on static spacetimes. These are a popular paradigm for studying quantum field theory on curved spacetimes.

Despite their supposed simplicity, these systems are interesting for at least two reasons. First, they provide examples where the basic concepts and methods of quantum field theory can be explained, understood and tested. Second, they provide the building blocks for the study of more complicated systems in quantum field theory and (non-equilibrium) statistical mechanics, where one or more such fields are (nonlinearly) coupled to each other or to other, possibly finite dimensional systems. Bose fields are for example a popular tool for modelling heath baths. The much studied spin-bose model and more generally the Pauli-Fierz models are all of this type.

In Sect. 2, I shall first briefly describe the classical mechanics of such systems in a unified way. This will then allow us in Sect. 3 to write down the corresponding quantum mechanical systems – the free bose fields – in a straightforward manner, for *both infinite and finite dimensional* systems. In particular, if you are familiar with the quantum mechanical description of finite dimensional systems, you should conclude after reading these two sections that the description of the infinite dimensional systems can be done quite analogously.

At that point, we will be ready to start studying the systems constructed, and to analyze their physical properties. The only issue I will address here, in Sect. 4, is not one that features prominently in quantum field theory books, but it has generated a fair amount of debate and even controversy. It is the one of local observables, and of local states, essential for the physical interpretation of the theory. Other topics will be discussed in [6]. I will adopt the definition of Knight of "strictly local excitation of the vacuum" (Definition 4), that I will refer to as a strictly local or a strictly localized state for brevity. I will then state and prove a generalization of Knight's Theorem [16] (Sect. 4.5) which asserts that finite particle states cannot be perfectly localized. It will furthermore be explained how Knight's a priori counterintuitive result can be readily understood if one remembers the analogy between finite and infinite dimensional harmonic systems alluded to above. I will also discuss the link between the above result and the so-called Newton-Wigner position operator thereby illuminating, I believe, the difficulties associated with the latter (Sect. 4.7). I will in particular argue that those difficulties do not find their origin in special relativity or in any form of causality violation, as is usually claimed. It will indeed be seen that the Newton-Wigner position operator has an immediate analog for a finite or infinite system of oscillators, and that it makes absolutely no sense there since it is at odds with basic physical intuition and since it is not compatible with the physically reasonable definition of Knight. The conclusion I will draw is that *the Newton-Wigner operator does not provide an appropriate tool to describe the strict localization properties of the states of extended systems of the type discussed here.* It shows up only because of an understandable but ill-fated desire to force too stringent a *particle interpretation* with all its usual attributes on the states of a *field.* The right notion of a (strictly) localized state is the one given by Knight. These issues have generated some debate in the context of relativistic quantum field theory over the years, upon which I shall comment in Sect. 4.7.

The text is written at the graduate level and is aimed at an audience of mathematicians with a taste for physics and of physicists with a taste for mathematics. A background in the classical and quantum theory of *finite* dimensional systems is assumed, although the text is basically self-contained. The approach to the subject chosen here differs both from the usual "second quantization" and "canonical quantization" treatments of quantum field theory prevalent in the physics literature (although it is very close to the latter). It is not axiomatic either. I feel it is fruitful because it allows one to apply the intuition gained from the study of finite dimensional systems in the infinite dimensional case. This helps in developing a good understanding of the basic physics of quantum field theory, and in particular to do away with some of the confusion surrounding even some of the simplest aspects of this theory, as I hope to illustrate with the discussion of "localization" in this context.

Although my approach here is resolutely non-relativistic, I hope to show it still sheds an interesting and illuminating light on relativistic theories as well. Indeed, the main feature of the systems under consideration is their infinite spatial extension, and it is this feature that distinguishes them from systems with a finite number of particles such as atomic or molecular systems, that have a finite spatial extension.

Related topics will be discussed in a much extended version of this manuscript, which is in preparation [6].

2 Classical Free Harmonic Systems

2.1 The Hamiltonian Structure

Let us now turn to the systems described briefly in the Preface. My first goal is to describe in detail the Hamiltonian structure underlying

$$\ddot{q} + \Omega^2 q = 0 \,. \tag{1}$$

For finite dimensional systems, it is well known how to view (1) as a Hamiltonian system, and we will now show how to do this for infinite dimensional systems using as only ingredient the positive operator Ω^2 on $\mathcal{D}(\Omega^2) \subset \mathcal{K}$. We need to identify a phase space on which the solutions to this equation define a Hamiltonian flow for a suitable Hamiltonian. For that purpose, note that, formally at least, (1) is equivalent to

$$\dot{q} = p, \ \dot{p} = -\Omega^2 q \,,$$

which are Hamilton's equations of motion for the Hamiltonian ($X = (q, p)$)

$$H(X) = \frac{1}{2} p \cdot p + \frac{1}{2} q \cdot \Omega^2 q \,, \tag{2}$$

with respect to the symplectic structure

$$s(X, X') = q \cdot p' - q' \cdot p \,.$$

Note that I use $\cdot$ for the inner product on $\mathcal{K}$. The Poisson bracket of two functions f and g on $\mathcal{K} \oplus \mathcal{K}$ is neatly expressed in terms of s by

$$\{f, g\} = s(\nabla_X f, \nabla_X g) \,,$$

where $\nabla_X f = (\nabla_q f, \nabla_p f)$. Solving Hamilton's equations of motion one obtains the Hamiltonian flow which in this case can simply be written

$$\Phi_t = \cos \Omega t I_2 - \sin \Omega t J \,, \tag{3}$$

where

$$I_2 = \begin{pmatrix} 1\,0 \\ 0\,1 \end{pmatrix}, \quad J = \begin{pmatrix} 0 & -\Omega^{-1} \\ \Omega & 0 \end{pmatrix}. \tag{4}$$

For later purposes, we remark that the corresponding Hamiltonian vector field X_H defined by

$$\frac{\mathrm{d}\Phi_t}{\mathrm{d}t} = X_H \Phi_t \,. \tag{5}$$

can be written

$$X_H = -J\Omega \,. \tag{6}$$

Of course, this is sloppy, because whereas s defines a symplectic structure on $\mathcal{K}\oplus\mathcal{K}$, the operator J is not a bounded operator on $\mathcal{K}\oplus\mathcal{K}$, so that the flow is not globally defined on this space! In other words, in the infinite dimensional case, we have to remember that both Ω and Ω^{-1} may be unbounded operators (think of the wave equation, for example) and therefore we have to carefully identify a suitable phase space on which both the symplectic structure and the flow Φ_t are globally well-defined. For that purpose, we introduce the scale of spaces ($\lambda \in \mathbb{R}$):

$$\mathcal{K}_\lambda = [\mathcal{D}(\Omega^\lambda)] \,.$$

Here the notation [ ] means that we completed $\mathcal{D}$ in the topology induced by $\| \Omega^\lambda q \|$ where $\| \cdot \|$ is the Hilbert space norm of $\mathcal{K}$: note that we have supposed that Ω has a trivial kernel, so that $\| \Omega^\lambda q \|$ defines a norm (and not just a semi-norm). Explicit examples are developed in Sect. 2.2.

It is easy to check that J and hence Φ_t are globally well defined on

$$\mathcal{H} = \mathcal{K}_{1/2} \oplus \mathcal{K}_{-1/2} \,.$$

Moreover, the symplectic form can also defined on this space via

$$s(X, X') = \Omega^{1/2}q \cdot \Omega^{-1/2}p' - \Omega^{1/2}q' \cdot \Omega^{-1/2}p \,. \tag{7}$$

Actually, it can be checked that $\mathcal{H}$ is the only space of the form $\mathcal{K}_\lambda \oplus \mathcal{K}_\mu$ with these properties. In what follows, I shall refer to $\mathcal{H}$ as the (real) phase space of the system. Note that, from now on, whenever $w \in \mathcal{K}_\lambda, w' \in \mathcal{K}_{-\lambda}$, we will write $w \cdot w' = \Omega^\lambda w \cdot \Omega^{-\lambda} w'$. With these notations, one easily checks that, for $a \in \mathcal{K}_{1/2}, b \in \mathcal{K}_{-1/2}$,

$$\{b \cdot q, a \cdot p\} = a \cdot b \,. \tag{8}$$

Here $\{\cdot,\cdot\}$ denotes the Poisson bracket.

Note that the phase space $\mathcal{H}$ may depend on Ω, for fixed $\mathcal{K}$. As long as both Ω and Ω^{-1} are bounded operators, one has clearly $\mathcal{H}(\Omega) = \mathcal{K}\oplus\mathcal{K}$. This is of course always the case when $\mathcal{K}$ is finite dimensional. So for systems with a finite number of degrees of freedom, the phase space is fixed a priori to be $\mathcal{K} \oplus \mathcal{K}$, and the dynamics can be defined a posteriori on this fixed phase space. However, whenever either Ω or Ω^{-1} are unbounded, $\mathcal{H}(\Omega)$ differs from $\mathcal{K} \oplus \mathcal{K}$ and depends explicitly on Ω. In other words, one cannot first choose the phase space, and then study various different dynamics on it. Instead, the phase space and the dynamics are intimately linked: changing the dynamics on a given fixed phase space may not make sense.

To conclude, so far, we have shown how the solutions of (1) define a (linear) Hamiltonian flow Φ_t on a (real) symplectic vector space $(\mathcal{H}, s)$.

As far as the classical mechanics of the system is concerned, this is really all we need. In order to construct the corresponding quantum theory (Sect. 3), and in particular the quantum Hilbert space, we do however need to exploit the structures underlying the classical theory some more. This I will do in Sect. 2.3. If we were only interested in the finite dimensional case, this would be of some interest, but not necessary. For the infinite dimensional case it is essential. Indeed, for finite dimensional harmonic systems, the usual Schrödinger quantum mechanics is of course perfectly adequate, and the formalism developed here is quite useless. It is however not possible to straightforwardly adapt the Schrödinger formulation to the infinite dimensional situation, and so we need to exploit the additional structures a little more. To understand the following developments, it is helpful to have some examples in mind.

2.2 Examples

Coupled Oscillators: Finite Dimension

Systems of point masses connected by springs have Hamiltonians of the type

$$H(X) = \frac{1}{2}(p^2 + q \cdot \Omega^2 q)$$

where $X = (q, p) \in \mathbb{R}^{2n}$, so that here $\mathcal{K} = \mathbb{R}^n$, and Ω^2 is a positive definite $n \times n$ matrix. More generally, this Hamiltonian arises when linearizing any potential about a stable equilibrium point. An instructive example is the

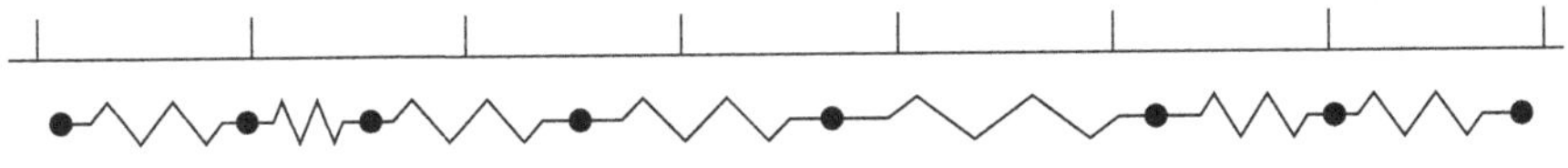

Fig. 1. A schematic representation of a chain of 8 oscillators moving horizontally. Linking the first to the last, you get a ring. The tick marks indicate their equilibrium positions. In the figure $\omega_{\mathrm{w}} = 0$

finite *oscillator chain* with periodic boundary conditions (see Fig. 1). There, n particles, constrained to move in one dimension only, are placed on a ring. They interact with their nearest neighbours only, through a force that is linear in the relative displacement of the particles and that is characterized by a frequency $\omega_{\rm n}$. In addition they are each subjected to a harmonic force with frequency $\omega_{\rm w}$. Assuming all the particles have identical masses, set equal to 1, the Hamiltonian for this system reads

$$H(X) = \frac{1}{2}\left(\sum_{i=1}^{n}(p(i)^2 + \omega_{\rm w}^2 q(i)^2 + \omega_{\rm n}^2(q(i+1) - q(i))^2\right) .$$

Note that in the sum the index is to be taken periodically, so that $q(n+1) = q(1)$, etc. I have adopted here and will continue to use the somewhat unusual notation $v(i)$ for the ith component of a vector $v \in \mathbb{R}^n$ or $\mathbb{C}^n$. This will prove very convenient later on. Introducing $\omega_0^2 = \omega_{\rm w}^2 + 2\omega_{\rm n}^2 > 0$ and

$$0 \leq \nu = \frac{\omega_{\rm n}^2}{\omega_0^2} \leq 1/2 ,$$

the equation of motion is, for all $j = 1, \ldots n$,

$$\ddot{q}(j) = -\omega_0^2\left[q(j) - \nu(q(j+1) + q(j-1))\right] = -(\Omega^2 q)(j) . \tag{9}$$

One readily finds the eigenvalues of Ω^2: they are given by

$$\omega^2(k) = \omega_0^2[1 - 2\nu\cos 2\pi k] \quad k = 1/n, 2/n, \ldots 1 .$$

Note that the eigenvalues are indeed positive, but, in order to make sure that 0 is not an eigenvalue, we have to impose $\nu < 1/2$, which amounts to requiring that $\omega_{\rm w} \neq 0$. This is intuitively clear: if $\omega_{\rm w} = 0$, the system allows for stationary solutions in which all oscillators are displaced by the same amount, so that the springs between the oscillators are not stretched. These are are referred to as a "zero modes". The above Hamiltonian provides the simplest model possible for a harmonic crystal, and is discussed in all books on solid state physics both from the classical and the quantum mechanical point of view.

Oscillator Chains and Lattices

Having understood the finite oscillator chain, it is easy to understand the first infinite dimensional system we shall consider, which is an infinite linear chain of oscillators, each one linked to its neighbours and to a wall with identical springs, so that the system is translationally invariant. The Hamiltonian and equation of motion of this system are the same as in the case of the ring, except that the sums now run over $\mathbb{Z}$. We now have $\mathcal{K} = \ell^2(\mathbb{Z}, \mathbb{R})$ and Ω^2, defined precisely as in (9) is a bounded operator. It has a purely absolutely continuous spectrum $\{\omega^2(k) \mid k \in [0,1]\}$, for all values of $\nu \in [0, 1/2]$. Indeed, even if $\nu = 1/2$, 0 is not an eigenvalue of Ω^2, since η_0 does not belong to $\ell^2(\mathbb{Z}, \mathbb{R})$.

It is instructive to identify the spaces $\mathcal{K}_\lambda$ explicitly in this case. For that purpose, note that the Fourier series transform

$$\hat{q}(k) = \sum_{j\in\mathbb{Z}} q(j)\mathrm{e}^{-\mathrm{i}2\pi jk}$$

identifies the real Hilbert space $\ell^2(\mathbb{Z},\mathbb{R})$ with the real subspace of the complex Hilbert space $L^2(\mathbb{R}/\mathbb{Z}, dk, \mathbb{C})$ for which $\overline{\hat{q}(k)} = \hat{q}(-k)$. It follows that $\mathcal{K}_\lambda$ can be identified with the space of locally integrable functions $\hat{q}$ for which $\overline{\hat{q}(k)} = \hat{q}(-k)$ and, more importantly, $\omega(k)^\lambda \hat{q}(k)$ belongs to $L^2(\mathbb{R}/\mathbb{Z}, dk, \mathbb{C})$.

First of all, consider $0 \le \nu < 1/2$. Then the spectrum is bounded away from zero, which means that both $\omega(k)$ and $\omega(k)^{-1}$ are bounded functions of k. As a result, then, for all $\lambda \in \mathbb{R}$, $\mathcal{K}_\lambda = \ell^2(\mathbb{Z},\mathbb{R})$. In particular, then $\mathcal{H} = \ell^2(\mathbb{Z},\mathbb{R}) \times \ell^2(\mathbb{Z},\mathbb{R})$ and does *not* depend on the value of ν in the range considered.

Something interesting happens, however, if we consider the case $\nu = 1/2$. Remember that this corresponds to setting $\omega_{\mathrm{w}} = 0$, which was not allowed in the finite ring because of the existence of the zero mode. Some remnant of this problem shows up here. Indeed, consider $\mathcal{K}_\lambda$, for $\lambda < 0$. Since

$$\omega^2(k) = (2\pi\omega_0)^2 k^2 + \mathrm{o}(k^2) ,$$

$q \in \mathcal{K}_\lambda$ if and only if $|k|^\lambda \hat{q}(k)$ belongs to $L^2(\mathbb{R}/\mathbb{Z}, dk, \mathbb{C})$ (and of course satisfies $\overline{\hat{q}(k)} = \hat{q}(-k)$). But, for $\lambda = -1/2$, this is not true for all $q \in \ell^2(\mathbb{Z},\mathbb{R})$. As a result, $\mathcal{K} = \ell^2(\mathbb{Z},\mathbb{R})$ is not a subspace of $\mathcal{K}_{-1/2}$ and similarly $\mathcal{K}_{1/2}$ is not a subspace of $\mathcal{K} = \ell^2(\mathbb{Z},\mathbb{R})$. Hence the phase space $\mathcal{H}$ is now different, as a set, from the phase space when $\nu \neq 1/2$ and in addition, one phase space is not included in the other. To see this has noticeable physical consequences, note the following. It seems like a reasonable thing to wish to study the motion of the chain when initially only one of the degrees of freedom is excited. Suppose therefore you wish to pick the initial condition $q(i) = p(i) = 0$, for all $i \neq 0$, $q(0) = 0 \neq p(0)$. In other words, the oscillator at the origin starts from its equilibrium position with a non-zero initial speed, while all other oscillators are at rest at their equilibrium positions. The trouble is that, when $\nu = 1/2$, this initial condition does not belong to the phase space! So it should be remembered that the choice of phase space I made here, which is reasonable from many a point of view, seems to nevertheless be somewhat too restrictive in this particular case, since it excludes certain very reasonable initial conditions from the state space of the system. This is one aspect of the so-called infrared problem and it will be relevant when discussing local observables in Sect. 4.

The generalization of the preceding considerations to d-dimensional translationally invariant lattices of oscillators is immediate. One has $\mathcal{K} = \ell^2(\mathbb{Z}^d,\mathbb{R})$ and, for all $j \in \mathbb{Z}^d$,

$$(\Omega^2 q)(j) = \omega_{\mathrm{w}}^2 q(j) - \omega_{\mathrm{n}}^2 \sum_{i\in\mathrm{nn}(j)} (q(i) - q(j)) = \omega_0^2 q(j) - \omega_{\mathrm{n}}^2 \sum_{i\in\mathrm{nn}(j)} q(i) , \quad (10)$$

where $\mathrm{nn}(j)$ designates the set of nearest neighbours of j and where this time

$$\omega_0^2 = \omega_{\mathrm{w}}^2 + 2d\omega_{\mathrm{n}}^2 \qquad \text{and} \qquad 0 \leq \nu = \frac{\omega_{\mathrm{n}}^2}{\omega_0^2} \leq \frac{1}{2d} .$$

Using the Fourier transform to diagonalize Ω^2 one finds the dispersion relation

$$\omega(k)^2 = \omega_0^2 \left[1 - 2\nu \left(\sum_{i=1}^{d} \cos 2\pi k_i\right)\right] .$$

This time the critical value of ν is $1/2d$ but it leads to less severe infrared behaviour. Indeed, if $\nu = 1/2d$, then

$$\omega(k)^2 = \omega_0^2 \frac{1}{d} \sum_{i=1}^{d} (2\pi k_i)^2 + \mathrm{o}(|k|^2) = \frac{(2\pi\omega_0)^2}{d} |k|^2 + \mathrm{o}(|k|^2) .$$

But now all compactly supported q belong to $\mathcal{K}_\lambda$, for all $-d/2 < \lambda$, as is easily checked. As a result, this time the phase space $\mathcal{H}$ contains all such initial conditions as soon as $d \geq 2$. We shall refer to them as *strictly local perturbations from equilibrium* and study their quantum analogues in Sect. 4. To be more precise, if $d \geq 2$, and if we denote by $C_{\mathrm{c}}(\mathbb{Z}^d)$ the space of compactly supported sequences, then $C_{\mathrm{c}}(\mathbb{Z}^d) \times C_{\mathrm{c}}(\mathbb{Z}^d) \subset \mathcal{H}$, for all possible values of ν. If $X = (q,p) \in C_{\mathrm{c}}(\mathbb{Z}^d) \times C_{\mathrm{c}}(\mathbb{Z}^d)$, then X describes an initial state in which only a finite number of oscillators is displaced from their equilibrium position and/or moving. So for this rather large and very natural class of initial conditions, the dynamics can be investigated as a function of ν, for all possible values of ν.

Lattices of oscillators are used to describe the thermal and acoustic properties of various solids, such as metals, crystals of all sorts, amorphous materials etc. Putting $\omega_{\mathrm{n}} = 0$ in the expressions above, one obtains the so-called Einstein model, in which the oscillators representing the ions of the solid are not coupled. The case where $\omega_{\mathrm{n}} \neq 0$ is the Debye model. In more sophisticated models still, different geometries may appear (hexagonal lattices, body or face centered cubic lattices etc.), and the spring constants may vary from site to site in periodic, quasi-periodic or random ways.

Wave and Klein-Gordon Equations

The wave equation

$$\partial_t^2 q(x,t) = \Delta q(x,t)$$

on a domain $K \subset \mathbb{R}^d$ with Dirichlet boundary conditions is another example of a free oscillator field where $\mathcal{K} = L^2(K,\mathbb{R})$ and $\mathcal{D}(\Omega)$ is the domain of the square root of the Dirichlet Laplacian. When K is a bounded set, the spectrum of the Dirichlet Laplacian is discrete. No infrared problem then arises, reflecting the fact that no arbitrary long wavelengths can occur in the system.

The case where $K = \mathbb{R}^d$ is instructive and easy to work out thanks to its translational invariance. The situation is completely analogous with the one in Sect. 2.2. Writing $\omega(k) = \sqrt{k^2}$, the space $\mathcal{K}_\lambda$ is for each real λ naturally isomorphic to the real subspace of $L^2(\mathbb{R}^d, \omega(k)^{2\lambda} dk, \mathbb{C})$ given by the condition $\overline{\hat{q}(k)} = q(-k)$. If $d \geq 2$, the Schwartz space is a subspace of $\mathcal{K}_{\pm\frac{1}{2}}$.

One can also consider the more general case where K is a Riemannian manifold with metric γ and $-\Delta$ the corresponding Laplace-Beltrami operator. Replacing $-\Delta$ by $-\Delta + m^2$ ($m > 0$) in the above, one obtains the Klein-Gordon equation. It plays an important role in the relativistic quantum field theory on flat or curved spacetimes.

2.3 A Preferred Complex Structure on the *Real* Classical Phase Space $\mathcal{H}$

The simple linear systems we are dealing with here have some extra structure that is encoded in the matrix J defined in (4). Noticing that $J^2 = -I_2$, one sees J defines an s-compatible (i.e. $s(JX, JY) = s(X,Y)$) and positive definite (i.e. $s(X, JX) \geq 0$ and $s(X, JY) = 0, \forall Y \in \mathcal{H}$ implies $X = 0$) complex structure on $\mathcal{H}$. As a result, $\mathcal{H}$ can first of all be viewed as a real Hilbert space, with inner product

$$g_\Omega(X,Y) \stackrel{\text{def}}{=} s(X, JY) = \sqrt{\Omega} q \cdot \sqrt{\Omega} q' + \sqrt{\Omega}^{-1} p \cdot \sqrt{\Omega}^{-1} p' , \qquad (11)$$

where $Y = (q', p')$. Of course, we recognize here the natural inner product on $\mathcal{H} = \mathcal{K}_{1/2} \oplus \mathcal{K}_{-1/2}$, written in terms of the symplectic form and J.

In addition, J can be used to equip $\mathcal{H}$ with a *complex* Hilbert space structure, where multiplication with the complex number $a + ib \in \mathbb{C}$ is defined by

$$(a + \mathrm{i}b)X \stackrel{\text{def}}{=} (a + bJ)X, \quad \forall X \in \mathcal{H}$$

and with the inner product

$$\langle X, Y \rangle_+ = \frac{1}{2}(g_\Omega(X,Y) + \mathrm{i}s(X,Y)) . \qquad (12)$$

Note that, when $\mathcal{H}$ has $2n$ *real* dimensions, the complex vector space $(\mathcal{H}, J)$ has only n complex dimensions.

Since Φ_t is symplectic and commutes with J, one easily checks that

$$g_\Omega(\Phi_t X, \Phi_t Y) = g_\Omega(X,Y) \quad \text{and} \quad \langle \Phi_t X, \Phi_t Y \rangle_+ = \langle X, Y \rangle_+ ,$$

so that Φ_t is a unitary operator on the complex Hilbert space $(\mathcal{H}, J, \langle \cdot, \cdot \rangle_+)$. As a result, $X_H = -J\Omega$, the generating Hamiltonian vector field is necessarily anti-self-adjoint and one can check that in addition

$$\mathrm{i}\langle X, X_H X\rangle_+ = H(X)\,. \tag{13}$$

It is natural to wonder if there exist many complex structures on $\mathcal{H}$ with these properties. In fact, J is the unique s-compatible, positive complex structure on $\mathcal{H}$ so that Φ_t is unitary on the corresponding complex Hilbert space [6]. In other words, the phase space $\mathcal{H}$ of an oscillator field, which is a *real* symplectic space, carries a natural, flow-invariant complex Hilbert space structure!

The ensuing complex Hilbert space seems a somewhat abstract object, but it can be naturally identified with $\mathcal{K}^{\mathbb{C}}$, the complexification of $\mathcal{K}$, as I now explain. In the following, whenever V is a real vector space, $V^{\mathbb{C}} = V \oplus \mathrm{i}V$ will denote its complexification. In the concrete examples I have in mind, where $V = \mathcal{K} = \mathbb{R}^n, \ell^2(\mathbb{Z}^d, \mathbb{R})$ or $L^2(\mathbb{R}^d, \mathbb{R})$, one finds $V^{\mathbb{C}} = \mathcal{K}^{\mathbb{C}} = \mathbb{C}^n, \ell^2(\mathbb{Z}^d, \mathbb{C})$ or $L^2(\mathbb{R}^d, \mathbb{C})$, respectively. The identification goes as follows :

$$z_\Omega : X = (q,p) \in \mathcal{H} \mapsto z_\Omega(X) = \frac{1}{\sqrt{2}}\left(\sqrt{\Omega}q + \mathrm{i}\frac{1}{\sqrt{\Omega}}p\right) \in \mathcal{K}^{\mathbb{C}}\,. \tag{14}$$

The following proposition is then easily proven.

Proposition 1. *The map z_Ω defines an isomorphism between the complex Hilbert spaces $(\mathcal{H}, J, \langle\cdot,\cdot\rangle_+)$ and $\mathcal{K}^{\mathbb{C}}$, intertwining the dynamics Φ_t with $\mathrm{e}^{-\mathrm{i}\Omega t}$. More precisely,*

$$z_\Omega(JX) = \mathrm{i}z_\Omega(X) \quad \overline{z_\Omega(X)}\cdot z_\Omega(X') = \langle X, X'\rangle_+\,. \tag{15}$$

and

$$z_\Omega(\Phi_t X) = \mathrm{e}^{-\mathrm{i}\Omega t} z_\Omega(X)\,. \tag{16}$$

Note that $\cdot$ has been extended to $\mathcal{K}^{\mathbb{C}}$ by linearity in each variable so that the inner product on $\mathcal{K}^{\mathbb{C}}$ is given by $\bar{z}\cdot z'$, for $z, z' \in \mathcal{K}^{\mathbb{C}}$. The choice of the unnatural looking factor $1/\sqrt{2}$ in the definition of z_Ω and of the matching factor $1/2$ in $\langle X, X'\rangle_+$ are conventions chosen to make comparison to the physics literature simple, as we will see further on. Similarly, for later purposes, we define

$$z^\dagger_\Omega : X = (q,p) \in \mathcal{H} \mapsto z^\dagger_\Omega(X) = \frac{1}{\sqrt{2}}(\sqrt{\Omega}q - \mathrm{i}\frac{1}{\sqrt{\Omega}}p) \in \mathcal{K}^{\mathbb{C}}\,, \tag{17}$$

which is complex anti-linear

$$z^\dagger_\Omega(JX) = -\mathrm{i}z^\dagger_\Omega(X) \tag{18}$$

and

$$\overline{z^\dagger_\Omega(X)}\cdot z^\dagger_\Omega(X') = \overline{\langle X, X'\rangle_+}\,. \tag{19}$$

The linear map z_Ω is readily inverted and one has, in obvious notations

$$q = \frac{1}{\sqrt{2\Omega}}(z_\Omega(X) + z^\dagger_\Omega(X)) \text{ and } p = \frac{\sqrt{\Omega}}{\mathrm{i}\sqrt{2}}(z_\Omega(X) - z^\dagger_\Omega(X))\,, \tag{20}$$

and

$$H(X) = z_\Omega(X)^\dagger \cdot \Omega z_\Omega(X)\,. \tag{21}$$

In conclusion, we established that, having started with a *real* Hilbert space $\mathcal{K}$ and a positive self-adjoint operator Ω, the classical phase space $\mathcal{H}$ of the corresponding oscillator equation $\ddot{q} = -\Omega^2 q$ can be identified naturally with the *complex* Hilbert space $\mathcal{K}^{\mathbb{C}}$, on which the dynamics is simply the unitary group generated by Ω, the symplectic structure is the imaginary part of the inner product and the Hamiltonian is given by $H(z) = \overline{z} \cdot \Omega z$. We therefore ended up with a *mathematically* completely equivalent description of the original phase space $\mathcal{H}$, its symplectic structure and the dynamics Φ_t generated by the Hamiltonian H in (2).

It is however important to understand that the *physical* interpretation of this new formulation should be done carefully, as I explain in Sect. 2.4.

2.4 Physical Interpretation

It is instructive to first look at what the formalism of the Sect. 2.3 yields for *finite dimensional* systems of coupled oscillators, such as the oscillator ring. In that case $\mathcal{K} = \mathbb{R}^n$ and hence $\mathcal{K}^{\mathbb{C}} = \mathbb{C}^n$. Note however that the identification of $\mathbb{R}^{2n}$ with $\mathbb{C}^n$ depends in a non-trivial way on Ω which makes a direct interpretation of points of $\mathbb{C}^n$ difficult. In particular, let $X = (q, p) \in \mathbb{R}^{2n} = \mathcal{H}$. Then the components of q and p have a direct physical interpretation as the displacements and momenta of the different oscillators. The ith component of the corresponding vector $z = z_\Omega(X) \in \mathbb{C}^n$ does not have such a direct simple interpretation since it is not a function of the diplacement q_i and momentum p_i of the ith oscillator alone, but it is a function of the diplacements q_j and momenta p_j of all the oscillators. This is so because in general, the matrix $\Omega^{1/2}$ has no (or few) zero off-diagonal entries, even if Ω^2 is tri-diagonal, as in the oscillator chain. Indeed, in that case, Ω^2 is a difference operator, but $\Omega^{1/2}$ is not. Conversely, as is clear from (20), q_i and p_i depend on all components of $z_\Omega(X)$, not only on the ith one. This explains why the alternative formulation of the problem in terms of the complex space $\mathcal{K}^{\mathbb{C}} = \mathbb{C}^n$ is not found in classical mechanics textbooks. Indeed, one is typically interested in questions concerning the displacements of the different oscillators, the energy distribution over the oscillators when the system is in a normal mode, energy propagation along the oscillators when originally only one oscillator is excited, etc. Such questions are obviously more easily addressed in the original formulation.

Another way to see why the alternative formulation leads to interpretational problems is as follows. Suppose we are studying two oscillator systems, one with potential $\frac{1}{2} q \cdot \Omega^2 q$ and another with $\frac{1}{2} q \cdot {\Omega'}^2 q$, where $\Omega^2 \neq {\Omega'}^2$. To fix ideas, we can think of ${\Omega'}^2$ as being a perturbation of Ω^2 which is obtained by changing just one spring constant. Suppose now that the state of the first system is $z \in \mathcal{K}^{\mathbb{C}}$, and of the second is $z' \in \mathcal{K}^{\mathbb{C}}$. Suppose $z = z' = z_0 \in \mathbb{C}^n$.

Would you say the two systems are in the same state? Certainly not in general! Indeed, as a result of what precedes, and in particular of (20), the same point $z_0 \in \mathcal{K}^{\mathbb{C}}$ yields entirely different values for the displacements q_i, q_i' and the momenta p_i, p_i' of the two oscillator systems! Indeed, we would normally say that the two systems are in the same state if the positions q_i, q_i' and momenta p_i, p_i' of the different degrees of freedom take the same values, that is to say if $X = X'$. But that is not the same as saying $z = z'$. In other words, if you decide to say $\mathcal{K}^{\mathbb{C}}$ is the phase space of your system, you should always remember that the physical interpretation of its points *depends on the dynamics,* i.e. on Ω. A similar phenomenon produces itself in the quantum mechanical description of oscillator systems as we will see in Sect. 3.3.

Suppose now we deal with an infinite dimensional oscillator field, such as an oscillator chain or a wave equation. As in the finite dimensional case, the elements of $\mathcal{H}$ then have a direct interpretation in terms of oscillator displacements, wave propagation etc., whereas those of $\mathcal{K}^{\mathbb{C}}$ don't. But now an additional complicating phenomenon that we already pointed out occurs: starting with a fixed $\mathcal{K}$, different choices of Ω may lead to different phase spaces $\mathcal{H}$! We gave an example for the oscillator chain in Sect. 2.2. Talking about "the same state" for different systems now becomes very difficult, since the state space $\mathcal{H}$ depends on the system considered. It is then tempting to prefer the alternative formulation where the phase space $\mathcal{K}^{\mathbb{C}}$ is independent of the dynamics, but at that point it should always be remembered that the same point in $\mathcal{K}^{\mathbb{C}}$ has a different interpretation depending on which system you consider.

In spite of those interpretational difficulties, the alternative formulation of the classical mechanics of oscillator systems will turn out to be useful (and even crucial) in the quantum mechanical description of oscillator fields. Indeed, the quantum Hilbert space for the free oscillator field will be seen to be the symmetric Fock space over $(\mathcal{H}, J, \langle\cdot,\cdot\rangle_+)$ (see Section 3). But identifying the latter with $\mathcal{K}^{\mathbb{C}}$ allows one to conveniently identify the quantum Hilbert space as the symmetric Fock space over $\mathcal{K}^{\mathbb{C}}$. This way, one can work on a fixed Hilbert space, while changing the dynamics by perturbing Ω, for example. This is very convenient. Still, the rather obvious, seemingly trivial and innocuous remarks above concerning the interpretation of the *classical* field theory are at the origin of further, more subtle interpretational difficulties with the quantum field theory of infinite dimensional oscillator fields as well, to which I shall come back in Sects. 3.3 and 3.5.

2.5 Creation and Annihilation Functions on $\mathcal{H}$

For the purposes of quantum mechanics, it will turn out to be convenient to develop the previous considerations somewhat further. Everybody is familiar with creation and annihilation operators in quantum mechanics. These objects are usually described as typically quantum mechanical in nature, but they have a perfectly natural classical analog, that I will call the creation and annihilation functions, and that are defined as follows.

For all $\xi \in \mathcal{K}^{\mathbb{C}}$,

$$a_{\mathrm{c}}(\xi) : X \in \mathcal{H} \mapsto \bar{\xi} \cdot z_{\Omega}(X) \in \mathbb{C} ,$$

and

$$a_{\mathrm{c}}^{\dagger}(\xi) : X \in \mathcal{H} \mapsto \xi \cdot z_{\Omega}^{\dagger}(X) \in \mathbb{C} .$$

Note that $a_{\mathrm{c}}(\xi)$ is anti-linear in ξ, whereas $a_{\mathrm{c}}^{\dagger}(\xi)$ is linear. The index "c" stands for "classical", so that the notation distinguishes between the classical creation/annihilation functions and the quantum creation/annihilation operators, to be introduced later. A direct computation now yields

$$\{a_{\mathrm{c}}(\xi_1), a_{\mathrm{c}}^{\dagger}(\xi_2)\} = -\mathrm{i}\bar{\xi}_1 \cdot \xi_2$$

and

$$a_{\mathrm{c}}(\xi) \circ \Phi_t = a_{\mathrm{c}}(\mathrm{e}^{\mathrm{i}\Omega t}\xi), \quad a_{\mathrm{c}}^{\dagger}(\xi) \circ \Phi_t = a_{\mathrm{c}}^{\dagger}(\mathrm{e}^{\mathrm{i}\Omega t}\xi) .$$

Also, for all $\eta \in \mathcal{K}_{-1/2}^{\mathbb{C}}$

$$\eta \cdot q = \frac{1}{\sqrt{2}}(a_{\mathrm{c}}(\Omega^{-1/2}\bar{\eta}) + a_{\mathrm{c}}^{\dagger}(\Omega^{-1/2}\eta)) = \frac{-\mathrm{i}}{\sqrt{2}}(a_{\mathrm{c}}^{\dagger}(\mathrm{i}\Omega^{-1/2}\eta) - a_{\mathrm{c}}(\mathrm{i}\Omega^{-1/2}\bar{\eta})) , \tag{22}$$

and, similarly, for all $\eta \in \mathcal{K}_{1/2}^{\mathbb{C}}$

$$\eta \cdot p = \frac{\mathrm{i}}{\sqrt{2}}(a_{\mathrm{c}}^{\dagger}(\Omega^{1/2}\eta) - a_{\mathrm{c}}(\Omega^{1/2}\bar{\eta})) . \tag{23}$$

In the language of the physics literature, these two equations express the oscilator field $\eta \cdot q$ and its conjugate field $\eta \cdot p$ *viewed as functions on phase space* in terms of the creation and annihilation functions.

It is finally instructive to write H explicitly in terms of the annihilation and creation functions. This is easily done when Ω has pure point spectrum, i.e. when there exists a basis of normalized eigenvectors for Ω on $\mathcal{K}^{\mathbb{C}}$:

$$\Omega\eta_i = \omega_i\eta_i, \ i \in \mathbb{N} .$$

Then, from (21)

$$H = \frac{1}{2}\sum_i \omega_i \left(a_{\mathrm{c}}^{\dagger}(\eta_i)a_{\mathrm{c}}(\eta_i) + a_{\mathrm{c}}(\bar{\eta}_i)a_{\mathrm{c}}^{\dagger}(\bar{\eta}_i)\right) . \tag{24}$$

Note that both sides of this equation are functions on (a suitable subset of) $\mathcal{H}$. Correspondingly, in quantum mechanics, both sides will be operators on the quantum Hilbert space of states.

3 The Quantum Theory of Free Harmonic Systems

3.1 Finite Dimensional Harmonic Systems: The Schrödinger Representation

How to give a quantum mechanical description of the classical free oscillator fields studied in Sect. 2? I shall proceed in two steps. I will first recall the quantum description of a system of a finite number of coupled oscillators, and then rewrite it in a manner suitable for immediate adaptation to infinite dimension.

The quantum Hamiltonian for a system with n degrees of freedom having a classical Hamiltonian given by

$$H = \frac{1}{2}p^2 + V(q)\ ,$$

where the potential V is a (smooth) real-valued function on $\mathbb{R}^n$ is, in the so-called position (or Schrödinger) representation given by

$$H = \frac{1}{2}P^2 + V(Q)\ ,$$

where $P = -\mathrm{i}\partial/\partial x$ and $Q = x$ are the usual momentum and position operators which are self-adjoint on there natural domains in the "quantum state space" $L^2(\mathbb{R}^n, \mathrm{d}x)$. Note that, just as in the classical description, the state space is independent of the dynamics, which makes it easy to compare the dynamics generated by two different Hamiltonians H and H', with potentials V and V'. To put it differently, just as a given point X in the classical phase space $\mathbb{R}^{2n}$ corresponds to the same state of the system, whatever its dynamics, so a given ψ in $L^2(\mathbb{R}^n)$ yields the same position and momentum distributions for the system, whatever the dynamics to which it is subjected.

Consequently, for an n-dimensional system of coupled oscillators with classical configuration space $\mathcal{K} = \mathbb{R}^n$ and phase space $\mathcal{H} = \mathbb{R}^{2n}$ the quantum Hamiltonian reads

$$H = \frac{1}{2}(P^2 + Q\cdot\Omega^2 Q)\ .$$

Unfortunately, these expressions stop making sense when $\mathcal{K}$ is an infinite dimensional space, in particular since it is not possible to make sense out of $L^2(\mathcal{K})$ in that case. So to describe the quantum mechanics of infinite dimensional harmonic systems, I will first rewrite the above Hamiltonian differently, in a manner allowing for immediate generalization to infinite dimension. This rewriting is, as we shall see, very analogous to the rewriting of the classical mechanics on $\mathcal{K}^{\mathbb{C}} = \mathbb{C}^n$, explained in Sect. 2.3, and is therefore also affected by the interpretational difficulties mentioned in Sect. 2.4. It is nevertheless very efficient and essential.

Let's define, for any $\xi \in \mathcal{K}^{\mathbb{C}} = \mathbb{C}^n$, the so-called creation and annihilation operators

$$\tilde{a}(\xi) = \overline{\xi} \cdot \frac{1}{\sqrt{2}}(\Omega^{1/2}Q + \mathrm{i}\Omega^{-1/2}P), \quad \tilde{a}^\dagger(\xi) = \xi \cdot \frac{1}{\sqrt{2}}(\Omega^{1/2}Q - \mathrm{i}\Omega^{-1/2}P)\,. \quad (25)$$

Note that those are first order differential operators, and that they depend on Ω, although the notation does not bring this dependence out. One checks easily that

$$[\tilde{a}(\xi), \tilde{a}^\dagger(\xi')] = \overline{\xi} \cdot \xi'\,, \quad (26)$$

all other commutators vanishing. In addition, for any $\eta \in \mathbb{C}^n$,

$$\eta \cdot Q = \frac{1}{\sqrt{2}}(\tilde{a}(\Omega^{-1/2}\overline{\eta}) + \tilde{a}^\dagger(\Omega^{-1/2}\eta))\,, \quad (27)$$

and, similarly,

$$\eta \cdot P = \frac{\mathrm{i}}{\sqrt{2}}(\tilde{a}^\dagger(\Omega^{1/2}\eta) - \tilde{a}(\Omega^{1/2}\overline{\eta}))\,. \quad (28)$$

The analogy of this and of the rest of this section with the developments of Sect. 2.5 should be self-evident. In particular, it is clear that the creation and annihilation operators are the "quantization" of the creation and annihilation functions $a_c, a_c^\dagger$ introduced earlier.

Furthermore, let $\eta_i \in \mathcal{K}^{\mathbb{C}} = \mathbb{C}^n, i = 1 \dots n$ be an orthonormal basis of eigenvectors of Ω^2 with eigenvalues $\omega_1^2 \leq \omega_2^2 \leq \dots \leq \omega_n^2$. Then it is easily checked that

$$H = \frac{1}{2}\sum_{i=1}^n \omega_i \left(\tilde{a}^\dagger(\eta_i)\tilde{a}(\eta_i) + \tilde{a}(\overline{\eta}_i)\tilde{a}^\dagger(\overline{\eta}_i)\right) = \sum_{i=1}^n \omega_i \tilde{a}^\dagger(\eta_i)\tilde{a}(\eta_i) + \frac{1}{2}\sum_{i=1}^n \omega_i\,.$$

The spectral analysis of H is now straightforwardly worked out, and described in any textbook on quantum mechanics. Let me recall the essentials.

It is first of all readily checked that there exists a unit vector $|0, \Omega\rangle$ in $L^2(\mathbb{R}^n)$ (unique up to a global phase), for which

$$\tilde{a}(\xi)|0, \Omega\rangle = 0, \forall \xi \in \mathbb{C}^n\,.$$

This common eigenvector of all the annihilation operators $\tilde{a}(\xi)$ is called the "vacuum". Remark that, as a vector in $L^2(\mathbb{R}^n)$, the vacuum $|0, \Omega\rangle$ obviously depends on Ω. One has indeed very explicitly

$$\langle x|0, \Omega\rangle = \frac{(\det\Omega)^{1/4}}{\pi^{n/4}} \exp -\frac{1}{2} x \cdot \Omega x\,. \quad (29)$$

Clearly $H|0, \Omega\rangle = \frac{1}{2}\sum_{i=1}^n \omega_i |0, \Omega\rangle$, so that the vacuum $|0, \Omega\rangle$ is actually the ground state of H. Writing for brevity $\tilde{a}_i = \tilde{a}(\eta_i), \tilde{a}_i^\dagger = \tilde{a}^\dagger(\eta_i)$ it follows (after some work) that the vectors

$$\frac{1}{\sqrt{m_1! m_2! m_3! \dots m_n!}} \left(\tilde{a}_1^\dagger\right)^{m_1} \left(\tilde{a}_2^\dagger\right)^{m_2} \left(\tilde{a}_3^\dagger\right)^{m_3} \dots \left(\tilde{a}_n^\dagger\right)^{m_n} |0, \Omega\rangle\,, \quad (30)$$

for all possible choices $(m_1, \ldots, m_n) \in \mathbb{N}^n$ form an orthonormal basis of eigenvectors for H.

Note that the position and momentum distributions of the ground state evidently depend on Ω and are in fact not totally trivial to compute, despite the apparent simplicity of the Gaussian expression above. Indeed, if you want to know, for example, $\langle 0, \Omega | Q_7^2 | 0, \Omega \rangle$ you actually need to be able to diagonalize Ω^2 explicitly, and you need in particular an explicit description of the normal modes. This can be done in simple cases, such as the oscillator ring, but not in general.

One can also introduce the "number operator"

$$\tilde{N} = \sum_{i=1}^{n} \tilde{a}_i^\dagger \tilde{a}_i \ ,$$

which commutes with H. The spectrum of $\tilde{N}$ is easily seen to equal to $\mathbb{N}$. Writing $\mathcal{E}_m$ for the eigenspace of $\tilde{N}$ with eigenvalue m, one has evidently

$$L^2(\mathbb{R}^n) = \sum_{m \in \mathbb{N}}^{\oplus} \mathcal{E}_m \ . \tag{31}$$

Each vector in (30) is readily checked to be an eigenvector of $\tilde{N}$ with eigenvalue $\sum_{k=1}^n m_k$. The preceding considerations will be the starting point for an equivalent reformulation of the quantum theory of finite dimensional oscillator systems in a manner suitable for generalization to infinite dimensional systems. This reformulation is based in an essential manner on the notion of Fock space, which I therefore first briefly recall in the next section.

3.2 Fock Spaces

The basic theory of symmetric and anti-symmetric Fock spaces can be found in many places ([2, 23] are two examples) and I will not detail it here, giving only the bare essentials, mostly for notational purposes. More information on this subject can also be found in the contribution of Jan Dereziński in this volume [4].

Let $\mathcal{V}$ be a complex Hilbert space, then the Fock space $\mathcal{F}(\mathcal{V})$ over $\mathcal{V}$ is

$$\mathcal{F}(\mathcal{V}) = \overline{\oplus_{m \in \mathbb{N}} \mathcal{F}_m(\mathcal{V})} \ ,$$

where $\mathcal{F}_m(\mathcal{V})$ is the m-fold tensor product of $\mathcal{V}$ with itself. Moreover $\mathcal{F}_0(\mathcal{V}) = \mathbb{C}$. An element $\psi \in \mathcal{F}(\mathcal{V})$ can be thought of as a sequence

$$\psi = (\psi_0, \psi_1, \ldots, \psi_m, \ldots),$$

where $\psi_m \in \mathcal{F}_m(\mathcal{V})$. I will also use the notation $\mathcal{F}^{\text{fin}}(\mathcal{V}) = \oplus_{m \in \mathbb{N}} \mathcal{F}_m(\mathcal{V})$, which is the dense subspace of $\mathcal{F}(\mathcal{V})$ made up of elements of the type

$$\psi = (\psi_0, \psi_1, \dots, \psi_N, 0, 0, \dots)$$

for some integer $N \geq 0$. Elements of $\mathcal{F}^{\text{fin}}(\mathcal{V})$ will be referred to as states *with a finite number of quanta*, a terminology that I will explain later.

I will freely use the Dirac notation for Hilbert space calculations. So I will write $|\psi\rangle \in \mathcal{F}(\mathcal{V})$ as well as $\psi \in \mathcal{F}(\mathcal{V})$, depending on which one seems more convenient at any given time. Also, when no confusion can arise, I will write $\mathcal{F}_m = \mathcal{F}_m(\mathcal{V})$.

Let $\mathcal{P}_m$ be the permutation group of m elements, then for each $\sigma \in \mathcal{P}_m$, we define the unitary operator $\hat{\sigma}$ on $\mathcal{F}_m(\mathcal{V})$ by

$$\hat{\sigma}\xi_1 \otimes \xi_2 \otimes \cdots \otimes \xi_m = \xi_{\sigma^{-1}(1)} \otimes \xi_{\sigma^{-1}(2)} \otimes \cdots \otimes \xi_{\sigma^{-1}(m)} ,$$

($\xi_j \in \mathbb{C}^n, j = 1, \dots, m$) and the projectors

$$P_{+,m} = \frac{1}{m!} \sum_{\sigma \in \mathcal{P}_m} \hat{\sigma}, \qquad P_{-,m} = \frac{1}{m!} \sum_{\sigma \in \mathcal{P}_m} \text{sgn}(\sigma)\hat{\sigma} .$$

Now we can define the (anti-)symmetric tensor product as

$$\mathcal{F}_m^{\pm}(\mathcal{V}) = P_{\pm,m}\mathcal{F}_m(\mathcal{V})$$

whereas the (anti-)symmetric Fock space $\mathcal{F}^{\pm}(\mathcal{V})$ over $\mathcal{V}$ is

$$\mathcal{F}^{\pm}(\mathcal{V}) = \overline{\oplus_{m \in \mathbb{N}} \mathcal{F}_m^{\pm}(\mathcal{V})} .$$

Introducing the projector $P_{\pm} = \sum_{m \in \mathbb{N}} P_{\pm,m}$, we also have

$$\mathcal{F}^{\pm}(\mathcal{V}) = P_{\pm}\mathcal{F}(\mathcal{V}) \quad \text{and} \quad \mathcal{F}^{\text{fin},\pm}(\mathcal{V}) = P_{\pm}\mathcal{F}^{fin}(\mathcal{V}) .$$

One refers to $\mathcal{F}^+(\mathcal{V})$ as the symmetric or bosonic Fock space and to $\mathcal{F}^-(\mathcal{V})$ as the anti-symmetric or fermionic Fock space. I will only deal with the former here.

Computations in Fock space are greatly simplified through the use of "creation" and "annihilation" operators, which are abstract versions of the operators $\tilde{a}(\xi)$ and $\tilde{a}^\dagger(\xi)$ introduced in Sect. 3.1.

Define, for any $\xi \in \mathcal{V}$,

$$d(\xi)\xi_1 \otimes \xi_2 \cdots \otimes \xi_m = (\overline{\xi} \cdot \xi_1)\, \xi_2 \otimes \cdots \otimes \xi_m .$$

This extends by linearity and yields a well-defined bounded operator from $\mathcal{F}_m$ to $\mathcal{F}_{m-1}$ which extends to a bounded operator on all of $\mathcal{F}(\mathcal{V})$, denoted by the same symbol.

Note that I use the notation $\overline{\xi} \cdot \eta$ for the inner product on the abstract space $\mathcal{V}$ because in the applications in these notes $\mathcal{V}$ will be $\mathcal{K}^{\mathbb{C}}$, in which case this notation is particularly transparent. Of course, on a general abstract $\mathcal{V}$, there is no natural definition of "the complex conjugate $\bar{\xi}$", but that does not mean we can't use $\bar{\xi} \cdot \eta$ as a notation for the inner product.

One has $\| d(\xi) \| = \| \xi \|$. Similarly, define

$$c(\xi)\xi_1 \otimes \xi_2 \cdots \otimes \xi_m = \xi \otimes \xi_1 \otimes \xi_2 \otimes \cdots \otimes \xi_m \,.$$

This again yields a well-defined bounded operator from $\mathcal{F}_m$ to $\mathcal{F}_{m+1}$ which extends to a bounded operator on all of $\mathcal{F}(\mathcal{V})$, denoted by the same symbol. One has $\| c(\xi) \| = \| \xi \|$ and $d(\xi)^* = c(\xi)$.

Introducing the self-adjoint "number operator" N by

$$N\psi = (0, \psi_1, 2\psi_2, \dots m\psi_m \dots) \,,$$

we can then define, on $\mathcal{F}^{\mathrm{fin}}$,

$$a_\pm(\xi) = P_\pm \sqrt{N+1} d(\xi) P_\pm, \qquad \text{and} \qquad a_\pm^\dagger(\xi) = P_\pm \sqrt{N} c(\xi) P_\pm \,.$$

The $a_\pm(\xi)$ are called "annihilation operators" and the $a_\pm^\dagger(\xi)$ creation operators. I will think of $a_-(\xi)$ as an operator on $\mathcal{F}^-$ and of $a_+(\xi)$ as an operator on $\mathcal{F}^+$. Direct computation (on $\mathcal{F}^{fin}$, for example) yields the following crucial commutation and anti-commutation relations between those operators:

$$[a_+(\xi_1), a_+(\xi_2)] = 0 = [a_+^\dagger(\xi_1), a_+^\dagger(\xi_2)], \qquad [a_+(\xi_1), a_+^\dagger(\xi_2)] = \overline{\xi_1} \cdot \xi_2 \,, \quad (32)$$

Those are referred to as the canonical commutation relations or CCR. You should compare (32) to (26) and be amazed.

Working in the bosonic Fock space $\mathcal{F}^+$ and using the above relations one establishes through direct computation that

$$\sqrt{m!} P_+ \xi_1 \otimes \xi_2 \otimes \cdots \otimes \xi_m = a_+^\dagger(\xi_1) a_+^\dagger(\xi_2) \dots a_+^\dagger(\xi_m) |0\rangle \,.$$

Here I introduced the notation $|0\rangle = (1, 0, 0, \dots) \in \mathcal{F}_0 \subset \mathcal{F}$. This vector is usually referred to as the Fock vacuum or simply as the vacuum. It can be characterized as being the unique vector in $\mathcal{F}^+$ for which

$$a_+(\xi)|\psi\rangle = 0, \qquad \forall \xi \in \mathcal{V} \,.$$

For explicit computations and in order to understand the physics literature, it is a Good Thing to have a convenient basis at hand. So suppose you have an orthonormal basis η_j of $\mathcal{V}$ (with $j = 1, 2, \dots \dim\mathcal{V}$). Then you can define, for any positive integer $k \leq \dim\mathcal{V}$ and for any choice of $(m_1, m_2, m_3, \dots, m_k) \in \mathbb{N}^k$, the vector

$$|m_1, m_2, m_3, \dots, m_k\rangle := (m_1! m_2! \dots m_k!)^{-1/2} \times \left(a_+^\dagger(\eta_1)\right)^{m_1} \left(a_+^\dagger(\eta_2)\right)^{m_2} \left(a_+^\dagger(\eta_3)\right)^{m_3} \dots \left(a_+^\dagger(\eta_k)\right)^{m_k} |0\rangle \,. \quad (33)$$

Those vectors are now easily checked to form an orthonormal basis of $\mathcal{F}^+$. The numbers m_j are often referred to as the "occupation numbers" of the states η_j. Note that each of them is an eigenvector of the number operator with eigenvalue given by $\sum_{j=1}^k m_j$.

It is a good exercise to prove that $N_+ = P_+NP_+$, the restriction of the number operator to $\mathcal{F}^+$ can be written

$$N_+ = \sum_j a_+^\dagger(\eta_j)a_+(\eta_j) .$$

If U is a unitary operator on $\mathcal{V}$, the unitary operator $\Gamma(U)$ on $\mathcal{F}^+$ is defined as $\otimes_{k=1}^m U$ when restricted to $\mathcal{F}_m^+$. When A is a self-adjoint operator on $\mathcal{V}$, $\mathrm{d}\Gamma(A)$ is the self-adjoint operator on $\mathcal{F}^+$ defined as

$$A \otimes \mathbb{1} \otimes \cdots \otimes \mathbb{1} + \mathbb{1} \otimes A \otimes \cdots \otimes \mathbb{1} + \cdots + \mathbb{1} \otimes \cdots \otimes \mathbb{1} \otimes A$$

on (a suitable domain) in $\mathcal{F}_m^+$, for each $m > 0$. Also $\mathrm{d}\Gamma(A)\mathcal{F}_0^+ = 0$. It is a good exercise to check that, if A has a basis of eigenvectors

$$A\eta_j = \alpha_j \eta_j$$

then

$$\mathrm{d}\Gamma(A) = \sum_i \alpha_i \, a_+^\dagger(\eta_i)a_+(\eta_i) .$$

3.3 The Fock Representation: Finite Dimensional Fields

It is now straightforward to reformulate the quantum description of the oscillator system in Sect. 3.1 as follows. First of all, in view of (31) and the considerations of the previous section, it is clear that there exists a unitary map T_Ω

$$T_\Omega : L^2(\mathbb{R}^n) \to \mathcal{F}^+(\mathbb{C}^n)$$

satisfying

$$T_\Omega \mathcal{E}_m = \mathcal{F}_m^+(\mathbb{C}^n), \quad T_\Omega H T_\Omega^{-1} = \mathrm{d}\Gamma(\Omega) + \frac{1}{2}\sum_{i=1}^n \omega_i ,$$

and

$$T_\Omega \tilde{a}(\xi) T_\Omega^{-1} = a_+(\xi) ,$$

for all $\xi \in \mathbb{C}^n$. In fact, quite explicitly, one has, for all $\xi_1, \xi_2, \dots \xi_m \in \mathbb{C}^n$,

$$\begin{aligned} T_\Omega : \tilde{a}^\dagger(\xi_1)\dots\tilde{a}^\dagger(\xi_m)|0,\Omega\rangle &\in \mathcal{E}_m \subset L^2(\mathbb{R}^n) \\ &\mapsto a_+^\dagger(\xi_1)\dots a_+^\dagger(\xi_m)|0\rangle \in \mathcal{F}_m^+(\mathbb{C}^n) \subset \mathcal{F}^+(\mathbb{C}^n) . \end{aligned}$$

The unitary map T_Ω transports each object of the theory from $L^2(\mathbb{R}^n)$ to the symmetric Fock space over $\mathbb{C}^n$ and provides in this manner an equivalent quantum mechanical description of the oscillator system, that goes under the name of Fock representation.

Note that in the left hand side of the above equations, the operators $\tilde{a}(\xi)$ or $\tilde{a}^\dagger(\xi)$ are the concrete differential operators on $L^2(\mathbb{R}^n)$ that were defined in (25) and that depend explicitly on Ω. In the right hand side, you find the abstract creation and annihilation operators defined in Sect. 3.2. Note that those do not depend on Ω at all. Similarly, the ground state vector $|0,\Omega\rangle$ of H appearing in the left hand side is of course Ω-dependent, whereas the Fock vacuum $|0\rangle$ in the right hand side is not. This is somewhat paradoxical. Indeed, since the vacuum is the ground state of the Hamiltonian, should it not depend on this Hamiltonian? The answer to this conundrum goes as follows, and is very similar to the discussion in Sect. 2.4 in the classical context. Recall that it is customary to say that each physical state of the system is represented by a vector in a Hilbert space. Consider for example the vacuum vector $|0\rangle$ in Fock space. To find out to which physical state of the system it corresponds, one has to compute the expectation value of physical observables in this state. Now, for a system of coupled oscillators, the most relevant observables are arguably the coordinates of position and momentum. In view of (27) and (28) it is now clear that

$$T_\Omega\eta\cdot QT_\Omega^{-1} = \frac{1}{\sqrt{2}}(a_+(\Omega^{-1/2}\overline{\eta}) + a_+^\dagger(\Omega^{-1/2}\eta))\,, \tag{34}$$

and, similarly,

$$T_\Omega\eta\cdot PT_\Omega^{-1} = \frac{i}{\sqrt{2}}(a_+^\dagger(\Omega^{1/2}\eta) - a_+(\Omega^{1/2}\overline{\eta}))\,. \tag{35}$$

I will in the following not hesitate to write $T_\Omega\eta\cdot QT_\Omega^{-1} = \eta\cdot Q$ and $T_\Omega\eta\cdot PT_\Omega^{-1} = \eta\cdot P$, in agreement with the usual convention that consists of not making the identification operator T_Ω notationally explicit. But it is now clear that, contrary to what happens on $L^2(\mathbb{R}^n)$, the explicit expression of the position and momentum observables as operators on Fock space depends on the dynamics, via Ω! Hence the expectation values of those operators, and of polynomial expressions in these operators will also depend on Ω. In this sense, *the same mathematical object*, namely the vector $|0\rangle \in \mathcal{F}^+(\mathbb{C}^n)$ *corresponds to a different physical state of the system* of n coupled oscillators for different choices of Ω, *i.e.* of the spring constants. Also, the *same physical quantity*, such as the displacement of the seventh oscillator, is represented by a different mathematical operator, namely the operator in the right hand side of (34), with $\eta(j) = \delta_{j7}$. In particular, if you are interested in the mean square displacement of the seventh oscillator when the system is in the ground state, *i.e.* $\langle 0|Q_7^2|0\rangle$, you will need a detailed spectral analysis of Ω and in particular a good understanding of the spatial distribution of its normal modes over the

n degrees of freedom of the system, as I already pointed out. The result you find will of course depend on Ω.

In the same manner, any other given fixed vector in the Fock space, such as for example a state of the form $a_+^\dagger(\xi)|0\rangle$, for some fixed choice of $\xi \in \mathbb{C}^n$, represents a *different* physical state depending on Ω.

In short, the interpretation of a given vector in Fock space as a state of a physical system depends on the dynamics of the system under consideration because the representation of the physical observables of the system by operators on Fock space is dynamics dependent.

To avoid confusion, these simple remarks need to be remembered when dealing with the infinite dimensional theory, where only the Fock representation survives. In particular, the name "vacuum vector" or "vacuum state" given to the Fock vacuum conveys the wrong idea that, somehow, when the system state is represented by this vector, space is empty, there is "nothing there" and therefore this state should have trivial physical properties that in fact should be independent of the system under consideration and in particular of the dynamics.

3.4 The Fock Representation: General Free Fields

Summing up, we have now reformulated the quantum mechanical description of a finite dimensional coupled oscillator system in a way that will be seen to carry over immediately – with only one moderate change – to the infinite dimensional case. Indeed, given a free oscillator field determined by $\mathcal{K}$ and Ω, it is now perfectly natural to choose as the quantum Hilbert space of such a system the Fock space $\mathcal{F}^+(\mathcal{K}^\mathbb{C})$, and as quantum Hamiltonian $H = \mathrm{d}\Gamma(\Omega)$. Note that this is a positive operator and that the Fock vacuum is its ground state, with eigenvalue 0. Proceeding in complete analogy with the finite dimensional case, the quantization of the classical creation and annhilition functions $a_\mathrm{c}(\xi), a_\mathrm{c}^\dagger(\xi)$ are the creation and annihilation operators $a_+(\xi), a_+^\dagger(\xi)$. In terms of those the quantized fields and their conjugates are then *defined* precisely as before ($\eta \in \mathcal{K}_{-1/2}^\mathbb{C}$):

$$\eta \cdot Q := \frac{1}{\sqrt{2}}(a_+(\Omega^{-1/2}\eta) + a_+^\dagger(\Omega^{-1/2}\eta)) , \tag{36}$$

and, similarly ($\eta \in \mathcal{K}_{1/2}^\mathbb{C}$),

$$\eta \cdot P := \frac{\mathrm{i}}{\sqrt{2}}(a_+^\dagger(\Omega^{1/2}\eta) - a_+(\Omega^{1/2}\overline{\eta})) . \tag{37}$$

It is often convenient to think of "the field Q" as the map that associates to each $\eta \in \mathcal{K}_{-1/2}$ the self-adjoint operator in the right hand side of (36), and similarly for "the conjugate field P", defined on $\mathcal{K}_{1/2}$. With this language, the field operator $\eta \cdot Q$ is the value of the field Q at $\eta \in \mathcal{K}_{-1/2}$. This notation is reasonable since the field is a linear function of its argument.

The moderate change to which I referred to above is the fact that, if I compare the above quantization prescription for the case $\mathcal{K} = \mathbb{R}^n$ to the one of Sect. 3.1 and Sect. 3.3, then it is clear that I substracted from the Hamiltonian the "zero-point energy", $\sum_{i=1}^n \omega_i$. It is argued in all quantum field theory texts that this constitutes an innocuous change, for two distinct reasons. First, adding a constant to the Hamiltonian does not change the dynamics in any fundamental way. Second only energy differences count in physics, so tossing out an additive constant in the definition of the energy should not change anything fundamentally. As a result, since the expression $\sum_{i=1}^n \omega_i$ makes no sense in general in infinite dimensions, where it is formally typically equal to $+\infty$, it seems like a good idea to toss it out from the very beginning! This means you calibrate the energy so that the ground state of the system, which is represented by the Fock vacuum, has zero *total* energy, independently of Ω, and leads to the choice of $H = \mathrm{d}\Gamma(\Omega)$ as the Hamiltonian.

While this is the reasoning found in all physics and mathematical physics texts the tossing out of the zero-point energy is not such an innocent operation after all. For the physics of the zero-point energy, I refer to [20]. See also [6] for further comments.

It is instructive to compute the evolution of the field and the conjugate field under the dynamics. Since

$$\mathrm{e}^{-\mathrm{i}Ht} = \Gamma(\mathrm{e}^{-\mathrm{i}\Omega t}) ,$$

it is easy to check that

$$\mathrm{e}^{\mathrm{i}Ht} a_+(\xi) \mathrm{e}^{-\mathrm{i}Ht} = a_+(\mathrm{e}^{\mathrm{i}\Omega t}\xi) .$$

Define then the evolved field $Q(t)$ as the map that associates to each $\eta \in \mathcal{K}_{-1/2}$ the self-adjoint operator $\eta \cdot Q(t)$ defined as follows:

$$\eta \cdot Q(t) \equiv \mathrm{e}^{\mathrm{i}Ht}(\eta \cdot Q)\mathrm{e}^{-\mathrm{i}Ht} .$$

A simple computation then yields

$$\eta \cdot Q(t) = \frac{1}{\sqrt{2}}(a_+(\Omega^{-1/2}\mathrm{e}^{\mathrm{i}\Omega t}\overline{\eta}) + a_+^\dagger(\Omega^{-1/2}\mathrm{e}^{\mathrm{i}\Omega t}\eta)) .$$

Hence

$$\frac{\mathrm{d}^2}{\mathrm{d}t^2}\eta \cdot Q(t) = -\Omega^2 \eta \cdot Q(t) .$$

One defines similarly $\eta \cdot P(t)$, which obeys the same equation. In fact, $\eta \cdot Q(t)$ and $\eta \cdot P(t)$ are operator-valued solutions of this with $\eta \cdot Q(t)$ satisfying the *equal time commutation relations*. They are called the Heisenberg field and conjugate field in the physics literature.

We are now in a position to further study these systems, a task I turn to next. First, a word on the "particle interpretation of the field states" is in order.

3.5 Particle Interpretation of the Field States

Physicists refer to $\mathcal{F}_m$ as the m particle sector of the Fock space ($m \geq 1$) and to $\mathcal{F}_0$ as the vacuum sector. This terminology comes from the following remark. As any beginners' text in quantum mechanics will tell you, whenever the quantum Hilbert space of a single particle (or a single system) is $\mathcal{V}$, the Hilbert space of states for m (identical) particles (or systems) is the m-fold tensor product of $\mathcal{V}$. The simplest case is the one where $\mathcal{V} = L^2(\mathbb{R}^d)$. Then the m-fold tensor product can be naturally identified with $L^2(\mathbb{R}^d \times \cdots \times \mathbb{R}^d = \mathbb{R}^{dm})$, which is isomorphic to $\otimes_m \mathcal{V}$. The same quantum mechanics course will teach you that, when the particles are indistinguishable, the state space needs to be restricted either to the symmetric or anti-symmetric tensor product. In the first case, which is the one we are dealing with here, the particles are said to be bosons, otherwise they are fermions. In the case where $\mathcal{V} = L^2(\mathbb{R}^d)$, the m-fold symmetric tensor product of $\mathcal{V}$ consists of all symmetric L^2-functions of m variables.

The above considerations suggest that, *conversely*, whenever the quantum state space of a physical system turns out to be a Fock space over some Hilbert space $\mathcal{V}$, one may think of $\mathcal{V}$ as a one-particle space, and of $\mathcal{F}_m(\mathcal{V})$ as the corresponding m-particle space. An arbitrary state of the system can then be thought of as a superposition of states with 0, 1, 2, $\ldots m$, $\ldots$ particles. These ideas emerged very quickly after the birth of quantum mechanics, as soon as physicists attacked the problem of analyzing the quantum mechanical behaviour of systems with an infinite number of degrees of freedom, such as the electromagnetic field. The Fock space structure of the Hilbert space of states describing the *field* immediately lead to such an interpretation in terms of *particles*. For the electromagnetic field, the particles were baptized "photons", and in complete analogy, the quantum mechanical description of lattice vibrations in solid state physics lead to the notion of "phonons". The idea that one can associate a particle interpretation to the states of a Fock space is further corroborated by the observation that those states carry energy and momentum in "lumps". This can be seen as follows. Suppose, in our notations, that Ω has a pure point spectrum:

$$\Omega\eta_j = \omega_j \eta_j, \; j \in \mathbb{N} \, .$$

Then the quantum Hamiltonian is

$$H = \mathrm{d}\Gamma(\Omega) = \sum_j \omega_j a_j^\dagger a_j \, ,$$

where I wrote $a_j^\dagger = a^\dagger(\eta_j)$. Note that I have dropped the index $+$ on the creation and annihilation operators, a practice that I shall stick to in what follows since I will at any rate be working on the symmetric Fock space all the time. Now consider for example the state

$$a_1^\dagger (a_5^\dagger)^3 a_{10}^\dagger |0\rangle \, .$$

This is a 5-particle state, and an eigenvector of the Hamiltonian with eigenvalue $\omega_1 + 3\omega_5 + \omega_{10}$. It is natural to think of it intuitively as being a state "containing" 3 particles of energy ω_5, and one particle of energy ω_1 and ω_{10} each. Similarly, in translationally invariant systems, such states can be seen to carry a total momentum which is the sum of "lumps" of momentum corresponding to its individual constituents. Of course, the particle interpretation of the states of the field is a very important feature of the theory since it is essential for the interpretation of high energy experiments, and so it has quite naturally received a lot of attention.

Despite its undeniable value, the suggestive interpretation of the states of Fock space in terms of particles may lead (and has lead) to some amount of confusion and has to be taken with a (large) grain of salt. Some of those problems seem to have been brought out clearly only when physicists started to investigate quantum field theory on curved space-times. A critical discussion of this issue can be found throughout [10]. Although Fulling does adopt the second quantization viewpoint, he stresses repeatedly the need to escape "from the tyranny of the particle concept" in order to "come to a completely field theoretic understanding of quantum field theory." Similarly, Wald, who does indeed adopt a field theoretic viewpoint throughout in [29], gives a critical analysis of the merits and limitations of the particle concept in quantum field theory. He actually stresses the need to "unlearn" some of the familiar concepts of quantum field theory on flat space times to understand the curved space time version of the theory.

There are in fact several sources of problems with the particle interpretation of the states in quantum field theory. The first one was already hinted at in Sect. 3.3: the use of the word "vacuum" to describe the ground state of the system invites one to think that when the system is in this state, there is "nothing there". Actually, one may be tempted to think the system itself is simply not there! But to see that makes no sense, it is enough to think of an oscillator lattice. Certainly, when this system is in its ground state, all oscillators are there! It is just that the system is not excited, so there are no "particles" in the above (Fock space) sense of the word, and this in spite of the fact that the mechanical particles making up the lattice are certainly present. Also, if one thinks of the vacuum state as empty space, it becomes impossible to understand how its properties can depend on the system considered via Ω. In fact, it is quite baffling to think "empty space" could have any properties at all. In particular, the mean square displacement of the field, for example, given by

$$\langle 0|(\eta \cdot Q)^2|0\rangle$$

is a function of Ω, as is easily seen even in finite dimensional oscillator systems. This quantity is an example of a so-called "vacuum fluctuation". Of course, for systems with a finite number of degrees of freedom, we find this phenomenon perfectly natural, but if you study the Klein-Gordon field, for example, and call the ground state the vacuum, you end up being surprised to see vacuum expectation values depend on the mass of particles that are not there!

A second source of confusion is that the notion of "particle" evokes a localized entity, carrying not only momentum and energy, but that one should also be able to localize in space, preferably with the help of a position operator. I will show in Sect. 4 that there is no reasonable notion of "position" that can be associated to the one-particle states of Fock space, contrary to what happens in the usual non-relativistic quantum mechanics of systems with a finite number of particles. In particular, there is no reasonable "position operator". This has nothing to do with relativity, but is true for large classes of Ω and in particular for all examples given so far. So even if the particles of field theory share a certain number of properties with the usual point particles of classical and quantum mechanical textbooks, they have some important features that make them quite different. They are analogous objects, but not totally similar ones. This, I will argue, has nothing to do either with special or general relativity, but is clear if one remembers systematically the analogy with finite dimensional oscillator systems.

As a constant reminder of the fact that the so-called particles of quantum field theory are nothing but excitations of its ground state, it is a good idea to use the older physics terminology and to talk systematically of "quasi-particles", "quanta", "field quanta" or of "elementary excitations of the field" rather than simply of particles when describing the states of Fock space. I will adhere as much as possible to this prudent practice.

Moreover, when testing your understanding of a notion in quantum field theory, try to see what it gives for a finite system of oscillators. If it looks funny there, it is likely to be a bad idea to use it in the infinite dimensional case.

The remaining parts of this section develop material that will be needed in Sect. 4. It is perhaps a good idea to start reading the latter, coming back to this material only as I refer to it.

3.6 Weyl Operators and Coherent States

Given a Hilbert space $\mathcal{V}$ and the corresponding symmetric Fock space $\mathcal{F}^+(\mathcal{V})$, we can first define, for any $\xi \in \mathcal{V}$, the *Weyl operator*

$$W_{\mathrm{F}}(\xi) = \mathrm{e}^{a^\dagger(\xi) - a(\xi)} .$$

A coherent state is then defined as a vector of $\mathcal{F}^+(\mathcal{V})$ of the form

$$|\xi\rangle \overset{\mathrm{def}}{=} W_{\mathrm{F}}(\xi)|0\rangle ,$$

for some $\xi \in \mathcal{V}$. Note that the map

$$\xi \in \mathcal{V} \mapsto |\xi\rangle \in \mathcal{F}^+(\mathcal{V})$$

provides a *nonlinear* imbedding of $\mathcal{V}$ into $\mathcal{F}^+(\mathcal{V})$ which is not to be confused with the trivial linear imbedding $\mathcal{V} \cong \mathcal{F}_1(\mathcal{V}) \subset \mathcal{F}^+(\mathcal{V})$. Given an arbitrary $0 \neq \psi \in \mathcal{V}$, one can likewise consider the family $W_{\mathrm{F}}(\xi)\psi$, and those vectors are also referred to as a family of coherent states.

Coherent states play an important role in the semi-classical analysis of quantum systems and in various branches of theoretical physics [17, 22]. We describe them here in the abstract context of symmetric Fock spaces. They are very simple objects to define but nevertheless have an seemingly inexhaustable set of interesting properties. I will only mention those I need.

To compute with the coherent states, we need a number of formulas that are listed below and that can all be obtained easily, if one remembers first of all that, if A and B are bounded operators so that $C = [A, B]$ commutes with both A and B, then

$$\mathrm{e}^{A+B} = \mathrm{e}^{A}\mathrm{e}^{B}\mathrm{e}^{-\frac{C}{2}}, C = [A, B] \,.$$

Computing with $a^\dagger(\xi)$ and $a(\xi)$ as if they were bounded operators, all formulas below follow from this and some perseverance in computing. Taking care of the domain problems to make them completely rigorous is tedious but character building and can be done using the techniques described in [3] or [23]. First of all, we have, for all $\xi_1, \xi_2 \in \mathcal{V}$,

$$W_{\mathrm{F}}(\xi_1)W_{\mathrm{F}}(\xi_2) = W_{\mathrm{F}}(\xi_1 + \xi_2)\mathrm{e}^{-\mathrm{i}\mathrm{Im}(\overline{\xi}_1\cdot\xi_2)} \,.$$

As a result

$$W_{\mathrm{F}}(\xi_1)W_{\mathrm{F}}(\xi_2) = W_{\mathrm{F}}(\xi_2)W_{\mathrm{F}}(\xi_1)\mathrm{e}^{-2\mathrm{i}\mathrm{Im}(\overline{\xi}_1\cdot\xi_2)} \,,$$

and

$$[W_{\mathrm{F}}(\xi), W_{\mathrm{F}}(\xi')] = W_{\mathrm{F}}(\xi')W_{\mathrm{F}}(\xi)\left(\mathrm{e}^{\mathrm{i}2\mathrm{Im}(\overline{\xi}'\cdot\xi)} - 1\right) \,.$$

Furthermore

$$W_{\mathrm{F}}(s\xi)W_{\mathrm{F}}(\xi')W_{\mathrm{F}}(t\xi) = W_{\mathrm{F}}((s+t)\xi)W_{\mathrm{F}}(\xi')\mathrm{e}^{2\mathrm{i}t\mathrm{Im}(\overline{\xi}\cdot\xi')}$$

and hence

$$W_{\mathrm{F}}(-\zeta)W_{\mathrm{F}}(\xi)W_{\mathrm{F}}(\zeta) = W_{\mathrm{F}}(\xi)\mathrm{e}^{\mathrm{i}2\mathrm{Im}(\overline{\zeta}\cdot\xi)} \,,$$

or

$$W_{\mathrm{F}}(-\zeta)W_{\mathrm{F}}(\xi)W_{\mathrm{F}}(\zeta) = \mathrm{e}^{\left[(a^\dagger(\xi)+\xi\cdot\overline{\zeta})-(a(\xi)+\overline{\xi}\cdot\zeta)\right]} \,.$$

One then finds

$$W_{\mathrm{F}}(-\zeta)\left[a^\dagger(\xi)\right]^n W_{\mathrm{F}}(\zeta) = (a^\dagger(\xi)+\xi\cdot\overline{\zeta})^n, \; W_{\mathrm{F}}(-\zeta)a^n(\xi)W_{\mathrm{F}}(\zeta) = (a(\xi)+\overline{\xi}\cdot\zeta)^n \,.$$

It is often convenient to write

$$W_{\mathrm{F}}(\xi) = \mathrm{e}^{a^\dagger(\xi)}\mathrm{e}^{-a(\xi)}\mathrm{e}^{-\frac{1}{2}\|\xi\|^2} = \mathrm{e}^{-a(\xi)}\mathrm{e}^{a^\dagger(\xi)}\mathrm{e}^{\frac{1}{2}\|\xi\|^2} \,.$$

Also, remark that, for all $\xi \neq 0$,

$$\| W_{\mathrm{F}}(\xi) - 1\!\!1 \| = 2 \text{ and s} - \lim_{t \to 0} W_{\mathrm{F}}(t\xi) = 1\!\!1 \,.$$

Using what precedes, one easily finds the following formulas involving the vacuum.

$$|\xi\rangle = \mathrm{e}^{-\frac{1}{2}\|\xi\|^2} \mathrm{e}^{a^\dagger(\xi)} |0\rangle \,, \tag{38}$$

$$\langle 0| a^n(\xi) |\xi'\rangle = \mathrm{e}^{-\frac{1}{2}\|\xi'\|^2} (\overline{\xi} \cdot \xi')^n \,, \tag{39}$$

and

$$\langle \zeta | W_{\mathrm{F}}(\xi) | \zeta \rangle = \mathrm{e}^{-\frac{1}{2}\|\xi\|^2} \mathrm{e}^{\mathrm{i}2\mathrm{Im}(\overline{\zeta}\cdot\xi)} \,. \tag{40}$$

3.7 Observables and Observable Algebras

Physically measurable quantities of a system are, in its classical description, represented by functions on phase space. Consider first finite dimensional systems. An example, in the case of an oscillator ring, is "the displacement of the ninth oscillator", represented by $q_9 : X = (q, p) \in \mathcal{H} \to q_9 \in \mathbb{R}$. Some interesting observables are represented by linear functions (such as position and momentum) or by quadratic functions (such as energy or angular momentum). More generally, they may be polynomial. To discuss the linear functions, it is helpful to notice that the *topological* dual space of $\mathcal{H}$ can conveniently be identified with $\mathcal{H}$ itself using the symplectic form: to each $Y \in \mathcal{H}$, we associate the linear map

$$X \in \mathcal{H} \mapsto s(Y, X) \in \mathbb{R} \,.$$

One has, from (8), for every $Y_1, Y_2 \in \mathcal{H}$,

$$\{s(Y_1, \cdot),\ s(Y_2, \cdot)\} = s(Y_1, Y_2) \,. \tag{41}$$

It is then convenient to introduce

$$V_{\mathrm{c}}(Y) = \mathrm{e}^{-\mathrm{i}s(Y,\cdot)} \tag{42}$$

which serves as a generating function for monomials of the type

$$s(Y_1, \cdot) s(Y_2, \cdot) \ldots s(Y_n, \cdot) = \frac{(i\partial)^n}{\partial t_1 \partial t_2 \ldots \partial t_n} V_{\mathrm{c}}(t_1 Y_1 + \ldots t_n Y_n)_{|t_1 = 0 = t_2 \cdots = t_n} \,.$$

It is immediate from the definition of the $V_{\mathrm{c}}(Y)$ that

$$V_{\mathrm{c}}(Y) \circ \Phi_t = V_{\mathrm{c}}(\Phi_{-t} Y) \,.$$

Working in the Schrödinger representation, the quantum mechanical analogues of the $V_{\mathrm{c}}(Y)$ are the Weyl operators

$$V(Y) = \mathrm{e}^{-\mathrm{i}(a \cdot P - b \cdot Q)}, \text{ where } Y = (a, b) \in \mathcal{H} \,. \tag{43}$$

The $V(Y)$ are clearly unitary operators on $L^2(\mathbb{R}^n)$ and satisfy the so-called Weyl relations

$$V(Y_1)V(Y_2) = \mathrm{e}^{-\frac{i}{2}s(Y_1,Y_2)}V(Y_1+Y_2), \qquad \forall\, Y_1, Y_2 \in \mathcal{H}\,.$$

In a Fock representation (determined by a choice of Ω), one has, with the notation of Sect. 3.3

$$T_\Omega V(Y) T_\Omega^{-1} = W_{\mathrm{F}}(z_\Omega(Y))\,.$$

Here the $W_{\mathrm{F}}(z_\Omega(Y))$ are the Weyl operators on the symmetric Fock space $\mathcal{F}^+(\mathbb{C}^n)$, as introduced in Sect. 3.6.

In the algebraic approach to quantum theory, one postulates that the interesting observables of the theory include at least those that can be written as finite sums of $V(Y)$. One therefore considers the algebra

$$\mathrm{CCR}_0(\mathbb{R}^{2n}) = \mathrm{span}\,\{W_{\mathrm{F}}(z_\Omega(Y)) \mid Y \in \mathbb{R}^{2n}\} = \mathrm{span}\,\{W_{\mathrm{F}}(\xi) \mid \xi \in \mathbb{C}^n\}\,.$$

This algebra is irreducible. This means that the only closed subspaces of $L^2(\mathbb{R}^n) \cong \mathcal{F}^+(\mathbb{C}^n)$ invariant under the above algebra are the trivial ones and is equivalent, via Schur's Lemma, to the statement that the only bounded operators that commute with all F in the algebra are the multiples of the identity. For a simple proof of these facts one may consult [5]. This implies via a well known result in the theory of von Neumann algebras (see [2], for example) that its weak closure is all of $\mathcal{B}(\mathcal{F}^+(\mathbb{C}^n))$: in this sense, "any bounded operator on Fock space can be approximated (in the weak topology!) by a function of Q and P." This is clearly a way of saying that the original algebra is quite large. Note nevertheless that its operator norm closure (called the CCR-algebra over $\mathbb{R}^{2n}$ and denoted by $\mathrm{CCR}(\mathbb{R}^{2n})$) is much smaller, since it contains no compact operators. For the purposes of these notes, I will consider $\mathrm{CCR}(\mathbb{R}^{2n})$ or $\mathrm{CCR}_0(\mathbb{R}^{2n})$ as "the" observable algebra of the systems considered.

Remark that these algebras do, as sets, not depend on Ω. But again, in close analogy to what we observed in Sect. 3.3, given an operator on Fock space belonging to one of these algebras, its expression in terms of Q and P does depend on Ω, and so does therefore its physical interpretation as an observable. So it is not only the identification of the appropriate observable algebra which is important, but the labeling, within this algebra, of the elements that describe the relevant physical observables. This will be crucial once we discuss local observables in Section 4, and become hopefully quite a bit clearer then too.

It is obviously not of much interest to discuss observable algebras if one is not going to say how the observables evolve in time. In finite dimensional systems, one is given a Hamiltonian H, which is a self-adjoint operator on $L^2(\mathbb{R}^n) \cong \mathcal{F}^+(\mathbb{C}^n)$. It generates the so-called Heisenberg evolution of each observable F, which is defined by $\alpha_t(F) = \mathrm{e}^{\mathrm{ii}Ht} F \mathrm{e}^{-\mathrm{i}Ht}$. It has to be checked that the algebra of observables and H are such that this defines an automorphism of the algebra (i.e. so that $\mathrm{e}^{\mathrm{i}Ht}F\mathrm{e}^{-\mathrm{i}Ht}$ still belongs to the algebra if F does).

That α_t is an automorphism of the CCR algebra is not true in general. For example, it is proven in [11] that, when $H(\lambda) = \frac{1}{2}P^2 + \lambda V$, with V a bounded L^1 function, then the Heisenberg evolution leaves the CCR algebra invariant for all values of t and of λ if and only if $V = 0$. In other words, the CCR algebra cannot possibly be a suitable algebra to describe most standard quantum mechanical systems with a finite number of degrees of freedom.

An exception to this rule are systems described by quadratic hamiltonians, which are precisely the ones we are interested in here. An easy example is provided by quadratic Hamiltonians of the type $H = \frac{1}{2}P^2 + \frac{1}{2}Q \cdot \Omega^2 Q$ in view of

$$\mathrm{e}^{\mathrm{id}\Gamma(\Omega)t} W_{\mathrm{F}}(\xi)\mathrm{e}^{-\mathrm{id}\Gamma(\Omega)t} = W_{\mathrm{F}}(\mathrm{e}^{-\mathrm{i}\Omega t}\xi),\ \forall \xi \in \mathcal{K}^{\mathbb{C}}\,,$$

which follows immediately from the discussion in Sect. 3.4. This clearly implies that the dynamics leaves the CCR algebra $\mathrm{CCR}(\mathbb{R}^{2n})$ invariant. Note that this will work in infinite dimensional systems just as well as in finite dimensional ones.

The discussion carries over to the infinite dimensional case without change. One defines the algebra of observables in the quantum theory to be

$$\mathrm{CCR}_0(\mathcal{H}) = \mathrm{span}\,\{W_{\mathrm{F}}(z_\Omega(Y)) \mid Y \in \mathcal{H}\} = \mathrm{span}\,\{W_{\mathrm{F}}(\xi) \mid \xi \in \mathcal{K}^{\mathbb{C}}\}\,.$$

Again, this algebra is independent of Ω and turns out to be irreducible [2], so that its weak closure is the algebra of all bounded operators on Fock space. Its norm closure, which is much smaller, is the so-called CCR-algebra over $\mathcal{H}$, for which I will write $\mathrm{CCR}(\mathcal{H})$. Since we will only work with quadratic Hamiltonians, this algebra is adequate for the description of such systems since it is then invariant under the dynamics. Here also, to no one's surprise by now, I hope, the interpretation of a given operator in the algebra as an observable *will* depend on Ω, as we will see in more detail in Section 4.

For further reference, let me define also the algebra

$$\mathrm{CCR}_0(\mathcal{M}) = \mathrm{span}\,\{W_{\mathrm{F}}(z_\Omega(Y))|Y \in \mathcal{M}\}\,,$$

whenever $\mathcal{M}$ is a vector subspace of $\mathcal{H}$ (even if $\mathcal{M}$ is not symplectic). In many situations it is natural and elegant not to work with the norm closure of the $\mathrm{CCR}_0(\mathcal{M})$, but with their weak closure, for which I shall write $\mathrm{CCR}_{\mathrm{w}}(\mathcal{M})$. Further developments concerning the CCR can be found in the contribution of J. Dereziński in this volume [4].

4 Local Observables and Local States

4.1 Introduction

The issue of what are local observables, local states and local measurements has attracted a fair amount of attention and has generated some surprises and even some controversy in the mathematical physics literature on relativistic quantum field theory. The controversy has centered on the question of particle localization, of possible causality violations and of relativistic invariance. I will address these issues in the present section within the restricted context of the free oscillator fields under study here, some of which are relativistically invariant, while others are not. I will argue that there is not much reason to be surprised and certainly no ground for controversy.

After defining what is meant by a local observable (Sect. 4.2) and giving some examples (Sect. 4.3), the notion of "strictly local excitation of the vacuum" is introduced in Sect. 4.4. I will then state a generalization of a theorem of Knight asserting that, if Ω is a non-local operator, then states with a finite number of field excitations cannot be strictly local excitations of the vacuum (Sect. 4.5). It will be shown through examples (Sect. 4.6) that the above condition on Ω is typically satisfied in models of interest and I will explain the link between the above notion of localized excitation of the vacuum and the so-called Newton-Wigner localization (Sect. 4.7). It will be argued that the latter is not a suitable notion to discuss the local properties of the states of oscillator fields. The actual proof of Knight's theorem is deferred to Sect. 4.8.

4.2 Definition of a Local Structure

Among the interesting observables of the oscillator systems we are studying are certainly the "local" ones. I will give a precise definition in a moment, but thinking again of the oscillator chain, "the displacement q_7 of the seventh oscillator" is certainly a "local" observable. In the same way, if dealing with a wave equation, "the value $q(x)$ of the field at x" is a local observable. The Hamiltonian is on the other hand not a local observable, since it involves sums or integrals over all oscillator displacements and momenta. Generally, "local observables" are functions of the fields and conjugate fields in a bounded region of space. Of course, this notion does not make sense for all harmonic systems, defined by giving a positive operator Ω^2 on some abstract Hilbert space $\mathcal{K}$. So let me reduce the level of abstractness of the discussion, therefore hopefully increasing its level of pertinence, and define what I mean by a system with a local structure.

In view of what precedes, I will limit my attention to free oscillator fields over a real Hilbert space $\mathcal{K}$ of the form $\mathcal{K} = L^2_{\mathbb{R}}(K, \mathrm{d}\mu)$, where K is a topological space and μ a Borel measure on K. Here the subscript "$\mathbb{R}$" indicates that we are dealing with the real Hilbert space of real-valued functions. In fact, all examples I have given so far are of the above type.

Definition 1. *A local structure for the oscillator field determined by Ω and $\mathcal{K} = L^2_{\mathbb{R}}(K, \mathrm{d}\mu)$ is a subspace $\mathcal{S}$ of $\mathcal{K}$ with the following properties:*

1. $\mathcal{S} \subset \mathcal{K}_{1/2} \cap \mathcal{K}_{-1/2}$;

2. *Let B be a Borel subset of K, then $\mathcal{S}_B \equiv \mathcal{S} \cap L^2_{\mathbb{R}}(B, \mathrm{d}\mu)$ is dense in $L^2_{\mathbb{R}}(B, \mathrm{d}\mu)$.*

This is a pretty strange definition, and I will give some examples in a second, but let me first show how to use this definition to define what is meant by "local observables". Note that, thanks to the density condition above,

$$\mathcal{H}(B, \Omega) \stackrel{\mathrm{def}}{=} \mathcal{S}_B \times \mathcal{S}_B$$

is a symplectic subspace of $\mathcal{H}$ so that the restriction of $W_{\mathrm{F}} \circ z_\Omega$ to $\mathcal{H}(B, \Omega)$ is a representation of the CCR over $\mathcal{H}(B, \Omega)$.

Definition 2. *Let $\mathcal{K} = L^2_{\mathbb{R}}(K, \mathrm{d}\mu), \Omega, \mathcal{S}$ be as above and let B be a Borel subset of K. The algebra of local observables over B is the algebra*

$$\mathrm{CCR}_0(\mathcal{H}(B, \Omega)) = \mathrm{span}\ \{W_{\mathrm{F}}(z_\Omega(Y)) \mid Y \in \mathcal{S}_B \times \mathcal{S}_B\}\ .$$

Note that Ω plays a role in the definition of $\mathcal{S}$ through the appearance of the spaces $\mathcal{K}_\lambda$. The first condition on $\mathcal{S}$ guarantees that $\mathcal{S} \times \mathcal{S} \subset \mathcal{H}$ so that, in particular, for all $Y \in \mathcal{S} \times \mathcal{S}$, $s(Y, \cdot)$ is well defined as a function on $\mathcal{H}$ which is important for the definition of the local observables to make sense. In practice, one wants to be able to use the same spatial structure $\mathcal{S}$ for various choices of Ω, in order to be able to compare different systems built over the same space $\mathcal{K} = L^2_{\mathbb{R}}(K, \mathrm{d}\mu)$. Note nevertheless that even then, the algebras of local and of quasi-local observables, which are algebras of bounded operators on the Fock space $\mathcal{F}^+(\mathcal{K}^{\mathbb{C}})$ do, *as sets*, depend on Ω. This is in contrast to the algebra of "all" observables,

$$\mathrm{CCR}_0(\mathcal{H}) = \mathrm{span}\ \{W_{\mathrm{F}}(z_\Omega(Y)) \mid Y \in \mathcal{H}\} = \mathrm{span}\ \{W_{\mathrm{F}}(\xi) \mid \xi \in \mathcal{K}^{\mathbb{C}}\}\ ,$$

which is, as a set, independent of Ω, as pointed out before. In other words, some of the physics is hidden in the way the local algebras are imbedded in the CCR algebra over $\mathcal{H}$.

4.3 Examples of Local Structures

Oscillator Lattices – Klein-Gordon Equations

In the case of the translationally invariant oscillator lattices in dimension 2 or higher presented in Sect. 2.2, $\mathcal{S}$ can be taken to be the space of sequences q of finite support, even in the massless case, as is easily checked. Alternatively, you could take $\mathcal{S}$ to be the larger space of sequences of fast decrease. This has the advantage that then $\mathcal{S} \times \mathcal{S}$ is dynamics invariant. Note that in neither of these examples $\mathcal{S}\times\mathcal{S}$ is J invariant, though, so that $\mathcal{S}\times\mathcal{S}$ will not be a complex vector subspace of $(\mathcal{H}, J)$, just a real one. This is also true for $\mathcal{S}_B \times \mathcal{S}_B$ and will be crucial when discussing "local excitations of the vacuum" in quantum field theory.

Exercise 1. Check all of the above statements in detail.

As an example of a local observable, we have, with $\eta \in \mathcal{S}$ of bounded support in some set $B \subset \mathbb{Z}^d$,

$$\mathrm{e}^{\mathrm{i}\eta\cdot Q} = W_{\mathrm{F}}\left(\frac{\mathrm{i}}{\sqrt{2}}\Omega^{-1/2}\eta\right) .$$

Very explicitly, one may think of taking $\eta(j) = \delta_{j,k}$ and then this is $\mathrm{e}^{\mathrm{i}Q(k)}$, a simple function of the displacement of the oscillator at site $k \in \mathbb{Z}^d$. At the risk of boring the wits out of you, let me point out yet again that this *fixed observable* is represented on Fock space by a *different operator* for different choices of Ω.

Similarly

$$\mathrm{e}^{\mathrm{i}\eta\cdot P} = W_{\mathrm{F}}\left(-\frac{1}{\sqrt{2}}\Omega^{1/2}\eta\right)$$

is a function of the momenta of the oscillators in the support of η.

In the one-dimensional translationally invariant lattice a spatial structure does not exist when $\nu = 1/2$ because of the strong infrared singularity. Indeed, due to the density condition in the definition of the local structure, it is clear that $\mathcal{S}$ must contain all sequences of finite support, and those do not belong to $\mathcal{K}_{-1/2}$ in dimension 1, as we already pointed out in Sect. 2.2.

Similarly, the wave and Klein-Gordon equations on $\mathbb{R}^d$ admit for example $C_0(\mathbb{R}^d)$ or the space of Schwartz functions as a spatial structure in dimension 2 or higher, as follows from the discussion in Sect. 2.2.

The Finite Dimensional Case

I find this example personally most instructive. It forces one into an unusual point of view on a system of n coupled oscillators that is well suited to the infinite dimensional case. Think therefore of a system of n oscillators characterized by a positive n by n matrix Ω^2, as in Sect. 2.2. A local observable of such a system should be a function of the positions and momenta of a fixed finite set of oscillators. Does the definition given above correctly incorporate this intuition? Let's check.

In this case, $\mathcal{K} = \mathbb{R}^n$, which I view as $L^2_{\mathbb{R}}(K)$, where K is simply the set of n elements. Indeed, $q \in \mathbb{R}^n$ can be seen as a function $q : j \in \{1, \dots n\} \mapsto q(j) \in \mathbb{R}$, obviously square integrable for the counting measure. I already explained in detail the identification between the quantum state space $L^2(\mathbb{R}^n)$

and $\mathcal{F}^+(\mathbb{C}^n)$ (Sect. 3.3). Here $\mathbb{C}^n$ is the complexification of $\mathbb{R}^n$, and as such naturally identified with $L^2(K, \mathbb{C})$. So, finally

$$L^2(\mathbb{R}^n) \cong \mathcal{F}^+(L^2(K, \mathbb{C}))\,.$$

Consider now a subset B of K, say $B = \{1, 6, 9\} (n \geq 9)$. It is an excellent exercise to convince oneself that, unraveling the various identifications, a local observable over B is a finite linear combination of operators on $L^2(\mathbb{R}^n)$ of the form ($a_j, b_j \in \mathbb{R}, j \in B$):

$$\exp -\mathrm{i} \left(\sum_{j \in B} (a_j P_j - b_j Q_j) \right)\,.$$

Better yet, if you write (with $\sharp B$ denoting the cardinality of the set B)

$$L^2(\mathbb{R}^n) \cong L^2(\mathbb{R}^{\sharp B}, \prod_{j \in B} \mathrm{d}x_j) \otimes L^2(\mathbb{R}^{n - \sharp B} \prod_{j \notin B} \mathrm{d}x_j)\,,$$

then it is clear that the weak closure of the above algebra is

$$\mathcal{B}(L^2(\mathbb{R}^{\sharp B}, \prod_{j \in B} \mathrm{d}x_j)) \otimes \mathbb{1}\,.$$

So, indeed, a local observable is clearly one that involves only the degrees of freedom indexed by elements of B.

Exercise 2. Convince yourself all of this is true.

Unbounded Local Observables

To make contact with the physics literature, it will be convenient on occasion in the following to refer to polynomials in $\frac{\mathrm{d}}{\mathrm{d}t} W(z_\Omega(tY))|_{t=0}$ with $Y \in \mathcal{H}(B, \Omega)$ as local observables over B as well. These are sums of expressions of the form

$$\Pi_S(z_\Omega(Y_1)) \Pi_S(z_\Omega(Y_2)) \ldots \Pi_S(z_\Omega(Y_n))$$

where each $Y_j \in \mathcal{H}(B, \Omega)$. Alternatively and perhaps more suggestively, these are sums of expressions of the form

$$(\eta_1 \cdot Q) \ldots (\eta_m \cdot Q) \quad \text{and} \quad (\eta_1 \cdot P) \ldots (\eta_m \cdot P)\,,$$

or of products thereof, where each $\eta_j \in \mathcal{S}_B$. Again, for lattices, these are polynomials in the positions and momenta of the individual oscillators in some subset B of the lattice $\mathbb{Z}^d$.

4.4 Strictly Localized Vacuum Excitations

I now want to give meaning to the notion of "local excitation of the vacuum" for general free oscillator fields with a local structure $\mathcal{S}$. So in this section $\mathcal{K} = L^2_{\mathrm{r}}(K, \mathrm{d}\mu)$, and $\mathcal{S}$ satisfies the conditions of Definition 1.

The equivalent classical notion is readily described and was already discussed in Sect. 2.2. The vacuum, being the ground state of the system, is the quantum mechanical equivalent of the global equilibrium $X = 0$, which belongs of course to the phase space $\mathcal{H}$, and a local perturbation of this equilibrium is an initial condition $X = (q, p) \in \mathcal{S} \times \mathcal{S}$ with the support of q and of p contained in a (typically bounded) subset B of K. An example of a local perturbation of an oscillator lattice is a state $X \in \mathcal{H}$ where only q_0 and p_0 differ from 0. In the classical theory, local perturbations of the equilibrium are therefore states that differ from the equilibrium state only inside a bounded subset B of K. It is this last formulation that is readily adapted to the quantum context, through the use of the notion of "local observable" introduced previously.

For that purpose, we first introduce the following notion, which is due to Knight [16].

Definition 3. *Let $\psi, \psi' \in \mathcal{F}^+(\mathcal{K}^{\mathbb{C}})$. We will say that ψ and ψ' are indistinguishable inside a Borel set $B \subset K$ if, for all $X \in \mathcal{H}(B, \Omega)$,*

$$\langle\psi|W_{\mathrm{F}}(z_\Omega(X))|\psi\rangle = \langle\psi'|W_{\mathrm{F}}(z_\Omega(X))|\psi'\rangle\,. \tag{44}$$

Note that, given ψ and B, it is easy to construct many states that are locally indistinguishable from ψ in B. Indeed, one may consider $W_{\mathrm{F}}(z_\Omega(X))|\psi\rangle$, for any $X \in \mathcal{H}(B^c, \Omega)$.

We are now ready to define what we mean by a strictly local excitation of the vacuum.

Definition 4. *If B is a Borel subset of K, a strictly local excitation of the vacuum with support in B is a normalized vector $\psi \in \mathcal{F}^+(\mathcal{K}^{\mathbb{C}})$, different from the vacuum itself, which is indistinguishable from the vacuum outside of B. In other words,*

$$\langle\psi|W_{\mathrm{F}}(z_\Omega(Y))|\psi\rangle = \langle 0|W_{\mathrm{F}}(z_\Omega(Y))|0\rangle \tag{45}$$

for all $Y = (q, p) \in \mathcal{H}(B^c, \Omega)$.

For brevity, I will occasionally call such states "local states", although this terminology conjures up images that are misleading. In view of what precedes, the coherent states $W_{\mathrm{F}}(z_\Omega(X))|0\rangle$, for any $X \in \mathcal{H}(B, \Omega)$ are strictly local excitations of the vacuum in B. The use of the adjective "strictly" is motivated by the possibility of relaxing condition (45) to allow for states that are only approximately localized in B, but for which the expectation values of observables located far from B converge more or less rapidly to the corresponding vacuum expectation values. I refer to [6] for details.

4.5 Knight's Theorem Revisited

Recall that states with a finite number of field quanta, *i.e.* states belonging to $\mathcal{F}^{\text{fin},+}(\mathcal{K}^{\mathbb{C}})$, are interpreted as states describing a finite number of quasi-particles (see Sect. 3.5). Hence one natural question is whether such a state can be a strictly local excitation of the vacuum in a set B. Theorem 1 below gives a necessary and sufficient condition for this to happen.

First, I need a definition:

Definition 5. *Ω is said to be strongly non-local on B if there does not exist a non-vanishing $h \in \mathcal{K}_{1/2}$ with the property that both h and Ωh vanish outside B.*

Here I used the further definition:

Definition 6. *Let $h \in \mathcal{K}_{\pm 1/2}$ and $B \subset K$. Then h is said to vanish in B if for all $\eta \in \mathcal{S}_B$, $\eta \cdot h = 0$. Similarly, it is said to vanish outside B, if for all $\eta \in \mathcal{S}_{B^c}$, $\eta \cdot h = 0$.*

Note that this definition uses the density of S_B in $L^2(B)$ implicitly, because without this property, it would not make much sense. Intuitively, a strongly non-local operator is one that does not leave the support of any function h invariant.

Theorem 1. *Let B be a Borel subset of K. Then the following are equivalent:*

(i) Ω is strongly non-local on B;

(ii) There do not *exist states in $\mathcal{F}^{fin,+}(\mathcal{K}^{\mathbb{C}})$ which are strictly strictly local excitations of the vacuum with support in $B \subset K$;*

I will give the proof of this result in Sect. 4.8.

Statement (i) of the theorem gives a more or less easily checked neccessary and sufficient condition for the non-existence of states with a finite number of field quanta that are localized in a region B. I will show in the examples developed in the following sections that this condition is so to speak always satisfied when B is a bounded set: I mean, it is satisfied in the various models that are typically studied in solid state physics, in relativistic quantum field theory, or in the theory of free quantum fields on curved space-times. Indeed, in these examples, Ω^2 is a finite difference or (second order elliptic) differential operator, so that it is local: it preserves the support. But its positive square root, Ω, is more like a pseudo-differential operator, and therefore does not preserve supports. This will be shown in several cases below. The upshot is that states with a finite number of particles, and a fortiori, one-particle states, are never strictly localized in a bounded set B. This gives a precise sense in which the elementary excitations of the vacuum in a bosonic field theory (relativistic or not) differ from the ordinary point particles of non-relativistic mechanics: their Hilbert space of states contains no states in which they are perfectly localized.

So, to sum it all up, you could put it this way. To the question

Why is there no sharp position observable for particles?

the answer is

It is the non-locality of Ω, stupid!

Should all this make you feel uncomfortable, I hope the further discussion in Sects. 4.7 of the history of the quest for a "position observable" in relativistic field theory will be of some help.

4.6 Examples

As a warm-up, here is my favourite example.

Exercise 3. Let $\mathcal{K} = \mathbb{R}^2$ so that Ω^2 is a two by two matrix and, as explained in Sect. 4.3, $K = \{1, 2\}$. Show that in this case, a state with a finite number of quanta can be a strictly localized excitation of the vacuum on $B = \{1\}$ only if Ω^2 is diagonal. In other words, this can happen only if the two oscillators are not coupled.

For typical translationally invariant systems, it is easy to see Ω is strongly non-local over bounded sets, so that we can conclude there are no strictly localized finite particle states. This is the content of the following results.

Theorem 2. *Let $\mathcal{K} = L^2_{\mathbb{R}}(\mathbb{R}^d, \mathrm{d}x)$ and let ω be a positive function belonging to $L^\infty_{\mathrm{loc}}(\mathbb{R}^d, \mathrm{d}k)$ with $\omega^{-1} \in L^1_{\mathrm{loc}}(\mathbb{R}^d, \mathrm{d}k)$. Suppose both ω and ω^{-1} are polynomially bounded at infinity. Let $\Omega = \omega(|\nabla|)$. Then $\mathcal{S} = \mathcal{S}(\mathbb{R}^d)$ is a local structure for this system. If ω does not extend to a holomorphic function on the complex plane, then Ω is strongly non-local on any bounded open set B. Consequently, there exist no states with a finite number of quasi-particles that are strictly localized excitations of the vacuum in such a set B.*

The proof is a simple application of the Paley-Wiener theorem together with Theorem 1. Note that the theorem applies to the Klein-Gordon equation: so we recover in this way Knight's original result. Pushing the use of the Paley-Wiener theorem a little further, one can also prove:

Theorem 3. *Let $\mathcal{K} = L^2_{\mathbb{R}}(\mathbb{R}^d, \mathrm{d}x)$ and $\Omega^2 = -\Delta + m^2$, with $d \geq 1, m > 0$, or $d \geq 2, m \geq 0$. Then $\mathcal{S} = \mathcal{S}(\mathbb{R}^d)$ is a local structure for this system and there exist no states with a finite number of quasi-particles that are strictly localized excitations of the vacuum in any set B with non-empty open complement.*

The result one needs here is proven in [26]: for $\Omega = \sqrt{-\Delta + m^2}$, h and Ωh cannot both vanish on the same open set. Via Theorem 1 this implies the above result.

An analogous result holds for the translationally invariant lattices discussed in Sect. 2.2. In particular, with Ω^2 as in (10), it is very easy to see that there are no states with a finite number of quanta that are perfectly localized perturbations of the vacuum on a finite number of lattice sites. The spatial structure is given here by the sequences of finite support, as discussed in Sect. 4.3.

Similarly, for the wave and Klein-Gordon equations the operator Ω is typically also strictly non-local, but I will not go into this here.

It is clear from these examples that Knight's theorem has less to do with relativity than with coupled oscillators, which is the point I wanted to make all along.

4.7 Newton-Wigner Localization

Knight's result appears counterintuitive. Indeed, we argued first that the Fock space structure of the Hilbert space of states of the field invites a particle interpretation (Sect. 3.5), we then introduced what looks like a perfectly reasonable notion of "strictly localized excitation of the vacuum", only to end up discovering that states with a finite number of particles cannot be strictly localized. Since the notion of a particle evokes an entity that is localized in space, this may seem paradoxical. My point of view is simple: the way out of this paradox is, as I have suggested before (Sect. 3.5), that one has to keep in mind that the particles under discussion here are just *excited states of an extended system* and that, just like in an oscillator ring, chain, or lattice, the analogy with the point particles of elementary classical or quantum mechanics courses should not be pushed too far. Calling those excitations particles amounts to nothing more than an occasionally confusing abuse of language. The lesson to be learned from Knight's result is therefore that such field quanta may carry momentum and energy, but they cannot be perfectly localized. Viewed from the angle I have chosen, this is not even surprising. The examples showed indeed this statement is true in a system with two oscillators, and in oscillator lattices. One should in particular not hope to associate a position operator with those quanta, having all the usual properties familiar from the description of point particles in ordinary Schrödinger quantum mechanics.

I could end the story there. But a very different point of view, based precisely on the use of a position operator (the so-called Newton-Wigner position operator) to locate the particles, was developed well before Knight's work, in the context of (free) relativistic quantum field theory of which the Klein-Gordon field is a particular example. Since this alternative point of view has met with a certain amount of popularity, it cannot be dismissed too lightly. Below I will explain it has an obvious analog for the oscillator systems under study here and I will show why, although it seems at first sight perfectly natural, it is clearly ill-conceived. The implication of this remark for the debate about supposed causality problems in relativistic quantum field theory and a further overview of some other issues related to "particle localization" in that context will also be given.

Newton-Wigner Localization: The Definition

Let us therefore turn again to an oscillator field with spatial structure so that $\mathcal{K}^{\mathbb{C}} = L^2(K, \mathrm{d}\mu, \mathbb{C})$. The state space of this system is the bosonic Fock space $\mathcal{F}^+(\mathcal{K}^{\mathbb{C}})$ of which $\mathcal{K}^{\mathbb{C}} = L^2(K, \mathrm{d}\mu, \mathbb{C})$ represents the one-particle sector. Now, if the system is in the state $\psi \in \mathcal{K}^{\mathbb{C}} \subset \mathcal{F}^+(\mathcal{K}^{\mathbb{C}})$, it is in view of the particle interpretation of the field states explained in Sect. 3.5 very tempting to interpret $\mid \psi \mid^2 (y)\mathrm{d}\mu$ as the probability for finding the "particle" in a volume $\mathrm{d}\mu$ around y, or in a preciser manner, to say that the probability for finding the particle in $B \subset K$ is given by

$$\int_B \mid \psi \mid^2 (y)\mathrm{d}\mu \, .$$

This seems like a quite reasonable thing to do because it is completely analogous to what is done in the non-relativistic Schrödinger quantum mechanics of particle systems. I will call the projection valued measure $B \mapsto \chi_B$, where χ_B is the operator of multiplication by the characteristic function of B the Newton-Wigner position observable. If $\psi \in L^2(K, \mathrm{d}\mu, \mathbb{C})$ is supported in $B \subset K$, we say ψ is "Newton-Wigner localized in B". This terminology is inspired by the observation that, when considering the particular example of an oscillator field given by the wave or Klein-Gordon equation, one has $\mathcal{K} = L^2(\mathbb{R}^3, \mathrm{d}x, \mathbb{C})$ and in that case the above measure is indeed the joint spectral measure of the usual Newton-Wigner position operator of relativistic quantum field theory [21]. This choice of position observable may seem reasonable, but it is only based on an analogy, and as I will now show, it is not reasonable at all.

For that purpose, let us go back to the particular example of the oscillator ring treated before (Sects. 2.2 and 4.3) and see what the Newton-Wigner position operator means in that case. Remember, this is just a system of n coupled oscillators. So the quantum Hilbert space can on the one hand be seen as $L^2(\mathbb{R}^n, \mathrm{d}x)$ (Schrödinger representation) and the system can be studied through the displacements and momenta of those oscillators. This is the usual point of view. Alternatively, it can be identified with the bosonic Fock space $\mathcal{F}^+(\mathcal{K}^{\mathbb{C}})$ (Fock representation), where now the one-particle subspace is $\mathcal{K}^{\mathbb{C}} = \mathbb{C}^n$. The latter, as explained in Sect. 4.3, can be thought of as $L^2(\mathbb{Z}/n\mathbb{Z}, \mathbb{C})$. In other words, it is tempting to interpret $\psi \in \mathcal{K}^{\mathbb{C}}$ as the quantum mechanical state of a "particle" hopping along n sites! Its probability of being at site i is then given by $|\psi(i)|^2$. More generally, any state of the n oscillators can be seen as a superposition of $0, 1, 2, \ldots$ "particle" states, where now "particle" refers to an imagined entity hopping along the sites of the chain. Speaking like this, we are pushing the particle interpretation maximally. The state $a^\dagger(\delta_i)|0\rangle$ is then thought of as particle perfectly localized on the site i.

But does this make sense? Certainly, whatever picture used, the mean square displacement of the oscillator at site j is a relevant physical observable in this system. The problem is that this mean square displacement will differ from its vacuum value if $j \neq i$:

$$\langle 0|a(\delta_i)Q_j^2 a^\dagger(\delta_i)|0\rangle \neq \langle 0|Q_j^2|0\rangle .$$

So the idea that the system contains only one particle, and that the latter is localized perfectly at i, the rest of the sites being "empty", is not tenable. Indeed, if the particle is at site i, and if space (here represented by the n sites) is otherwise "empty", how can any observable at site j take a value different from its vacuum value? The problem is of course readily solved if one stops trying to interpret the quantity $|\psi(i)|^2$ as a probability of presence for a particle.

The same analysis carries immediately over to the oscillator chains or lattices discussed before. It is perhaps even more telling there. Now the one particle space is $\ell^2(\mathbb{Z}^d, \mathbb{C})$ and so the idea of thinking of states in this space as describing a particle hopping on the sites of the lattice $\mathbb{Z}^d$ may seem even more reasonable. Models of this type are used in solid state physics to describe lattice vibrations, and the quanta are then called phonons. They are excitations of the oscillator lattice and – as Knight's theorem tells us – cannot be perfectly localized in the sense that, if the system is in a one-phonon state $\psi \in \ell^2(\mathbb{Z}^d)$, then it cannot coincide with the vacuum outside a finite subset of the lattice. This does not lead to any interpretational difficulties, *as long as one does not try to interpret* $|\psi(i)|^2$ *as the probability of finding the particle at site* i *of the lattice.*

Finally, without any change whatsoever, the same analysis carries over to the Klein-Gordon equation. Let B be a bounded subset of $\mathbb{R}^3$ and $\psi \in L^2(\mathbb{R}^3, \mathrm{d}x, \mathbb{C})$ be supported in B. As Knight's theorem tells us, the corresponding one-particle state of the field is not an excitation of the vacuum localized inside B.

The conclusion I draw from all this is that *the Newton-Wigner operator does not provide an appropriate tool to describe the strict localization properties of the states of extended systems of the type discussed here.* It shows up only because of an understandable but ill-fated desire to force a *particle interpretation* with all its usual attributes on the states of a *field.* The right notion of a (strictly) localized state is the one given by Knight (Definition 4). This has lead to some debate in the context of relativistic quantum field theory, upon which I shall comment below. Anticipating on the discussion there, I would like to stress that my line of argument here, and in particular my criticism of the use of the Newton-Wigner operator has nothing to do with relativity, or with causality, but is related instead to the fact that we are dealing with extended systems.

Causality Problems

In the early days of relativistic quantum field theory, and well before anything like Knight's theorem was formulated or proven, the particle interpretation of the field states made it perfectly natural to search for a position operator with the usual properties familiar from non-relativistic quantum mechanics. In other words, if the field quanta are particles, one would want to answer the question: "Where is the particle?" It should therefore not come as a surprise that a fair amount of literature was devoted to this problem. The theory received its definite form in [21] and a slightly more rigorous treatment was subsequently given in [30]. References to earlier work can be found in those two papers and in [25]. The discussion in [21] centers on the question how to identify, inside a relativistic elementary system (i.e. inside a unitary irreducible representation of the Poincaré group), a "position operator $\hat{x} = (\hat{x}_1, \hat{x}_2, \hat{x}_3)$", using only natural requirements – formulated as axioms – on the transformation properties of this operator under rotations and translations. The upshot of this analysis is that such an operator exists (for most values of spin and mass) and that it is unique. It is called the Newton-Wigner position operator in the literature. As an example, there exists such an operator in the one field quantum sector of the quantized Klein-Gordon field, which carries an irreducible representation of the Poincaré group of zero spin and it is precisely the one discussed in the previous subsection. Now, the joint spectral measure of the three components of $\hat{x}$ defines a projection valued measure P_B, where B is a Borel subset of $\mathbb{R}^3$. If the interpretation of $\hat{x}$ as a position operator along the lines of the usual interpretational rules of quantum mechanics is to make sense, then eigenstates of P_B with eigenvalue 1 are to be thought of as states "perfectly localized inside B". This is referred to as NW-localization. This is precisely the interpretation given to the Newton-Wigner operator in the literature which is, as explained before, at odds with Knight's notion of local excitation of the vacuum. Nevertheless, the axiomatic derivation of the Newton-Wigner operator, and its perfect analogy with the familiar situation in the quantum mechanics of non-relativistic particles gives it something very compelling, which probably explains its success. As a result, some authors have written that the Newton-Wigner operator is the only possible position operator for relativistic quantum particles. In [30], one reads the following claim: "I venture to say that any notion of localizability in three-dimensional space which does not satisfy [the axioms] will represent a radical departure from present physical ideas." Newton and Wigner say something similar, but do not put it so forcefully: "It seems to us that the above postulates are a reasonable expression for the localization of the system to the extent that one would naturally call a system unlocalizable if it should prove to be impossible to satisfy these requirements. In [25] one can read: "One either accepts the Newton-Wigner position operator when it exists, or abandons his axioms. We believe the first alternative is well worth investigation and adopt it here." I of course have argued above that one should abandon it, and that this neither constitutes a departure from standard physical ideas, nor means that one abandons the notion of localizability.

Still, even among those that have advocated the use of Newton-Wigner localization, this notion has stirred up a fair amount of debate, since it violates causality, as I now briefly explain.

Indeed, first of all, a one-particle state of the Klein-Gordon field perfectly NW-localized in some bounded set B at an initial time, is easily seen to have a non-zero probability to be found arbitrarily far away from B, at any later time, violating causality. Since the theory is supposed to be relativistic, this is a real problem that has received much attention. Actually, replacing the projection operators P_B of the NW-position operator by any other positive operators transforming correctly under space translations, Hegerfeldt proved that the causality problem remains (see [13, 14] and for a more recent overview, [15]). In addition, and directly linked to the previous observation, a state perfectly localized in one Lorentz frame is not in another one. These difficulties, while well known and widely stressed, are often dismissed with a vague appeal to one of the following somewhat related ideas. Although the Newton-Wigner derivation does not refer to any underlying field theory, these arguments all involve remembering that the "particles" in relativistic field theory are excitations of the field.

The first such argument goes as follows. In a field theory a position measurement of a particle would lead to pair creation (see [25]) and so the appearance of particles far away is not paradoxical. This line of reasoning is not very satisfactory (as already pointed out in [21]), since it seems to appeal to a (non-specified) theory of interacting fields to deal with the a priori simple non-interacting field. An alternative argument stresses that in a theory which allows for multi-particle states, the observation of exactly one particle inside a bounded set B entails the observation of the absence of particles everywhere else, and is therefore not really a local measurement. As such, the appearance later on of particles far away does not violate causality (see [10]). This argument is certainly correct. But it is again qualitative and nothing guarantees that it can correctly account for the "amount" of causality violation generated by the Newton-Wigner position.

All in all, it seems considerably simpler to adopt the notion of "strictly localized vacuum excitation" introduced by Knight, which is perfectly adapted to the study of the extended systems under consideration here and to accept once and for all that the particles of field theory are elementary excitations of the system (or field quanta) that do not have all the usual attributes of the point particles of our first mechanics and quantum mechanics courses. This seems to be the point of view implicitly prevalent among physicists, although it is never clearly spelled out in the theoretical physics textbooks for example, as I will discuss in more detail in [6]. It also has the advantage that no causality problems arise. Although traces of this argument can occasionally be found in the more mathematically oriented literature, Knight's definition of a strictly local excitation of the vacuum and his result on the non-localizability of finite particle states seem to be mostly ignored in discussions of the issue of the localizability of particles in field theory, of which there continue to be many [1, 7, 8, 12, 27, 28].

Having advocated Knight's definition of "local state", it remains to prove the extension of his theorem given above.

4.8 Proof of Theorem 1

The theorem is reduced to abstract nonsense through the following proposition. Note that, for any subset $\mathcal{M}$ of a Hilbert space $\mathcal{V}$, $\mathcal{M}^\perp$ denotes its orthogonal complement, which is a complex subspace of $\mathcal{V}$.

Proposition 2. *Let $\mathcal{K} = L^2_{\mathbb{R}}(K, \mathrm{d}\mu)$, Ω and $\mathcal{S}$ be as before. Let $B \subset K$. Then the following statements are equivalent.*

(i) Ω is strongly non-local over B.

(ii) $(z_\Omega(\mathcal{H}(B^c, \Omega)))^\perp = \{0\}$ or, equivalently,

$$(\overline{\mathrm{span}}_{\mathbb{C}} z_\Omega(\mathcal{H}(B^c, \Omega)))^\perp = \{0\} . \tag{46}$$

Indeed, that Theorem 1 (i) and (ii) are equivalent now follows from Theorem 4 below.

Proof. (Proposition 2) It is easy to see that $\xi \in (z_\Omega(\mathcal{H}(B^c, \Omega)))^\perp$ if and only if

$$\overline{\xi} \cdot \Omega^{1/2}\eta = 0 = \overline{\xi} \cdot \Omega^{-1/2}\eta ,$$

for all $\eta \in \mathcal{S}_{B^c}$. We can suppose without loss of generality that ξ is real. Now, if $\xi \in \mathcal{K}$, then $\Omega^{1/2}\xi \in \mathcal{K}_{-1/2}$ and hence

$$0 = \xi \cdot \Omega^{1/2}\eta = \Omega^{1/2}\xi \cdot \eta$$

which proves $\Omega^{1/2}\xi$ vanishes outside B. Similarly $\Omega^{-1/2}\xi$ vanishes outside B. Setting $h = \Omega^{-1/2}\xi$ the result follows.

Theorem 4. *Let $\mathcal{W}$ be a real subspace of $\mathcal{V}$.*

(i) If $\psi \in \mathcal{F}^+((\overline{\mathrm{span}}_{\mathbb{C}}\mathcal{W})^\perp) \subset \mathcal{F}^+(\mathcal{V})$, $\parallel \psi \parallel = 1$, then

$$\langle\psi|W_{\mathrm{F}}(\xi)|\psi\rangle = \langle 0|W_{\mathrm{F}}(\xi)|0\rangle, \qquad \forall \xi \in \mathcal{W} . \tag{47}$$

(ii) If $\mathrm{span}_{\mathbb{C}}\mathcal{W}$ is dense in $\mathcal{V}$ then there exist no $\psi \in \mathcal{F}^{\mathrm{fin},+}(\mathcal{V})$ other than $|0\rangle$ itself so that (47) holds.

Clearly, the equivalence of (i) and (ii) in Theorem 1 is obtained by taking $\mathcal{W} = z_\Omega(\mathcal{H}(B^c, \Omega))$ in the above theorem and applying Proposition 2.

Proof. As a warm-up, let us prove that, if $|\psi\rangle = a^\dagger(\xi')|0\rangle$, for some $\xi' \in \mathcal{V}$, then (47) holds if and only if $\xi' \in (\overline{\mathrm{span}}_{\mathbb{C}}\mathcal{W})^\perp$ and $\parallel \xi' \parallel = 1$. Indeed, for all $\xi \in \mathcal{W}$,

$$\begin{aligned}\langle\psi|W_{\mathrm{F}}(\xi)|\psi\rangle &= \langle 0|a(\xi')[\mathbb{1}+a^\dagger(\xi)][\mathbb{1}-a(\xi)]a^\dagger(\xi')|0\rangle \mathrm{e}^{-\frac{1}{2}\|\xi\|^2} \\ &= \langle 0|W_{\mathrm{F}}(\xi)|0\rangle \parallel \xi' \parallel^2 - \langle 0|a(\xi')a^\dagger(\xi)a(\xi)a^\dagger(\xi')|0\rangle \mathrm{e}^{-\frac{1}{2}\|\xi\|^2} \\ &= \langle 0|W_{\mathrm{F}}(\xi)|0\rangle \left[\parallel \xi' \parallel^2 - (\overline{\xi}'\cdot\xi)(\overline{\xi}\cdot\xi')\right] .\end{aligned}$$

Supposing (47) holds, this clearly implies $\parallel \xi' \parallel = 1$ and $\xi' \in (\overline{\mathrm{span}}_{\mathbb{C}}\mathcal{W})^\perp$. The converse is equally obvious. This proves the theorem for the very particular case of states containing exactly one quantum. Note that this completely characterizes the states with exactly one field quantum that are "localized".

To prove part (i), we can now proceed as follows. Recall that

$$W_{\mathrm{F}}(\xi) = \mathrm{e}^{-\frac{1}{2}\|\xi\|^2}\mathrm{e}^{a^\dagger(\xi)}\mathrm{e}^{-a(\xi)} .$$

Let $\xi \in \mathcal{W}$. Suppose $\psi = (\psi_0, \psi_1, \psi_2, \dots, \psi_N, 0, 0, \dots) \in \mathcal{F}^{\mathrm{fin},+}((\overline{\mathrm{span}}_{\mathbb{C}}\mathcal{W})^\perp)$. Then

$$\langle\psi|W_{\mathrm{F}}(\xi)|\psi\rangle = \mathrm{e}^{-\frac{1}{2}\|\xi\|^2}\langle\psi,\psi\rangle = \langle 0|W_{\mathrm{F}}(\xi)|0\rangle .$$

Indeed, as a result of the fact that $\xi \in \mathcal{W}$ and $\psi \in \mathcal{F}^{\mathrm{fin},+}((\overline{\mathrm{span}}_{\mathbb{C}}\mathcal{W})^\perp)$, it follows that $a(\xi)\psi = 0$ so that $\mathrm{e}^{-a(\xi)}\psi = \psi$. From this one can conclude as follows. For any $\psi = (\psi_0, \dots, \psi_n, \dots) \in \mathcal{F}^+((\overline{\mathrm{span}}_{\mathbb{C}}\mathcal{W})^\perp)$ and for any $N \in \mathbb{N}$, we can write

$$\psi = \psi_{<N} + \psi_{>N}$$

where $\psi_{<N} = (\psi_0, \dots, \psi_N, 0, \dots,)$. Then, for any $\epsilon > 0$, there exists $N_\epsilon \in \mathbb{N}$ so that

$$\langle\psi|W_{\mathrm{F}}(\xi)|\psi\rangle = \langle\psi_{<N_\epsilon}|W_{\mathrm{F}}(\xi)|\psi_{<N_\epsilon}\rangle + \mathcal{O}(\epsilon) = \langle 0|W_{\mathrm{F}}(\xi)|0\rangle + \mathcal{O}(\epsilon) ,$$

where the error term is uniform in ξ. Taking ϵ to 0, the result now follows.

In order to prove part (ii), I start with the following preliminary computation. Let $N \in \mathbb{N}$ and consider $\psi = (\psi_0, \psi_1, \psi_2, \dots, \psi_N, 0, 0, \dots) \in \mathcal{F}^{\mathrm{fin},+}(\mathcal{V})$ with $\psi_N \neq 0$. We wish to compute, for any $t \in \mathbb{R}$, for any $\xi \in \mathcal{W}$,

$$\langle\psi|W_{\mathrm{F}}(t\xi)|\psi\rangle = \sum_{n,m=0}^{N}\langle\psi_n|W_{\mathrm{F}}(t\xi)|\psi_m\rangle .$$

We will first establish that

$$\langle\psi|W_{\mathrm{F}}(t\xi)|\psi\rangle \mathrm{e}^{\frac{1}{2}t^2\|\xi\|^2}$$

is a polynomial of degree at most $2N$ in t, for fixed ξ. For that purpose, it is enough to notice that any term of the type

$$\langle\psi_n|W_{\mathrm{F}}(t\xi)|\psi_m\rangle \mathrm{e}^{\frac{1}{2}t^2\|\xi\|^2}$$

is a polynomial of degree at most $n+m$. This follows from

$$\langle\psi_n|W_{\mathrm{F}}(t\xi)|\psi_m\rangle e^{\frac{1}{2}t^2\|\xi\|^2} = \langle\psi_n|e^{a^\dagger(t\xi)}e^{-a(t\xi)}|\psi_m\rangle$$
$$= \sum_{\ell_1=0}^{n}\sum_{\ell_2=0}^{m}\frac{1}{\ell_1!\ell_2!}\langle\psi_n|(a^\dagger(t\xi))^{\ell_1}(-a(t\xi))^{\ell_2}|\psi_m\rangle$$

It is clear that this is a polynomial of degree at most $n+m$. Also, the sum can actually be restricted to those ℓ_1, ℓ_2 for which

$$m - \ell_2 = n - \ell_1 \,.$$

The term of degree $2N$ of the above polynomial is now easily identified:

$$\langle\psi_N|W_{\mathrm{F}}(t\xi)|\psi_N\rangle e^{\frac{1}{2}\|t\xi\|^2} = \langle\psi_N|e^{a^\dagger(t\xi)}e^{-a(t\xi)}|\psi_N\rangle$$
$$= \sum_{\ell_1,\ell_2=0}^{N}\frac{1}{\ell_1!\ell_2!}\langle\psi_N|(a^\dagger(t\xi))^{\ell_1}(-a(t\xi))^{\ell_2}|\psi_N\rangle$$
$$= \frac{(-1)^N t^{2N}}{N!N!}\langle\psi_N|(a^\dagger(\xi))^N(a(\xi))^N|\psi_N\rangle + \mathcal{O}(t^{2N-1})\,.$$

Suppose now (47) holds for ψ. Then this polynomial actually has to be a constant, so, if $N \geq 1$,

$$(a(\xi))^N|\psi_N\rangle = 0$$

for all $\xi \in \mathcal{W}$. Now let $\xi_1 \dots \xi_N \in \mathcal{W}$ and consider the polynomial

$$(a(t_1\xi_1 + \cdots + t_N\xi_N))^N|\psi_N\rangle = 0$$

in the variables $t_1, \dots, t_N \in \mathbb{R}$. Since each of its coefficients must vanish, we conclude that

$$a(\xi_1)a(\xi_2)\dots a(\xi_N)|\psi_N\rangle = 0\,,$$

for any choice of the $\xi_1 \dots \xi_N \in \mathcal{W}$. Consequently, this is also true for any choice of $\xi_1 \dots \xi_N \in \mathrm{span}_{\mathbb{C}}\mathcal{W}$. Introduce now an orthonormal basis η_i, $i \in \mathbb{N}$, of $\mathcal{V}$, with each $\eta_i \in \mathrm{span}_{\mathbb{C}}\mathcal{W}$. Then, in view of the above,

$$a(\eta_{i_1})a(\eta_{i_2})\dots a(\eta_{i_N})|\psi_N\rangle = 0\,,$$

for any choice $i_1 \dots i_N \in \mathbb{N}$. It is then clear that $\psi_N = 0$. Since by hypothesis $\psi_N \neq 0$, it follows that $N = 0$, so that ψ belongs to the zero-particle subspace $\mathcal{F}_0^+(\mathcal{V})$.

References

1. H. Bacry, *Localizability and space in quantum physics*, Lecture Notes in Physics 308, Springer Verlag (1988).
2. O. Bratteli, D.W. Robinson, *Operator algebras and quantum statistical mechanics*, Vol. 1, Springer-Verlag (1987).

3. O. Bratteli, D.W. Robinson, *Operator algebras and quantum statistical mechanics*, Vol. 2, Springer-Verlag (1996).
4. J. Dereziński, *Introduction to representations of canonical commutation and anti-commutation relations.* Lect. Notes Phys. **695**, 65–145 (2006).
5. S. De Bièvre, *Quantum chaos: a brief first visit*, Contempary Mathematics, **289**, 161–218 (2001).
6. S. De Bièvre, *Classical and quantum harmonic systems*, in preparation.
7. G. Fleming, *Hyperplane dependent quantized fields and Lorentz invariance*, in *Philosophical foundations of quantum field theory*, Ed. H. R. Brown and R. Harré, Clarendon Press Oxford, 93–115 (1988).
8. G. Fleming, J. Butterfield, *Strange positions*, in "From physics to philosophy", Ed. J. Butterfield, C. Pagonis, Cambridge University Press, 108–165 (1999).
9. Folland, *Harmonic analysis on phase space*, Princeton University Press, Princeton (1988).
10. S. Fulling, *Aspects of quantum field theory on curved space-time*, Cambridge UP (1989).
11. M. Fannes, A. Verbeure, *On the time-evolution automorphisms of the CCR-algebra for quantum mechanics*, Commun. Math. Phys. **35**, 257–264 (1974).
12. R. Haag, *Local quantum physics*, Springer (1996).
13. G. C. Hegerfeldt, *Remark on causality and particle localization*, Phys. Rev. D **10**, 10, 3320–3321 (1974).
14. G. C. Hegerfeldt, *Violation of causality in relativistic quantum theory?*, Phys. Rev. Letters **54**, 22, 2395–2398 (1985).
15. G. C. Hegerfeldt, *Causality, particle localization and positivity of the energy*, in "Irreversibility and Causality (Goslar 1996)", Lecture Notes in Physics 504, Springer Verlag, 238–245 (1998).
16. J.M. Knight, *Strict localization in quantum field theory*, Journal of Mathematical Physics **2**, 4, 459–471 (1961).
17. J. Klauder and B.-S. Skagerstam, *Coherent states: applications in physics and mathematical physics*, World Scientific, Singapore (1985).
18. A. L. Licht, *Strict localization*, Journal of Mathematical Physics **4**, 11, 1443–1447 (1963).
19. A. L. Licht, *Local states*, Journal of Mathematical Physics, **7**, 9, 1656–1669 (1966).
20. P. W. Milonni, *The quantum vacuum: an introduction to quantum electrodynamics*, Academic Press (1992).
21. T. D. Newton and E. P. Wigner, *Localized states for elementary systems*, Rev. Mod. Phys. **21**, 3, 400–406 (1949).
22. A. Perelomov, *Generalized coherent states and their applications*, Springer Berlin-New York NY-Paris, Texts and monographs in physics (1986).
23. M. Reed and B. Simon, *A course in mathematical physics*, Volume II, Academic Press, London (1972).
24. D. Robert, *Autour de l'approximation semi-classique*, Birkhaüser (1987).
25. S.S. Schweber, A.S. Wightman, *Configuration space methods in relativistic quantum field theory I*, Physical Review **98**, 3, 812–831.
26. I. E. Segal, R. W. Goodman, *Anti-locality of certain Lorentz-invariant operators*, Journal of Mathematics and Mechanics, **14**, 4, 629–638 (1965).
27. R. F. Streater, *Why should anyone want to axiomatize quantum field theory?*, in *Philosophical foundations of quantum field theory*, Ed. H. R. Brown and R. Harré, Clarendon Press Oxford, 137–148 (1988).

28. P. Teller, *An interpretive introduction to quantum field theory*, Princeton University Press (1995).
29. R.M. Wald, *Quantum field theory in curved spacetime and black hole thermodynamics*, University of Chicago Press, Chicago Lecture Notes in Physics (1994).
30. A.S. Wightman, *On the localizability of quantum mechanical systems*, Rev. Mod. Phys. **34**, 1, 845–872 (1962).

Introduction to Representations of the Canonical Commutation and Anticommutation Relations

J. Dereziński

Department of Mathematical Methods in Physics, Warsaw University
Hoża 74, 00-682, Warszawa, Poland
jan.derezinski@fuw.edu.pl

J. Dereziński: *Introduction to Representations of the Canonical Commutation and Anticommutation Relations*, Lect. Notes Phys. **695**, 63–143 (2006)
www.springerlink.com

1 Introduction

Since the early days of quantum mechanics it has been noted that the position operator x and the momentum operator $D := -\mathrm{i}\nabla$ satisfy the following commutation relation:

$$[x, D] = \mathrm{i}\,. \tag{1}$$

Similar commutation relation hold in the context of the second quantization. The bosonic creation operator a^* and the annihilation operator a satisfy

$$[a, a^*] = 1\,. \tag{2}$$

If we set $a^* = \frac{1}{\sqrt{2}}(x - \mathrm{i}D)$, $a = \frac{1}{\sqrt{2}}(x + \mathrm{i}D)$, then (1) implies (2), so we see that both kinds of commutation relations are closely related.

Strictly speaking the formulas (1) and (2) are ill defined because it is not clear how to interpret the commutator of unbounded operators. Weyl proposed to replace (1) by

$$\mathrm{e}^{\mathrm{i}\eta x}\mathrm{e}^{\mathrm{i}qD} = \mathrm{e}^{-\mathrm{i}q\eta}\mathrm{e}^{\mathrm{i}qD}\mathrm{e}^{\mathrm{i}\eta x},\ \ \eta, q \in \mathbb{R}, \tag{3}$$

which has a clear mathematical meaning [75]. Equation (1) is often called the canonical commutation relation (CCR) in the Heisenberg form and (3) in the Weyl form.

It is natural to ask whether the commutation relations fix the operators x and D uniquely up to the unitary equivalence. If we assume that we are given two self-adjoint operators x and D acting irreducibly on a Hilbert space and satisfying (3), then the answer is positive, as proven by Stone and von Neumann [51], see also [68].

It is useful to generalize the relations (1) and (2) to systems with many degrees of freedom. They were used by Dirac to describe quantized electromagnetic field in [30].

In systems with many degrees of freedom it is often useful to use a more abstract setting for the CCR. One can consider n self-adjoint operators $\phi_1, \dots, \phi_n$ satisfying relations

$$[\phi_j, \phi_k] = \mathrm{i}\omega_{jk} , \tag{4}$$

where ω_{jk} is an antisymmetric matrix. Alternatively one can consider the Weyl (exponentiated) form of (4) satisfied by the so-called Weyl operators $\mathrm{e}^{\mathrm{i}(y_1\phi_1+\cdots+y_n\phi_n)}$, where $(y_1, \dots, y_n) \in \mathbb{R}^n$.

The Stone-von Neumann Theorem about the uniqueness can be extended to the case of regular representations of the CCR in the Weyl form if ω_{jk} is a finite dimensional symplectic matrix. Note that in this case the relations (4) are invariant with respect to the symplectic group. This invariance is implemented by a projective unitary representation of the symplectic group, as has been noted by Segal [63]. It can be expressed in terms of a representation of the two-fold covering of the symplectic group – the so-called metaplectic representation, [61, 74].

The symplectic invariance of the CCR plays an important role in many problems of quantum theory and of partial differential equations. An interesting and historically perhaps first nontrivial application is due to Bogolubov, who used it in the theory of superfluidity of the Bose gas [15]. Since then, the idea of using symplectic transformations in bosonic systems often goes in the physics literature under the name of the Bogolubov method, see e.g. [34].

The Canonical Anticommutation Relations (CAR) appeared in mathematics before quantum theory in the context of Clifford algebras [19]. A Clifford algebra is the associative algebra generated by elements $\phi_1, \dots, \phi_n$ satisfying the relations

$$[\phi_i, \phi_j]_+ = 2\delta_{ij}, \tag{5}$$

where $[\cdot, \cdot]_+$ denotes the anticommutator. It is natural to assume that the ϕ_i are self-adjoint. It is not difficult to show that if the representation (5) is irreducible, then it is unique up to the unitary equivalence for n even and that there are two inequivalent representations for an odd n.

In quantum physics, the CAR appeared in the description of fermions [45]. If $a_1^*, \dots, a_m^*$ are fermionic creation and $a_1, \dots, a_m$ fermionic annihilation operators, then they satisfy

$$[a_i^*, a_j^*]_+ = 0, \quad [a_i, a_j]_+ = 0, \quad [a_i^*, a_j]_+ = \delta_{ij} .$$

If we set $\phi_{2j-1} := a_j^* + a_j$, $\phi_{2j} := -\mathrm{i}(a_j^* - a_j)$, then we see that they satisfy the relations (5) with $n = 2m$.

Another application of the CAR in quantum physics are Pauli [52] and Dirac [31] matrices used in the description of spin $\frac{1}{2}$ particles.

Clearly, the relations (5) are preserved by orthogonal transformations applied to $(\phi_1, \ldots, \phi_n)$. The orthogonal invariance of the CAR is implemented by a projective unitary representation. It can be also expressed in terms of a representation of the double covering of the orthogonal group, called the Pin group. The so-called spinor representations of orthogonal groups were studied by Cartan [18], and Brauer and Weyl [12].

The orthogonal invariance of the CAR relations appears in many disguises in algebra, differential geometry and quantum physics. In quantum physics it is again often called the method of Bogolubov transformations. A particularly interesting application of this method can be found in the theory of superfluidity (a version of the BCS theory that can be found e.g. in [34]).

The notion of a representation of the CCR and CAR gives a convenient framework to describe Bogolubov transformations and their unitary implementations. Analysis of Bogolubov transformations becomes especially interesting in the case of an infinite number of degrees of freedom. In this case there exist many inequivalent representations of the CCR and CAR, as noticed in the 50's, e.g. by Segal [64] and Gaarding and Wightman [38].

The most commonly used representations of the CCR/CAR are the so-called Fock representations, defined in bosonic/fermionic Fock spaces. These spaces have a distinguished vector Ω called the vacuum killed by the annihilation operators and cyclic with respect to creation operators. They were introduced in quantum physics by Fock [35] to describe systems of many particle systems with the Bose/Fermi statistics. Their mathematical structure, and also the essential self-adjointness of bosonic field operators, was established by Cook [21].

The passage from a one particle system to a system with an arbitrary number of particles subject to the Bose/Fermi statistics is usually called second quantization. Early mathematical research on abstract aspects of second quantization was done by Friedrichs [37] and Segal [63, 64].

In the case of an infinite number of degrees of freedom, the symplectic/orthogonal invariance of representations of the CCR/CAR becomes much more subtle. The unitary implementability of symplectic/orthogonal transformations in the Fock space is described by the Shale/Shale-Stinespring Theorem. These theorems say that implementable symplectic/orthogonal transformation belong to a relatively small group $Sp_2(\mathcal{Y})/O_2(\mathcal{Y})$, [61]/ [62]. In the case of an infinite number of degrees of freedom there also exists an analogue of the metaplectic/Pin representation. This seems to have been first noted by Lundberg [49].

Among early works describing these results let us mention the book by Berezin [14]. It gives concrete formulas for the implementation of Bogolubov transformations in bosonic and fermionic Fock spaces. Related problems were discussed, often independently, by other researchers, such as Ruijsenaars [59, 60].

As stressed by Segal [64], it is natural to apply the language of C^*-algebras in the description of the CCR and CAR. This is easily done in the case of the CAR, where there exists an obvious candidate for the C^*-algebra of the CAR over a given Euclidean space [17]. If this Euclidean space is of countably infinite dimension, the C^*-algebra of the CAR is isomorphic to the so called $UHF(2^\infty)$ algebra studied by Glimm. Using representations of this C^*-algebra one can construct various non-isomorphic kinds of factors (W^*-algebras with a trivial center), studied in particular by Powers [54] and Powers and Stœrmer [55].

In the case of the CCR, the choice of the corresponding C^*-algebra is less obvious. The most popular choice is the C^*-algebra generated by the Weyl operators, studied in particular by Slawny [66]. One can, however, argue that the "Weyl algebra" is not very physical and that there are other more natural choices of the C^*-algebra of the CCR. Partly to avoid discussing such (quite academic) topics, in our lecture notes we avoid the language of C^*-algebras. On the other hand, we will use the language of W^*-algebras, which seems natural in this context.

One class of representations of the CCR and CAR – the quasi-free representations – is especially widely used in quantum physics. In mathematical literature they have been first identified by Robinson [58] and Shale and Stinespring [62]. Quasi-free representations were extensively studied, especially by Araki [2,4,5,7,10] and van Daele [22].

A concrete realization of quasi-free representations is furnished by the so-called Araki-Woods representations [8] in the bosonic and Araki-Wyss representations [9] in the fermionic case. We describe these representations in detail. From the physical point of view, they can be viewed as a kind of a thermodynamical limit of representations for a finite number of degrees of freedom. From the mathematical point of view, they provide interesting and physically well motivated examples of factors of type II and III. It is very instructive to use the Araki-Woods and Araki-Wyss representations as illustrations for the Tomita-Takesaki theory and for the so-called standard form of a W^*-algebra [42] (see also [4,16,20,29,67]). They are quite often used in recent works on quantum statistical physics, see e.g. [27,44].

It is interesting to note that the famous paper of Haag, Hugenholtz and Winnink [41] about the KMS condition was directly inspired by the Araki-Woods representation.

Araki-Woods/Araki-Wyss representations can be considered also in the case of a finite number of degrees of freedom. In this case, they are equivalent to a multiple of the usual representations of the CCR/CAR. This equivalence can be described by the GNS representation with respect to a quasi-free state composed with an appropriate unitarily implemented Bogolubov transformation. We discuss this topic in the section devoted to "confined" Bose/Fermi gas.

It is easy to see that real subspaces of a complex Hilbert space form a complete complemented lattice, where the complementation is given by the symplectic orthogonal complement. It is also clear that von Neumann algebras on a given Hilbert space form a complete complemented lattice with the commutant as the complementation. It was proven by Araki [1] (see also [32]) that von Neumann algebras on a bosonic Fock space associated to real subspaces of the classical phase space also form a complemented complete lattice isomorphic to the corresponding lattice of real subspaces. We present this result, used often in algebraic quantum field theory. We also describe the fermionic analog of this result (which seems to have been overlooked in the literature).

In the last section we describe a certain class of operators that we call Pauli-Fierz operators, which are used to describe a small quantum system interacting with a bosonic reservoir, see [11, 25–27] and references therein. These operators have interesting mathematical and physical properties, which have been studied in recent years by various authors. Pauli-Fierz operators provide a good opportunity to illustrate the use of various representations of the CCR.

The concepts discussed in these lectures, in particular representations of the CCR and CAR, constitute, in one form or another, a part of the standard language of mathematical physics. More or less explicitly they are used in any textbook on quantum field theory. Usually the authors first discuss quantum fields "classically" – just the relations they satisfy without specifying their representation. Only then one introduces their representation in a Hilbert space. In the zero temperature, it is usually the Fock representation determined by the requirement that the Hamiltonian should be bounded from below, see e.g. [24]. In positive temperatures one usually chooses the GNS representation given by an appropriate thermal state.

The literature devoted to topics contained in our lecture notes is quite large. Let us mention some of the monographs. The exposition of the C^*-algebraic approach to the CCR and CAR can be found in [17]. This monograph provides also extensive historical remarks. One could also consult an older monograph [33]. Modern exposition of the mathematical formalism of second quantization can be also found e.g. in [13, 39]. We would also like to mention the book by Neretin [50], which describes infinite dimensional metaplectic and Pin groups, and review articles by Varilly and Gracia-Bondia [72, 73]. A very comprehensive article devoted to the CAR C^*-algebras was written by Araki [6]. Introductions to Clifford algebras can be found in [47, 48, 71]. In this collection of lecture notes De Bièvre discusses the localizability for bosonic fields [24].

The theory of the CCR and CAR involves a large number of concepts coming from algebra, analysis and physics. This is why the literature about this subject is very scattered and uses various conventions, notations and terminology. Even the meaning of the expressions "a representation of the CCR" and "a representation of the CAR" depends slightly on the author.

In our lectures we want to stress close analogies between the CCR and CAR. Therefore, we tried to present both formalisms in a possibly parallel way.

We also want to draw the reader's attention to W^*-algebraic aspects of the theory. They shed a lot of light onto some aspects of mathematical physics. The CAR and CCR are also a rich source of illustrations for various concepts of the theory of W^*-algebras.

We often refrain from giving proofs. Most of them are quite elementary and can be easily provided by the interested reader.

Acknowledgments

The research of the author was partly supported by the Postdoctoral Training Program HPRN-CT-2002-0277 and the Polish KBN grants SPUB127 and 2 P03A 027 25. Part of the work was done when the author visited the Erwin Schrödinger Institute as the Senior Research Fellow.

The author profited from discussions with V. Jakšić and S. De Bièvre. He would also like to express his gratitude to H. Araki for reading a previous version of the manuscript and pointing out some errors.

2 Preliminaries

In this section we review our basic notation, mostly about vector spaces and linear operators.

2.1 Bilinear Forms

Let α be a bilinear form on a vector space $\mathcal{Y}$. The action of α on a pair of vectors $y_1, y_2 \in \mathcal{Y}$ will be written as $y_1\alpha y_2$. We say that a linear map r on $\mathcal{Y}$ preserves α iff

$$(ry_1)\alpha(ry_2) = y_1\alpha y_2, \quad y_1, y_2 \in \mathcal{Y}.$$

We say that α is nondegenerate, if for any non-zero $y_1 \in \mathcal{Y}$ there exists $y_2 \in \mathcal{Y}$ such that $y_2\alpha y_1 \neq 0$.

An antisymmetric nondegenerate form is called symplectic. A symmetric nondegenerate form is called a scalar product.

2.2 Operators in Hilbert Spaces

The scalar product of two vectors Φ, Ψ in a Hilbert space will be denoted by $(\Phi|\Psi)$. It will be antilinear in the first argument and linear in the second.

If $\mathcal{H}_1, \mathcal{H}_2$ are Hilbert spaces, then $B(\mathcal{H}_1, \mathcal{H}_2)$, resp. $U(\mathcal{H}_1, \mathcal{H}_2)$ denotes bounded, resp. unitary operators from $\mathcal{H}_1$ to $\mathcal{H}_2$.

A^* denotes the hermitian adjoint of the operator A.

An operator $U : \mathcal{H}_1 \to \mathcal{H}_2$ is called antiunitary iff it is antilinear, bijective and $(U\Phi|U\Psi) = \overline{(\Phi|\Psi)}$.

$B^2(\mathcal{H}_1, \mathcal{H}_2)$ denotes Hilbert-Schmidt operators from $\mathcal{H}_1$ to $\mathcal{H}_2$, that is $A \in B^2(\mathcal{H}_1, \mathcal{H}_2)$ iff $\mathrm{Tr} A^*A < \infty$. Note that $B^2(\mathcal{H}_1, \mathcal{H}_2)$ has a natural structure of the Hilbert space with the scalar product

$$(A|B) := \mathrm{Tr} A^* B \, .$$

$B^1(\mathcal{H}_1, \mathcal{H}_2)$ denotes trace class operators from $\mathcal{H}_1$ to $\mathcal{H}_2$, that is $A \in B^1(\mathcal{H}_1, \mathcal{H}_2)$ iff $\mathrm{Tr}(A^*A)^{1/2} < \infty$.

For a single space $\mathcal{H}$, we will write $B(\mathcal{H}) = B(\mathcal{H}, \mathcal{H})$, etc. $B_{\mathrm{h}}(\mathcal{H})$ will denote bounded self-adjoint operators on $\mathcal{H}$ (the subscript h stands for "hermitian"). $B_+(\mathcal{H})$ denotes positive bounded operators on $\mathcal{H}$. Similarly, $B^2_+(\mathcal{H})$ and $B^1_+(\mathcal{H})$ stand for positive Hilbert-Schmidt and trace class operators on $\mathcal{H}$ respectively.

By saying that A is an operator from $\mathcal{H}_1$ to $\mathcal{H}_2$ we will mean that the domain of A, denoted by $\mathrm{Dom} A$ is a subspace of $\mathcal{H}_1$ and A is a linear map from $\mathrm{Dom} A$ to $\mathcal{H}_2$.

The spectrum of an operator A is denoted by $\mathrm{sp} A$.

2.3 Tensor Product

We assume that the reader is familiar with the concept of the algebraic tensor product of two vector spaces. The tensor product of a vector $z \in \mathcal{Z}$ and a vector $w \in \mathcal{W}$ will be, as usual, denoted by $z \otimes w$.

Let $\mathcal{Z}$ and $\mathcal{W}$ be two Hilbert spaces. The notation $\mathcal{Z} \otimes \mathcal{W}$ will be used to denote the tensor product of $\mathcal{Z}$ and $\mathcal{W}$ in the sense of Hilbert spaces. Thus $\mathcal{Z} \otimes \mathcal{W}$ is a Hilbert space equipped with the operation

$$\mathcal{Z} \times \mathcal{W} \ni (z, w) \mapsto z \otimes w \in \mathcal{Z} \otimes \mathcal{W} \, ,$$

and with a scalar product satisfying

$$(z_1 \otimes w_1 | z_2 \otimes w_2) = (z_1|z_2)(w_1|w_2) \, . \tag{6}$$

$\mathcal{Z} \otimes \mathcal{W}$ is the completion of the algebraic tensor product of $\mathcal{Z}$ and $\mathcal{W}$ in the norm given by (6).

2.4 Operators in a Tensor Product

Let a and b be (not necessarily everywhere defined) operators on $\mathcal{Z}$ and $\mathcal{W}$. Then we can define a linear operator $a \otimes b$ with the domain equal to the algebraic tensor product of $\mathrm{Dom}\, a$ and $\mathrm{Dom}\, b$, satisfying

$$(a \otimes b)(z \otimes w) := (az) \otimes (bw) .$$

If a and b are densely defined, then so is $a \otimes b$.

If a and b are closed, then $a \otimes b$ is closable. To see this, note that the algebraic tensor product of $\mathrm{Dom}\, a^*$ and $\mathrm{Dom}\, b^*$ is contained in the domain of $(a \otimes b)^*$. Hence $(a \otimes b)^*$ has a dense domain.

We will often denote the closure of $a \otimes b$ with the same symbol.

2.5 Conjugate Hilbert Spaces

Let $\mathcal{H}$ be a complex vector space. The space $\overline{\mathcal{H}}$ conjugate to $\mathcal{H}$ is a complex vector space together with a distinguished antilinear bijection

$$\mathcal{H} \ni \Psi \mapsto \overline{\Psi} \in \overline{\mathcal{H}} . \tag{7}$$

The map (7) is called a conjugation on $\mathcal{H}$. It is convenient to denote the inverse of the map (7) by the same symbol. Thus $\overline{\overline{\Psi}} = \Psi$.

Assume in addition that $\mathcal{H}$ is a Hilbert space. Then we assume that $\overline{\mathcal{H}}$ is also a Hilbert space and (7) is antiunitary, so that the scalar product on $\overline{\mathcal{H}}$ satisfies

$$(\overline{\Phi}|\overline{\Psi} = \overline{(\Phi|\Psi)} .$$

For $\Psi \in \mathcal{H}$, let $(\Psi|$ denote the operator in $B(\mathcal{H}, \mathbb{C})$ given by

$$\mathcal{H} \ni \Phi \mapsto (\Psi|\Phi) \in \mathbb{C} .$$

We will write $|\Psi) := (\Psi|^*$.

By the Riesz lemma, the map

$$\overline{\mathcal{H}} \ni \overline{\Psi} \mapsto (\Psi| \in B(\mathcal{H}, \mathbb{C})$$

is an isomorphism between $\overline{\mathcal{H}}$ and the dual of $\mathcal{H}$, that is $B(\mathcal{H}, \mathbb{C})$.

If $A \in B(\mathcal{H})$, then $\overline{A} \in B(\overline{\mathcal{H}})$ is defined by

$$\overline{\mathcal{H}} \ni \overline{\Psi} \mapsto \overline{A}\, \overline{\Psi} := \overline{A\Psi} \in \overline{\mathcal{H}} .$$

We will identify $\overline{B(\mathcal{H})}$ with $B(\overline{\mathcal{H}})$.

If $\mathcal{H}$ is a real vector space, we always take $\Psi \mapsto \overline{\Psi}$ to be the identity.

Let $\mathcal{Z}, \mathcal{W}$ be Hilbert spaces. We will often use the identification of the set of Hilbert-Schmidt operators $B^2(\mathcal{Z}, \mathcal{W})$ with $\mathcal{W} \otimes \overline{\mathcal{Z}}$, so that $|\Phi)(\Psi| \in B^2(\mathcal{Z}, \mathcal{W})$ corresponds to $\Phi \otimes \overline{\Psi} \in \mathcal{W} \otimes \overline{\mathcal{Z}}$.

We identify $\overline{\mathcal{Z}\otimes\mathcal{W}}$ with $\overline{\mathcal{W}}\otimes\overline{\mathcal{Z}}$.

If $A\in B(\mathcal{Z},\mathcal{W})$, then $A^{\#}\in B(\overline{\mathcal{W}},\overline{\mathcal{Z}})$ is defined as $A^{\#}:=\overline{A}^{*}$ (recall that $*$ denotes the hermitian conjugation). $A^{\#}$ is sometimes called the transpose of A.

This is especially useful if $A\in B(\overline{\mathcal{Z}},\mathcal{Z})$. Then we say that A is symmetric iff $A^{\#}=A$ and antisymmetric if $A^{\#}=-A$. In other words, A is symmetric if

$$(z_1|A\overline{z}_2)=(z_2|A\overline{z}_1),\quad z_1,z_2\in\mathcal{Z}\ ;$$

antisymmetric if

$$(z_1|A\overline{z}_2)=-(z_2|A\overline{z}_1),\quad z_1,z_2\in\mathcal{Z}\ .$$

The space of symmetric and antisymmetric bounded operators from $\overline{\mathcal{Z}}$ to $\mathcal{Z}$ is denoted by $B_{\mathrm{s}}(\overline{\mathcal{Z}},\mathcal{Z})$ and $B_{\mathrm{a}}(\overline{\mathcal{Z}},\mathcal{Z})$ resp. The space of Hilbert-Schmidt symmetric and antisymmetric operators from $\overline{\mathcal{Z}}$ to $\mathcal{Z}$ is denoted by $B^2_{\mathrm{s}}(\overline{\mathcal{Z}},\mathcal{Z})$ and $B^2_{\mathrm{a}}(\overline{\mathcal{Z}},\mathcal{Z})$ resp.

Remark 1. Note that, unfortunately, in the literature, e.g. in [56], the word "symmetric" is sometimes used in a different meaning ("hermitian but not necessarily self-adjoint").

2.6 Fredholm Determinant

Let $\mathcal{Y}$ be a (real or complex) Hilbert space.

Let $1+B^1(\mathcal{Y})$ denote the set of operators of the form $1+a$ with $a\in B^1(\mathcal{Y})$.

Theorem 1. *There exists a unique function* $1+B^1(\mathcal{Y})\ni r\mapsto \det r$ *that satisfies*

1) *If* $\mathcal{Y}=\mathcal{Y}_1\oplus\mathcal{Y}_2$ *with* $\dim\mathcal{Y}_1<\infty$ *and* $r=r_1\oplus 1$, *then* $\det r=\det r_1$, *where* $\det r_1$ *is the usual determinant of a finite dimensional operator* r_1*;*
2) $1+B^1(\mathcal{Y})\ni r\mapsto\det r$ *is continuous in the trace norm.*

$\det r$ is called the Fredholm determinant of r, see e.g. [57] Sect. XIII.17.

3 Canonical Commutation Relations

In this section we introduce one of the basic concept of our lectures – a representation of the canonical commutation relations (CCR). We choose the exponential form of the CCR – often called the Weyl form of the CCR.

In the literature the terminology related to CCR often depends on the author, [13, 17, 25, 33]. What we call representations of the CCR is known also as Weyl or Heisenberg-Weyl systems.

3.1 Representations of the CCR

Let $\mathcal{Y}$ be a real vector space equipped with an antisymmetric form ω. (Note that ω does not need to be nondegenerate). Let $\mathcal{H}$ be a Hilbert space. Recall that $U(\mathcal{H})$ denotes the set of unitary operators on $\mathcal{H}$. We say that a map

$$\mathcal{Y} \ni y \mapsto W^\pi(y) \in U(\mathcal{H}) \tag{8}$$

is a *representation of the CCR over $\mathcal{Y}$ in $\mathcal{H}$* if

$$W^\pi(y_1)W^\pi(y_2) = \mathrm{e}^{-\frac{\mathrm{i}}{2}y_1\omega y_2}W^\pi(y_1+y_2), \qquad y_1, y_2 \in \mathcal{Y}. \tag{9}$$

Note that (9) implies

Theorem 2. *Let $y, y_1, y_2 \in \mathcal{Y}$. Then*

$$W^\pi(y_1)W^\pi(y_2) = \mathrm{e}^{-\mathrm{i}y_1\omega y_2}W^\pi(y_2)W^\pi(y_1)\,, \tag{10}$$

$$W^{\pi*}(y) = W^\pi(-y), \qquad W^\pi(0) = 1\,, \tag{11}$$

$$W^\pi(t_1y)W^\pi(t_2y) = W^\pi((t_1+t_2)y), \quad t_1, t_2 \in \mathbb{R}\,. \tag{12}$$

Equation (10) is known as the canonical commutation relation in the Weyl form.

We say that a subset $K \subset \mathcal{H}$ is cyclic for (8) if

$$\mathrm{Span}\{W^\pi(y)\Psi \ : \ \Psi \in K,\ y \in \mathcal{Y}\}$$

is dense in $\mathcal{H}$. We say that $\Psi_0 \in \mathcal{H}$ is cyclic if $\{\Psi_0\}$ is cyclic.

We say that the representation (8) is irreducible if the only closed subspace of $\mathcal{H}$ preserved by $W^\pi(y)$ for all $y \in \mathcal{Y}$ is $\{0\}$ and $\mathcal{H}$. Clearly, in the case of an irreducible representation, all nonzero vectors in $\mathcal{H}$ are cyclic.

Suppose we are given two representations of the CCR over the same space $(\mathcal{Y}, \omega)$:

$$\mathcal{Y} \ni y \mapsto W^{\pi_1}(y) \in U(\mathcal{H}_1)\,, \tag{13}$$

$$\mathcal{Y} \ni y \mapsto W^{\pi_2}(y) \in U(\mathcal{H}_2)\,. \tag{14}$$

We say that (13) is unitarily equivalent to (14) iff there exists a unitary operator $U \in U(\mathcal{H}_1, \mathcal{H}_2)$ such that

$$UW^{\pi_1}(y) = W^{\pi_2}(y)U, \quad y \in \mathcal{Y}\,.$$

Clearly, given a representation of the CCR (8) and a linear transformation r on $\mathcal{Y}$ that preserves ω,

$$\mathcal{Y} \ni y \mapsto W^{\pi}(ry) \in U(\mathcal{H})$$

is also a representation of the CCR.

If we have two representations of the CCR

$$\mathcal{Y}_1 \ni y_1 \mapsto W^{\pi_1}(y_1) \in U(\mathcal{H}_1),$$

$$\mathcal{Y}_2 \ni y_2 \mapsto W^{\pi_2}(y_2) \in U(\mathcal{H}_2),$$

then

$$\mathcal{Y}_1 \oplus \mathcal{Y}_2 \ni (y_1, y_2) \mapsto W^{\pi_1}(y_1) \otimes W^{\pi_2}(y_2) \in U(\mathcal{H}_1 \otimes \mathcal{H}_2)$$

is also a representation of the CCR.

By (12), for any representation of the CCR,

$$\mathbb{R} \ni t \mapsto W^{\pi}(ty) \in U(\mathcal{H}) \tag{15}$$

is a 1-parameter group. We say that a representation of the CCR (8) is *regular* if (15) is strongly continuous for each $y \in \mathcal{Y}$. Representations of the CCR that appear in applications are usually regular.

3.2 Schrödinger Representation of the CCR

Let $\mathcal{X}$ be a finite dimensional real space. Let $\mathcal{X}^{\#}$ denote the space dual to $\mathcal{X}$. Then the form

$$(\eta_1, q_1)\omega(\eta_2, q_2) = \eta_1 q_2 - \eta_2 q_1 \tag{16}$$

on $\mathcal{X}^{\#} \oplus \mathcal{X}$ is symplectic.

Let x be the generic name of the variable in $\mathcal{X}$, and simultaneously the operator of multiplication by the variable x in $L^2(\mathcal{X})$. More precisely for any $\eta \in \mathcal{X}^{\#}$ the symbol ηx denotes the self-adjoint operator on $L^2(\mathcal{X})$ acting on its domain as

$$(\eta x \Psi)(x) := (\eta x)\Psi(x)\ .$$

Let $D := \frac{1}{\mathrm{i}}\nabla_x$ be the momentum operator on $L^2(\mathcal{X})$. More precisely, for any $q \in \mathcal{X}$ the symbol qD denotes the self-adjoint operator on $L^2(\mathcal{X})$ acting on its domain as

$$(qD\Psi)(x) := \frac{1}{\mathrm{i}} q\nabla_x \Psi(x)\ .$$

It is easy to see that

Theorem 3. *The map*

$$\mathcal{X}^{\#} \oplus \mathcal{X} \ni (\eta, q) \mapsto \mathrm{e}^{\mathrm{i}(\eta x + qD)} \in U(L^2(\mathcal{X})) \tag{17}$$

is an irreducible regular representation of the CCR over $\mathcal{X}^{\#} \oplus \mathcal{X}$ *in* $L^2(\mathcal{X})$.

Equation (17) is called the *Schrödinger representation.*

Conversely, suppose that $\mathcal{Y}$ is a finite dimensional real vector space equipped with a symplectic form ω. Let

$$\mathcal{Y} \ni y \mapsto W^\pi(y) \in U(\mathcal{H}) \tag{18}$$

be a representation of the CCR. There exists a real vector space $\mathcal{X}$ such that the symplectic space $\mathcal{Y}$ can be identified with $\mathcal{X}^\# \oplus \mathcal{X}$ equipped with the symplectic form (16). Thus we can rewrite (18) as

$$\mathcal{X}^\# \oplus \mathcal{X} \ni (\eta, q) \mapsto W^\pi(\eta, q) \in U(\mathcal{H})$$

satisfying

$$W^\pi(\eta_1, q_1)W^\pi(\eta_2, q_2) = \mathrm{e}^{-\frac{\mathrm{i}}{2}(\eta_1 q_2 - \eta_2 q_1)} W^\pi(\eta_1 + \eta_2, q_1 + q_2) .$$

In particular,

$$\mathcal{X}^\# \ni \eta \mapsto W^\pi(\eta, 0) \in U(\mathcal{H}) , \tag{19}$$

$$\mathcal{X} \ni q \mapsto W^\pi(0, q) \in U(\mathcal{H}) \tag{20}$$

are unitary representations satisfying

$$W^\pi(\eta, 0)W^\pi(0, q) = \mathrm{e}^{-\mathrm{i}\eta q} W^\pi(0, q)W^\pi(\eta, 0) .$$

If (18) is regular, then (19) and (20) are strongly continuous.

The following classic result says that a representation of the CCR over a symplectic space with a finite number of degrees of freedom is essentially unique up to the multiplicity (see e.g. [17,33]):

Theorem 4 (The Stone–von Neumann theorem). *Suppose that $(\mathcal{Y}, \omega)$ is a finite dimensional symplectic space and (18) a regular representation of the CCR. Suppose that we fix an identification of $\mathcal{Y}$ with $\mathcal{X}^\# \oplus \mathcal{X}$. Then there exists a Hilbert space $\mathcal{K}$ and a unitary operator $U : L^2(\mathcal{X}) \otimes \mathcal{K} \to \mathcal{H}$ such that*

$$W^\pi(\eta, q)U = U\left(\mathrm{e}^{\mathrm{i}(\eta x + qD)} \otimes 1_\mathcal{K}\right) .$$

The representation of the CCR (18) is irreducible iff $\mathcal{K} = \mathbb{C}$.

Corollary 1. *Suppose that $\mathcal{Y}$ is a finite dimensional symplectic space. Let $\mathcal{Y} \ni y \mapsto W^{\pi_1}(y) \in U(\mathcal{H})$ and $\mathcal{Y} \ni y \mapsto W^{\pi_2}(y) \in U(\mathcal{H})$ be two regular irreducible representations of the CCR. Then they are unitarily equivalent.*

3.3 Field Operators

In this subsection we assume that we are given a regular representation $\mathcal{Y} \ni y \mapsto W^\pi(y) \in U(\mathcal{H})$. Recall that $\mathbb{R} \ni t \mapsto W^\pi(ty)$ is a strongly continuous unitary group. By the Stone theorem, for any $y \in \mathcal{Y}$, we can define its self-adjoint generator

$$\phi^\pi(y) := -\mathrm{i}\frac{\mathrm{d}}{\mathrm{d}t} W^\pi(ty)\Big|_{t=0} .$$

$\phi^\pi(y)$ will be called the *field operator* corresponding to $y \in \mathcal{Y}$.

Remark 2. Sometimes, the operators $\phi^\pi(y)$ are called *Segal field operators.*

Theorem 5. *Let $y, y_1, y_2 \in \mathcal{Y}$.*
1) *Let $\Psi \in \mathrm{Dom}\,\phi^\pi(y_1) \cap \mathrm{Dom}\,\phi^\pi(y_2)$, $c_1, c_2 \in \mathbb{R}$. Then*

$$\Psi \in \mathrm{Dom}\,\phi^\pi(c_1y_1 + c_2y_2),$$

$$\phi^\pi(c_1y_1 + c_2y_2)\Psi = c_1\phi^\pi(y_1)\Psi + c_2\phi^\pi(y_2)\Psi.$$

2) *Let $\Psi_1, \Psi_2 \in \mathrm{Dom}\,\phi^\pi(y_1) \cap \mathrm{Dom}\,\phi^\pi(y_2)$. Then*

$$(\phi^\pi(y_1)\Psi_1|\phi^\pi(y_2)\Psi_2) - (\phi^\pi(y_2)\Psi_1|\phi^\pi(y_1)\Psi_2) = \mathrm{i}y_1\omega y_2(\Psi_1|\Psi_2)\,. \tag{21}$$

3) *$\phi^\pi(y_1) + \mathrm{i}\phi^\pi(y_2)$ is a closed operator on the domain*

$$\mathrm{Dom}\,\phi^\pi(y_1) \cap \mathrm{Dom}\,\phi^\pi(y_2)\,.$$

Equation (21) can be written somewhat imprecisely as

$$[\phi^\pi(y_1), \phi^\pi(y_2)] = \mathrm{i}y_1\omega y_2 \tag{22}$$

and is called the canonical commutation relation in the Heisenberg form.

3.4 Bosonic Bogolubov Transformations

Let $(\mathcal{Y}, \omega)$ be a finite dimensional symplectic space. Linear maps on $\mathcal{Y}$ preserving ω are then automatically invertible. They form a group, which will be called the *symplectic group of* $\mathcal{Y}$ and denoted $Sp(\mathcal{Y})$.

Let $\mathcal{Y} \ni y \mapsto W(y) \in U(\mathcal{H})$ be a regular irreducible representation of canonical commutation relations. The following theorem is an immediate consequence of Corollary 1:

Theorem 6. *For any $r \in Sp(\mathcal{Y})$ there exists $U \in U(\mathcal{H})$, defined uniquely up to a phase factor (a complex number of absolute value 1), such that*

$$UW(y)U^* = W(ry)\,. \tag{23}$$

Let $\mathcal{U}_r$ be the class of unitary operators satisfying (23). Then

$$Sp(\mathcal{Y}) \ni r \mapsto \mathcal{U}_r \in U(\mathcal{H})/U(1)$$

is a group homomorphism, where $U(1)$ denotes the group of unitary scalar operators on $\mathcal{H}$.

One can ask whether one can fix uniquely the phase factor appearing in the above theorem and obtain a group homomorphism of $Sp(\mathcal{Y})$ into $U(\mathcal{H})$ satisfying (23). This is impossible, the best what one can do is the following improvement of Theorem 6:

Theorem 7. *For any $r \in Sp(\mathcal{Y})$ there exists a unique pair of operators $\{U_r, -U_r\} \subset U(\mathcal{H})$ such that*

$$U_r W(y) U_r^* = W(ry) ,$$

and such that we have a group homomorphism

$$Sp(\mathcal{Y}) \ni r \mapsto \pm U_r \in U(\mathcal{H})/\{1, -1\} . \tag{24}$$

The representation (24) is called the *metaplectic representation of* $Sp(\mathcal{Y})$.

Note that the homotopy group of $Sp(\mathcal{Y})$ is $\mathbb{Z}$. Hence, for any $n \in \{1, 2, \dots, \infty\}$, we can construct the n-fold covering of $Sp(\mathcal{Y})$. The image of the metaplectic representation is isomorphic to the double covering of $Sp(\mathcal{Y})$. It is often called the metaplectic group.

In the physics literature the fact that symplectic transformations can be unitarily implemented is generally associated with the name of Bogolubov, who successfully applied this idea to the superfluidity of the Bose gas. Therefore, in the physics literature, the transformations described in Theorems 6 and 7 are often called *Bogolubov transformations.*

The proofs of Theorems 6 and 7 are most conveniently given in the Fock representation, where one has simple formulas for U_r (see e.g. [14, 36]). We will describe these formulas later on (in the more general context of an infinite number of degrees of freedom).

4 Canonical Anticommutation Relations

In this section we introduce the second basic concept of our lectures, that of a representation of the canonical anticommutation relations (CAR). Again, there is no uniform terminology in this domain [13, 17, 33]. What we call a representation of the CAR is often called Clifford relations, which is perhaps more justified historically. Our terminology is intended to stress the analogy between the CCR and CAR.

4.1 Representations of the CAR

Let $\mathcal{Y}$ be a real vector space with a positive scalar product α. Let $\mathcal{H}$ be a Hilbert space. Recall that $B_{\mathrm{h}}(\mathcal{H})$ denotes the set of bounded self-adjoint operators on $\mathcal{H}$. We say that a linear map

$$\mathcal{Y} \ni y \mapsto \phi^\pi(y) \in B_{\mathrm{h}}(\mathcal{H}) \tag{25}$$

is a *representation of the CAR over* $\mathcal{Y}$ *in* $\mathcal{H}$ iff $\phi^\pi(y)$ satisfy

$$[\phi^\pi(y_1), \phi^\pi(y_2)]_+ = 2 y_1 \alpha y_2, \quad y_1, y_2 \in \mathcal{Y} . \tag{26}$$

We will often drop π, and write $|y|_\alpha := (y \alpha y)^{1/2}$.

Remark 3. The reason of putting the factor 2 in (26) is the identity $\phi(y)^2 = y\alpha y$. Note, however, that in (22) there is no factor of 2, and therefore at some places the treatment of the CCR and CAR will be not as parallel as it could be.

Theorem 8. 1) $\mathrm{sp}\,\phi(y) = \{-|y|_\alpha, |y|_\alpha\}$.
2) *Let* $t \in \mathbb{C}$, $y \in \mathcal{Y}$. *Then* $\|t + \phi(y)\| = \max\{|t + |y|_\alpha|, |t - |y|_\alpha|\}$.
3) $\mathrm{e}^{\mathrm{i}\phi(y)} = \cos|y|_\alpha + \mathrm{i}\frac{\sin|y|_\alpha}{|y|_\alpha}\phi(y)$.
4) *Let* $\mathcal{Y}^{\mathrm{cpl}}$ *be the completion of* $\mathcal{Y}$ *in the norm* $|\cdot|_\alpha$. *Then there exists a unique extension of (25) to a continuous map*

$$\mathcal{Y}^{\mathrm{cpl}} \ni y \mapsto \phi^{\pi^{\mathrm{cpl}}}(y) \in B_{\mathrm{h}}(\mathcal{H}) . \tag{27}$$

Moreover, (27) is a representation of the CAR.

Motivated by the last statement, henceforth we will assume that $\mathcal{Y}$ is complete – that is, $\mathcal{Y}$ is a real Hilbert space.

By saying that $(\phi_1, \ldots, \phi_n)$ is a representation of the CAR on $\mathcal{H}$ we will mean that we have a representation of the CAR $\mathbb{R}^n \ni y \mapsto \phi(y) \in B_{\mathrm{h}}(\mathcal{H})$, where $\mathbb{R}^n$ is equipped with the canonical scalar product, e_i is the canonical basis of $\mathbb{R}^n$ and $\phi_i = \phi^\pi(e_i)$. Clearly, this is equivalent to the relations $[\phi_i, \phi_j]_+ = 2\delta_{ij}$.

We say that a subset $K \subset \mathcal{H}$ is cyclic for (25) if $\mathrm{Span}\{\phi^\pi(y_1)\cdots\phi^\pi(y_n)\Psi : \Psi \in K,\ y_1, \ldots, y_n \in \mathcal{Y},\ n = 1, 2, \ldots\}$ is dense in $\mathcal{H}$. We say that $\Psi_0 \in \mathcal{H}$ is cyclic for (25) if $\{\Psi_0\}$ is cyclic.

We say that (25) is irreducible if the only closed subspace of $\mathcal{H}$ preserved by $\phi^\pi(y)$ for all $y \in \mathcal{Y}$ is $\{0\}$ and $\mathcal{H}$. Clearly, in the case of an irreducible representation, all nonzero vectors in $\mathcal{H}$ are cyclic.

Suppose we are given two representations of the CAR over the same space $(\mathcal{Y}, \alpha)$:

$$\mathcal{Y} \ni y \mapsto \phi^{\pi_1}(y) \in B_{\mathrm{h}}(\mathcal{H}_1) , \tag{28}$$

$$\mathcal{Y} \ni y \mapsto \phi^{\pi_2}(y) \in B_{\mathrm{h}}(\mathcal{H}_2) , \tag{29}$$

then we say that (28) is unitarily equivalent to (29) iff there exists a unitary operator $U \in U(\mathcal{H}_1, \mathcal{H}_2)$ such that

$$U\phi^{\pi_1}(y) = \phi^{\pi_2}(y)U, \quad y \in \mathcal{Y} .$$

Let $\mathcal{Y}_1$, $\mathcal{Y}_2$ be two real Hilbert spaces. Suppose that I is a self-adjoint operator on $\mathcal{H}_1$ and

$$\mathcal{Y}_1 \oplus \mathbb{R} \ni (y_1, t) \mapsto \phi^{\pi_1}(y_1) + tI \in B_{\mathrm{h}}(\mathcal{H}_1) ,$$

$$\mathcal{Y}_2 \ni y_2 \mapsto \phi^{\pi_2}(y_2) \in B_{\mathrm{h}}(\mathcal{H}_2)$$

are representations of the CAR. Then

$$\mathcal{Y}_1 \oplus \mathcal{Y}_2 \ni (y_1, y_2) \mapsto \phi^{\pi_1}(y_1) \otimes 1 + I \otimes \phi^{\pi_2}(y_2) \in B(\mathcal{H}_1 \otimes \mathcal{H}_2)$$

is a representation of the CAR.

If $r \in B(\mathcal{Y})$ preserves the scalar product (is isometric), and we are given a representation of the CAR (25), then

$$\mathcal{Y} \ni y \mapsto \phi^{\pi}(ry) \in B_{\mathrm{h}}(\mathcal{H})$$

is also a representation of the CAR.

Most of the above material was very similar to its CCR counterpart. The following construction, however, has no analog in the context of the CCR:

Theorem 9. *Suppose that $\mathbb{R}^n \ni y \mapsto \phi(y)$ is a representation of the CAR. Let $y_1, \dots, y_n$ be an orthonormal basis in $\mathbb{R}^n$. Set*

$$Q := \mathrm{i}^{n(n-1)/2} \phi(y_1) \cdots \phi(y_n) \, .$$

Then the following is true:

1) *Q depends only on the orientation of the basis (it changes the sign under the change of the orientation).*
2) *Q is unitary and self-adjoint, moreover, $Q^2 = 1$.*
3) *$Q\phi(y) = (-1)^n \phi(y) Q$, for any $y \in \mathcal{Y}$.*
4) *If $n = 2m$, then $Q = \mathrm{i}^m \phi(y_1) \cdots \phi(y_{2m})$ and*

$$\mathbb{R}^{2m+1} \ni (y, t) \mapsto \phi(y) \pm tQ$$

are two representations of the CAR.

5) *If $n = 2m+1$, then $Q = (-\mathrm{i})^m \phi(y_1) \cdots \phi(y_{2m+1})$ and $\mathcal{H} = \mathrm{Ker}(Q-1) \oplus \mathrm{Ker}(Q+1)$ gives a decomposition of $\mathcal{H}$ into a direct sum of subspaces preserved by our representation.*

4.2 Representations of the CAR in Terms of Pauli Matrices

In the space $\mathbb{C}^2$ we introduce the usual Pauli spin matrices σ_1, σ_2 and σ_3. This means

$$\sigma_1 = \begin{bmatrix} 0 & 1 \\ 1 & 0 \end{bmatrix}, \quad \sigma_2 = \begin{bmatrix} 0 & -\mathrm{i} \\ \mathrm{i} & 0 \end{bmatrix}, \quad \sigma_3 = \begin{bmatrix} 1 & 0 \\ 0 & -1 \end{bmatrix} .$$

Note that $\sigma_i^2 = 1$, $\sigma_i^* = \sigma_i$, $i = 1, 2, 3$, and

$$\begin{aligned} \sigma_1\sigma_2 &= -\sigma_2\sigma_1 = \mathrm{i}\sigma_3, \\ \sigma_2\sigma_3 &= -\sigma_3\sigma_2 = \mathrm{i}\sigma_1 \\ \sigma_3\sigma_1 &= -\sigma_1\sigma_3 = \mathrm{i}\sigma_2. \end{aligned} \tag{30}$$

Moreover, $B(\mathbb{C}^2)$ has a basis $(1, \sigma_1, \sigma_2, \sigma_3)$. Clearly, $(\sigma_1, \sigma_2, \sigma_3)$ is a representation of the CAR over $\mathbb{R}^3$.

In the algebra $B(\otimes^n \mathbb{C}^2)$ we introduce the operators

$$\sigma_i^{(j)} := 1^{\otimes(j-1)} \otimes \sigma_i \otimes 1^{\otimes(n-j)}, \quad i = 1, 2, 3, \quad j = 1, \ldots, n\,.$$

Note that $\sigma_i^{(j)}$ satisfy (30) for any j and commute for distinct j. Moreover, $B(\otimes^n \mathbb{C}^2)$ is generated as an algebra by $\{\sigma_i^{(j)} \ : \ j = 1, \ldots, n, \ i = 1, 2\}$. Set $I_j := \sigma_3^{(1)} \cdots \sigma_3^{(j)}$. In order to transform spin matrices into a representation of the CAR we need to apply the so-called *Jordan-Wigner construction.* According to this construction,

$$\left(\sigma_1^{(1)}, \sigma_2^{(1)}, I_1\sigma_1^{(2)}, I_1\sigma_2^{(2)}, \ldots, I_{n-1}\sigma_1^{(n)}, I_{n-1}\sigma_2^{(n)}\right)$$

is a representation of the CAR over $\mathbb{R}^{2n}$. By adding $\pm I_n$ we obtain a representation of the CAR over $\mathbb{R}^{2n+1}$.

The following theorem can be viewed as a fermionic analog of the Stone-von Neumann Theorem 4. It is, however, much easier to prove and belongs to standard results about Clifford algebras [47].

Theorem 10. 1) *Let $(\phi_1, \phi_2, \ldots, \phi_{2n})$ be a representation of the CAR over $\mathbb{R}^{2n}$ in a Hilbert space $\mathcal{H}$. Then there exists a Hilbert space $\mathcal{K}$ and a unitary operator*

$$U : \otimes^n \mathbb{C}^2 \otimes \mathcal{K} \to \mathcal{H}$$

such that

$$U\left(I_{j-1}\sigma_1^{(j)} \otimes 1_{\mathcal{K}}\right) = \phi_{2j-1} U,$$

$$U\left(I_{j-1}\sigma_2^{(j)} \otimes 1_{\mathcal{K}}\right) = \phi_{2j} U, \quad j = 1, \ldots, n.$$

The representation is irreducible iff $\mathcal{K} = \mathbb{C}$.

2) *Let $(\phi_1, \phi_2, \ldots, \phi_{2n+1})$ be a representation of the CAR over $\mathbb{R}^{2n+1}$ in a Hilbert space $\mathcal{H}$. Then there exist Hilbert spaces $\mathcal{K}_-$ and $\mathcal{K}_+$ and a unitary operator*

$$U : \otimes^n \mathbb{C}^2 \otimes (\mathcal{K}_+ \oplus \mathcal{K}_-) \to \mathcal{H}$$

such that

$$U\left(I_{j-1}\sigma_1^{(j)} \otimes 1_{\mathcal{K}_+ \oplus \mathcal{K}_-}\right) = \phi_{2j-1} U,$$

$$U\left(I_{j-1}\sigma_2^{(j)} \otimes 1_{\mathcal{K}_+ \oplus \mathcal{K}_-}\right) = \phi_{2j} U, \quad j = 1, \ldots, n,$$

$$U\left(I_n \otimes (1_{\mathcal{K}_+} \oplus -1_{\mathcal{K}_-})\right) = \phi_{2n+1} U.$$

Corollary 2. *Suppose that $\mathcal{Y}$ is an even dimensional real Hilbert space. Let $\mathcal{Y} \ni y \mapsto \phi^{\pi_1}(y) \in B_{\mathrm{h}}(\mathcal{H})$ and $\mathcal{Y} \ni y \mapsto \phi^{\pi_2}(y) \in B_{\mathrm{h}}(\mathcal{H})$ be two irreducible representations of the CAR. Then they are unitarily equivalent.*

4.3 Fermionic Bogolubov Transformations

Let $(\mathcal{Y}, \alpha)$ be a finite dimensional real Hilbert space (a Euclidean space). Linear transformations on $\mathcal{Y}$ that preserve the scalar product are invertible and form a group, which will be called the orthogonal group of $\mathcal{Y}$ and denoted $O(\mathcal{Y})$.

Let

$$\mathcal{Y} \ni y \mapsto \phi(y) \in B_{\mathrm{h}}(\mathcal{H}) \tag{31}$$

be an irreducible representation of the CAR. The following theorem is an immediate consequence of Corollary 2:

Theorem 11. *Let* $\dim \mathcal{Y}$ *be even. For any* $r \in O(\mathcal{Y})$ *there exists* $U \in U(\mathcal{H})$ *such that*

$$U\phi(y)U^* = \phi(ry) \; . \tag{32}$$

The unitary operator U *in (32) is defined uniquely up to a phase factor. Let* $\mathcal{U}_r$ *denote the class of such operators. Then*

$$O(\mathcal{Y}) \ni r \mapsto \mathcal{U}_r \in U(\mathcal{H})/U(1)$$

is a group homomorphism.

One can ask whether one can fix uniquely the phase factor appearing in the above theorem and obtain a group homomorphism of $O(\mathcal{Y})$ into $U(\mathcal{H})$ satisfying (32). This is impossible, the best one can do is the following improvement of Theorem 11:

Theorem 12. *Let* $\dim \mathcal{Y}$ *be even. For any* $r \in O(\mathcal{Y})$ *there exists a unique pair* $\{U_r, -U_r\} \subset U(\mathcal{H})$ *such that*

$$U_r\phi(y)U_r^* = \phi(ry) \; ,$$

and such that we have a group homomorphism

$$O(\mathcal{Y}) \ni r \mapsto \pm U_r \in U(\mathcal{H})/\{1, -1\} \; . \tag{33}$$

The map (33) is called the *Pin representation of* $O(\mathcal{Y})$.

Note that for $\dim \mathcal{Y} > 2$, the homotopy group of $O(\mathcal{Y})$ is $\mathbb{Z}_2$. Hence the double covering of $O(\mathcal{Y})$ is its universal covering. The image of the Pin representation in $U(\mathcal{H})$ is isomorphic to this double covering and is called the Pin group.

In the physics literature the fact that orthogonal transformations can be unitarily implemented is again associated with the name of Bogolubov and the transformations described in Theorems 11 and 12 are often called (fermionic) Bogolubov transformations. They are used e.g. in the theory of the superconductivity.

Theorems 11 and 12 are well known in mathematics in the context of theory of Clifford algebras. They are most conveniently proven by using, what we call, the Fock representation, where one has simple formulas for U_r. We will describe these formulas later on (in a more general context of the infinite number of degrees of freedom).

5 Fock Spaces

In this section we fix the notation for bosonic and fermionic Fock spaces. Even though these concepts are widely used, there seem to be no universally accepted symbols for many concepts in this area.

5.1 Tensor Algebra

Let $\mathcal{Z}$ be a Hilbert space. Let $\otimes^n \mathcal{Z}$ denote the n-fold tensor product of $\mathcal{Z}$. We set

$$\otimes \mathcal{Z} = \mathop{\oplus}_{n=0}^{\infty} \otimes^n \mathcal{Z} .$$

Here $\oplus$ denotes the direct sum in the sense of Hilbert spaces, that is the completion of the algebraic direct sum. $\otimes \mathcal{Z}$ is sometimes called the *full Fock space*. The element $1 \in \mathbb{C} = \otimes^0 \mathcal{Z} \subset \otimes \mathcal{Z}$ is often called the *vacuum* and denoted Ω.

Sometimes we will need $\otimes^{\mathrm{fin}} \mathcal{Z}$ which is the subspace of $\otimes \mathcal{Z}$ with a finite number of particles, that means the algebraic direct sum of $\otimes^n \mathcal{Z}$.

$\otimes \mathcal{Z}$ and $\otimes^{\mathrm{fin}} \mathcal{Z}$ are associative algebras with the operation $\otimes$ and the identity Ω.

5.2 Operators dΓ and Γ in the Full Fock Space

If p is a closed operator from $\mathcal{Z}_1$ to $\mathcal{Z}_2$, then we define the closed operator $\Gamma^n(p)$ from $\otimes^n \mathcal{Z}_1$ to $\otimes^n \mathcal{Z}_2$ and $\Gamma(p)$ from $\otimes \mathcal{Z}_1$ to $\otimes \mathcal{Z}_2$:

$$\Gamma^n(p) := p^{\otimes n},$$

$$\Gamma(p) := \mathop{\oplus}_{n=0}^{\infty} \Gamma^n(p) .$$

$\Gamma(p)$ is bounded iff $\|p\| \leq 1$. $\Gamma(p)$ is unitary iff p is.

Likewise, if h is a closed operator on $\mathcal{Z}$, then we define the closed operator $\mathrm{d}\Gamma^n(h)$ on $\otimes^n \mathcal{Z}$ and $\mathrm{d}\Gamma(h)$ on $\otimes \mathcal{Z}$:

$$\mathrm{d}\Gamma^n(h) = \sum_{j=1}^{n} 1_{\mathcal{Z}}^{\otimes(j-1)} \otimes h \otimes 1_{\mathcal{Z}}^{\otimes(n-j)} ,$$

$$\mathrm{d}\Gamma(h) := \mathop{\oplus}_{n=0}^{\infty} \mathrm{d}\Gamma^n(h) .$$

$\mathrm{d}\Gamma(h)$ is self-adjoint iff h is.

The *number operator* is defined as $N = \mathrm{d}\Gamma(1)$. The *parity operator* is

$$I := (-1)^N = \Gamma(-1) . \tag{34}$$

Let us give a sample of properties of operators on full Fock spaces.

Theorem 13. 1) *Let* $h, h_1, h_2 \in B(\mathcal{Z})$, $p_1 \in B(\mathcal{Z}, \mathcal{Z}_1)$, $p_2 \in B(\mathcal{Z}_1, \mathcal{Z}_2)$, $\|p_i\| \leq 1$. *We then have*

$$\Gamma(\mathrm{e}^{\mathrm{i}h}) = \exp(\mathrm{d}\Gamma(\mathrm{i}h)) , \tag{35}$$

$$\Gamma(p_2)\Gamma(p_1) = \Gamma(p_2 p_1) ,$$

$$[\mathrm{d}\Gamma(h_1), \mathrm{d}\Gamma(h_2)] = \mathrm{d}\Gamma([h_1, h_2]) .$$

2) *Let* $\Phi, \Psi \in \otimes^{\mathrm{fin}}\mathcal{Z}$, $h \in B(\mathcal{Z})$, $p \in B(\mathcal{Z}, \mathcal{Z}_1)$. *Then*

$$\Gamma(p)\ (\Phi \otimes \Psi) = (\Gamma(p)\Phi) \otimes (\Gamma(p)\Psi) ,$$

$$\mathrm{d}\Gamma(h)\ (\Phi \otimes \Psi) = (\mathrm{d}\Gamma(h)\Phi) \otimes \Psi + \Phi \otimes (\mathrm{d}\Gamma(h)\Psi) .$$

Of course, under appropriate technical conditions, similar statements are true for unbounded operators. In particular (35) is true for any self-adjoint h.

5.3 Symmetric and Antisymmetric Fock Spaces

Let $S^n \ni \sigma \mapsto \Theta(\sigma) \in U(\otimes^n \mathcal{Z})$ be the natural representation of the permutation group S^n given by

$$\Theta(\sigma) z_1 \otimes \cdots \otimes z_n := z_{\sigma^{-1}(1)} \otimes \cdots \otimes z_{\sigma^{-1}(n)} .$$

We define

$$\Theta_{\mathrm{s}}^n := \frac{1}{n!} \sum_{\sigma \in S^n} \Theta(\sigma) ,$$

$$\Theta_{\mathrm{a}}^n := \frac{1}{n!} \sum_{\sigma \in S^n} (\operatorname{sgn} \sigma)\, \Theta(\sigma) .$$

Θ_{s}^n and Θ_{a}^n are orthogonal projections in $\otimes^n \mathcal{Z}$.

We will write s/a as a subscript that can mean either s or a. We set

$$\Theta_{\mathrm{s/a}} := \mathop{\oplus}_{n=0}^{\infty} \Theta_{\mathrm{s/a}}^n .$$

Clearly, $\Theta_{\mathrm{s/a}}$ is an orthogonal projection in $\otimes \mathcal{Z}$.

Define

$$\Gamma_{\mathrm{s/a}}^{n}(\mathcal{Z}) := \Theta_{\mathrm{s/a}}^{n}(\otimes^{n}\mathcal{Z}),$$

$$\Gamma_{\mathrm{s/a}}(\mathcal{Z}) := \Theta_{\mathrm{s/a}}(\otimes\mathcal{Z}) = \mathop{\oplus}_{n=0}^{\infty} \Gamma_{\mathrm{s/a}}^{n}(\mathcal{Z}).$$

$\Gamma_{\mathrm{s/a}}(\mathcal{Z})$ is often called the *bosonic/fermionic* or *symmetric/antisymmetric Fock space.*

We also introduce the finite particle Fock spaces

$$\Gamma_{\mathrm{s/a}}^{\mathrm{fin}}(\mathcal{Z}) = (\otimes^{\mathrm{fin}}\mathcal{Z}) \cap \Gamma_{\mathrm{s/a}}(\mathcal{Z}) .$$

$\Gamma_{\mathrm{s/a}}(\mathcal{Z})$ is a Hilbert space (as a closed subspace of $\otimes\mathcal{Z}$).

Note that $\Gamma_{\mathrm{s/a}}^{0}(\mathcal{Z}) = \mathbb{C}$ and $\Gamma_{\mathrm{s/a}}^{1}(\mathcal{Z}) = \mathcal{Z}$. $\mathcal{Z}$ is often called the *1-particle space* and $\Gamma_{\mathrm{s/a}}(\mathcal{Z})$ the second quantization of $\mathcal{Z}$.

The following property of bosonic Fock spaces is often useful:

Theorem 14. *The span of elements of the form $z^{\otimes n}$, $z \in \mathcal{Z}$, is dense in $\Gamma_{\mathrm{s}}^{n}(\mathcal{Z})$.*

5.4 Symmetric and Antisymmetric Tensor Product

If $\Psi, \Phi \in \Gamma_{\mathrm{s/a}}^{\mathrm{fin}}(\mathcal{Z})$, then we define

$$\Psi \otimes_{\mathrm{s/a}} \Phi := \Theta_{\mathrm{s/a}}(\Psi \otimes \Phi) \in \Gamma_{\mathrm{s/a}}^{\mathrm{fin}}(\mathcal{Z}) .$$

$\Gamma_{\mathrm{s/a}}^{\mathrm{fin}}(\mathcal{Z})$ is an associative algebra with the operation $\otimes_{\mathrm{s/a}}$ and the identity Ω.

Note that $z^{\otimes n} = z^{\otimes_{\mathrm{s}} n}$.

Instead of $\otimes_{\mathrm{a}}$ one often uses the wedge product, which for $\Psi \in \Gamma_{\mathrm{a}}^{p}(\mathcal{Z})$, $\Phi \in \Gamma_{\mathrm{a}}^{r}(\mathcal{Z})$ is defined as

$$\Psi \wedge \Phi := \frac{(p+r)!}{p!r!} \Psi \otimes_{\mathrm{a}} \Phi \in \Gamma_{\mathrm{a}}^{\mathrm{fin}}(\mathcal{Z}) .$$

It is also an associative operation. Its advantage over $\otimes_{\mathrm{a}}$ is visible if we compare the following identities:

$$v_1 \wedge \cdots \wedge v_p = \sum_{\sigma \in S^p} (\mathrm{sgn}\sigma)\ v_{\sigma(1)} \otimes \cdots \otimes v_{\sigma(p)},$$

$$v_1 \otimes_{\mathrm{a}} \cdots \otimes_{\mathrm{a}} v_p = \frac{1}{p!} \sum_{\sigma \in S^p} (\mathrm{sgn}\sigma)\ v_{\sigma(1)} \otimes \cdots \otimes v_{\sigma(p)}, \qquad v_1, \cdots, v_p \in \mathcal{Z}.$$

The advantage of $\otimes_{\mathrm{a}}$ is that it is fully analogous to $\otimes_{\mathrm{s}}$.

5.5 dΓ and Γ Operations

If p is a closed operator from $\mathcal{Z}$ to $\mathcal{W}$, then $\Gamma^{n}(p)$, defined in Subsect. 5.2, maps $\Gamma_{\mathrm{s/a}}^{n}(\mathcal{Z})$ into $\Gamma_{\mathrm{s/a}}^{n}(\mathcal{W})$. Hence $\Gamma(p)$ maps $\Gamma_{\mathrm{s/a}}(\mathcal{Z})$ into $\Gamma_{\mathrm{s/a}}(\mathcal{W})$. We will use the same symbols $\Gamma^n(p)$ and $\Gamma(p)$ to denote the corresponding restricted operators.

If h is a closed operator on $\mathcal{Z}$, then $\mathrm{d}\Gamma^n(h)$ maps $\Gamma^n_{\mathrm{s/a}}(\mathcal{Z})$ into itself. Hence, $\mathrm{d}\Gamma(h)$ maps $\Gamma_{\mathrm{s/a}}(\mathcal{Z})$ into itself. We will use the same symbols $\mathrm{d}\Gamma^n(h)$ and $\mathrm{d}\Gamma(h)$ to denote the corresponding restricted operators.

$\Gamma(p)$ is called the 2nd quantization of p. Similarly, $\mathrm{d}\Gamma(h)$ is sometimes called the 2nd quantization of h.

Theorem 15. *Let $p \in B(\mathcal{Z}, \mathcal{Z}_1)$, $h \in B(\mathcal{Z})$, $\Psi, \Phi \in \Gamma^{\mathrm{fin}}_{\mathrm{s/a}}(\mathcal{Z})$. Then*

$$\Gamma(p)\left(\Psi \otimes_{\mathrm{s/a}} \Phi\right) = (\Gamma(p)\Psi) \otimes_{\mathrm{s/a}} (\Gamma(p)\Phi),$$

$$\mathrm{d}\Gamma(h)\left(\Psi \otimes_{\mathrm{s/a}} \Phi\right) = (\mathrm{d}\Gamma(h)\Psi) \otimes_{\mathrm{s/a}} \Phi + \Psi \otimes_{\mathrm{s/a}} (\mathrm{d}\Gamma(h)\Phi).$$

5.6 Tensor Product of Fock Spaces

In this subsection we describe the so-called *exponential law for Fock spaces.*

Let $\mathcal{Z}_1$ and $\mathcal{Z}_2$ be Hilbert spaces. We introduce the identification

$$U : \Gamma^{\mathrm{fin}}_{\mathrm{s/a}}(\mathcal{Z}_1) \otimes \Gamma^{\mathrm{fin}}_{\mathrm{s/a}}(\mathcal{Z}_2) \to \Gamma^{\mathrm{fin}}_{\mathrm{s/a}}(\mathcal{Z}_1 \oplus \mathcal{Z}_2)$$

as follows. Let $\Psi_1 \in \Gamma^n_{\mathrm{s/a}}(\mathcal{Z}_1)$, $\Psi_2 \in \Gamma^m_{\mathrm{s/a}}(\mathcal{Z}_2)$. Let j_i be the imbedding of $\mathcal{Z}_i$ in $\mathcal{Z}_1 \oplus \mathcal{Z}_2$. Then

$$U(\Psi_1 \otimes \Psi_2) := \sqrt{\tfrac{(n+m)!}{n!m!}}(\Gamma(j_1)\Psi_1) \otimes_{\mathrm{s/a}} (\Gamma(j_2)\Psi_2) \, . \tag{36}$$

Theorem 16. 1) *$U(\Omega_1 \otimes \Omega_2) = \Omega$.*

2) *U extends to a unitary operator $\Gamma_{\mathrm{s/a}}(\mathcal{Z}_1) \otimes \Gamma_{\mathrm{s/a}}(\mathcal{Z}_2) \to \Gamma_{\mathrm{s/a}}(\mathcal{Z}_1 \oplus \mathcal{Z}_2)$.*

3) *If $h_i \in B(\mathcal{Z}_i)$, then*

$$U\big(\mathrm{d}\Gamma(h_1) \otimes 1 + 1 \otimes \mathrm{d}\Gamma(h_2)\big) = \mathrm{d}\Gamma(h_1 \oplus h_2)U.$$

4) *If $p_i \in B(\mathcal{Z}_i)$, then*

$$U\left(\Gamma(p_1) \otimes \Gamma(p_2)\right) = \Gamma(p_1 \oplus p_2)U.$$

5.7 Creation and Annihilation Operators

Let $\mathcal{Z}$ be a Hilbert space and $w \in \mathcal{Z}$. We consider the bosonic/fermionic Fock space $\Gamma_{\mathrm{s/a}}(\mathcal{Z})$.

Let $w \in \mathcal{Z}$. We define two operators with the domain $\Gamma^{\mathrm{fin}}_{\mathrm{s/a}}(\mathcal{Z})$. The *creation operator* is defined as

$$a^*(w)\Psi := \sqrt{n+1}\, w \otimes_{\mathrm{s/a}} \Psi, \quad \Psi \in \Gamma^n_{\mathrm{s/a}}(\mathcal{Z})$$

In the fermionic case, $a^*(w)$ is bounded. In the bosonic case, $a^*(w)$ is densely defined and closable. In both cases we define denote the closure of $a^*(w)$ by the same symbol. Likewise, in both cases we define the *annihilation operator* by

$$a(w) := a^*(w)^* .$$

Note that

$$a(w)\Psi = \sqrt{n}\big((w| \otimes 1\big)\Psi, \quad \Psi \in \Gamma^n_{\mathrm{s/a}}(\mathcal{Z}) .$$

Theorem 17. 1) *In the bosonic case we have*

$$[a^*(w_1), a^*(w_2)] = 0, \quad [a(w_1), a(w_2)] = 0,$$
$$[a(w_1), a^*(w_2)] = (w_1|w_2).$$

2) *In the fermionic case we have*

$$[a^*(w_1), a^*(w_2)]_+ = 0, \quad [a(w_1), a(w_2)]_+ = 0,$$
$$[a(w_1), a^*(w_2)]_+ = (w_1|w_2).$$

In both bosonic and fermionic cases the following is true:

Theorem 18. *Let $p, h \in B(\mathcal{Z})$ and $w \in \mathcal{Z}$. Then*
1) $\Gamma(p)a(w) = a(p^{*-1}w)\Gamma(p)$,
2) $[\mathrm{d}\Gamma(h), a(w)] = -a(h^*w)$,
3) $\Gamma(p)a^*(w) = a^*(pw)\Gamma(p)$,
4) $[\mathrm{d}\Gamma(h), a^*(w)] = a^*(hw)$.

The exponential law for creation/annihilation operators is slightly different in the bosonic and fermionic cases:

Theorem 19. *Let $\mathcal{Z}_1$ and $\mathcal{Z}_2$ be Hilbert spaces and $(w_1, w_2) \in \mathcal{Z}_1 \oplus \mathcal{Z}_2$. Let U be the defined for these spaces as in Theorem 16.*
1) *In the bosonic case we have*

$$a^*(w_1, w_2)U = U(a^*(w_1) \otimes 1 + 1 \otimes a^*(w_2)) ,$$
$$a(w_1, w_2)U = U(a(w_1) \otimes 1 + 1 \otimes a(w_2)) .$$

2) *In the fermionic case, if I_1 denotes the parity operator for $\Gamma_{\mathrm{a}}(\mathcal{Z}_1)$ (see (34)), then*

$$a^*(w_1, w_2)U = U(a^*(w_1) \otimes 1 + I_1 \otimes a^*(w_2)) ,$$
$$a(w_1, w_2)U = U(a(w_1) \otimes 1 + I_1 \otimes a(w_2)) .$$

Set

$$\Lambda := (-1)^{N(N-1)/2}. \tag{37}$$

The following property is valid both in the bosonic and fermionic cases:

$$\begin{aligned} \Lambda a^*(z)\Lambda &= -Ia^*(z) = a^*(z)I, \\ \Lambda a(z)\Lambda &= Ia(z) = -a(z)I. \end{aligned} \tag{38}$$

In the fermionic case, (38) allows to convert the anticommutation relations into commutation relations

$$[\Lambda a^*(z)\Lambda, a^*(w)] = [\Lambda a(z)\Lambda, a(w)] = 0,$$

$$[\Lambda a^*(z)\Lambda, a(w)] = I(w|z).$$

Theorem 20. *Let the assumptions of Theorem 19 be satisfied. Let N_i, I_i, Λ_i be the operators defined as above corresponding to $\mathcal{Z}_i$, $i = 1, 2$. Then*

1) $\Lambda U = U(\Lambda_1 \otimes \Lambda_2)(-1)^{N_1 \otimes N_2}$.

2) *In the fermionic case,*

$$\Lambda a^*(w_1, w_2)\Lambda\, U = U(a^*(w_1)I_1 \otimes I_2 + 1 \otimes a^*(w_2)I_2),$$

$$\Lambda a(w_1, w_2)\Lambda\, U = U(-a(w_1)I_1 \otimes I_2 - 1 \otimes a(w_2)I_2).$$

Proof. To prove 2) we use 1) and $(-1)^{N_1 \otimes N_2} a(w) \otimes 1 (-1)^{N_1 \otimes N_2} = a(w) \otimes I_2$.
□

5.8 Multiple Creation and Annihilation Operators

Let $\Phi \in \Gamma_{\mathrm{s/a}}^m(\mathcal{Z})$. We define the operator of creation of Φ with the domain $\Gamma_{\mathrm{s/a}}^{\mathrm{fin}}(\mathcal{Z})$ as follows:

$$a^*(\Phi)\Psi := \sqrt{(n+1)\cdots(n+m)}\,\Phi \otimes_{\mathrm{s/a}} \Psi, \quad \Psi \in \Gamma_{\mathrm{s/a}}^n .$$

$a^*(\Phi)$ is a densely defined closable operator. We denote its closure by the same symbol. We set

$$a(\Phi) := (a^*(\Phi))^* .$$

$a(\Phi)$ is called the operator of annihilation of Φ. For $w_1, \dots, w_m \in \mathcal{Z}$ we have

$$a^*(w_1 \otimes_{\mathrm{s/a}} \cdots \otimes_{\mathrm{s/a}} w_m) = a^*(w_1) \cdots a^*(w_m) ,$$

$$a(w_1 \otimes_{\mathrm{s/a}} \cdots \otimes_{\mathrm{s/a}} w_m) = a(w_m) \cdots a(w_1) .$$

Recall from Subsect. 2.5 that we can identify the space $B^2(\overline{\mathcal{Z}}, \mathcal{Z})$ with $\otimes^2 \mathcal{Z}$. Hence, we have an identification of $B_{\mathrm{s/a}}^2(\overline{\mathcal{Z}}, \mathcal{Z})$ with $\Gamma_{\mathrm{s/a}}^2(\mathcal{Z})$.

Thus if $c \in B_{\mathrm{s/a}}^2(\overline{\mathcal{Z}}, \mathcal{Z})$, then by interpreting c as an element of $\Gamma_{\mathrm{s/a}}^2(\mathcal{Z})$, we can use the notation $a^*(c)$ / $a(c)$ for the corresponding two-particle creation/annihilation operators.

6 Representations of the CCR in Fock Spaces

6.1 Field Operators

Let $\mathcal{Z}$ be a (complex) Hilbert space. Define the real vector space

$$\mathrm{Re}(\mathcal{Z} \oplus \overline{\mathcal{Z}}) := \{(z, \overline{z}) \ : \ z \in \mathcal{Z}\} . \tag{39}$$

Clearly, $\mathrm{Re}(\mathcal{Z} \oplus \overline{\mathcal{Z}})$ is a real subspace of $\mathcal{Z} \oplus \overline{\mathcal{Z}}$. For shortness, we will usually write $\mathcal{Y}$ for $\mathrm{Re}(\mathcal{Z} \oplus \overline{\mathcal{Z}})$. In this section we will treat $\mathcal{Y}$ as a symplectic space equipped with the symplectic form

$$(z, \overline{z})\omega(w, \overline{w}) = 2\mathrm{Im}(z|w) .$$

Consider the creation/annihilation operators $a^*(z)$ and $a(z)$ acting on the bosonic Fock space $\Gamma_{\mathrm{s}}(\mathcal{Z})$. For $y = (w, \overline{w}) \in \mathcal{Y}$ we define

$$\phi(y) := a^*(w) + a(w) .$$

Note that $\phi(y)$ is essentially self-adjoint on $\Gamma_{\mathrm{s}}^{\mathrm{fin}}(\mathcal{Z})$. We use the same symbol $\phi(y)$ for its self-adjoint extension.

We have the following commutation relations

$$[\phi(y_1), \phi(y_2)] = \mathrm{i} y_1 \omega y_2, \quad y_1, y_2 \in \mathcal{Y} ,$$

as an identity on $\Gamma_{\mathrm{s}}^{\mathrm{fin}}(\mathcal{Z})$.

It is well known that in every bosonic Fock space we have a natural representation of the CCR:

Theorem 21. *The map*

$$\mathcal{Y} \ni y \mapsto W(y) := \mathrm{e}^{\mathrm{i}\phi(y)} \in U(\Gamma_{\mathrm{s}}(\mathcal{Z})) \tag{40}$$

is a regular irreducible representation of the CCR.

(40) is called *the Fock representation of the CCR.*

One often identifies the spaces $\mathcal{Y}$ and $\mathcal{Z}$ through

$$\mathcal{Z} \ni z \mapsto \frac{1}{\sqrt{2}}(z, \overline{z}) \in \mathcal{Y} . \tag{41}$$

With this identification, one introduces the field operators for $w \in \mathcal{Z}$ as

$$\phi(w) := \frac{1}{\sqrt{2}} \left(a^*(w) + a(w)\right) .$$

The converse identities are

$$\begin{aligned} a^*(w) &= \tfrac{1}{\sqrt{2}} \left(\phi(w) - \mathrm{i}\phi(\mathrm{i}w)\right), \\ a(w) &= \tfrac{1}{\sqrt{2}} \left(\phi(w) + \mathrm{i}\phi(\mathrm{i}w)\right). \end{aligned}$$

Note that in the fermionic case a different identification seems more convenient (see (54)). In this section we will avoid to identify $\mathcal{Z}$ with $\mathcal{Y}$.

Note that the physical meaning of $\mathcal{Z}$ and $\mathcal{Y}$ is different: $\mathcal{Z}$ is the one-particle Hilbert space of the system, $\mathcal{Y}$ is its classical phase space and $\mathcal{Z} \oplus \overline{\mathcal{Z}}$ can be identified with the complexification of the classical phase space, that is $\mathbb{C}\mathcal{Y}$. For instance, if we are interested in a real scalar quantum field theory, then $\mathcal{Z}$ is the space of positive energy solutions of the Klein-Gordon equation, $\mathcal{Y}$ is the space of real solutions and $\mathcal{Z} \oplus \overline{\mathcal{Z}}$ is the space of complex solutions. See e.g. [24] in this collection of lecture notes, where this point is dicussed in more detail.

6.2 Bosonic Gaussian Vectors

Let $c \in B_{\mathrm{s}}^2(\overline{\mathcal{Z}}, \mathcal{Z})$. Recall that c can be identified with an element of $\Gamma_{\mathrm{s}}^2(\mathcal{Z})$. Recall from Subsect. 5.8 that we defined an unbounded operator $a^*(c)$ on $\Gamma_{\mathrm{s}}(\mathcal{Z})$ such that for $\Psi_n \in \Gamma_{\mathrm{s}}^n(\mathcal{Z})$

$$a^*(c)\Psi_n := \sqrt{(n+2)(n+1)}\, c \otimes_{\mathrm{s}} \Psi_n \in \Gamma_{\mathrm{s}}^{n+2}(\mathcal{Z}). \tag{42}$$

Theorem 22. *Assume that* $\|c\| < 1$.
1) $\mathrm{e}^{\frac{1}{2}a^*(c)}$ *is a closable operator on* $\Gamma_{\mathrm{s}}^{\mathrm{fin}}(\mathcal{Z})$.
2) $\det(1 - cc^*) > 0$, *so that we can define the vector*

$$\Omega_c := (\det(1 - cc^*))^{\frac{1}{4}} \exp\left(\tfrac{1}{2}a^*(c)\right) \Omega \tag{43}$$

It is the unique vector in $\Gamma_{\mathrm{s}}(\mathcal{Z})$ *satisfying*

$$\|\Omega_c\| = 1, \quad (\Omega_c|\Omega) > 0, \quad (a(z) - a^*(c\overline{z}))\Omega_c = 0, \quad z \in \mathcal{Z}\,.$$

In the Schrödinger representation the vectors Ω_c are normalized Gaussians with an arbitrary dispersion – hence they are often called *squeezed states.*

6.3 Complex Structures Compatible with a Symplectic Form

Before analyzing Bogolubov transformations on a Fock space it is natural to start with a little linear algebra of symplectic vector spaces.

We can treat $\mathcal{Y}$ as a real Hilbert space. In fact, we have a natural scalar product

$$(z, \overline{z})\alpha(w, \overline{w}) := \mathrm{Re}(z|w)\,.$$

This scalar product will have a fundamental importance in the next section, when we will discuss fermions. In this section we need it only to define bounded and trace class operators.

We define $Sp(\mathcal{Y})$ to be the set of all bounded invertible linear maps on $\mathcal{Y}$ preserving ω. (This extends the definition of $Sp(\mathcal{Y})$ from the case of a finite dimensional symplectic space $\mathcal{Y}$ to the present context).

A linear map j is called a *complex structure* (or an *antiinvolution*) iff $j^2 = -1$. We say that it is *compatible with a symplectic form* ω iff $j \in Sp(\mathcal{Y})$ and the symmetric form $y_1 \omega j y_2$, where $y_1, y_2 \in \mathcal{Y}$, is positive definite. (One also says that j is Kähler with respect to ω).

On $\mathcal{Y}$ we introduce the linear map

$$j(z, \overline{z}) := (\mathrm{i}z, -\mathrm{i}\overline{z}) .$$

It is easy to see that j is a complex structure compatible with ω.

Note that fixing the complex structure j on the symplectic space $\mathcal{Y}$ compatible with the symplectic form ω is equivalent to identifying $\mathcal{Y}$ with $\mathrm{Re}(\mathcal{Z} \oplus \overline{\mathcal{Z}})$ for some complex Hilbert space $\mathcal{Z}$.

Let $r \in B(\mathcal{Y})$. We can extend r to $\mathcal{Z} \oplus \overline{\mathcal{Z}}$ by complex linearity. On $\mathcal{Z} \oplus \overline{\mathcal{Z}}$ we can write r as a 2 by 2 matrix

$$r = \begin{bmatrix} p & q \\ \overline{q} & \overline{p} \end{bmatrix} ,$$

where $p \in B(\mathcal{Z}, \mathcal{Z})$, $q \in B(\overline{\mathcal{Z}}, \mathcal{Z})$. Now $r \in Sp(\mathcal{Y})$ iff

$$p^*p - q^\# \overline{q} = 1, \;\; p^\# \overline{q} - q^* p = 0,$$

$$pp^* - qq^* = 1, \;\; pq^\# - qp^\# = 0.$$

We have

$$pp^* \geq 1, \quad p^*p \geq 1 .$$

Hence p^{-1} exists and $\|p^{-1}\| \leq 1$.

We define the operators $c, d \in B(\overline{\mathcal{Z}}, \mathcal{Z})$

$$c := p^{-1} q = q^\# (p^\#)^{-1}, \quad d := q\overline{p}^{-1} = (p^*)^{-1} q^\# .$$

Note that d, c are symmetric (in the sense defined in Sect. 2), $\|d\| \leq 1$, $\|c\| \leq 1$,

$$r = \begin{bmatrix} 1 & d \\ 0 & 1 \end{bmatrix} \begin{bmatrix} (p^*)^{-1} & 0 \\ 0 & \overline{p} \end{bmatrix} \begin{bmatrix} 1 & 0 \\ \overline{c} & 1 \end{bmatrix} , \tag{44}$$

$$1 - cc^* = (p^*p)^{-1}, \quad 1 - dd^* = (\overline{p}^* \overline{p})^{-1} .$$

The decomposition (44) plays an important role in the description of Bogolubov transformations.

In the following theorem we introduce a certain subgroup of $Sp(\mathcal{Y})$, which will play an important role in Shale's Theorem on the implementabilty of Bogolubov transformations.

Theorem 23. *Let $r \in Sp(\mathcal{Y})$. The following conditions are equivalent:*

0) $j - rjr^{-1} \in B^2(\mathcal{Y})$, 1) $rj - jr \in B^2(\mathcal{Y})$,

2) $\mathrm{Tr} q^* q < \infty$, 3) $\mathrm{Tr}(pp^* - 1) < \infty$,

4) $d \in B^2(\overline{\mathcal{Z}}, \mathcal{Z})$, 5) $c \in B^2(\overline{\mathcal{Z}}, \mathcal{Z})$.

Define $Sp_2(\mathcal{Y})$ to be the set of $r \in Sp(\mathcal{Y})$ satisfying the above conditions. Then $Sp_2(\mathcal{Y})$ is a group.

6.4 Bosonic Bogolubov Transformations in the Fock Representation

Consider now the Fock representation $\mathcal{Y} \ni y \mapsto W(y) \in U(\Gamma_{\rm s}(\mathcal{Z}))$ defined in (40).

The following theorem describes when a symplectic transformation is implementable by a unitary transformation. Part 1) was originally proven in [61]. Proof of 1) and 2) can be found in [14, 60].

Theorem 24 (Shale Theorem).
1) *Let $r \in Sp(\mathcal{Y})$. Then the following conditions are equivalent:*
  a) There exists $U \in U(\Gamma_{\rm s}(\mathcal{Z}))$ such that

$$UW(y)U^* = W(ry), \quad y \in \mathcal{Y}. \tag{45}$$

  b) $r \in Sp_2(\mathcal{Y})$.

2) *If the above conditions are satisfied, then U is defined uniquely up to a phase factor. Moreover, if we define*

$$U_r^j = |\det pp^*|^{-\frac{1}{4}} \mathrm{e}^{-\frac{1}{2}a^*(d)} \Gamma((p^*)^{-1}) \mathrm{e}^{\frac{1}{2}a(c)}, \tag{46}$$

then U_r^j is the unique unitary operator satisfying (45) and

$$(\Omega | U_r^j \Omega) > 0. \tag{47}$$

3) *If $\mathcal{U}_r = \{\lambda U_r^j \ : \ \lambda \in \mathbb{C}, \ |\lambda| = 1\}$, then*

$$Sp_2(\mathcal{Y}) \ni r \mapsto \mathcal{U}_r \in U(\Gamma_{\rm s}(\mathcal{Z}))/U(1)$$

is a homomorphism of groups.

6.5 Metaplectic Group in the Fock Representation

$r \mapsto U_r^j$ is not a representation of $Sp(\mathcal{Y})$, it is only a projective representation. By taking a certain subgroup of $Sp_2(\mathcal{Y})$ we can obtain a representation analogous to the metaplectic representation described in Theorem 7.

Define $Sp_1(\mathcal{Y}) := \{r \in Sp(\mathcal{Y}) \ : \ r - 1 \in B^1(\mathcal{Y})\}$. (Recall that $B^1(\mathcal{Y})$ are trace class operators).

Theorem 25. 1) *$Sp_1(\mathcal{Y})$ is a subgroup of $Sp_2(\mathcal{Y})$.*
2) *$r \in Sp_1(\mathcal{Y})$ iff $p - 1 \in B^1(\mathcal{Z})$.*

For $r \in Sp_1(\mathcal{Y})$, define

$$\pm U_r = \pm(\det p^*)^{-\frac{1}{2}} e^{-\frac{1}{2}a^*(d)} \Gamma((p^*)^{-1}) e^{\frac{1}{2}a(c)} . \tag{48}$$

(We take both signes of the square root, thus $\pm U_r$ denotes a pair of operators differing by a sign).

Theorem 26. 1) $\pm U_r \in U(\Gamma_s(\mathcal{Z}))/\{1 - 1\}$;
2) $U_r W(y) U_r^* = W(ry)$.
3) *The following map is a group homomorphism.*

$$Sp_1(\mathcal{Y}) \ni r \mapsto \pm U_r \in U(\mathcal{H})/\{1, -1\} . \tag{49}$$

Clearly, the operators $\pm U_r$ differ by a phase factor from U_r^j from Theorem 24.

6.6 Positive Symplectic Transformations

Special role is played by positive symplectic transformations. It is easy to show that $r \in Sp_2(\mathcal{Y})$ is a positive self-adjoint operator on $\mathcal{Y}$ iff it is of the form

$$r = \begin{bmatrix} (1 - cc^*)^{-1/2} & (1 - cc^*)^{-1/2} c \\ (1 - c^*c)^{-1/2} c^* & (1 - c^*c)^{-1/2} \end{bmatrix} , \tag{50}$$

for some $c \in B_s^2(\overline{\mathcal{Z}}, \mathcal{Z})$.

The following theorem describes Bogolubov transformations associated with positive symplectic transformations.

Theorem 27. 1) *The formula*

$$R_c := \det(1 - cc^*)^{\frac{1}{4}} \exp\left(-\tfrac{1}{2}a^*(c)\right) \Gamma\left((1 - cc^*)^{\frac{1}{2}}\right) \exp\left(\tfrac{1}{2}a(c)\right) \tag{51}$$

defines a unitary operator on $\Gamma_s(\mathcal{Z})$.
2) *If* Ω_c *is defined in (43), then* $\Omega = R_c \Omega_c$.
3)

$$R_c a^*(z) R_c^* = a^*\left((1 - cc^*)^{-1/2} z\right) + a\left((1 - cc^*)^{-1/2} c\overline{z}\right) ,$$

$$R_c a(z) R_c^* = a^*\left((1 - cc^*)^{-1/2} c\overline{z}\right) + a\left((1 - cc^*)^{-1/2} z\right) .$$

4) *If* r *is related to* c *by (50), then* R_c *coincides with* U_r^j *defined in (46).*
5) R_c *coincides with* U_r *defined in (48), where we take the plus sign and the positive square root.*

7 Representations of the CAR in Fock Spaces

7.1 Field Operators

Let $\mathcal{Z}$ be a Hilbert space. As in the previous section, let $\mathcal{Y} := \mathrm{Re}(\mathcal{Z} \oplus \overline{\mathcal{Z}})$. This time, however, we treat it as a real Hilbert space equipped with the scalar product

$$(z, \overline{z})\alpha(w, \overline{w}) = \mathrm{Re}(z|w) .$$

For $w \in \mathcal{Z}$, consider the creation/annihilation operators $a^*(w)$ and $a(w)$ acting on the fermionic Fock space $\Gamma_{\mathrm{a}}(\mathcal{Z})$. For $y = (w, \overline{w}) \in \mathcal{Y}$ we define

$$\phi(y) := a^*(w) + a(w) .$$

Note that $\phi(y)$ are bounded and self-adjoint for any $y \in \mathcal{Y}$. Besides,

$$[\phi(y_1), \phi(y_2)]_+ = 2y_1\alpha y_2, \quad y_1, y_2 \in \mathcal{Y} .$$

Thus we have

Theorem 28.

$$\mathcal{Y} \ni y \mapsto \phi(y) \in B_{\mathrm{h}}(\Gamma_{\mathrm{a}}(\mathcal{Z})) \tag{52}$$

is an irreducible representation of the CAR over the space $(\mathcal{Y}, \alpha)$.

The map (52) is called the *Fock representation of the CAR.*

Let $w_1, \dots, w_m$ be an orthonormal basis of the complex Hilberts space $\mathcal{Z}$. Then

$$(w_1, \overline{w}_1), (-\mathrm{i}w_1, \mathrm{i}\overline{w}_1), \dots (w_m, \overline{w}_m), (-\mathrm{i}w_m, \mathrm{i}\overline{w}_m) \tag{53}$$

is an orthonormal basis of the real Hilbert space $\mathcal{Y} = \mathrm{Re}(\mathcal{Z} \oplus \overline{\mathcal{Z}})$. It is easy to see that the orientation of (53) does not depend on the choice of $w_1, \dots, w_m$.

The operator Q defined as in Theorem 9 for this orientation equals the parity operator $I = \Gamma(-1) = (-1)^N$. In fact, using Theorem 9 4), we can compute

$$\begin{aligned} Q &= \mathrm{i}^m \prod_{j=1}^{m} \phi(w_j, \overline{w}_j)\phi(-\mathrm{i}w_j, \mathrm{i}\overline{w}_j) \\ &= \prod_{j=1}^{m} (-a^*(w_j)a(w_j) + a(w_j)a^*(w_j)) = \Gamma(-1). \end{aligned}$$

In the fermionic case, one often identifies the spaces $\mathcal{Y}$ and $\mathcal{Z}$ through

$$\mathcal{Z} \ni w \mapsto (w, \overline{w}) \in \mathcal{Y} . \tag{54}$$

With this identification, one introduces the field operators for $w \in \mathcal{Z}$ as

$$\phi(w) := a^*(w) + a(w) .$$

The converse identities are

$$a^*(w) = \frac{1}{2}\left(\phi(w) - \mathrm{i}\phi(\mathrm{i}w)\right),$$

$$a(w) = \frac{1}{2}\left(\phi(w) + \mathrm{i}\phi(\mathrm{i}w)\right).$$

Using these identifications, we have for $z, w \in \mathcal{Z}$ the identities

$$[\phi(z), \phi(w)]_+ = 2\mathrm{Re}(w|z),$$

$$\Lambda\phi(w)\Lambda = -\mathrm{i}\phi(\mathrm{i}w)I = \mathrm{i}I\phi(\mathrm{i}w),$$

$$[\Lambda\phi(z)\Lambda, \phi(w)] = 2\mathrm{Im}(w|z)I,$$

where I is the parity operator and Λ was introduced in (37).

Note that the identification (54) is different from the one used in the bosonic case (41). In this section we will avoid identifying $\mathcal{Z}$ with $\mathcal{Y}$.

7.2 Fermionic Gaussian vectors

Let $c \in \Gamma_{\mathrm{a}}^2(\mathcal{Z})$. Note that it can be identified with an element of $B_{\mathrm{a}}^2(\overline{\mathcal{Z}}, \mathcal{Z})$. cc^* is trace class, so $\det(1 + cc^*)$ is well defined.

Theorem 29. *Define a vector in* $\Gamma_{\mathrm{a}}(\mathcal{Z})$ *by*

$$\Omega_c := (\det(1 + cc^*))^{-\frac{1}{4}} \exp\left(\tfrac{1}{2}a^*(c)\right)\Omega. \tag{55}$$

It is the unique vector satisfying

$$\|\Omega_c\| = 1, \quad (\Omega_c|\Omega) > 0, \quad (a(z) + a^*(c\overline{z}))\Omega_c = 0, \quad z \in \mathcal{Z}.$$

Vectors of the form Ω_c are often used in the many body quantum theory. In particular, they appear as convenient variational states in theory of superconductivity that goes back to the work of Bardeen-Cooper-Schrieffer, see e.g. [34].

7.3 Complex Structures Compatible with a Scalar Product

Similarly as for bosons, it is convenient to study some abstract properties of orthogonal transformations on a real Hilbert space as a preparation for the analysis of fermionic Bogolubov transformations.

Let $O(\mathcal{Y})$ denote the group of orthogonal transformations on $\mathcal{Y}$.

We say that a complex structure j is *compatible with the scalar product* α (or is Kähler with respect to α) if $j \in O(\mathcal{Y})$.

Recall that on $\mathcal{Y}$ we have a distinguished complex structure

$$j(z, \overline{z}) := (\mathrm{i}z, -\mathrm{i}\overline{z}).$$

It is easy to see that j is compatible with α.

Note that fixing the complex structure j on a real Hilbert space $\mathcal{Y}$ compatible with the scalar product α is equivalent with identifying $\mathcal{Y}$ with $\mathrm{Re}(\mathcal{Z}\oplus\overline{\mathcal{Z}})$ for some complex Hilbert space $\mathcal{Z}$.

Let $r \in B(\mathcal{Y})$. Recall that we can extend r to $\mathcal{Z}\oplus\overline{\mathcal{Z}}$ by complex linearity and write it as

$$r = \begin{bmatrix} p & q \\ \overline{q} & \overline{p} \end{bmatrix},$$

where $p \in B(\mathcal{Z},\mathcal{Z})$, $q \in B(\overline{\mathcal{Z}},\mathcal{Z})$. Now $r \in O(\mathcal{Y})$ iff

$$p^*p + q^\#\overline{q} = 1, \;\; p^\#\overline{q} + q^*p = 0,$$

$$pp^* + qq^* = 1, \;\; pq^\# + qp^\# = 0.$$

It is convenient to distinguish a certain class of orthogonal transformations given by the following theorem:

Theorem 30. *Let $r \in O(\mathcal{Y})$. Then the following conditions are equivalent:*
1) $\mathrm{Ker}(rj + jr) = \{0\}$*;*
2) $\mathrm{Ker}(r^*j + jr^*) = \{0\}$*;*
3) $\mathrm{Ker}\, p = \{0\}$*;*
4) $\mathrm{Ker}\, p^* = \{0\}$*.*

If the conditions of Theorem 30 are satisfied, then we say that r is j-*nondegenerate.* Let us assume that this is the case. Then p^{-1} and p^{*-1} are densely defined operators. Set

$$d = q\overline{p}^{-1} = -(p^*)^{-1}q^\#, \quad c = p^{-1}q = -q^\#(p^\#)^{-1},$$

and assume that they are bounded. Then $d, c \in B_{\mathrm{a}}(\overline{\mathcal{Z}},\mathcal{Z})$. The following factorization of r plays an important role in the description of fermionic Bogolubov transformations:

$$r = \begin{bmatrix} 1 & d \\ 0 & 1 \end{bmatrix} \begin{bmatrix} (p^*)^{-1} & 0 \\ 0 & \overline{p} \end{bmatrix} \begin{bmatrix} 1 & 0 \\ \overline{c} & 1 \end{bmatrix}.$$

We also have

$$1 + cc^* = (p^*p)^{-1}, \quad 1 + d^*d = (\overline{pp}^*)^{-1}.$$

The following group will play an important role in the Shale-Stinespring Theorem on the implementability of fermionic Bogolubov transformations:

Proposition 1. *Let $r \in O(\mathcal{Y})$. The following conditions are equivalent:*
1) $j - rjr^{-1} \in B^2(\mathcal{Y})$,
2) $rj - jr \in B^2(\mathcal{Y})$,
3) $q \in B^2(\overline{\mathcal{Z}},\mathcal{Z})$.

Define $O_2(\mathcal{Y})$ to be the set of $r \in O(\mathcal{Y})$ satisfying the above conditions. Then $O_2(\mathcal{Y})$ is a group.

Note that if r is j-nondegenerate, then it belongs to $O_2(\mathcal{Y})$ iff $c \in B^2(\overline{\mathcal{Z}},\mathcal{Z})$, or equivalently, $d \in B^2(\overline{\mathcal{Z}},\mathcal{Z})$.

7.4 Fermionic Bogolubov Transformations in the Fock Representation

Consider now the Fock representation of the CAR, $\mathcal{Y} \ni y \mapsto \phi(y) \in B_{\mathrm{h}}(\Gamma_{\mathrm{a}}(\mathcal{Z}))$.

Theorem 31. 1) *Let $r \in O(\mathcal{Y})$. Then the following conditions are equivalent:*

a) There exists $U \in U(\Gamma_{\mathrm{a}}(\mathcal{Z}))$ such that

$$U\phi(y)U^* = \phi(ry), \quad y \in \mathcal{Y}\,. \tag{56}$$

b) $r \in O_2(\mathcal{Y})$

2) *For $r \in O_2(\mathcal{Y})$, the unitary operator U satisfying (56) is defined uniquely up to a phase factor. Let $\mathcal{U}_r$ denote the class of these operators. Then*

$$O_2(\mathcal{Y}) \ni r \mapsto \mathcal{U}_r \in U(\Gamma_{\mathrm{a}}(\mathcal{Z}))/U(1)$$

is a homomorphism of groups.

3) *Let $r \in O_2(\mathcal{Y})$ be j-nondegenerate. Let p, c, d be defined as in the previous subsection. Set*

$$U_r^j = |\det pp^*|^{\frac{1}{4}} \mathrm{e}^{\frac{1}{2}a^*(d)} \Gamma((p^*)^{-1}) \mathrm{e}^{-\frac{1}{2}a(c)}\,. \tag{57}$$

Then U_r^j is the unique unitary operator satisfying (56) and

$$(\Omega|U_r^j\Omega) > 0\,. \tag{58}$$

7.5 Pin Group in the Fock Representation

Define $O_1(\mathcal{Y}) := \{r \in O(\mathcal{Y}) \ :\ r - 1 \in B^1(\mathcal{Y})\}$.

Theorem 32. 1) *$O_1(\mathcal{Y})$ is a subgroup of $O_2(\mathcal{Y})$.*

2) *$r \in O_1(\mathcal{Y})$ iff $p - 1 \in B^1(\mathcal{Z})$.*

The following theorem describes the Pin representation for an arbitrary number of degrees of freedom:

Theorem 33. *There exists a group homomorphism*

$$O_1(\mathcal{Y}) \ni r \mapsto \pm U_r \in U(\Gamma_{\mathrm{a}}(\mathcal{Z}))/\{1, -1\} \tag{59}$$

satisfying $U_r\phi(y)U_r^* = \phi(ry)$.

In order to give a formula for $\pm U_r$, which is analogous to the bosonic formula (48), we have to restrict ourselves to j-nondegenerate transformations.

Theorem 34. *Suppose that $r \in O_1(\mathcal{Y})$ is j-nondegenerate. Then*

$$\pm\, U_r = \pm(\det p^*)^{\frac{1}{2}} \mathrm{e}^{\frac{1}{2}a^*(d)} \Gamma((p^*)^{-1}) \mathrm{e}^{-\frac{1}{2}a(c)}. \tag{60}$$

Similarly as in the bosonic case, it is easy to see that the operators $\pm U_r$ differ by a phase factor from U_r^j from Theorem 31.

7.6 j-Self-Adjoint Orthogonal Transformations

Special role is played by $r \in O_2(\mathcal{Y})$ satisfying $rj = j^* r$. Such transformations will be called *j-self-adjoint.*

One can easily show that $r \in O_2(\mathcal{Y})$ is j-self-adjoint if

$$r = \begin{bmatrix} (1+cc^*)^{-1/2} & (1+cc^*)^{-1/2}c \\ -(1+c^*c)^{-1/2}c^* & (1+c^*c)^{-1/2} \end{bmatrix} \tag{61}$$

for some $c \in B^2_{\mathrm{a}}(\mathcal{Y})$.

Theorem 35. 1) *The formula*

$$R_c := \det(1+cc^*)^{-\frac{1}{4}} \exp\left(\tfrac{1}{2}a^*(c)\right) \Gamma((1+cc^*)^{-\frac{1}{2}}) \exp\left(-\tfrac{1}{2}a(c)\right) \tag{62}$$

defines a unitary operator on $\Gamma_{\mathrm{a}}(\mathcal{Z})$.

2) *If Ω_c is defined in (55), then $\Omega = R_c \Omega_c$.*

3)

$$R_c a^*(z) R_c^* = a^*\left((1+cc^*)^{-1/2}z\right) + a\left((1+cc^*)^{-1/2}c\overline{z}\right),$$

$$R_c a(z) R_c^* = a^*\left((1+cc^*)^{-1/2}c\overline{z}\right) + a\left((1+cc^*)^{-1/2}z\right).$$

4) *If r and c are related by (61), then the operator R_c coincides with the operator U_r^j defined in (57).*

5) *R_c coincides with U_r defined in (34) with the plus sign and the positive square root.*

8 W^*-Algebras

In this section we review some elements of the theory of W^*-algebras needed in our paper. For more details we refer the reader to [29], and also [16, 17, 67, 69, 70].

$\mathfrak{M}$ is a *W^*-algebra* if it is a C^*-algebra, possessing a *predual.* (This means that there exists a Banach space $\mathcal{Y}$ such that $\mathfrak{M}$ is isomorphic as a Banach space to the dual of $\mathcal{Y}$. This Banach space $\mathcal{Y}$ is called a predual of $\mathfrak{M}$).

One can show that a predual of a W^*-algebra is defined uniquely up to an isomorphism. The topology on $\mathfrak{M}$ given by the functionals in the predual (the $*$-weak topology in the terminology of theory of Banach spaces) will be called the *σ-weak topology.* The set σ-weakly continuous linear functionals coincides with the predual of $\mathfrak{M}$.

$\mathbb{R} \ni t \mapsto \tau^t$ is called a *W^*-dynamics* if it is a 1-parameter group with values in $*$-automorphisms of $\mathfrak{M}$ and, for any $A \in \mathfrak{M}$, $t \mapsto \tau^t(A)$ is σ-weakly continuous. The pair $(\mathfrak{M}, \tau)$ is called a *W^*-dynamical system.*

$$\mathfrak{M} \cap \mathfrak{M}' := \{B \in \mathfrak{M} \ : \ AB = BA, \ A \in \mathfrak{M}\}$$

is called the *center of the algebra* $\mathfrak{M}$. A W^*-algebra with a trivial center is called a *factor.*

If $\mathfrak{A}$ is a subset of $B(\mathcal{H})$ for some Hilbert space $\mathcal{H}$, then

$$\mathfrak{A}' := \{B \ : \ AB = BA, \ A \in \mathfrak{A}\}$$

is called the *commutant of* $\mathfrak{A}$.

8.1 Standard Representations

We say that $\mathcal{H}^+$ is a *self-dual cone* in a Hilbert space $\mathcal{H}$ if

$$\mathcal{H}^+ = \{\Phi \in \mathcal{H} \ : \ (\Phi|\Psi) \geq 0, \ \Psi \in \mathcal{H}^+\} \, .$$

We say that a quadruple $(\pi, \mathcal{H}, J, \mathcal{H}^+)$ is a *standard representation of a W^*-algebra* $\mathfrak{M}$ if $\pi : \mathfrak{M} \to B(\mathcal{H})$ is an injective σ-weakly continuous $*$-representation, J is an antiunitary involution on $\mathcal{H}$ and $\mathcal{H}^+$ is a self-dual cone in $\mathcal{H}$ satisfying the following conditions:
1) $J\pi(\mathfrak{M})J = \pi(\mathfrak{M})'$;
2) $J\pi(A)J = \pi(A)^*$ for A in the center of $\mathfrak{M}$;
3) $J\Psi = \Psi$ for $\Psi \in \mathcal{H}^+$;
4) $\pi(A)J\pi(A)\mathcal{H}^+ \subset \mathcal{H}^+$ for $A \in \mathfrak{M}$.

Every W^*-algebra has a unique (up to the unitary equivalence) standard representation, [42] (see also [4,17,20,29,70]).

The standard representation has several important properties. First, every σ-weakly continuous state ω has a unique vector representative in $\mathcal{H}^+$ (in other words, there is a unique normalized vector $\Omega \in \mathcal{H}^+$ such that $\omega(A) = (\Omega|\pi(A)\Omega)$). Secondly, for every $*$-automorphism τ of $\mathfrak{M}$ there exists a unique unitary operator $U \in B(\mathcal{H})$ such that

$$\pi(\tau(A)) = U\pi(A)U^*, \quad U\mathcal{H}^+ \subset \mathcal{H}^+ \, .$$

Finally, for every W^*-dynamics $\mathbb{R} \ni t \mapsto \tau^t$ on $\mathfrak{M}$ there is a unique self-adjoint operator L on $\mathcal{H}$ such that

$$\pi(\tau^t(A)) = \mathrm{e}^{\mathrm{i}tL}\pi(A)\mathrm{e}^{-\mathrm{i}tL}, \quad \mathrm{e}^{\mathrm{i}tL}\mathcal{H}^+ = \mathcal{H}^+. \tag{63}$$

The operator L is called the *standard Liouvillean of* $t \mapsto \tau^t$.

Given a standard representation $(\pi, \mathcal{H}, J, \mathcal{H}^+)$ we also have the *right representation* $\pi_{\mathrm{r}} : \overline{\mathfrak{M}} \to B(\mathcal{H})$ given by $\pi_{\mathrm{r}}(\overline{A}) := J\pi(A)J$. Note that the image of π_{r} is $\pi(\mathfrak{M})'$. We will often write π_{l} for π and call it the *left representation.*

8.2 Tomita-Takesaki Theory

Let $\pi : \mathfrak{M} \to B(\mathcal{H})$ be an injective σ-weakly continuous $*$-representation and Ω a cyclic and separating vector for $\pi(\mathfrak{M})$. One then proves that the formula

$$S\pi(A)\Omega = \pi(A)^*\Omega$$

defines a closable antilinear operator S on $\mathcal{H}$. The *modular operator* Δ and the *modular conjugation* J are defined by the polar decomposition of S:

$$S = J\Delta^{1/2} .$$

If we set

$$\mathcal{H}^+ = \{\pi(A)J\pi(A)\Omega \ : \ A \in \mathfrak{M}\}^{\mathrm{cl}} ,$$

then $(\pi, \mathcal{H}, J, \mathcal{H}^+)$ is a standard representation of $\mathfrak{M}$. Given $\mathfrak{M}$, π and $\mathcal{H}$, it is the unique standard representation with the property $\Omega \in \mathcal{H}^+$.

8.3 KMS States

Let $(\mathfrak{M}, \tau)$ be a W^*-dynamical system. Let β be a positive number (having the physical interpretation of the inverse temperature). A σ-weakly continuous state ω on $\mathfrak{M}$ is called a *(τ, β)-KMS state* (or a *β-KMS state for τ*) iff for all $A, B \in \mathfrak{M}$ there exists a function $F_{A,B}$, analytic inside the strip $\{z : 0 < \mathrm{Im} z < \beta\}$, bounded and continuous on its closure, and satisfying the KMS boundary conditions

$$F_{A,B}(t) = \omega(A\tau^t(B)), \qquad F_{A,B}(t + \mathrm{i}\beta) = \omega(\tau^t(B)A).$$

A KMS-state is τ-invariant. If $\mathfrak{M}$ is a factor, then $(\mathfrak{M}, \tau)$ can have at most one β-KMS state.

If $\mathfrak{M} \subset B(\mathcal{H})$ and $\Phi \in \mathcal{H}$, we will say that Φ is a *(τ, β)-KMS vector* iff $(\Phi | \cdot \Phi)$ is a (τ, β)-KMS state.

The acronym KMS stands for Kubo-Martin-Schwinger.

8.4 Type I Factors – Irreducible Representation

The most elementary example of a factor is the so-called type I factor – this means the algebra of all bounded operators on a given Hilbert space. In this and the next two subsections we describe various concepts of theory of W^*-algebras on this example.

The space of σ-weakly continuous functionals on $B(\mathcal{H})$ (the predual of $B(\mathcal{H})$) can be identified with $B^1(\mathcal{H})$ (trace class operators) by the formula

$$\psi(A) = \mathrm{Tr}\gamma A, \qquad \gamma \in B^1(\mathcal{H}), \ \ A \in B(\mathcal{H}) . \tag{64}$$

In particular, σ-weakly continuous states are determined by positive trace one operators, called density matrices. A state given by a density matrix γ is faithful iff $\mathrm{Ker}\gamma = \{0\}$.

If τ is a $*$-automorphism of $B(\mathcal{H})$, then there exists $W \in U(\mathcal{H})$ such that

$$\tau(A) = WAW^*, \quad A \in B(\mathcal{H}) . \tag{65}$$

If $t \mapsto \tau^t$ is a W^*-dynamics, then there exists a self-adjoint operator H on $\mathcal{H}$ such that

$$\tau^t(A) = \mathrm{e}^{\mathrm{i}tH} A \mathrm{e}^{-\mathrm{i}tH}, \quad A \in B(\mathcal{H}) .$$

See e.g. [16].

A state given by (64) is invariant with respect to the W^*-dynamics (65) iff H commutes with γ.

There exists a (β, τ)–KMS state iff $\mathrm{Tr}\,\mathrm{e}^{-\beta H} < \infty$, and then it has the density matrix $\mathrm{e}^{-\beta H}/\mathrm{Tr}\,\mathrm{e}^{-\beta H}$.

8.5 Type I Factor – Representation in Hilbert-Schmidt Operators

Clearly, the representation of $B(\mathcal{H})$ in $\mathcal{H}$ is not in a standard form. To construct a standard form of $B(\mathcal{H})$, consider the Hilbert space of Hilbert-Schmidt operators on $\mathcal{H}$, denoted $B^2(\mathcal{H})$, and two injective representations:

$$\begin{aligned} B(\mathcal{H}) \ni A &\mapsto \pi_{\mathrm{l}}(A) \in B(B^2(\mathcal{H})), \quad \pi_{\mathrm{l}}(A)B := AB, \quad B \in B^2(\mathcal{H}) ; \\ \overline{B(\mathcal{H})} \ni \overline{A} &\mapsto \pi_{\mathrm{r}}(\overline{A}) \in B(B^2(\mathcal{H})), \quad \pi_{\mathrm{r}}(\overline{A})B := BA^*, \quad B \in B^2(\mathcal{H}) . \end{aligned} \tag{66}$$

Set $J_{\mathcal{H}}B := B^*$, $B \in B^2(\mathcal{H})$. Then $J_{\mathcal{H}}\pi_{\mathrm{l}}(A)J_{\mathcal{H}} = \pi_{\mathrm{r}}(\overline{A})$ and

$$(\pi_l, B^2(\mathcal{H}), J_{\mathcal{H}}, B^2_+(\mathcal{H}))$$

is a standard representation of $B(\mathcal{H})$.

If a state on $B(\mathcal{H})$ is given by a density matrix $\gamma \in B^1_+(\mathcal{H})$, then its standard vector representative is $\gamma^{\frac{1}{2}} \in B^2_+(\mathcal{H})$. The standard implementation of the $*$-authomorphism $\tau(A) = WAW^*$ equals $\pi_{\mathrm{l}}(W)\pi_r(\overline{W})$. If the W^*-dynamics $t \mapsto \tau^t$ is given by a self-adjoint operator H, then its standard Liouvillean is $\pi_{\mathrm{l}}(H) - \pi_{\mathrm{r}}(\overline{H})$.

8.6 Type I Factors – Representation in $\mathcal{H} \otimes \overline{\mathcal{H}}$

An alternative formalism, which can be used to describe a standard form of type I factors, uses the notion of a conjugate Hilbert space.

Recall that $B^2(\mathcal{H})$ has a natural identification with $\mathcal{H} \otimes \overline{\mathcal{H}}$. Under the identification the representations (66) become

$$\begin{aligned} B(\mathcal{H}) \ni A &\mapsto A \otimes 1_{\overline{\mathcal{H}}} \in B(\mathcal{H} \otimes \overline{\mathcal{H}}) ; \\ \overline{B(\mathcal{H})} \ni \overline{A} &\mapsto 1_{\mathcal{H}} \otimes \overline{A} \in B(\mathcal{H} \otimes \overline{\mathcal{H}}) . \end{aligned} \tag{67}$$

(Abusing the notation, sometimes the above representations will also be denoted by π_{l} and π_{r}).

Note that the standard unitary implementation of the automorphism $\tau(A) = WAW^*$ is then equal to $W \otimes \overline{W}$. The standard Liouvillean for $\tau^t(A) = \mathrm{e}^{\mathrm{i}tH} A \mathrm{e}^{-\mathrm{i}tH}$ equals $L = H \otimes 1 - 1 \otimes \overline{H}$. The modular conjugation is $J_{\mathcal{H}}$ defined by

$$J_{\mathcal{H}}\left(\Psi_1 \otimes \overline{\Psi_2}\right) := \Psi_2 \otimes \overline{\Psi_1} . \tag{68}$$

The positive cone is then equal to

$$(\mathcal{H} \otimes \overline{\mathcal{H}})_+ := \mathrm{Conv}\{\Psi \otimes \overline{\Psi} \ : \ \Psi \in \mathcal{H}\}^{\mathrm{cl}} ,$$

where Conv denotes the convex hull.

8.7 Perturbations of W^*-Dynamics and Liouvilleans

The material of this subsection will be needed only in the last section devoted to Pauli-Fierz systems.

Let τ be a W^*-dynamics on a W^*-algebra $\mathfrak{M}$ and let $(\pi, \mathcal{H}, J, \mathcal{H}_+)$ be a standard representation of $\mathfrak{M}$. Let L be the standard Liouvillean of τ.

The following theorem is proven in [29]:

Theorem 36. *Let V be a self-adjoint operator on $\mathcal{H}$ affiliated to $\mathfrak{M}$. (That means that all spectral projections of V belong to $\pi(\mathfrak{M})$). Let $L+V$ be essentially self-adjoint on* $\mathrm{Dom}(L) \cap \mathrm{Dom}(V)$ *and $L_V := L+V-JVJ$ be essentially self-adjoint on* $\mathrm{Dom}(L) \cap \mathrm{Dom}(V) \cap \mathrm{Dom}(JVJ)$. *Set*

$$\tau_V^t(A) := \pi^{-1}\left(\mathrm{e}^{\mathrm{i}t(L+V)} \pi(A) \mathrm{e}^{-\mathrm{i}t(L+V)}\right) .$$

Then $t \mapsto \tau_V^t$ is a W^*-dynamics on $\mathfrak{M}$ and L_V is its standard Liouvillean.

9 Quasi-Free Representations of the CCR

9.1 Bosonic Quasi-Free Vectors

Let $(\mathcal{Y}, \omega)$ be a real vector space with an antisymmetric form. Let

$$\mathcal{Y} \ni y \mapsto W(y) \in U(\mathcal{H}) \tag{69}$$

be a representation of the CCR. We say that $\Psi \in \mathcal{H}$ is a *quasi-free vector* for (69) iff there exists a quadratic form η such that

$$(\Psi | W(y) \Psi) = \exp\left(-\tfrac{1}{4} y \eta y\right) . \tag{70}$$

Note that η is necessarily positive, that is $y\eta y \geq 0$ for $y \in \mathcal{Y}$.

A representation (69) is called *quasi-free* if there exists a cyclic quasi-free vector in $\mathcal{H}$.

The following fact is easy to see:

Theorem 37. *A quasi-free representation is regular.*

Therefore, in a quasi-free representation we can define the corresponding field operators, denoted $\phi(y)$.

If m is an integer, we say that σ is a *pairing of* $\{1,\dots,2m\}$ if it is a permutation of $\{1,\dots,2m\}$ satisfying

$$\sigma(1)<\sigma(3)<\cdots<\sigma(2m-1),\quad \sigma(2j-1)<\sigma(2j),\ j=1,\dots,m\,.$$

$P(2m)$ will denote the set of pairings of $\{1,\dots,2m\}$.

Theorem 38. *Suppose we are given a regular representation of the CCR*

$$\mathcal{Y}\ni y\mapsto \mathrm{e}^{\mathrm{i}\phi(y)}\in U(\mathcal{H})\,.$$

Let $\Psi\in\mathcal{H}$. Then the following statements are equivalent:

1) *For any $n=1,2,\dots$, $y_1,\dots y_n\in\mathcal{Y}$, $\Psi\in\mathrm{Dom}\,(\phi(y_1)\cdots\phi(y_n))$, and*

$$\begin{aligned}(\Psi|\phi(y_1)\cdots\phi(y_{2m-1})\Psi)&=0,\\ (\Psi|\phi(y_1)\cdots\phi(y_{2m})\Psi)&=\sum_{\sigma\in P(2m)}\prod_{j=1}^{m}(\Psi|\phi(y_{\sigma(2j-1)})\phi(y_{\sigma(2j)})\Psi).\end{aligned}$$

2) *Ψ is a quasi-free vector.*

Theorem 39. *Suppose that Ψ is a quasi-free vector with η satisfying (70). Then*

1) $y_1\left(\eta+\frac{\mathrm{i}}{2}\omega\right)y_2=(\Psi|\phi(y_1)\phi(y_2)\Psi)$;

2) $|y_1\omega y_2|\leq 2|y_1\eta y_1|^{1/2}|y_2\eta y_2|^{1/2},\quad y_1,y_2\in\mathcal{Y}$.

Proof. Note that $(\Psi|\phi(y)^2\Psi)=y\eta y$. This implies

$$\frac{1}{2}\Big((\Psi|\phi(y_1)\phi(y_2)\Psi)+(\Psi|\phi(y_2)\phi(y_1)\Psi)\Big)=y_1\eta y_2\,.$$

From the canonical commutation relations we get

$$\frac{1}{2}\Big((\Psi|\phi(y_1)\phi(y_2)\Psi)-(\Psi|\phi(y_2)\phi(y_1)\Psi)\Big)=\frac{\mathrm{i}}{2}y_1\omega y_2\,.$$

This yields 1).

From

$$\|(\phi(y_1)\pm\mathrm{i}\phi(y_2))\Psi\|^2\geq 0\,.$$

we get

$$|y_2\omega y_1|\leq y_1\eta y_1+y_2\eta y_2\,. \tag{71}$$

This implies 2). □

9.2 Classical Quasi-Free Representations of the CCR

Let us briefly discuss quasi-free representations for the trivial antisymmetric form. In this case the fields commute and can be interpreted as classical random variables, hence we will call such representations *classical.*

Consider a real vector space $\mathcal{Y}$ equipped with a positive scalar product η. Consider the probabilistic Gaussian measure given by the covariance η. That means, if $\dim \mathcal{Y} = n < \infty$, then it is the measure $\mathrm{d}\mu = (\det \eta)^{1/2}(2\pi)^{-n/2}\mathrm{e}^{-y\eta y/2}\mathrm{d}y$, where $\mathrm{d}y$ denotes the Lebesgue measure on $\mathcal{Y}$. If $\dim \mathcal{Y} = \infty$, see e.g. [65].

Consider the Hilbert space $L^2(\mu)$. Note that a dense subspace of of $L^2(\mu)$ can be treated as functions on $\mathcal{Y}$. For $y \in \mathcal{Y}$, let $\phi(y)$ denote the function $\mathcal{Y} \ni v \mapsto y\eta v \in \mathbb{R}$. $\phi(y)$ can be treated as a self-adjoint operator on $L^2(\mu)$.

We equip $\mathcal{Y}$ with the antisymmetric form $\omega = 0$. Then

$$\mathcal{Y} \ni y \mapsto \mathrm{e}^{\mathrm{i}\phi(y)} \in U(L^2(\mu)) \tag{72}$$

is a representation of the CCR.

Let $\Psi \in L^2(\mu)$ be the constant function equal to 1. Then Ψ is a cyclic quasi-free vector for (72).

In the remaining part of this section we will discuss quasi-free representations that are fully "quantum" – whose CCR are given by a non-degenerate antisymmetric form ω.

9.3 Araki-Woods Representation of the CCR

In this subsection we describe the *Araki-Woods representations of the CCR* and the corresponding W^*-algebras. These representations were introduced in [8]. They are examples of quasi-free representations. In our presentation we follow [27].

Let $\mathcal{Z}$ be a Hilbert space and consider the Fock space $\Gamma_{\mathrm{s}}(\mathcal{Z} \oplus \overline{\mathcal{Z}})$. We will identify the symplectic space $\mathrm{Re}\big((\mathcal{Z} \oplus \overline{\mathcal{Z}}) \oplus \overline{(\mathcal{Z} \oplus \overline{\mathcal{Z}})}\big)$ with $\mathcal{Z} \oplus \overline{\mathcal{Z}}$, as in (41). Therefore, for $(z_1, \overline{z}_2) \in \mathcal{Z} \oplus \overline{\mathcal{Z}}$, the operator

$$\phi(z_1, \overline{z}_2) := \frac{1}{\sqrt{2}}\big(a^*(z_1, \overline{z}_2) + a(z_1, \overline{z}_2)\big)$$

is the corresponding field operator and $W(z_1, \overline{z}_2) = \mathrm{e}^{\mathrm{i}\phi(z_1, \overline{z}_2)}$ is the corresponding Weyl operator.

We will parametrize the Araki-Woods representation by a self-adjoint operator γ on $\mathcal{Z}$ satisfying $0 \le \gamma \le 1$, $\mathrm{Ker}(\gamma - 1) = \{0\}$. Another important object associated to the Araki-Woods representation is a positive operator ρ on $\mathcal{Z}$ called the "1-particle density". It is related to γ by

$$\gamma := \rho(1 + \rho)^{-1}, \quad \rho = \gamma(1 - \gamma)^{-1} \,. \tag{73}$$

(Note that in [27] we used ρ to parametrize Araki-Woods representations).

For $z \in \mathrm{Dom}(\rho^{\frac{1}{2}})$ we define two unitary operators on $\Gamma_{\mathrm{s}}(\mathcal{Z} \oplus \overline{\mathcal{Z}})$ as:

$$W_{\gamma,\mathrm{l}}^{\mathrm{AW}}(z) := W\big((\rho+1)^{\frac{1}{2}} z, \overline{\rho}^{\frac{1}{2}} \overline{z}\big),$$
$$W_{\gamma,\mathrm{r}}^{\mathrm{AW}}(\overline{z}) := W\big(\rho^{\frac{1}{2}} z, (\overline{\rho}+1)^{\frac{1}{2}} \overline{z}\big).$$

We denote by $\mathfrak{M}_{\gamma,\mathrm{l}}^{\mathrm{AW}}$ and $\mathfrak{M}_{\gamma,\mathrm{r}}^{\mathrm{AW}}$ the von Neumann algebras generated by $\{W_{\gamma,\mathrm{l}}^{\mathrm{AW}}(z) : z \in \mathrm{Dom}(\rho^{\frac{1}{2}})\}$ and $\{W_{\gamma,\mathrm{r}}^{\mathrm{AW}}(\overline{z}) : z \in \mathrm{Dom}(\rho^{\frac{1}{2}})\}$ respectively. We will be call them respectively the *left* and the *right Araki-Woods algebra.* We drop the superscript AW until the end of the section.

The operators τ and ϵ, defined by

$$\mathcal{Z} \oplus \overline{\mathcal{Z}} \ni (z_1, \overline{z}_2) \mapsto \tau(z_1, \overline{z}_2) := (\overline{z}_2, z_1) \in \overline{\mathcal{Z}} \oplus \mathcal{Z}\ , \tag{74}$$

$$\mathcal{Z} \oplus \overline{\mathcal{Z}} \ni (z_1, \overline{z}_2) \mapsto \epsilon(z_1, \overline{z}_2) := (z_2, \overline{z}_1) \in \mathcal{Z} \oplus \overline{\mathcal{Z}}\ , \tag{75}$$

will be useful. Note that τ is linear, ϵ antilinear, and

$$\epsilon(z_1, \overline{z}_2) = \overline{\tau(z_1, \overline{z}_2)}. \tag{76}$$

In the following theorem we will describe some basic properties of the Araki-Woods algebras.

Theorem 40. 1) $\mathcal{Z} \supset \mathrm{Dom}(\rho^{\frac{1}{2}}) \ni z \mapsto W_{\gamma,\mathrm{l}}(z) \in U(\Gamma_{\mathrm{s}}(\mathcal{Z} \oplus \overline{\mathcal{Z}}))$ *is a regular representation of the CCR. In particular,*

$$W_{\gamma,\mathrm{l}}(z_1) W_{\gamma,\mathrm{l}}(z_2) = \mathrm{e}^{-\frac{\mathrm{i}}{2}\mathrm{Im}(z_1|z_2)} W_{\gamma,\mathrm{l}}(z_1 + z_2)\ .$$

It will be called the left Araki-Woods representation of the CCR associated to the pair $(\mathcal{Z}, \gamma)$. The corresponding field, creation and annihilation operators are affiliated to $\mathfrak{M}_{\gamma,\mathrm{l}}$ and are given by

$$\phi_{\gamma,\mathrm{l}}(z) = \phi\Big((\rho+1)^{\frac{1}{2}} z, \overline{\rho}^{\frac{1}{2}} \overline{z}\Big),$$

$$a_{\gamma,\mathrm{l}}^*(z) = a^*\Big((\rho+1)^{\frac{1}{2}} z, 0\Big) + a\Big(0, \overline{\rho}^{\frac{1}{2}} \overline{z}\Big),$$

$$a_{\gamma,\mathrm{l}}(z) = a\Big((\rho+1)^{\frac{1}{2}}\ z, 0\Big) + a^*\Big(0, \overline{\rho}^{\frac{1}{2}} \overline{z}\Big).$$

2) $\overline{\mathcal{Z}} \supset \mathrm{Dom}(\overline{\rho}^{\frac{1}{2}}) \ni \overline{z} \mapsto W_{\gamma,\mathrm{r}}(\overline{z}) \in U(\Gamma_{\mathrm{s}}(\mathcal{Z} \oplus \overline{\mathcal{Z}}))$ *is a regular representation of the CCR. In particular*

$$W_{\gamma,\mathrm{r}}(\overline{z}_1) W_{\gamma,\mathrm{r}}(\overline{z}_2) = \mathrm{e}^{-\frac{\mathrm{i}}{2}\mathrm{Im}(\overline{z}_1|\overline{z}_2)} W_{\gamma,\mathrm{r}}(\overline{z}_1 + \overline{z}_2) = \mathrm{e}^{\frac{\mathrm{i}}{2}\mathrm{Im}(z_1|z_2)} W_{\gamma,\mathrm{r}}(\overline{z}_1 + \overline{z}_2)\ .$$

It will be called the right Araki-Woods representation of the CCR associated to the pair $(\mathcal{Z}, \gamma)$. The corresponding field, creation and annihilation operators are affiliated to $\mathfrak{M}_{\gamma,\mathrm{r}}$ and are given by

$$\phi_{\gamma,\mathrm{r}}(\overline{z}) = \phi\Big(\rho^{\frac{1}{2}} z, (\overline{\rho}+1)^{\frac{1}{2}}\overline{z}\Big),$$

$$a^*_{\gamma,\mathrm{r}}(\overline{z}) = a\Big(\rho^{\frac{1}{2}} z, 0\Big) + a^*\Big(0, (\overline{\rho}+1)^{\frac{1}{2}}\overline{z}\Big),$$

$$a_{\gamma,\mathrm{r}}(\overline{z}) = a^*\Big(\rho^{\frac{1}{2}} z, 0\Big) + a\Big(0, (\overline{\rho}+1)^{\frac{1}{2}}\overline{z}\Big).$$

3) *Set*

$$J_{\mathrm{s}} = \Gamma(\epsilon) \tag{77}$$

Then we have

$$J_{\mathrm{s}} W_{\gamma,\mathrm{l}}(z) J_{\mathrm{s}} = W_{\gamma,\mathrm{r}}(\overline{z}),$$

$$J_{\mathrm{s}} \phi_{\gamma,\mathrm{l}}(z) J_{\mathrm{s}} = \phi_{\gamma,\mathrm{r}}(\overline{z}),$$

$$J_{\mathrm{s}} a^*_{\gamma,\mathrm{l}}(z) J_{\mathrm{s}} = a^*_{\gamma,\mathrm{r}}(\overline{z}),$$

$$J_{\mathrm{s}} a_{\gamma,\mathrm{l}}(z) J_{\mathrm{s}} = a_{\gamma,\mathrm{r}}(\overline{z}).$$

4) *The vacuum Ω is a bosonic quasi-free vector for $W_{\gamma,\mathrm{l}}$, its expectation value for the Weyl operators (the "generating function") is equal to*

$$\big(\Omega|W_{\gamma,\mathrm{l}}(z)\Omega\big) = \exp\Big(-\tfrac{1}{4}(z|z) - \tfrac{1}{2}(z|\rho z)\Big) = \exp\Big(-\tfrac{1}{4}\Big(z|\tfrac{1+\gamma}{1-\gamma}z\Big)\Big)$$

and the "two-point functions" are equal to

$$\big(\Omega|\phi_{\gamma,\mathrm{l}}(z_1)\phi_{\gamma,\mathrm{l}}(z_2)\Omega\big) = \tfrac{1}{2}(z_1|z_2) + \mathrm{Re}(z_1|\rho z_2),$$

$$\big(\Omega|a_{\gamma,\mathrm{l}}(z_1)a^*_{\gamma,\mathrm{l}}(z_2)\Omega\big) = (z_1|(1+\rho)z_2) = (z_1|(1-\gamma)^{-1}z_2),$$

$$\big(\Omega|a^*_{\gamma,\mathrm{l}}(z_1)a_{\gamma,\mathrm{l}}(z_2)\Omega\big) = (z_2|\rho z_1) = (z_2|\gamma(1-\gamma)^{-1}z_1),$$

$$\big(\Omega|a^*_{\gamma,\mathrm{l}}(z_1)a^*_{\gamma,\mathrm{l}}(z_2)\Omega\big) = 0,$$

$$\big(\Omega|a_{\gamma,\mathrm{l}}(z_1)a_{\gamma,\mathrm{l}}(z_2)\Omega\big) = 0.$$

5) $\mathfrak{M}_{\gamma,\mathrm{l}}$ *is a factor.*
6) $\mathrm{Ker}\gamma = \{0\}$ *iff Ω is separating for $\mathfrak{M}_{\gamma,\mathrm{l}}$ iff Ω is cyclic for $\mathfrak{M}_{\gamma,\mathrm{l}}$. If this is the case, then the modular conjugation for Ω is given by (77) and the modular operator for Ω is given by*

$$\Delta = \Gamma\left(\gamma \oplus \overline{\gamma}^{-1}\right). \tag{78}$$

7) *We have*

$$\mathfrak{M}'_{\gamma,\mathrm{l}} = \mathfrak{M}_{\gamma,\mathrm{r}}\,. \tag{79}$$

8) *Define*

$$\Gamma_{\mathrm{s},\gamma}^{+}(\mathcal{Z}\oplus\overline{\mathcal{Z}}) := \{AJ_{\mathrm{s}}A\Omega \ : \ A\in\mathfrak{M}_{\gamma,\mathrm{l}}\}^{\mathrm{cl}} . \tag{80}$$

Then $(\mathfrak{M}_{\gamma,\mathrm{l}}, \Gamma_{\mathrm{s}}(\mathcal{Z}\oplus\overline{\mathcal{Z}}), J_{\mathrm{s}}, \Gamma_{\mathrm{s},\gamma}^{+}(\mathcal{Z}\oplus\overline{\mathcal{Z}}))$ *is a* W^**-algebra in the standard form.*

9) *If* γ *has some continuous spectrum, then* $\mathfrak{M}_{\gamma,\mathrm{l}}$ *is a factor of type* III_1 *[69].*

10) *If* $\gamma=0$*, then* $\mathfrak{M}_{\gamma,\mathrm{l}}$ *is a factor of type* I.

11) *Let* h *be a self-adjoint operator on* $\mathcal{Z}$ *commuting with* γ *and*

$$\tau^t(W_{\gamma,\mathrm{l}}(z)) := W_{\gamma,\mathrm{l}}(\mathrm{e}^{\mathrm{i}th}z) .$$

Then $t\mapsto\tau^t$ *extends to a* W^**-dynamics on* $\mathfrak{M}_{\gamma,\mathrm{l}}$ *and*

$$L = \mathrm{d}\Gamma(h\oplus(-\overline{h}))$$

is its standard Liouvillean.

12) Ω *is a* (τ,β)*-KMS vector iff* $\gamma=\mathrm{e}^{-\beta h}$.

Proof. 1)–4) follow by straightforward computations.

Let us prove 5). We have

$$W_{\gamma,\mathrm{l}}(z_1)W_{\gamma,\mathrm{r}}(\overline{z}_2) = W_{\gamma,\mathrm{r}}(\overline{z}_2)W_{\gamma,\mathrm{l}}(z_1), \quad z_1,z_2\in\mathrm{Dom}\rho^{\frac{1}{2}} .$$

Consequently, $\mathfrak{M}_{\gamma,\mathrm{l}}$ and $\mathfrak{M}_{\gamma,\mathrm{r}}$ commute with one another.

Now

$$\begin{aligned}
&\left(\mathfrak{M}_{\gamma,\mathrm{l}}\cup\mathfrak{M}_{\gamma,\mathrm{l}}'\right)' \subset \left(\mathfrak{M}_{\gamma,\mathrm{l}}\cup\mathfrak{M}_{\gamma,\mathrm{r}}\right)' \\
&= \left\{W\big((\rho+1)^{\frac{1}{2}}z_1+\rho^{\frac{1}{2}}z_2, \overline{\rho}^{\frac{1}{2}}\overline{z}_1+(\overline{\rho}+1)^{\frac{1}{2}}\overline{z}_2\big) \ : \ z_1,z_2\in\mathcal{Z}\right\}' \\
&= \{W(w_1,\overline{w}_2) \ : \ w_1,w_2\in\mathcal{Z}\}' = \mathbb{C}1 ,
\end{aligned}$$

because

$$\{\big((\rho+1)^{\frac{1}{2}}z_1+\rho^{\frac{1}{2}}z_2, \overline{\rho}^{\frac{1}{2}}\overline{z}_1+(\overline{\rho}+1)^{\frac{1}{2}}\overline{z}_2\big) \ : \ z_1,z_2\in\mathcal{Z}\}$$

is dense in $\mathcal{Z}\oplus\overline{\mathcal{Z}}$, and Weyl operators depend strongly continuously on their parameters and act irreducibly on $\Gamma_{\mathrm{s}}(\mathcal{Z}\oplus\overline{\mathcal{Z}})$. Therefore,

$$\left(\mathfrak{M}_{\gamma,\mathrm{l}}\cup\mathfrak{M}_{\gamma,\mathrm{l}}'\right)' = \mathbb{C}1 ,$$

which means that $\mathfrak{M}_{\gamma,\mathrm{l}}$ is a factor and proves 5).

Let us prove the $\Rightarrow$ part of 6). Assume first that $\mathrm{Ker}\gamma=\{0\}$. Set $\tau^t(A) := \Gamma(\gamma,\overline{\gamma}^{-1})^{\mathrm{i}t}A\Gamma(\gamma,\overline{\gamma}^{-1})^{-\mathrm{i}t}$. We first check that τ^t preseves $\mathfrak{M}_{\gamma,\mathrm{l}}$. Therefore, it is a W^*-dynamics on $\mathfrak{M}_{\gamma,\mathrm{l}}$.

Next we check that $(\Omega|\cdot\Omega)$ satisfies the $(\tau,-1)$-KMS condition. This is straightforward for the Weyl operators $W_{\gamma,\mathrm{l}}(z)$. Therefore, it holds for the $*-$algebra $\mathfrak{M}_{\gamma,\mathrm{l},0}$ of finite linear combinations of $W_{\gamma,\mathrm{l}}(z)$. By the Kaplansky

Theorem, the unit ball of $\mathfrak{M}_{\gamma,\mathrm{l},0}$ is σ-weakly dense in the unit ball of $\mathfrak{M}_{\gamma,\mathrm{l}}$. Using this we extend the KMS condition to $\mathfrak{M}_{\gamma,\mathrm{l}}$.

A KMS state on a factor is always faithful. By 5), $\mathfrak{M}_{\gamma,\mathrm{l}}$ is a factor. Hence Ω is separating.

Let $\mathcal{H}$ be the closure of $\mathfrak{M}_{\gamma,\mathrm{l}}\Omega$. $\mathcal{H}$ is an invariant subspace for $\mathfrak{M}_{\gamma,\mathrm{l}}$, moreover Ω is cyclic and separating for $\mathfrak{M}_{\gamma,\mathrm{l}}$ on $\mathcal{H}$. Let us compute the operators S, Δ and J of the modular theory for Ω on $\mathcal{H}$.

Clearly $\mathcal{H}$ is spanned by vectors of the form $\Psi_z := \mathrm{e}^{\mathrm{i}a^*\left((1+\rho)^{\frac{1}{2}}z,\overline{\rho}^{\frac{1}{2}}\overline{z}\right)}\Omega$. Let us compute:

$$\Gamma(\gamma,\overline{\gamma}^{-1})^{\frac{1}{2}}\Psi_z = \mathrm{e}^{\mathrm{i}a^*\left(\rho^{\frac{1}{2}}z,(1+\overline{\rho})^{\frac{1}{2}}\overline{z}\right)}\Omega\,,$$

$$\begin{aligned} S\Psi_z &= \mathrm{e}^{-\mathrm{i}a^*\left((1+\rho)^{\frac{1}{2}}z,\overline{\rho}^{\frac{1}{2}}\overline{z}\right)}\Omega \\ &= J_{\mathrm{s}}\Gamma(\gamma,\overline{\gamma}^{-1})^{\frac{1}{2}}\Psi_z\,. \end{aligned}$$

We have

$$\Gamma(\gamma,\overline{\gamma}^{-1})^{\mathrm{i}t}\Psi_z = \Psi_{\gamma^{\mathrm{i}t}z}\,.$$

Hence $\Gamma(\gamma,\overline{\gamma}^{-1})$ preserves $\mathcal{H}$ and vectors Ψ_z form an essential domain for its restriction to $\mathcal{H}$. Besides, $\Gamma(\gamma,\overline{\gamma}^{-1})\mathcal{H}$ is dense in $\mathcal{H}$ and S preserves $\mathcal{H}$ as well. Therefore, J_{s} preserves $\mathcal{H}$. Thus

$$S = J_{\mathrm{s}}\Big|_{\mathcal{H}}\;\Gamma(\gamma,\overline{\gamma}^{-1})\Big|_{\mathcal{H}}$$

is the the polar decomposition of S and defines the modular operator and the modular conjugation. Next we see that

$$W_{\gamma,\mathrm{l}}(z_1)J_{\mathrm{s}}W_{\gamma,\mathrm{l}}(z_2)\Omega = W((\rho+1)^{\frac{1}{2}}z_1+\rho^{\frac{1}{2}}z_2,\overline{\rho}^{\frac{1}{2}}\overline{z}_1+(\overline{\rho}+1)^{\frac{1}{2}}\overline{z}_2)\Omega.$$

Therefore, $\mathfrak{M}_{\gamma,\mathrm{l}}J_{\mathrm{s}}\mathfrak{M}_{\gamma,\mathrm{l}}\Omega$ is dense in $\Gamma_{\mathrm{s}}(\mathcal{Z}\oplus\overline{\mathcal{Z}})$. But $\mathfrak{M}_{\gamma,\mathrm{l}}J_{\mathrm{s}}\mathfrak{M}_{\gamma,\mathrm{l}}\Omega\subset\mathcal{H}$. Hence $\mathcal{H}=\Gamma_{\mathrm{s}}(\mathcal{Z}\oplus\overline{\mathcal{Z}})$ and Ω is cyclic. This proves the $\Rightarrow$ part of 6).

To prove 7), we first assume that $\mathrm{Ker}\gamma=\{0\}$. By 6), we can apply the modular theory, which gives $\mathfrak{M}_{\gamma,\mathrm{l}}' = J_{\mathrm{s}}\mathfrak{M}_{\gamma,\mathrm{l}}J_{\mathrm{s}}$. By 3) we have $J_{\mathrm{s}}\mathfrak{M}_{\gamma,\mathrm{l}}J_{\mathrm{s}} = \mathfrak{M}_{\gamma,\mathrm{r}}$.

For a general γ, we decompose $\mathcal{Z}=\mathcal{Z}_0\oplus\mathcal{Z}_1$, where $\mathcal{Z}_0=\mathrm{Ker}\gamma$ and $\mathcal{Z}_1$ is equipped with a nondegenerate $\gamma_1:=\gamma\big|_{\mathcal{Z}_1}$. We then have $\mathfrak{M}_{\gamma,\mathrm{l}}\simeq B(\Gamma_{\mathrm{s}}(\mathcal{Z}_0))\otimes\mathfrak{M}_{\gamma_1,\mathrm{l}}$ and $\mathfrak{M}_{\gamma,\mathrm{r}}\simeq B(\Gamma_{\mathrm{s}}(\overline{\mathcal{Z}}_0))\otimes\mathfrak{M}_{\gamma_1,\mathrm{r}}$. This implies that $\mathfrak{M}_{\gamma,\mathrm{l}}'=\mathfrak{M}_{\gamma,\mathrm{r}}$ and ends the proof of 7).

From the decomposition $\mathfrak{M}_{\gamma,\mathrm{l}}\simeq B(\Gamma_{\mathrm{s}}(\mathcal{Z}_0))\otimes\mathfrak{M}_{\gamma_1,\mathrm{l}}$ we see that if $\mathrm{Ker}\gamma=\mathcal{Z}_0\neq\{0\}$, then Ω is neither cyclic nor separating. This completes the proof of 6) [32]. □

9.4 Quasi-Free Representations of the CCR as the Araki-Woods Representations

A large class of quasi-free representation is unitarily equivalent to the Araki-Woods representation for some γ.

Theorem 41. *Suppose that we are given a representation of the CCR*

$$\mathcal{Y}_0 \ni y \mapsto W(y) \in U(\mathcal{H}) , \tag{81}$$

with a cyclic quasi-free vector Ψ satisfying $(\Psi|W(y)\Psi) = \mathrm{e}^{-\frac{1}{4}y\eta y}$. Suppose that the symmetric form η is nondegenerate. Let $\mathcal{Y}$ be the completion of $\mathcal{Y}_0$ to a real Hilbert space with the scalar product given by η. By Theorem 39 2), ω extends to a bounded antisymmetric form on $\mathcal{Y}$, which we denote also by ω. Assume that ω is nondegenerate on $\mathcal{Y}$. Then there exists a Hilbert space $\mathcal{Z}$ and a positive operator γ on $\mathcal{Z}$, a linear injection of $\mathcal{Y}_0$ onto a dense subspace of $\mathcal{Z}$ and an isometric operator $U : \mathcal{H} \to \Gamma_{\mathrm{s}}(\mathcal{Z} \oplus \overline{\mathcal{Z}})$ such that

$$U\Psi = \Omega ,$$

$$UW(y) = W_{\gamma,\mathrm{l}}(y)U, \quad y \in \mathcal{Y}_0.$$

Proof. Without loss of generality we can assume that $\mathcal{Y}_0 = \mathcal{Y}$.

Working in the real Hilbert space $\mathcal{Y}$ equipped with the scalar product η and using Theorem 39 2), we see that ω is a bilinear form bounded by 2. Therefore there exist a bounded antisymmetric operator μ with a trivial kernel, $\|\mu\| \leq 1$, such that

$$y_1\omega y_2 = 2y_1\eta\mu y_2 .$$

Consider the polar decomposition

$$\mu = |\mu|j = j|\mu| .$$

Then j is an orthogonal operator satisfying $j^2 = -1$.

Let $\mathcal{Z}$ be the completion of $\mathcal{Y}$ with respect to the scalar product $\eta|\mu|$. Then j maps $\mathcal{Z}$ into itself and is an orthogonal operator for the scalar product $\eta|\mu|$ satisfying $j^2 = -1$. We can treat $\mathcal{Z}$ as a complex space, identifying $-j$ with the imaginary unit. We equip it with the (sesquilinear) scalar product

$$(y_1|y_2) := y_1\eta|\mu|y_2 + \mathrm{i}y_1\eta\mu y_2 = y_1\omega j y_2 + \mathrm{i}y_1\omega y_2 .$$

$\rho := |\mu|^{-1} - 1$ defines a positive operator on $\mathcal{Z}$ such that $\mathcal{Y} = \mathrm{Dom}\rho^{\frac{1}{2}}$. Now

$$\begin{aligned}(\Psi|\phi(y_1)\phi(y_2)\Psi) &= y_1\eta y_2 + \tfrac{\mathrm{i}}{2}y_1\omega y_2 \\ &= y_1\eta|\mu|y_2 + \mathrm{i}y_1\eta\mu y_2 + y_1\eta|\mu|(|\mu|^{-1} - 1)y_2 \\ &= (y_1|y_2) + \mathrm{Re}(y_1|\rho y_2) \\ &= (\Omega|\phi_{\gamma,\mathrm{l}}(y_1)\phi_{\gamma,\mathrm{l}}(y_2)\Omega),\end{aligned}$$

for γ as in (73). Therefore,

$$UW(y)\Psi := W_{\gamma,1}(y)\Omega$$

extends to an isometric map from $\mathcal{H}$ to $\Gamma_{\mathrm{s}}(\mathcal{Z} \oplus \overline{\mathcal{Z}})$ that intertwines the representation (81) with the Araki-Woods representation for $(\mathcal{Z}, \gamma)$. □

10 Quasi-Free Representations of the CAR

10.1 Fermionic Quasi-Free Vectors

Let $(\mathcal{Y}, \alpha)$ be a real Hilbert space. Let

$$\mathcal{Y} \ni y \mapsto \phi(y) \in B_{\mathrm{h}}(\mathcal{H}) \tag{82}$$

be a representation of the CAR. We say that $\Psi \in \mathcal{H}$ is a *quasi-free vector* for (82) iff

$$(\Psi|\phi(y_1)\cdots\phi(y_{2m-1})\Psi) = 0\ ,$$

$$(\Psi|\phi(y_1)\cdots\phi(y_{2m})\Psi) = \sum_{\sigma\in P(2m)} \mathrm{sgn}\sigma \prod_{j=1}^{m} (\Psi|\phi(y_{\sigma(2j-1)})\phi(y_{\sigma(2j)})\Psi)\ .$$

We say that (82) is a *quasi-free representation* if there exists a cyclic quasi-free vector Ψ in $\mathcal{H}$.

Define the antisymmetric form ω

$$y_1\omega y_2 := \frac{1}{\mathrm{i}}(\Psi|[\phi(y_1), \phi(y_2)]\Psi)\ . \tag{83}$$

Theorem 42. 1) $(\Psi|\phi(y_1)\phi(y_2)\Psi) = y_1\alpha y_2 + \frac{\mathrm{i}}{2}y_1\omega y_2$;
2) $|y_1\omega y_2| \leq 2|y_1\alpha y_1|^{\frac{1}{2}}|y_2\alpha y_2|^{\frac{1}{2}}$.

10.2 Araki-Wyss Representation of the CAR

In this subsection we describe *Araki-Wyss representations of the CAR* [9], see also [44]. They are examples of quasi-free representations of the CAR.

Let $\mathcal{Z}$ be a Hilbert space and consider the Fock space $\Gamma_{\mathrm{a}}(\mathcal{Z} \oplus \overline{\mathcal{Z}})$. We will identify the real Hilbert space $\mathrm{Re}\big((\mathcal{Z} \oplus \overline{\mathcal{Z}}) \oplus \overline{(\mathcal{Z} \oplus \overline{\mathcal{Z}})}\big)$ with $\mathcal{Z} \oplus \overline{\mathcal{Z}}$, as in (54). Therefore, for $(z_1, \overline{z}_2) \in \mathcal{Z} \oplus \overline{\mathcal{Z}}$,

$$\phi(z_1, \overline{z}_2) := a^*(z_1, \overline{z}_2) + a(z_1, \overline{z}_2)$$

are the corresponding field operators.

We will parametrize Araki-Wyss representation by a positive operator γ on $\mathcal{Z}$, possibly with a non-dense domain. We will also use the operator χ, called the "1-particle density", which satisfies $0 \leq \chi \leq 1$. The two operators are related to one another by

$$\gamma := \chi(1-\chi)^{-1}, \quad \chi = \gamma(\gamma+1)^{-1}. \tag{84}$$

For $z \in \mathcal{Z}$ we define the Araki-Wyss field operators on $\Gamma_{\mathrm{a}}(\mathcal{Z} \oplus \overline{\mathcal{Z}})$ as:

$$\begin{aligned}
\phi_{\gamma,\mathrm{l}}^{\mathrm{AW}}(z) &:= \phi\big((1-\chi)^{\frac{1}{2}} z, \overline{\chi}^{\frac{1}{2}} \overline{z}\big) , \\
\phi_{\gamma,\mathrm{r}}^{\mathrm{AW}}(\overline{z}) &:= \Lambda\phi\big(\chi^{\frac{1}{2}} z, (1-\overline{\chi})^{\frac{1}{2}} \overline{z}\big)\Lambda = \mathrm{i}I\phi\big(\mathrm{i}\chi^{\frac{1}{2}} z, \mathrm{i}(1-\overline{\chi})^{\frac{1}{2}} \overline{z}\big) ,
\end{aligned}$$

where recall that $\Lambda = (-1)^{N(N-1)/2}$ and $I = (-1)^N = \Gamma(-1)$. The maps $z \mapsto \phi_{\gamma,\mathrm{l}}(z)$, $\overline{z} \mapsto \phi_{\gamma,\mathrm{r}}(\overline{z})$, are called respectively the left and the right Araki-Wyss representation of the CAR associated to the pair $(\mathcal{Z}, \gamma)$. We denote by $\mathfrak{M}_{\gamma,\mathrm{l}}^{\mathrm{AW}}$ and $\mathfrak{M}_{\gamma,\mathrm{r}}^{\mathrm{AW}}$ the von Neumann algebras generated by $\{\phi_{\gamma,\mathrm{l}}(z) : z \in \mathcal{Z}\}$ and $\{\phi_{\gamma,\mathrm{r}}(\overline{z}) : z \in \mathcal{Z}\}$. They will be called respectively the *left* and the *right Araki-Wyss algebra.*

We drop the superscript AW until the end of the section.

In the following theorem we will describe some basic properties of the Araki-Wyss algebras.

Theorem 43. 1) $\mathcal{Z} \ni z \mapsto \phi_{\gamma,\mathrm{l}}(z) \in B_{\mathrm{h}}(\Gamma_{\mathrm{a}}(\mathcal{Z} \oplus \overline{\mathcal{Z}}))$ *is a representation of the CAR. In particular,*

$$[\phi_{\gamma,\mathrm{l}}(z_1), \phi_{\gamma,\mathrm{l}}(z_2)]_+ = 2\mathrm{Re}(z_1|z_2) .$$

The corresponding creation and annihilation operators belong to $\mathfrak{M}_{\gamma,\mathrm{l}}$ *and are given by*

$$\begin{aligned}
a_{\gamma,\mathrm{l}}^*(z) &= a^*\Big((1-\chi)^{\frac{1}{2}} z, 0\Big) + a\Big(0, \overline{\chi}^{\frac{1}{2}} \overline{z}\Big), \\
a_{\gamma,\mathrm{l}}(z) &= a\Big((1-\chi)^{\frac{1}{2}} z, 0\Big) + a^*\Big(0, \overline{\chi}^{\frac{1}{2}} \overline{z}\Big).
\end{aligned}$$

2) $\overline{\mathcal{Z}} \ni \overline{z} \mapsto \phi_{\gamma,\mathrm{r}}(\overline{z}) \in B_{\mathrm{h}}(\Gamma_{\mathrm{a}}(\mathcal{Z} \oplus \overline{\mathcal{Z}}))$ *is a representation of the CAR. In particular*

$$[\phi_{\gamma,\mathrm{r}}(\overline{z}_1), \phi_{\gamma,\mathrm{r}}(\overline{z}_2)]_+ = 2\mathrm{Re}(z_1|z_2) .$$

The corresponding creation and annihilation operators belong to $\mathfrak{M}_{\gamma,\mathrm{r}}$ *and are given by*

$$\begin{aligned}
a_{\gamma,\mathrm{r}}^*(z) &= \Lambda\Big(a(\chi^{\frac{1}{2}} z, 0) + a^*(0, (1-\overline{\chi})^{\frac{1}{2}} \overline{z})\Big)\Lambda, \\
a_{\gamma,\mathrm{r}}(z) &= \Lambda\Big(a^*(\chi^{\frac{1}{2}} z, 0) + a(0, (1-\overline{\chi})^{\frac{1}{2}} \overline{z})\Big)\Lambda.
\end{aligned}$$

3) *Set*

$$J_{\mathrm{a}} := \Lambda\Gamma(\epsilon) . \tag{85}$$

We have

$$J_{\rm a}\phi_{\gamma,{\rm l}}(z)J_{\rm a}=\phi_{\gamma,{\rm r}}(\overline{z})\ ,$$

$$J_{\rm a}a^*_{\gamma,{\rm l}}(z)J_{\rm a}=a^*_{\gamma,{\rm r}}(\overline{z})\ ,$$

$$J_{\rm a}a_{\gamma,{\rm l}}(z)J_{\rm a}=a_{\gamma,{\rm r}}(\overline{z}).$$

4) *The vacuum Ω is a fermionic quasi-free vector, the "two-point functions" are equal*

$$(\Omega|\phi_{\gamma,{\rm l}}(z_1)\phi_{\gamma,{\rm l}}(z_2)\Omega)=(z_1|z_2)-{\rm i}2{\rm Im}(z_1|\chi z_2),$$

$$\big(\Omega|a_{\gamma,{\rm l}}(z_1)a^*_{\gamma,{\rm l}}(z_2)\Omega\big)=(z_1|(1-\chi)z_2)=(z_1|(1+\gamma)^{-1}z_2)\ ,$$

$$\big(\Omega|a^*_{\gamma,{\rm l}}(z_1)a_{\gamma,{\rm l}}(z_2)\Omega\big)=(z_2|\chi z_1)=(z_2|\gamma(\gamma+1)^{-1}z_1)\ ,$$

$$\big(\Omega|a^*_{\gamma,{\rm l}}(z_1)a^*_{\gamma,{\rm l}}(z_2)\Omega\big)=0\ ,$$

$$\big(\Omega|a_{\gamma,{\rm l}}(z_1)a_{\gamma,{\rm l}}(z_2)\Omega\big)=0.$$

5) $\mathfrak{M}_{\gamma,{\rm l}}$ *is a factor.*
6) ${\rm Ker}\gamma={\rm Ker}\gamma^{-1}=\{0\}$ *(equivalently,* ${\rm Ker}\chi={\rm Ker}(1-\chi)=\{0\}$*) iff Ω is separating for $\mathfrak{M}_{\gamma,{\rm l}}$ iff Ω is cyclic for $\mathfrak{M}_{\gamma,{\rm l}}$. If this is the case, then the modular conjugation for Ω is given by (85) and the modular operator for Ω is given by*

$$\Delta=\Gamma\left(\gamma\oplus\overline{\gamma}^{-1}\right). \tag{86}$$

7) *We have*

$$\mathfrak{M}'_{\gamma,{\rm l}}=\mathfrak{M}_{\gamma,{\rm r}}\ . \tag{87}$$

8) *Set*

$$\Gamma^+_{{\rm a},\gamma}(\mathcal{Z}\oplus\overline{\mathcal{Z}}):=\{AJ_{\rm a}A\Omega\ :\ A\in\mathfrak{M}_{\gamma,{\rm l}}\}^{\rm cl}\ . \tag{88}$$

Then $(\mathfrak{M}_{\gamma,{\rm l}},\Gamma_{\rm a}(\mathcal{Z}\oplus\overline{\mathcal{Z}}),J_{\rm a},\Gamma^+_{{\rm a},\gamma}(\mathcal{Z}\oplus\overline{\mathcal{Z}}))$ is a W^*-algebra in the standard form.
9) *If γ has some continuous spectrum, then $\mathfrak{M}_{\gamma,{\rm l}}$ is a factor of type* ${\rm III}_1$.
10) *If $\mathcal{Z}$ is an infinite dimensional Hilbert space and $\gamma=\lambda$ or $\gamma=\lambda^{-1}$ with $\lambda\in]0,1[$, then $\mathfrak{M}_{\gamma,{\rm l}}$ is a factor of type* ${\rm III}_\lambda$, *[70].*
11) *If $\mathcal{Z}$ is an infinite dimensional Hilbert space and $\gamma=1$ (equivalently, $\chi=\frac{1}{2}$), then $\mathfrak{M}_{\gamma,{\rm l}}$ is a factor of type* ${\rm II}_1$.
12) *If $\gamma=0$ or $\gamma^{-1}=0$, (equivalently, $\chi=0$ or $\chi=1$), then $\mathfrak{M}_{\gamma,{\rm l}}$ is a factor of type* I.
13) *Let h be a self-adjoint operator on $\mathcal{Z}$ commuting with γ and*

$$\tau^t(\phi_{\gamma,{\rm l}}(z)):=\phi_{\gamma,{\rm l}}({\rm e}^{{\rm i}th}z)\ .$$

Then $t\mapsto\tau^t$ extends uniquely to a W^*-dynamics on $\mathfrak{M}_{\gamma,{\rm l}}$ and

$$L={\rm d}\Gamma(h\oplus(-\overline{h}))$$

is its standard Liouvillean.

14) *Ω is a (τ,β)-KMS vector iff* $\gamma = \mathrm{e}^{-\beta h}$.

Proof. 1)–4) follow by direct computations.

The proof of 5) will be divided into a number of steps.

Step 1. We have

$$\phi_{\gamma,\mathrm{l}}(z_1)\phi_{\gamma,\mathrm{r}}(\overline{z}_2) = \phi_{\gamma,\mathrm{r}}(\overline{z}_2)\phi_{\gamma,\mathrm{l}}(z_1) .$$

Consequently, $\mathfrak{M}_{\gamma,\mathrm{l}}$ and $\mathfrak{M}_{\gamma,\mathrm{r}}$ commute with one another.

Step 2. For simplicity, in this step we assume that $\mathcal{Z}$ is separable and $0 \leq \chi \leq \frac{1}{2}$; the generalization of the proof to the general case is easy. By a well known theorem (see e.g. [69], Ex. II.1.4), for any $\epsilon > 0$, we can find a self-adjoint operator ν such that $\mathrm{Tr}(\chi^{\frac{1}{2}} - \nu^{\frac{1}{2}})^*(\chi^{\frac{1}{2}} - \nu^{\frac{1}{2}}) < \epsilon^2$ and there exists an orthonormal basis $w_1, w_2, \ldots$ of eigenvectors of ν. Let $\nu w_i = \nu_i w_i$, $\nu_i \in \mathbb{R}$.

Introduce the operators

$$\begin{aligned}
A_j := {} & \phi\left((1-\nu_j)^{\frac{1}{2}} w_j, \overline{\nu}_j^{\frac{1}{2}} \overline{w}_j\right) \phi\left(\mathrm{i}(1-\nu_j)^{\frac{1}{2}} w_j, -\mathrm{i}\overline{\nu}_j^{\frac{1}{2}} \overline{w}_j\right) \\
& \times \Lambda\phi\left(\nu_j^{\frac{1}{2}} w_j, (1-\overline{\nu}_j)^{\frac{1}{2}} \overline{w}_j\right) \phi\left(\mathrm{i}\nu_j^{\frac{1}{2}} w_j, -\mathrm{i}(1-\overline{\nu}_j)^{\frac{1}{2}} \overline{w}_j\right) \Lambda \\
:= {} & \phi\left((1-\nu_j)^{\frac{1}{2}} w_j, \overline{\nu}_j^{\frac{1}{2}} \overline{w}_j\right) \phi\left(\mathrm{i}(1-\nu_j)^{\frac{1}{2}} w_j, -\mathrm{i}\overline{\nu}_j^{\frac{1}{2}} \overline{w}_j\right) \\
& \times \phi\left(\mathrm{i}\nu_j^{\frac{1}{2}} w_j, \mathrm{i}(1-\overline{\nu}_j)^{\frac{1}{2}} \overline{w}_j\right) \phi\left(-\nu_j^{\frac{1}{2}} w_j, (1-\overline{\nu}_j)^{\frac{1}{2}} \overline{w}_j\right) \\
= {} & \left(2a^*(w_j,0)a(w_j,0) - 1\right)\left(2a^*(0,\overline{w}_j)a(0,\overline{w}_j) - 1\right).
\end{aligned}$$

Note that A_j commute with one another and $\prod\limits_{j=1}^{\infty} A_j$ converges in the σ-weak topology to I.

Introduce also

$$\begin{aligned}
B_j := {} & \phi_{\gamma,\mathrm{l}}(w_j)\phi_{\gamma,\mathrm{l}}(\mathrm{i}w_j)\phi_{\gamma,\mathrm{r}}(\overline{w}_j)\phi_{\gamma,\mathrm{r}}(\mathrm{i}\overline{w}_j) \\
= {} & \left(2a^*\left((1-\chi)^{\frac{1}{2}} w_j, \overline{\chi}^{\frac{1}{2}} \overline{w}_j\right) a\left((1-\chi)^{\frac{1}{2}} w_j, \overline{\chi}^{\frac{1}{2}} \overline{w}_j\right) - 1\right) \\
& \times \left(2a^*\left(-\chi^{\frac{1}{2}} w_j, (1-\overline{\chi})^{\frac{1}{2}} \overline{w}_j\right) a\left(-\chi^{\frac{1}{2}} w_j, (1-\overline{\chi})^{\frac{1}{2}} \overline{w}_j\right) - 1\right)
\end{aligned}$$

Note that B_j belongs to the algebra generated by $\mathfrak{M}_{\gamma,\mathrm{l}}$ and $\mathfrak{M}_{\gamma,\mathrm{r}}$ and

$$\left\| \prod_{j=1}^{n} A_j - \prod_{j=1}^{n} B_j \right\| \leq c\epsilon .$$

This proves that

$$I \in \left(\mathfrak{M}_{\gamma,\mathrm{l}} \cup \mathfrak{M}_{\gamma,\mathrm{r}}\right)'' . \tag{89}$$

Step 3. We have

$$\left(\mathfrak{M}_{\gamma,\mathrm{l}} \cup \mathfrak{M}'_{\gamma,\mathrm{l}}\right)' \subset \left(\mathfrak{M}_{\gamma,\mathrm{l}} \cup \mathfrak{M}_{\gamma,\mathrm{r}}\right)' = \left(\mathfrak{M}_{\gamma,\mathrm{l}} \cup \mathfrak{M}_{\gamma,\mathrm{r}} \cup \{I\}\right)'$$
$$= \{\phi((1-\chi)^{\frac{1}{2}}w_1 - \chi^{\frac{1}{2}}w_2, \overline{\chi}^{\frac{1}{2}}\overline{w}_1 + (1-\chi)^{\frac{1}{2}}\overline{w}_2) \ : \ w_1, w_2 \in \mathcal{Z}\}'$$
$$= \{\phi(w_1, \overline{w}_2) \ : \ w_1, w_2 \in \mathcal{Z}\}' = \mathbb{C}1 ,$$

where at the beginning we used Step 1, then (89), next we used

$$\phi_{\gamma,\mathrm{r}}(\overline{z}) = \mathrm{i}\phi(-\chi^{\frac{1}{2}}\mathrm{i}z, -(1-\chi)^{\frac{1}{2}}\overline{\mathrm{i}z})I ,$$

$$\{(1-\chi)^{\frac{1}{2}}w_1 - \chi^{\frac{1}{2}}w_2, \overline{\chi}^{\frac{1}{2}}\overline{w}_1 + (1-\chi)^{\frac{1}{2}}\overline{w}_2) \ : \ w_1, w_2 \in \mathcal{Z}\} = \mathcal{Z} \oplus \overline{\mathcal{Z}} ,$$

and finally the irreducibility of fermionic fields. This shows that $\left(\mathfrak{M}_{\gamma,\mathrm{l}} \cup \mathfrak{M}'_{\gamma,\mathrm{l}}\right)' = \mathbb{C}1$, which means that $\mathfrak{M}_{\gamma,\mathrm{l}}$ is a factor and ends the proof of 5).

The proof of the $\Rightarrow$ part of 6) is similar to its bosonic analog. Assume that $\mathrm{Ker}\gamma = \{0\}$. Set

$$\tau^t(A) := \Gamma(\gamma, \overline{\gamma}^{-1})^{\mathrm{i}t} A \Gamma(\gamma, \overline{\gamma}^{-1})^{-\mathrm{i}t} .$$

We first check that τ^t preserves $\mathfrak{M}_{\gamma,\mathrm{l}}$. Therefore, it is a W^*-dynamics on $\mathfrak{M}_{\gamma,\mathrm{l}}$.

Next we check that $(\Omega| \cdot \Omega)$ satisfies the $(\tau, -1)$-KMS condition. This is straightforward for the Weyl operators $\phi_{\gamma,\mathrm{l}}(z)$. Therefore, it holds for the $*-$algebra $\mathfrak{M}_{\gamma,\mathrm{l},0}$ of polynomials in $\phi_{\gamma,\mathrm{l}}(z)$. By the Kaplansky Theorem, the unit ball of $\mathfrak{M}_{\gamma,\mathrm{l},0}$ is σ-weakly dense in the unit ball of $\mathfrak{M}_{\gamma,\mathrm{l}}$. Using this we extend the KMS condition to $\mathfrak{M}_{\gamma,\mathrm{l}}$.

A KMS state on a factor is always faithful. By 5), $\mathfrak{M}_{\gamma,\mathrm{l}}$ is a factor. Hence Ω is separating.

Let $\mathcal{H}$ be the closure of $\mathfrak{M}_{\gamma,\mathrm{l}}\Omega$. $\mathcal{H}$ is invariant for $\mathfrak{M}_{\gamma,\mathrm{l}}$, moreover Ω is cyclic and separating for $\mathfrak{M}_{\gamma,\mathrm{l}}$ on $\mathcal{H}$. Computing on polynomials in $\phi_{\gamma,\mathrm{l}}(z)$ acting on Ω, we check that $\Gamma(\gamma, \overline{\gamma}^{-1})$ and J_{a} preserve $\mathcal{H}$, the modular conjugation for Ω is given by $J_{\mathrm{a}}\Big|_{\mathcal{H}}$ and the modular operator equals $\Gamma(\gamma, \overline{\gamma}^{-1})\Big|_{\mathcal{H}}$. Now

$$\phi_{\gamma,\mathrm{l}}(z_1) \cdots \phi_{\gamma,\mathrm{l}}(z_n) J_{\mathrm{a}} \phi_{\gamma,\mathrm{l}}(z'_1) \cdots \phi_{\gamma,\mathrm{l}}(z'_m)\Omega$$
$$= \phi_{\gamma,\mathrm{l}}(z_1) \cdots \phi_{\gamma,\mathrm{l}}(z_n) \phi_{\gamma,\mathrm{r}}(\overline{z}'_1) \cdots \phi_{\gamma,\mathrm{r}}(\overline{z}'_m)\Omega .$$

Thus, $\mathfrak{M}_{\gamma,\mathrm{l}} J_{\mathrm{a}} \mathfrak{M}_{\gamma,\mathrm{l}}\Omega$ is dense in $\Gamma_{\mathrm{a}}(\mathcal{Z} \oplus \overline{\mathcal{Z}})$. But $\mathfrak{M}_{\gamma,\mathrm{l}} J_{\mathrm{a}} \mathfrak{M}_{\gamma,\mathrm{l}}\Omega \subset \mathcal{H}$. Hence Ω is cyclic and $\mathcal{H} = \Gamma_{\mathrm{a}}(\mathcal{Z} \oplus \overline{\mathcal{Z}})$. This proves 6).

In this section, we will prove 7) only under the assumption $\mathrm{Ker}\gamma = \mathrm{Ker}\gamma^{-1} = \{0\}$. By 5) this implies that $(\Omega| \cdot \Omega)$ is faithful and we can apply the modular theory, which gives $J_{\mathrm{a}}\mathfrak{M}_{\gamma,\mathrm{l}}J_{\mathrm{a}} = \mathfrak{M}'_{\gamma,\mathrm{l}}$. By 3) we have $J_{\mathrm{a}}\mathfrak{M}_{\gamma,\mathrm{l}}J_{\mathrm{a}} = \mathfrak{M}_{\gamma,\mathrm{r}}$.

7) for a general γ will follow from Theorem 55, proven later. □

10.3 Quasi-Free Representations of the CAR as the Araki-Wyss Representations

There is a simple condition, which allows to check whether a given quasi-free representation of the CAR is unitarily equivalent to an Araki-Wyss representation:

Theorem 44. *Suppose that* $(\mathcal{Y}, \alpha)$ *is a real Hilbert space,*

$$\mathcal{Y} \ni y \mapsto \phi(y) \in B_{\mathrm{h}}(\mathcal{H}) \tag{90}$$

is a representation of the CAR and Ψ *a cyclic quasi-free vector for (90). Let* ω *be defined by (83). Suppose that* $\mathrm{Ker}\omega$ *is even or infinite dimensional. Then there exists a complex Hilbert space* $\mathcal{Z}$*, an operator* γ *on* $\mathcal{Z}$ *satisfying* $0 \leq \gamma \leq 1$*, and an isometric operator* $U : \mathcal{H} \to \Gamma_{\mathrm{a}}(\mathcal{Z} \oplus \overline{\mathcal{Z}})$ *such that*

$$U\Psi = \Omega \,,$$

$$U\phi(y) = \phi_{\gamma,1}(y)U, \quad y \in \mathcal{Y}.$$

$\mathcal{Z}$ *equipped with the real part of its scalar product coincides with* $(\mathcal{Y}, \alpha)$.

Proof. By Theorem 42 2), there exists an antisymmetric operator μ such that $\|\mu\| \leq 1$ and

$$y_1 \omega y_2 = 2 y_1 \alpha \mu y_2 \,.$$

Let $\mathcal{Y}_0 := \mathrm{Ker}\mu$ and $\mathcal{Y}_1$ be its orthogonal complement. On $\mathcal{Y}_1$ we can make the polar decomposition

$$\mu = |\mu| j = j |\mu| \,,$$

j is an orthogonal operator such that $j^2 = -1$. Thus j is a complex structure. If $\dim \mathcal{Y}_0$ is even or infinite, then we can extend j to a complex structure on $\mathcal{Y}$. Interpreting $-j$ as the complex structure, we convert $\mathcal{Y}$ into a complex space, which will be denoted by $\mathcal{Z}$, and equip it with the (sesquilinear) scalar product

$$(y_1 | y_2) := y_1 \alpha y_2 + \mathrm{i} y_1 \alpha j y_2 \,.$$

Now $\chi := \frac{1}{2}(1 - |\mu|)$ and $\gamma := \chi(1-\chi)^{-1}$ define operators on $\mathcal{Z}$ such that $0 \leq \chi \leq \frac{1}{2}$ and $0 \leq \gamma \leq 1$. We have

$$\begin{aligned}
(\Psi | \phi(y_1)\phi(y_2)\Psi) &= y_1 \alpha y_2 + \tfrac{\mathrm{i}}{2} y_1 \omega y_2 \\
&= y_1 \alpha y_2 + \mathrm{i} y_1 \alpha j y_2 + \mathrm{i} y_1 \alpha j (|\mu| - 1) y_2 \\
&= (y_1 | y_2) - 2\mathrm{i}\mathrm{Im}(y_1 | \chi y_2) \\
&= (\Omega | \phi_{\gamma,1}(y_1)\phi_{\gamma,1}(y_2)\Omega).
\end{aligned}$$

Now we see that

$$U\phi(y)\Psi := \phi_{\gamma,1}(y)\Omega$$

extends to an isometric operator intertwining the representation (90) with the Araki-Wyss representation for $(\mathcal{Z}, \gamma)$. □

10.4 Tracial Quasi-Free Representations

Tracial representations of the CAR are the fermionic analogs of classical quasifree representations of the CCR.

Consider a real Hilbert space $\mathcal{V}$. Let $\mathcal{W} = \mathcal{V} \oplus \mathrm{i}\mathcal{V}$ be its complexification. Let κ denote the natural conjugation on $\mathcal{W}$, which means $\kappa(v_1 + \mathrm{i}v_2) := v_1 - \mathrm{i}v_2$, $v_1, v_2 \in \mathcal{V}$. Consider the pair of representations of the CAR

$$\begin{aligned} \mathcal{V} \ni v &\mapsto \phi_{\mathcal{V},\mathrm{l}}(v) := \phi(v) \in B_{\mathrm{h}}(\Gamma_{\mathrm{a}}(\mathcal{W}))\ , \\ \mathcal{V} \ni v &\mapsto \phi_{\mathcal{V},\mathrm{r}}(v) := \Lambda\phi(v)\Lambda \in B_{\mathrm{h}}(\Gamma_{\mathrm{a}}(\mathcal{W})). \end{aligned} \tag{91}$$

Let $\mathfrak{M}_{\mathcal{V},\mathrm{l}}$ and $\mathfrak{M}_{\mathcal{V},\mathrm{r}}$ be the von Neumann algebras generated by $\{\phi_{\mathcal{V},\mathrm{l}}(v) : v \in \mathcal{V}\}$ and $\{\phi_{\mathcal{V},\mathrm{r}}(v) : v \in \mathcal{V}\}$ respectively.

Theorem 45. 1) *(91) are two commuting representations of the CAR:*

$$\begin{aligned} [\phi_{\mathcal{V},\mathrm{l}}(v_1), \phi_{\mathcal{V},\mathrm{l}}(v_2)]_+ &= 2(v_1|v_2)\ , \\ [\phi_{\mathcal{V},\mathrm{r}}(v_1), \phi_{\mathcal{V},\mathrm{r}}(v_2)]_+ &= 2(v_1|v_2)\ , \\ [\phi_{\mathcal{V},\mathrm{l}}(v_1), \phi_{\mathcal{V},\mathrm{r}}(v_2)] &= 0. \end{aligned}$$

2) *Set*

$$J_{\mathrm{a}} := \Lambda\Gamma(\kappa)\ . \tag{92}$$

We have

$$J_{\mathrm{a}}\phi_{\mathcal{V},\mathrm{l}}(v)J_{\mathrm{a}} = \phi_{\mathcal{V},\mathrm{r}}(v)\ .$$

3) *Ω is a quasi-free vector for (91) with the 2-point function*

$$(\Omega|\phi_{\mathcal{V},\mathrm{l}}(v_1)\phi_{\mathcal{V},\mathrm{l}}(v_2)\Omega) = (v_1|v_2)\ .$$

4) *Ω is cyclic and separating on $\mathfrak{M}_{\mathcal{V},\mathrm{l}}$. $(\Omega|\cdot\,\Omega)$ is tracial, which means*

$$(\Omega|AB\Omega) = (\Omega|BA\Omega), \quad A, B \in \mathfrak{M}_{\mathcal{V},\mathrm{l}}\ .$$

The corresponding modular conjugation is given by (92) and the modular operator equals $\Delta = 1$.

5) *We have $\mathfrak{M}_{\mathcal{V},\mathrm{r}} = \mathfrak{M}_{\mathcal{V},\mathrm{l}}'$.*

6) *If* $\dim\mathcal{V}$ *is even or infinite, then the tracial representations of the CAR are unitarily equivalent to the Araki-Wyss representations with $\gamma = 1$ (equivalently, $\chi = \frac{1}{2}$).*

7) *If* $\dim\mathcal{V}$ *is odd, then the center of $\mathfrak{M}_{\mathcal{V},\mathrm{l}}$ is 2-dimensional: it is spanned by* 1 *and Q introduced in Theorem 9.*

10.5 Putting Together an Araki-Wyss and a Tracial Representation

A general quasifree representation of the CAR can be obtained by putting together an Araki-Wyss representation and a tracial representation. Actually, one can restrict oneself to a tracial representation with just one dimensional $\mathcal{V}$, but we will consider the general case.

Let $\mathcal{Z}$, γ be as in the subsection on Araki-Wyss representations and $\mathcal{V}, \mathcal{W}$ be as in the subsection on tracial representations. Define the following operators on $\Gamma_{\mathrm{a}}(\mathcal{Z} \oplus \overline{\mathcal{Z}} \oplus \mathcal{W})$

$$\begin{aligned} \mathcal{Z} \oplus \mathcal{V} \ni (z, v) &\mapsto \phi^{\mathrm{AW}}_{\mathcal{V},\gamma,\mathrm{l}}(z, v) := \phi((1-\chi)^{\frac{1}{2}} z, \overline{\chi}^{\frac{1}{2}} \overline{z}, v) , \\ \overline{\mathcal{Z}} \oplus \mathcal{V} \ni (\overline{z}, v) &\mapsto \phi^{\mathrm{AW}}_{\mathcal{V},\gamma,\mathrm{r}}(\overline{z}, v) := \Lambda \phi(\chi^{\frac{1}{2}} z, (1-\overline{\chi})^{\frac{1}{2}} \overline{z}, v) \Lambda , \end{aligned} \tag{93}$$

(We drop the superscript AW in what follows).

Theorem 46. 1) *(93) are two commuting representations of the CAR:*

$$\begin{aligned} [\phi_{\mathcal{V},\gamma,\mathrm{l}}(z_1, v_1), \phi_{\mathcal{V},\gamma,\mathrm{l}}(z_2, v_2)]_+ &= 2\mathrm{Re}(z_1|z_2) + 2(v_1|v_2) , \\ [\phi_{\mathcal{V},\gamma,\mathrm{r}}(\overline{z}_1, v_1), \phi_{\mathcal{V},\gamma,\mathrm{r}}(\overline{z}_2, v_2)]_+ &= 2\mathrm{Re}(z_1|z_2) + 2(v_1|v_2) , \\ [\phi_{\mathcal{V},\gamma,\mathrm{l}}(z_1, v_1), \phi_{\mathcal{V},\gamma,\mathrm{r}}(\overline{z}_2, v_2)] &= 0. \end{aligned}$$

2) *Set*

$$J_{\mathrm{a}} := \Lambda \Gamma(\epsilon \oplus \kappa). \tag{94}$$

We have

$$J_{\mathrm{a}} \phi_{\mathcal{V},\gamma,\mathrm{l}}(v) J_{\mathrm{a}} = \phi_{\mathcal{V},\gamma,\mathrm{r}}(v) .$$

3) *Ω is a quasi-free vector for with the 2-point function*

$$(\Omega|\phi_{\mathcal{V},\gamma,\mathrm{l}}(z_1, v_1)\phi_{\mathcal{V},\gamma,l}(z_2, v_2)\Omega) = (z_1|z_2) - 2\mathrm{i}\mathrm{Im}(z_1|\chi z_2) + (v_1|v_2) .$$

4) $\mathrm{Ker}\gamma = \mathrm{Ker}\gamma^{-1} = \{0\}$ *(equivalently,* $\mathrm{Ker}\chi = \mathrm{Ker}(1-\chi) = \{0\}$*) iff Ω is separating on $\mathfrak{M}_{\mathcal{V},\gamma,\mathrm{l}}$ iff Ω is cyclic on $\mathfrak{M}_{\mathcal{V},\gamma,\mathrm{l}}$. If this is the case, the corresponding modular conjugation is given by (94) and the modular operator equals $\Delta = \Gamma(\gamma \oplus \overline{\gamma}^{-1} \oplus 1)$.*
5) $\mathfrak{M}_{\mathcal{V},\gamma,\mathrm{r}} = \mathfrak{M}'_{\mathcal{V},\gamma,\mathrm{l}}$.
6) *If* $\dim \mathcal{V}$ *is even or infinite, then the representations (93) are unitarily equivalent to the Araki-Wyss representations.*
7) *If* $\dim \mathcal{Z}$ *is finite and* $\dim \mathcal{V}$ *is odd, then the center of $\mathfrak{M}_{\mathcal{V},\mathrm{l}}$ is 2-dimensional: it is spanned by* 1 *and Q introduced in Theorem 9.*

11 Confined Bose and Fermi Gas

Sometimes the Araki-Woods representation of the CCR and the Araki-Wyss representations of the CAR are equivalent to a multiple of the Fock representation and the corresponding W^*-algebra is type I. This happens e.g. in the case of a finite number of degrees of freedom. More generally, this holds if

$$\mathrm{Tr}\gamma < \infty \,. \tag{95}$$

Representations satisfying this condition will be called "confined".

Let us explain the name "confined". Consider free Bose or Fermi gas with the Hamiltonian equal to $\mathrm{d}\Gamma(h)$, where h is the 1-particle Hamiltonian. One can argue that in the physical description of this system stationary quasi-free states are of special importance. They are given by density matrices of the form

$$\Gamma(\gamma)/\mathrm{Tr}\Gamma(\gamma) \,, \tag{96}$$

with γ commuting with h. In particular, γ can have the form $\mathrm{e}^{-\beta h}$, in which case (96) is the Gibbs state at inverse temperature β.

For (96) to make sense, $\mathrm{Tr}\Gamma(\gamma)$ has to be finite. As we will see later on, $\mathrm{Tr}\Gamma(\gamma) < \infty$ is equivalent to (95).

A typical 1-particle Hamiltonian h of free Bose or Fermi gas is the Laplacian with, say, Dirichlet boundary conditions at the boundary of its domain. If the domain is unbounded, then, usually, the spectrum of h is continuous and, therefore, there are no non-zero operators γ that commute with h and satisfy (95). If the domain is bounded ("confined"), then the spectrum of h is discrete, and hence many such operators γ exist. In particular, $\gamma = \mathrm{e}^{-\beta h}$ has this property. This is the reason why we call "confined" the free Bose or Fermi gas satisfying (95).

In this section we will show how Araki-Woods and Araki-Wyss representations arise in confined systems. We will construct a natural intertwiner between the Araki-Woods/Araki-Wyss representations and the Fock representation. We will treat the bosonic and fermionic case parallel. Whenever possible, we will use the same formula to describe both the bosonic and fermionic case. Some of the symbols will denote different things in the bosonic/fermionic cases (e.g. the fields $\phi(z)$); others will have subscripts s/a indicating the two possible meanings. Sometimes there will be signs $\pm$ or $\mp$ indicating the two possible versions of the formula, the upper in the bosonic case, the lower in the fermionic case.

11.1 Irreducible Representation

In this subsection we consider the W^*-algebra $B(\Gamma_{\mathrm{s/a}}(\mathcal{Z}))$ acting in the obvious way on the Hilbert space $\Gamma_{\mathrm{s/a}}(\mathcal{Z})$. Recall that the W^*-algebra $B(\Gamma_{\mathrm{s}}(\mathcal{Z}))$ is generated by the representation of the CCR

$$\mathcal{Z} \ni z \mapsto W(z) = \mathrm{e}^{\mathrm{i}\phi(z)} \in U(\Gamma_{\mathrm{s}}(\mathcal{Z}))$$

and the W^*-algebra $B(\Gamma_{\mathrm{a}}(\mathcal{Z}))$ is generated by the representation of CAR

$$\mathcal{Z} \ni z \mapsto \phi(z) \in B_{\mathrm{h}}(\Gamma_{\mathrm{a}}(\mathcal{Z})) \; .$$

In both bosonic and fermionic cases we will also use a certain operator γ on $\mathcal{Z}$.

Recall that in the bosonic case, γ satisfies $0 \leq \gamma \leq 1$, $\mathrm{Ker}(1-\gamma) = \{0\}$, and we introduce the 1-particle density operator denoted ρ, as in (73).

In the fermionic case, γ is a positive operator, possibly with a non-dense domain, and we introduce the 1-particle density denoted χ, as in (84).

Throughout the section we assume that γ is trace class. In the bosonic case it is equivalent to assuming that ρ is trace class. We have

$$\mathrm{Tr}\Gamma(\gamma) = \det(1-\gamma)^{-1} = \det(1+\rho) \; .$$

In the fermionic case, if we assume that $\mathrm{Ker}\gamma^{-1} = \{0\}$ (or $\mathrm{Ker}(\chi - 1) = \{0\}$), γ is trace class iff χ is trace class. We have

$$\mathrm{Tr}\Gamma(\gamma) = \det(1+\gamma) = \det(1-\chi)^{-1} \; .$$

Define the state ω_γ on the W^*-algebra $B(\Gamma_{\mathrm{s/a}}(\mathcal{Z}))$ given by the density matrix

$$\Gamma(\gamma)/\mathrm{Tr}\Gamma(\gamma) \; .$$

Let h be another self-adjoint operator on $\mathcal{Z}$. Define the dynamics on $B(\Gamma_{\mathrm{s/a}}(\mathcal{Z}))$:

$$\tau^t(A) := \mathrm{e}^{\mathrm{i}t\mathrm{d}\Gamma(h)} A \mathrm{e}^{-\mathrm{i}t\mathrm{d}\Gamma(h)} \; , \qquad A \in B(\Gamma_{\mathrm{s}}(\mathcal{Z})) \; .$$

Clearly, ω_γ is τ-invariant iff h commutes with γ.

The state ω_γ is (β, τ)-KMS iff γ is proportional to $\mathrm{e}^{-\beta h}$.

11.2 Standard Representation

We need to identify the complex conjugate of the Fock space $\overline{\Gamma_{\mathrm{s/a}}(\mathcal{Z})}$ with the Fock space over the complex conjugate $\Gamma_{\mathrm{s/a}}(\overline{\mathcal{Z}})$. In the bosonic case this is straightforward. In the fermionic case, however, we will not use the naive identification, but the identification that "reverses the order of particles", consistent with the convention adopted in Subsect. 2.3. More precisely, if $z_1, \dots, z_n \in \mathcal{Z}$, then the identification looks as follows:

$$\begin{aligned} \overline{\Gamma_{\mathrm{a}}^n(\mathcal{Z})} \ni \overline{z_1 \otimes_{\mathrm{a}} \cdots \otimes_{\mathrm{a}} z_n} &\mapsto V\, \overline{z_1 \otimes_{\mathrm{a}} \cdots \otimes_{\mathrm{a}} z_n} \\ &:= \overline{z}_n \otimes_{\mathrm{a}} \cdots \otimes_{\mathrm{a}} \overline{z}_1 \in \Gamma_{\mathrm{a}}^n(\overline{\mathcal{Z}}). \end{aligned} \tag{97}$$

(Thus the identification $V : \overline{\Gamma_{\mathrm{a}}(\mathcal{Z})} \to \Gamma_{\mathrm{a}}(\overline{\mathcal{Z}})$ equals Λ times the naive, "non-reversing", identification).

Using (97) at the second step and the exponential law at the last step, we have the identification

$$\begin{aligned} B^2(\Gamma_{\mathrm{s/a}}(\mathcal{Z})) &\simeq \Gamma_{\mathrm{s/a}}(\mathcal{Z}) \otimes \overline{\Gamma_{\mathrm{s/a}}(\mathcal{Z})} \\ &\simeq \Gamma_{\mathrm{s/a}}(\mathcal{Z}) \otimes \Gamma_{\mathrm{s/a}}(\overline{\mathcal{Z}}) \simeq \Gamma_{\mathrm{s/a}}(\mathcal{Z} \oplus \overline{\mathcal{Z}}). \end{aligned} \tag{98}$$

As before, define

$$J_{\mathrm{s}} := \Gamma(\epsilon), \quad J_{\mathrm{a}} := \Lambda\Gamma(\epsilon)\ .$$

Theorem 47. *In the bosonic/fermionic case, under the above identification, the hermitian conjugation $*$ becomes $J_{\mathrm{s/a}}$.*

Proof. We restrict ourselves to the fermionic case. Consider

$$B = |z_1 \otimes_{\mathrm{a}} \cdots \otimes_{\mathrm{a}} z_n)(w_1 \otimes_{\mathrm{a}} \cdots \otimes_{\mathrm{a}} w_m| \in B^2(\Gamma_{\mathrm{a}}(\mathcal{Z}))\ .$$

It corresponds to

$$\begin{aligned} &\sqrt{(n+m)!}z_1 \otimes_{\mathrm{a}} \cdots \otimes_{\mathrm{a}} z_n \otimes_{\mathrm{a}} \overline{w_1 \otimes_{\mathrm{a}} \cdots \otimes_{\mathrm{a}} w_m} \\ = &\sqrt{(n+m)!}z_1 \otimes_{\mathrm{a}} \cdots \otimes_{\mathrm{a}} z_n \otimes_{\mathrm{a}} \overline{w}_m \otimes_{\mathrm{a}} \cdots \otimes_{\mathrm{a}} \overline{w}_1 \in \Gamma_{\mathrm{a}}(\mathcal{Z} \oplus \overline{\mathcal{Z}}). \end{aligned}$$

On the other hand,

$$B^* = |w_1 \otimes_{\mathrm{a}} \cdots \otimes_{\mathrm{a}} w_m)(z_1 \otimes_{\mathrm{a}} \cdots \otimes_{\mathrm{a}} z_n|$$

corresponds to

$$\begin{aligned} &\sqrt{(n+m)!}w_1 \otimes_{\mathrm{a}} \cdots \otimes_{\mathrm{a}} w_m \otimes_{\mathrm{a}} \overline{z_1 \otimes_{\mathrm{a}} \cdots \otimes_{\mathrm{a}} z_n} \\ = &\sqrt{(n+m)!}w_1 \otimes_{\mathrm{a}} \cdots \otimes_{\mathrm{a}} w_m \otimes_{\mathrm{a}} \overline{z}_n \otimes_{\mathrm{a}} \cdots \otimes_{\mathrm{a}} \overline{z}_1 \\ = &(-1)^{\frac{n(n-1)}{2} + \frac{m(m-1)}{2} + nm} \sqrt{(n+m)!}\overline{z}_1 \otimes_{\mathrm{a}} \cdots \otimes_{\mathrm{a}} \overline{z}_n \otimes_{\mathrm{a}} w_m \otimes_{\mathrm{a}} \cdots \otimes_{\mathrm{a}} w_1 \\ = &\Lambda\Gamma(\epsilon)\sqrt{(n+m)!}z_1 \otimes_{\mathrm{a}} \cdots \otimes_{\mathrm{a}} z_n \otimes_{\mathrm{a}} \overline{w}_m \otimes_{\mathrm{a}} \cdots \otimes_{\mathrm{a}} \overline{w}_1\ , \end{aligned}$$

where at the last step we used $\Gamma(\epsilon)z_i = \overline{z}_i$, $\Gamma(\epsilon)\overline{w}_i = w_i$ and

$$\tfrac{n(n-1)}{2} + \tfrac{m(m-1)}{2} + nm = \tfrac{(n+m)(n+m-1)}{2}.$$

□

The W^*-algebras $B(\Gamma_{\mathrm{s/a}}(\mathcal{Z}))$ and $\overline{B(\Gamma_{\mathrm{s/a}}(\mathcal{Z}))}$ have a natural standard representation in the Hilbert space $B^2(\Gamma_{\mathrm{s/a}}(\mathcal{Z}))$, as described in Subsect. 8.5.

They have also a natural representation in the Hilbert space $\Gamma_{s/a}(\mathcal{Z})\otimes\overline{\Gamma_{s/a}(\mathcal{Z})}$, as described in Subsect. 8.6. Using the identification (98) we obtain the representation θ_l of $B(\Gamma_{s/a}(\mathcal{Z}))$ and θ_r of $\overline{B(\Gamma_{s/a}(\mathcal{Z}))}$ in the space $\Gamma_{s/a}(\mathcal{Z}\oplus\overline{\mathcal{Z}})$.

Let us describe the last pair of representations in detail. Let

$$U : \Gamma_{s/a}(\mathcal{Z}) \otimes \Gamma_{s/a}(\overline{\mathcal{Z}}) \to \Gamma_{s/a}(\mathcal{Z} \oplus \overline{\mathcal{Z}}) \tag{99}$$

be the unitary map defined as in (36). Let V be defined in (97). Then

$$\begin{aligned} B(\Gamma_{s/a}(\mathcal{Z})) \ni A \mapsto \theta_l(A) &:= U\, A\otimes 1_{\Gamma_{s/a}(\overline{\mathcal{Z}})}\, U^* \in B(\Gamma_{s/a}(\mathcal{Z} \oplus \overline{\mathcal{Z}}))\,, \\ \overline{B(\Gamma_s(\mathcal{Z}))} \ni \overline{A} \mapsto \theta_r(\overline{A}) &:= U\, 1_{\Gamma_s(\mathcal{Z})}\otimes\overline{A}\, U^* \in B(\Gamma_s(\mathcal{Z} \oplus \overline{\mathcal{Z}}))\,, \\ \overline{B(\Gamma_a(\mathcal{Z}))} \ni \overline{A} \mapsto \theta_r(\overline{A}) &:= U\, 1_{\Gamma_a(\mathcal{Z})}\otimes(V\overline{A}V^*)\, U^* \in B(\Gamma_a(\mathcal{Z} \oplus \overline{\mathcal{Z}})). \end{aligned} \tag{100}$$

We have 2 commuting representations of the CCR

$$\mathcal{Z} \ni z \mapsto W(z,0) = \theta_l(W(z)) \in U(\Gamma_s(\mathcal{Z} \oplus \overline{\mathcal{Z}})), \tag{101}$$

$$\overline{\mathcal{Z}} \ni \overline{z} \mapsto W(0,\overline{z}) = \theta_r(\overline{W(z)}) \in U(\Gamma_s(\mathcal{Z} \oplus \overline{\mathcal{Z}}))\,. \tag{102}$$

The algebra $\theta_l(B(\Gamma_s(\mathcal{Z})))$ is generated by the image of (101) and the algebra $\theta_r(B(\Gamma_s(\overline{\mathcal{Z}})))$ is generated by the image of (102).

We have also 2 commuting representations of the CAR

$$\mathcal{Z} \ni z \mapsto \phi(z,0) = \theta_l(\phi(z)) \in B_h(\Gamma_a(\mathcal{Z} \oplus \overline{\mathcal{Z}})), \tag{103}$$

$$\overline{\mathcal{Z}} \ni \overline{z} \mapsto \Lambda\phi(0,\overline{z})\Lambda = \theta_r(\overline{\phi(z)}) \in B_h(\Gamma_a(\mathcal{Z} \oplus \overline{\mathcal{Z}})). \tag{104}$$

The algebra $\theta_l(B(\Gamma_a(\mathcal{Z})))$ is generated by the image of (103) and the algebra $\theta_r(B(\Gamma_a(\overline{\mathcal{Z}})))$ is generated by the image of (104).

Let $\Gamma^+_{s/a}(\mathcal{Z} \oplus \overline{\mathcal{Z}})$ be the image of $B^2_+(\Gamma_{s/a}(\mathcal{Z}))$ under the identification (98).

Theorem 48. 1) $\left(\theta_l, \Gamma_{s/a}(\mathcal{Z} \oplus \overline{\mathcal{Z}}), J_{s/a}, \Gamma^+_{s/a}(\mathcal{Z} \oplus \overline{\mathcal{Z}})\right)$ *is a standard representation of* $B(\Gamma_{s/a}(\mathcal{Z}))$.

2) $J_{s/a}\theta_l(A)J_{s/a} = \theta_r(\overline{A})$.

3) $\mathrm{d}\Gamma(h\oplus(-\overline{h}))$ *is the standard Liouvillean of* $t \mapsto \tau^t$ *in this representation*

4) *The standard vector representative of* ω_γ *in this representation is*

$$\Omega_\gamma := \det(1 \mp \gamma)^{\pm\frac{1}{2}} \exp\left(\frac{1}{2}a^*\left(\begin{bmatrix} 0 & \gamma^{\frac{1}{2}} \\ \pm\overline{\gamma}^{\frac{1}{2}} & 0 \end{bmatrix}\right)\right)\Omega. \tag{105}$$

Proof. 1), 2) and 3) are straightforward. Let us prove 4), which is a little involved, since we have to use various identifications we have introduced.

In the representation of Subsect. 8.5, the standard vector representative of ω_γ equals

$$
\begin{aligned}
&(\mathrm{Tr}\Gamma(\gamma))^{-\frac{1}{2}}\Gamma(\gamma^{\frac{1}{2}})\\
&=(\mathrm{Tr}\Gamma(\gamma))^{-\frac{1}{2}}\sum_{n=0}^{\infty}\Theta_{\mathrm{s/a}}^{n}(\gamma^{\frac{1}{2}})^{\otimes n}\Theta_{\mathrm{s/a}}^{n}\in B^2(\Gamma_{\mathrm{s/a}}(\mathcal{Z})).
\end{aligned}
$$

(Recall that $\Theta_{\mathrm{s/a}}^{n}$ denotes the orthogonal projection onto $\Gamma_{\mathrm{s/a}}^{n}(\mathcal{Z})$).

Clearly, $\gamma^{1/2}\in B^2(\mathcal{Z})$ corresponds to a certain vector $\Psi\in\mathcal{Z}\otimes\overline{\mathcal{Z}}$.

Let σ be the permutation of $(1,\dots,2n)$ given by

$$
\sigma(2j-1)=j,\quad \sigma(2j)=2n-j+1,\quad j=1,\dots,n\,.
$$

This permutation defines the unitary transformation

$$
\Theta(\sigma):(\mathcal{Z}\otimes\overline{\mathcal{Z}})^{\otimes n}\to(\otimes^{n}\mathcal{Z})\otimes\left(\otimes^{n}\overline{\mathcal{Z}}\right)\,.
$$

Now $(\gamma^{1/2})^{\otimes n}$ can be interpreted in two fashions. It can be interpreted as an element of $\otimes^{n}B^2(\mathcal{Z})$, and then it corresponds to the vector $\Psi^{\otimes n}\in(\mathcal{Z}\otimes\overline{\mathcal{Z}})^{\otimes n}$. It can be also interpreted as an element of $B^2(\mathcal{Z}^{\otimes n})$ and then it corresponds to

$$
\Theta(\sigma)\Psi^{\otimes n}\in(\otimes^{n}\mathcal{Z})\otimes\left(\otimes^{n}\overline{\mathcal{Z}}\right)\simeq\otimes^{n}\mathcal{Z}\otimes\left(\overline{\otimes^{n}\mathcal{Z}}\right)\,.
$$

(Note that we have taken into account the convention about the complex conjugate of the tensor product adopted in Subsect. 2.3).

Now $\Theta_{\mathrm{s/a}}^{n}(\gamma^{1/2})^{\otimes n}\Theta_{\mathrm{s/a}}^{n}\in B^2(\Gamma_{\mathrm{s/a}}^{n}(\mathcal{Z}))$ corresponds to

$$
\left(\Theta_{\mathrm{s/a}}^{n}\otimes\overline{\Theta_{\mathrm{s/a}}^{n}}\right)\,\Theta(\sigma)\,\Psi^{\otimes n}\in\Gamma_{\mathrm{s/a}}^{n}(\mathcal{Z})\otimes\Gamma_{\mathrm{s/a}}^{n}(\overline{\mathcal{Z}})\,. \tag{106}
$$

The identification

$$
U:\Gamma_{\mathrm{s/a}}^{n}(\mathcal{Z})\otimes\Gamma_{\mathrm{s/a}}^{n}(\overline{\mathcal{Z}})\to\Gamma_{\mathrm{s/a}}^{2n}(\mathcal{Z}\oplus\overline{\mathcal{Z}})
$$

is obtained by first treating $\Gamma_{\mathrm{s/a}}^{n}(\mathcal{Z})\otimes\Gamma_{\mathrm{s/a}}^{n}(\overline{\mathcal{Z}})$ as a subspace of $\otimes^{2n}(\mathcal{Z}\oplus\overline{\mathcal{Z}})$ and then applying $\frac{\sqrt{(2n)!}}{n!}\,\Theta_{\mathrm{s/a}}^{2n}$. Therefore, (106) is identified with

$$
\begin{aligned}
&\frac{\sqrt{(2n)!}}{n!}\Theta_{\mathrm{s/a}}^{2n}\left(\Theta_{\mathrm{s/a}}^{n}\otimes\overline{\Theta_{\mathrm{s/a}}^{n}}\right)\,\Theta(\sigma)\,\Psi^{\otimes n}\\
&=\frac{\sqrt{(2n)!}}{n!}\Theta_{\mathrm{s/a}}^{2n}\left(\Theta_{\mathrm{s/a}}^{2}\Psi\right)^{\otimes n}=\frac{\sqrt{(2n)!}}{n!}\left(\Theta_{\mathrm{s/a}}^{2}\Psi\right)^{\otimes_{\mathrm{s/a}}n}\in\Gamma_{\mathrm{s/a}}^{n}(\mathcal{Z}\oplus\overline{\mathcal{Z}})\,,
\end{aligned}
\tag{107}
$$

where we used the fact that

$$
\begin{aligned}
\Theta_{\mathrm{s/a}}^{2n}\left(\Theta_{\mathrm{s/a}}^{n}\otimes\overline{\Theta_{\mathrm{s/a}}^{n}}\right)\,\Theta(\sigma)&=\Theta_{\mathrm{s/a}}^{2n}\\
&=\Theta_{\mathrm{s/a}}^{2n}\left(\Theta_{\mathrm{s/a}}^{2}\otimes\cdots\otimes\Theta_{\mathrm{s/a}}^{2}\right).
\end{aligned}
\tag{108}
$$

(In the fermionic case, to see the first identity of (108) we need to note that the permutation σ is even). Now, if τ denotes the transposition, then $\Theta^2_{\mathrm{s/a}}\Psi = \frac{1}{2}(\Psi \pm \Theta(\tau)\Psi)$. Recall that $\Psi \in \mathcal{Z}\otimes\overline{\mathcal{Z}}$ corresponds to $\gamma^{1/2} \in B^2(\mathcal{Z})$. Therefore, $\Theta(\tau)\Psi$ corresponds to $(\gamma^{1/2})^{\#} = \overline{\gamma}^{1/2}$. Hence, $\Theta^2_{\mathrm{s/a}}\Psi \in \Gamma^2_{\mathrm{s/a}}(\mathcal{Z}\oplus\overline{\mathcal{Z}})$ is identified with

$$\frac{1}{2}c = \frac{1}{2}\begin{bmatrix} 0 & \gamma^{\frac{1}{2}} \\ \pm\overline{\gamma}^{\frac{1}{2}} & 0 \end{bmatrix} \in B^2_{\mathrm{s/a}}(\overline{\mathcal{Z}}\oplus\mathcal{Z}, \mathcal{Z}\oplus\overline{\mathcal{Z}}). \tag{109}$$

Thus, (107) corresponds to

$$(n!)^{-1}\left(\frac{1}{2}a^*(c)\right)^n \Omega \in \Gamma^{2n}_{\mathrm{s/a}}(\mathcal{Z}\oplus\overline{\mathcal{Z}}) .$$

So, finally, $\Gamma(\gamma^{\frac{1}{2}})$ corresponds to

$$\exp\left(\tfrac{1}{2}a^*(c)\right)\Omega.$$

Clearly,

$$cc^* = \begin{bmatrix} 0 & \gamma^{\frac{1}{2}} \\ \pm\overline{\gamma}^{\frac{1}{2}} & 0 \end{bmatrix}\begin{bmatrix} 0 & \pm\overline{\gamma}^{\frac{1}{2}} \\ \gamma^{\frac{1}{2}} & 0 \end{bmatrix} = \begin{bmatrix} \gamma & 0 \\ 0 & \overline{\gamma} \end{bmatrix} .$$

Therefore,

$$\det(1 \mp cc^*)^{\mp\frac{1}{2}} = \det(1 \mp \gamma)^{\mp 1} = \mathrm{Tr}\Gamma(\gamma) .$$

□

Note that the vector Ω_γ is an example of a bosonic/fermionic Gaussian state considered in (43) and (55), where it was denoted Ω_c:

$$\Omega_\gamma = \det(1 \mp cc^*)^{\pm\frac{1}{4}} \exp\left(\tfrac{1}{2}a^*(c)\right)\Omega.$$

11.3 Standard Representation in the Araki-Woods/Araki-Wyss Form

Define the following transformation on $\Gamma_{\mathrm{s/a}}(\mathcal{Z}\oplus\overline{\mathcal{Z}})$:

$$\begin{aligned} R_\gamma := {} & \det(1 \mp \gamma)^{\pm\frac{1}{2}} \exp\left(\mp\frac{1}{2}a^*\left(\begin{bmatrix} 0 & \gamma^{\frac{1}{2}} \\ \pm\overline{\gamma}^{\frac{1}{2}} & 0 \end{bmatrix}\right)\right) \\ & \times\Gamma((1 \mp \gamma)\oplus(1 \mp \overline{\gamma}))^{\pm\frac{1}{2}} \exp\left(\pm\frac{1}{2}a\left(\begin{bmatrix} 0 & \gamma^{\frac{1}{2}} \\ \pm\overline{\gamma}^{\frac{1}{2}} & 0 \end{bmatrix}\right)\right). \end{aligned} \tag{110}$$

Theorem 49. *R_γ is a unitary operator satisfying*

$$\begin{aligned} & R_\gamma\phi(z_1, \overline{z_2})R^*_\gamma \\ & = \phi((1 \mp \gamma)^{\pm\frac{1}{2}}z_1 \pm (\gamma \mp 1)^{\pm\frac{1}{2}}z_2, (\overline{\gamma} \mp 1)^{\pm\frac{1}{2}}\overline{z_1} + (1 \mp \overline{\gamma})^{\pm\frac{1}{2}}\overline{z_2}). \end{aligned} \tag{111}$$

Proof. Let c be defined as in (109). Using

$$\Gamma(1 \mp cc^*) = \Gamma((1 \mp \gamma) \oplus (1 \mp \overline{\gamma})) ,$$

we see that

$$R_\gamma := \det(1 \mp cc^*)^{\pm\frac{1}{4}} \exp\left(\mp\tfrac{1}{2}a^*(c)\right) \Gamma(1 \mp cc^*)^{\pm\frac{1}{2}} \exp\left(\pm\tfrac{1}{2}a(c)\right) .$$

Thus R_γ is in fact the transformation R_c considered in (51) and (62). □

Let $\phi_{\gamma,\mathrm{l}}(z)$, $\phi_{\gamma,\mathrm{r}}(z)$, $\Gamma^+_{\mathrm{s/a},\gamma}(\mathcal{Z} \oplus \overline{\mathcal{Z}})$, etc. be defined as in Theorems 40 and 43.

R_γ intertwines between the usual left/right representations and the left/right Araki-Woods/Araki-Wyss representations, which is expressed by the following identities:

$$\begin{aligned}
R_\gamma \phi(z, 0)) R_\gamma^* &= \phi_{\gamma,\mathrm{l}}(z) , \\
R_\gamma \phi(0, \overline{z}) R_\gamma^* &= \phi_{\gamma,\mathrm{r}}(\overline{z}), \text{ in the bosonic case} , \\
R_\gamma \Lambda \phi(0, \overline{z}) \Lambda R_\gamma^* &= \phi_{\gamma,\mathrm{r}}(\overline{z}), \text{ in the fermionic case} , \\
R_\gamma \theta_\mathrm{l}(B(\Gamma_{\mathrm{s/a}}(\mathcal{Z}))) R_\gamma^* &= \mathfrak{M}_{\gamma,\mathrm{l}} , \\
R_\gamma \theta_\mathrm{r}(\overline{B(\Gamma_{\mathrm{s/a}}(\mathcal{Z}))}) R_\gamma^* &= \mathfrak{M}_{\gamma,\mathrm{r}} , \\
R_\gamma J_{\mathrm{s/a}} R_\gamma^* &= J_{\mathrm{s/a}} , \\
R_\gamma \Gamma^+_{\mathrm{s/a}}(\mathcal{Z} \oplus \overline{\mathcal{Z}}) &= \Gamma^+_{\mathrm{s/a},\gamma}(\mathcal{Z} \oplus \overline{\mathcal{Z}}) , \\
R_\gamma \Omega_\gamma &= \Omega , \\
R_\gamma \mathrm{d}\Gamma(h, -\overline{h}) R_\gamma^* &= \mathrm{d}\Gamma(h, -\overline{h}).
\end{aligned}$$

For $A \in B(\Gamma_{\mathrm{s/a}}(\mathcal{Z}))$, set

$$\begin{aligned}
\theta_{\gamma,\mathrm{l}}(A) &:= R_\gamma \theta_\mathrm{l}(A) R_\gamma^* \in B(\Gamma_{\mathrm{s/a}}(\mathcal{Z} \oplus \overline{\mathcal{Z}})) , \\
\theta_{\gamma,\mathrm{r}}(\overline{A}) &:= R_\gamma \theta_\mathrm{r}(\overline{A}) R_\gamma^* \in B(\Gamma_{\mathrm{s/a}}(\mathcal{Z} \oplus \overline{\mathcal{Z}})).
\end{aligned} \tag{112}$$

Finally, we see that in the confined case the algebra $\mathfrak{M}_{\gamma,\mathrm{l}}$ is isomorphic to $B(\Gamma_{\mathrm{s/a}}(\mathcal{Z}))$:

Theorem 50. 1) $\left(\theta_{\gamma,\mathrm{l}}, \Gamma_{\mathrm{s/a}}(\mathcal{Z} \oplus \overline{\mathcal{Z}}), J_{\mathrm{s/a}}, \Gamma^+_{\mathrm{s/a},\gamma}(\mathcal{Z} \oplus \overline{\mathcal{Z}})\right)$ *is a standard representation of* $B(\Gamma_{\mathrm{s/a}}(\mathcal{Z}))$.

2) $J_{\mathrm{s/a}} \theta_{\gamma,\mathrm{l}}(A) J_{\mathrm{s/a}} = \theta_{\gamma,\mathrm{r}}(\overline{A})$.

3) $\mathrm{d}\Gamma(h \oplus (-\overline{h}))$ *is the standard Liouvillean of* $t \mapsto \tau^t$ *in this representation.*

4) Ω *is the standard vector representative of* ω_γ *in this representation.*

12 Lattice of von Neumann Algebras in a Fock Space

Let $\mathcal{Z}$ be a Hilbert space. With every real closed subspace of $\mathcal{Z}$ we can naturally associate a certain von Neumann subalgebra of $B(\Gamma_{\mathrm{s/a}}(\mathcal{Z}))$, both in the bosonic and fermionic case. These von Neumann subalgebras form a complete lattice. Properties of this lattice are studied in this section. They have important applications in quantum field theory.

12.1 Real Subspaces in a Complex Hilbert Space

In this subsection we analyze real subspaces in a complex Hilbert space. (For a similar analysis of two complex subspaces in a complex Hilbert space see [28,43].) This analysis will be then used both in the bosonic and fermionic case. We start with a simple fact which is true both in the complex and real case.

Lemma 1. *Let $\mathcal{V}_1, \mathcal{V}_2$ be closed subspaces of a (real or complex) Hilbert space. Let p_1, p_2 be the corresponding orthogonal projections. Then*

$$\mathcal{V}_1 \cap \mathcal{V}_2 + \mathcal{V}_1^{\perp} \cap \mathcal{V}_2^{\perp} = \mathrm{Ker}(p_1 - p_2)\,.$$

Proof. Let $(p_1 - p_2)z = 0$. Then $z = p_1 z + (1-p_1)z$, where $p_1 z = p_2 z \in \mathcal{V}_1 \cap \mathcal{V}_2$ and $(1-p_1)z = (1-p_2)z \in \mathcal{V}_1^{\perp} \cap \mathcal{V}_2^{\perp}$. □

Next suppose that $\mathcal{W}$ is a complex Hilbert space. Then it is at the same time a real Hilbert space with the scalar product given by the real part of the original scalar product. If $K \subset \mathcal{W}$, then $K^{\perp}$ will denote the orthogonal complement of K in the sense of the complex scalar product and K^{perp} will denote the orthogonal complement with respect to the real scalar product. That means

$$K^{\mathrm{perp}} := \{z \in \mathcal{Z} \ : \ \mathrm{Re}(v|z) = 0,\ v \in K\}\,.$$

Moreover,

$$\mathrm{i}K^{\mathrm{perp}} = \{z \in \mathcal{Z} \ : \ \mathrm{Im}(v|z) = 0,\ v \in K\}\,,$$

so $\mathrm{i}K^{\mathrm{perp}}$ can be called the symplectic complement of K. Note that if $\mathcal{V}$ is a closed real subspace of $\mathcal{W}$, then $(\mathcal{V}^{\mathrm{perp}})^{\mathrm{perp}} = \mathcal{V}$ and $\mathrm{i}(\mathrm{i}\mathcal{V}^{\mathrm{perp}})^{\mathrm{perp}} = \mathcal{V}$.

Theorem 51. *Let $\mathcal{V}$ be a closed real subspace of a complex Hilbert space $\mathcal{W}$. Let p, q be the orthogonal projections onto $\mathcal{V}$ and $\mathrm{i}\mathcal{V}$ respectively. Then the following conditions holds:*

1) $\mathcal{V} \cap \mathrm{i}\mathcal{V} = \mathcal{V}^{\mathrm{perp}} \cap \mathrm{i}\mathcal{V}^{\mathrm{perp}} = \{0\} \ \Leftrightarrow\ \mathrm{Ker}(p-q) = \{0\}$;
2) $\mathcal{V}^{\mathrm{perp}} \cap \mathrm{i}\mathcal{V} = \{0\} \Leftrightarrow\ \mathrm{Ker}(p+q-1) = \{0\}$.

Proof. By the previous lemma applied to the real Hilbert space $\mathcal{W}$ and its subspaces $\mathcal{V}$, $\mathrm{i}\mathcal{V}$ we get

$$\mathcal{V} \cap \mathrm{i}\mathcal{V} + \mathcal{V}^{\mathrm{perp}} \cap \mathrm{i}\mathcal{V}^{\mathrm{perp}} = \mathrm{Ker}(p-q)\,.$$

This gives 1). Applying this lemma to $\mathcal{V}$, $\mathrm{i}\mathcal{V}^{\mathrm{perp}}$ yields

$$\mathcal{V} \cap \mathrm{i}\mathcal{V}^{\mathrm{perp}} + \mathcal{V}^{\mathrm{perp}} \cap \mathrm{i}\mathcal{V} = \mathrm{Ker}(p + q - 1) .$$

Using $\mathcal{V} \cap \mathrm{i}\mathcal{V}^{\mathrm{perp}} = \mathrm{i}(\mathcal{V}^{\mathrm{perp}} \cap \mathrm{i}\mathcal{V})$ we obtain 2). □

We will say that a real subspace $\mathcal{V}$ of a complex Hilbert space is *in a general position* if it satisfies both conditions of the previous theorem. The following fact is immediate:

Theorem 52. *Let $\mathcal{V}$ be a closed real subspace of a complex Hilbert space $\mathcal{W}$. Set*

$$\mathcal{W}_+ := \mathcal{V} \cap \mathrm{i}\mathcal{V}, \qquad \mathcal{W}_- := \mathcal{V}^{\mathrm{perp}} \cap \mathrm{i}\mathcal{V}^{\mathrm{perp}} ,$$

$$\mathcal{W}_1 := \mathcal{V} \cap \mathrm{i}\mathcal{V}^{\mathrm{perp}} + \mathrm{i}\mathcal{V} \cap \mathcal{V}^{\mathrm{perp}}, \quad \mathcal{W}_0 := (\mathcal{W}_+ + \mathcal{W}_- + \mathcal{W}_0)^{\perp},$$

Then $\mathcal{W}_-, \mathcal{W}_+, \mathcal{W}_0, \mathcal{W}_1$ are complex subspaces of $\mathcal{W}$ and

$$\mathcal{W} = \mathcal{W}_+ \oplus \mathcal{W}_0 \oplus \mathcal{W}_1 \oplus \mathcal{W}_- .$$

We have

$$\mathcal{W}_+ = \mathcal{V} \cap \mathcal{W}_+, \quad \{0\} = \mathcal{V} \cap \mathcal{W}_- .$$

Set

$$\mathcal{V}_0 := \mathcal{V} \cap \mathcal{W}_0, \qquad \mathcal{V}_1 := \mathcal{V} \cap \mathcal{W}_1 = \mathcal{V} \cap \mathrm{i}\mathcal{V}^{\mathrm{perp}} .$$

Then $\mathcal{V}_0$ is a subspace of $\mathcal{W}_0$ in a general position and

$$\mathcal{V} = \mathcal{W}_+ \oplus \mathcal{V}_0 \oplus \mathcal{V}_1 \oplus \{0\} ,$$

$$\mathrm{i}\mathcal{V}^{\mathrm{perp}} = \{0\} \oplus \mathrm{i}\mathcal{V}_0^{\mathrm{perp}} \oplus \mathcal{V}_1 \oplus \mathcal{W}_- ,$$

(where $\mathcal{V}_0^{\mathrm{perp}}$ is the real orthogonal complement of $\mathcal{V}_0$ taken inside $\mathcal{W}_0$).

Theorem 53. *Let $\mathcal{V}$ be a closed real subspace of a complex Hilbert space $\mathcal{W}$ in a general position. Then the following is true:*

1) *There exists a closed complex subspace $\mathcal{Z}$ of $\mathcal{W}$, a antiunitary operator ϵ on $\mathcal{W}$ and a self-adjoint operator χ such that $\epsilon^2 = 1$, $\epsilon\mathcal{Z} = \mathcal{Z}^{\perp}$, $0 \leq \chi \leq \frac{1}{2}$, $\mathrm{Ker}\chi = \mathrm{Ker}(\chi - \frac{1}{2}) = \{0\}$ and*

$$\{(1-\chi)^{\frac{1}{2}} z + \epsilon\chi^{\frac{1}{2}} z \ : \ z \in \mathcal{Z}\} = \mathcal{V} ,$$

$$\{\chi^{\frac{1}{2}} z + \epsilon(1-\chi)^{\frac{1}{2}} z \ : \ z \in \mathcal{Z}\} = \mathrm{i}\mathcal{V}^{\mathrm{perp}}.$$

2) *Set $\rho := \chi(1-2\chi)^{-1}$. Then ρ is a positive operator on $\mathcal{Z}$ with $\mathrm{Ker}\rho = \{0\}$ and*

$$\{(1+\rho)^{\frac{1}{2}} z + \epsilon\rho^{\frac{1}{2}} z \ : \ z \in \mathrm{Dom}\rho^{1/2}\} \quad \textit{is dense in} \ \ \mathcal{V} ,$$

$$\{\rho^{\frac{1}{2}} z + \epsilon(1+\rho)^{\frac{1}{2}} z \ : \ z \in \mathrm{Dom}\rho^{1/2}\} \quad \textit{is dense in} \ \ \mathrm{i}\mathcal{V}^{\mathrm{perp}}.$$

Proof. Let p, q be defined as above. Clearly, $q = \mathrm{i}p\mathrm{i}^{-1}$.

Define the self-adjoint real-linear operators $m := p + q - 1$, $n := p - q$. Note that

$$n^2 = 1 - m^2 = p + q - pq - qp\ ,$$

$$mn = -nm \quad = qp - pq,$$

$$\mathrm{i}m = m\mathrm{i}, \qquad \mathrm{i}n = -n\mathrm{i}\ ,$$

$$\mathrm{Ker}\, m = \{0\}, \qquad \mathrm{Ker}\, n = \{0\}\ ,$$

$$\mathrm{Ker}(m \pm 1) = \{0\}, \quad \mathrm{Ker}(n \pm 1) = \{0\}\ ,$$

$$-1 \le m \le 1, \qquad -1 \le n \le 1.$$

We can introduce their polar decompositions

$$n = |n|\epsilon = \epsilon|n|, \quad m = w|m| = |m|w\ .$$

Clearly, ϵ and w are orthogonal operators satisfying

$$w^2 = \epsilon^2 = 1, \quad w\epsilon = -\epsilon w\ ,$$

$$w\mathrm{i} = \mathrm{i}w, \quad \mathrm{i}\epsilon = -\mathrm{i}\epsilon.$$

Set

$$\mathcal{Z} := \mathrm{Ker}(w - 1) = \mathrm{Ran} 1_{]0,1[}(m)\ .$$

Let $1_\mathcal{Z}$ denote the orthogonal projection from $\mathcal{W}$ onto $\mathcal{Z}$.

We have

$$\epsilon\mathcal{Z} = \mathrm{Ker}(w + 1) = \mathrm{Ran} 1_{]-\infty,0[}(m)\ ,$$

Clearly, we have the orthogonal direct sum $\mathcal{W} = \mathcal{Z} \oplus \epsilon\mathcal{Z}$.

Using $p = \frac{m+n+1}{2}$ we get

$$1_\mathcal{Z} p 1_\mathcal{Z} = \tfrac{m+1}{2} 1_\mathcal{Z}\ ,$$

$$\epsilon 1_\mathcal{Z} \epsilon p 1_\mathcal{Z} = \epsilon 1_\mathcal{Z} \epsilon \tfrac{n}{2} 1_\mathcal{Z} = \epsilon \tfrac{\sqrt{1-m^2}}{2} 1_\mathcal{Z}.$$

Therefore,

$$p 1_\mathcal{Z} = \tfrac{m+1}{2} 1_\mathcal{Z} + \epsilon \tfrac{\sqrt{1-m^2}}{2} 1_\mathcal{Z}.$$

Set $\chi := \frac{1}{2} 1_\mathcal{Z}(1 - m)$. Then

$$\{(1-\chi)^{\frac{1}{2}} z + \epsilon\chi^{\frac{1}{2}} z\ :\ z \in \mathcal{Z}\} = \{pz\ :\ z \in \mathcal{Z}\} \subset \mathcal{V}\ . \tag{113}$$

Suppose now that $v \in \mathcal{V} \cap \{pz\ :\ z \in \mathcal{Z}\}^{\mathrm{perp}}$. Then

$$0 = \mathrm{Re}(v|p1_\mathcal{Z} v) = \mathrm{Re}(v|1_\mathcal{Z} v) = \|1_\mathcal{Z} v\|^2\ .$$

Hence, $v \in \mathcal{Z}^\perp = \epsilon\mathcal{Z}$. Therefore, using $q = m + 1 - p$ we obtain

$$\mathrm{Re}(v|qv) = \mathrm{Re}(v|mv) \leq 0\,.$$

Hence $qv = 0$. Thus $v \in \mathrm{i}\mathcal{V}^{\mathrm{perp}} \cap \mathcal{V}$, which means that $v = 0$. Therefore, the left hand side of (113) is dense in $\mathcal{V}$.

To see that we have an equality in (113), we note that the operator

$$(1-\chi)^{1/2}1_{\mathcal{Z}} + \epsilon\chi^{1/2}1_{\mathcal{Z}} \tag{114}$$

is an isometry from $\mathcal{Z}$ to $\mathcal{W}$. Hence the range of (114) is closed. This ends the proof of 1).

2) follows easily from 1). □

Note that $\epsilon\mathcal{Z}$ can be identified with $\overline{\mathcal{Z}}$. Thus $\mathcal{W}$ can be identified with $\mathcal{Z} \oplus \overline{\mathcal{Z}}$. Under this identification, the operator ϵ coincides with ϵ defined in (75).

Theorem 53 gives 2 descriptions of a real subspace $\mathcal{V}$. The description 2) is used in the Araki-Woods representations of the CCR and 1) in the Araki-Wyss representations of the CAR.

12.2 Complete Lattices

In this subsection we recall some definitions concerning abstract lattices (see e.g. [53]). They provide a convenient language that can be used to express some properties of a class of von Neumann algebras acting on a Fock space.

Suppose that $(X, \leq)$ is an ordered set. Let $\{x_i \ : \ i \in I\}$ be a nonempty subset of X.

We say that u is a largest minorant of $\{x_i \ : \ i \in I\}$ if

1) $i \in I$ implies $u \leq x_i$;
2) $u_1 \leq x_i$ for all $i \in I$ implies $u_1 \leq u$.

If $\{x_i \ : \ i \in I\}$ possesses a largest minorant, then it is uniquely defined. The largest minorant of a set $\{x_i \ : \ i \in I\}$ is usually denoted

$$\mathop{\wedge}_{i\in I} x_i\,.$$

Analogously we define the smallest majorant of $\{x_i \ : \ i \in I\}$, which is usually denoted by

$$\mathop{\vee}_{i\in I} x_i\,.$$

We say that $(X, \leq)$ is a complete lattice, if every nonempty subset of X possesses the largest minorant and the smallest majorant. It is then equipped with the operations $\wedge$ and $\vee$.

We will say that the complete lattice is complemented if it is equipped with the operation $X \ni x \mapsto \sim x \in X$ such that

1) $\sim(\sim x) = x$;
2) $x_1 \leq x_2$ implies $\sim x_2 \leq \sim x_1$;
3) $\sim \mathop{\wedge}_{i\in I} x_i = \mathop{\vee}_{i\in I}(\sim x_i)$.

The operation $\sim$ will be called the complementation.

Let $\mathcal{W}$ be a topological vector space. For a family $\{\mathcal{V}_i\}_{i\in I}$ of closed subspaces of $\mathcal{W}$ we define

$$\mathop{\vee}_{i\in I} \mathcal{V}_i := \Big(\sum_{i\in I} \mathcal{V}_i\Big)^{\mathrm{cl}} .$$

Closed subspaces of $\mathcal{W}$ form a complete lattice with the order relation $\subset$ and the operations $\cap$ and $\vee$. If in addition $\mathcal{W}$ is a (real or complex) Hilbert space, then taking the orthogonal complement is an example of a complementation. In the case of a complex Hilbert space (or a finite dimensional symplectic space), taking the symplectic complement is also an example of a complementation.

Let $\mathcal{H}$ be a Hilbert space. For a family of von Neumann algebras $\mathfrak{M}_i \subset B(\mathcal{H})$, $i \in I$, we set

$$\mathop{\vee}_{i\in I} \mathfrak{M}_i := \Big(\bigcup_{i\in I} \mathfrak{M}_i\Big)'' .$$

(Recall that the prime denotes the commutant). Von Neumann algebras in $B(\mathcal{H})$ form a complete lattice with the order relation $\subset$ and the operations $\cap$ and $\vee$. Taking the commutant is an example of a complementation.

12.3 Lattice of Von Neumann Algebras in a Bosonic Fock Space

In this subsection we describe the result of Araki describing the lattice of von Neumann algebras naturally associated to a bosonic Fock space [1,32]. In the proof of this result it is convenient to use the facts about the Araki-Woods representation derived earlier.

Let $\mathcal{W}$ be a complex Hilbert space. We will identify $\mathrm{Re}(\mathcal{W}\oplus\overline{\mathcal{W}})$ with $\mathcal{W}$. Consider the Hilbert space $\Gamma_{\mathrm{s}}(\mathcal{W})$ and the corresponding Fock representation $\mathcal{W} \ni w \mapsto W(w) \in B(\Gamma_{\mathrm{s}}(\mathcal{W}))$.

For a real subspace $\mathcal{V} \subset \mathcal{W}$ we define the von Neumann algebra

$$\mathfrak{M}(\mathcal{V}) := \{W(w) \ : \ w \in \mathcal{V}\}'' \subset B(\Gamma_{\mathrm{s}}(\mathcal{W})) .$$

First note that it follows from the strong continuity of $\mathcal{W} \ni w \mapsto W(w)$ that $\mathfrak{M}(\mathcal{V}) = \mathfrak{M}(\mathcal{V}^{\mathrm{cl}})$. Therefore, in what follows it is enough to restrict ourselves to closed subspaces of $\mathcal{W}$.

The following theorem was proven by Araki [1], and then a simpler proof of the most difficult statement, the duality (6), was given by Eckmann and Osterwalder [32]:

Theorem 54. 1) $\mathfrak{M}(\mathcal{V}_1) = \mathfrak{M}(\mathcal{V}_2)$ *iff* $\mathcal{V}_1 = \mathcal{V}_2$.

2) $\mathcal{V}_1 \subset \mathcal{V}_2$ *implies* $\mathfrak{M}(\mathcal{V}_1) \subset \mathfrak{M}(\mathcal{V}_2)$.

3) $\mathfrak{M}(\mathcal{W}) = B(\Gamma_{\mathrm{s}}(\mathcal{W}))$ *and* $\mathfrak{M}(\{0\}) = \mathbb{C}1$.

4) $\mathfrak{M}\Big(\mathop{\vee}_{i\in I} \mathcal{V}_i\Big) = \mathop{\vee}_{i\in I} \mathfrak{M}(\mathcal{V}_i)$.

5) $\mathfrak{M}\Big(\bigcap\limits_{i\in I}\mathcal{V}_i\Big)=\bigcap\limits_{i\in I}\mathfrak{M}(\mathcal{V}_i)$.
6) $\mathfrak{M}(\mathcal{V})'=\mathfrak{M}(\mathrm{i}\mathcal{V}^{\mathrm{perp}})$.
7) $\mathfrak{M}(\mathcal{V})$ *is a factor iff* $\mathcal{V}\cap\mathrm{i}\mathcal{V}^{\mathrm{perp}}=\{0\}$.

Proof. To prove 1), assume that $\mathcal{V}_1$ and $\mathcal{V}_2$ are distinct closed subspaces. It is enough to assume that $\mathcal{V}_2\not\subset\mathcal{V}_1$. Then we can find $w\in\mathrm{i}\mathcal{V}_1^{\mathrm{perp}}\backslash\mathrm{i}\mathcal{V}_2^{\mathrm{perp}}$. Now $W(w)\in\mathfrak{M}(\mathcal{V}_1)'\backslash\mathfrak{M}(\mathcal{V}_2)'$. This implies $\mathfrak{M}(\mathcal{V}_1)'\neq\mathfrak{M}(\mathcal{V}_2)'$, which yields 1).

2) and 3) are immediate. The inclusion $\subset$ in 4) and the inclusion $\supset$ in 5) are immediate. The inclusion $\supset$ in 4) follows easily if we invoke the strong continuity of $\mathcal{W}\ni w\mapsto W(w)$.

If we know 6), then the remaining inclusion $\subset$ in 5) follows from $\supset$ in 4). 7) follows from 1), 5) and 6).

Thus what remains to be shown is (6). Its original proof was surprisingly involved, see [1]. We will give a somewhat simpler proof [32], which uses properties of the Araki-Woods representations, which in turn are based on the Tomita-Takesaki theory.

First, assume that $\mathcal{V}$ is in general position in $\mathcal{W}$. Then, according to Theorem 53 2), and identifying $\epsilon\mathcal{Z}$ with $\overline{\mathcal{Z}}$, we obtain a decomposition $\mathcal{V}=\mathcal{Z}\oplus\overline{\mathcal{Z}}$ and a positive operator ρ such that

$$\{(1+\rho)^{\frac{1}{2}}z+\overline{\rho}^{\frac{1}{2}}\overline{z}\ :\ z\in\mathcal{Z}\}\ \text{ is dense in }\ \mathcal{V}\,.$$

Then we see that $\mathfrak{M}(\mathcal{V})$ is the left Araki-Woods algebra $\mathfrak{M}_{\rho,\mathrm{l}}^{\mathrm{AW}}$. By Theorem 40, the commutant of $\mathfrak{M}_{\rho,\mathrm{l}}^{\mathrm{AW}}$ is $\mathfrak{M}_{\rho,\mathrm{r}}^{\mathrm{AW}}$. But

$$\{\rho^{\frac{1}{2}}z+(1+\overline{\rho})^{\frac{1}{2}}\overline{z}\ :\ z\in\mathcal{Z}\}\ \text{ is dense in }\ \mathrm{i}\mathcal{V}^{\mathrm{perp}}\,.$$

Therefore, $\mathfrak{M}_{\rho,\mathrm{r}}^{\mathrm{AW}}$ coincides with $\mathfrak{M}(\mathrm{i}\mathcal{V}^{\mathrm{perp}})$. This ends the proof of 6) in the case of $\mathcal{V}$ in a general position.

For an arbitrary $\mathcal{V}$, we decompose $\mathcal{W}=\mathcal{W}_+\oplus\mathcal{W}_0\oplus\mathcal{W}_1\oplus\mathcal{W}_-$ and $\mathcal{V}=\mathcal{W}_+\oplus\mathcal{V}_0\oplus\mathcal{V}_1\oplus\{0\}$, as in Theorem 52. Then we can write

$$\begin{aligned}B(\Gamma_{\mathrm{s}}(\mathcal{W}))&\simeq B(\Gamma_{\mathrm{s}}(\mathcal{W}_+))\otimes B(\Gamma_{\mathrm{s}}(\mathcal{W}_0))\otimes B(\Gamma_{\mathrm{s}}(\mathcal{W}_1))\otimes B(\Gamma_{\mathrm{s}}(\mathcal{W}_-))\,,\\ \mathfrak{M}(\mathcal{V})&\simeq B(\Gamma_{\mathrm{s}}(\mathcal{W}_+))\otimes\mathfrak{M}(\mathcal{V}_0)\otimes\mathfrak{M}(\mathcal{V}_1)\otimes 1.\end{aligned}$$

Clearly, $\mathrm{i}\mathcal{V}^{\mathrm{perp}}=\{0\}\oplus\mathrm{i}\mathcal{V}_0^{\mathrm{perp}}\oplus\mathcal{V}_1\oplus\mathcal{W}_-$ and the commutant of $\mathfrak{M}(\mathcal{V})$ equals

$$\begin{aligned}\mathfrak{M}(\mathcal{V})'&\simeq 1\otimes\mathfrak{M}(\mathcal{V}_0)'\otimes\mathfrak{M}(\mathcal{V}_1)\otimes B(\Gamma_{\mathrm{s}}(\mathcal{W}_-))\\ &=1\otimes\mathfrak{M}(\mathrm{i}\mathcal{V}_0^{\mathrm{perp}})\otimes\mathfrak{M}(\mathcal{V}_1)\otimes B(\Gamma_{\mathrm{s}}(\mathcal{W}_-))\\ &\simeq\mathfrak{M}(\mathrm{i}\mathcal{V}^{\mathrm{perp}}).\end{aligned}$$

□

Note that the above theorem can be interpreted as an isomorphism of the complete lattices of closed real subspaces of the complex Hilbert space $\mathcal{W}$ with the symplectic complement as the complementation, and the lattice of von Neumann algebras $\mathfrak{M}(\mathcal{V}) \subset B(\Gamma_{\mathrm{s}}(\mathcal{W}))$, with the complementation given by the commutant.

12.4 Lattice of Von Neumann Algebras in a Fermionic Fock Space

In this subsection we describe the fermionic analog of Araki's result about the lattice of von Neumann algebras in a bosonic Fock space.

Again, consider a complex Hilbert space $\mathcal{W}$. We will identify $\mathrm{Re}(\mathcal{W} \oplus \overline{\mathcal{W}})$ with $\mathcal{W}$. Consider the Hilbert space $\Gamma_{\mathrm{a}}(\mathcal{W})$ and the corresponding Fock representation $\mathcal{W} \ni w \mapsto \phi(w) \in B(\Gamma_{\mathrm{a}}(\mathcal{W}))$.

Consider the Hilbert space $\Gamma_{\mathrm{a}}(\mathcal{W})$ and the corresponding Fock representation. We will identify $\mathrm{Re}(\mathcal{W} \oplus \overline{\mathcal{W}})$ with $\mathcal{W}$. For a real subspace $\mathcal{V} \subset \mathcal{W}$ we define the von Neumann algebra

$$\mathfrak{M}(\mathcal{V}) := \{\phi(z) \ : \ z \in \mathcal{V}\}'' \subset B(\Gamma_{\mathrm{a}}(\mathcal{W}))\,.$$

Let the operator Λ defined in (37).

First note that it follows from the norm continuity of $\mathcal{W} \ni w \mapsto \phi(w)$ that $\mathfrak{M}(\mathcal{V}) = \mathfrak{M}(\mathcal{V}^{\mathrm{cl}})$. Therefore, in what follows it is enough to restrict ourselves to closed real subspaces of $\mathcal{W}$.

Theorem 55. 1) $\mathfrak{M}(\mathcal{V}_1) = \mathfrak{M}(\mathcal{V}_2)$ *iff* $\mathcal{V}_1 = \mathcal{V}_2$.
2) $\mathcal{V}_1 \subset \mathcal{V}_2$ *implies* $\mathfrak{M}(\mathcal{V}_1) \subset \mathfrak{M}(\mathcal{V}_2)$.
3) $\mathfrak{M}(\mathcal{W}) = B(\Gamma_{\mathrm{a}}(\mathcal{W}))$ *and* $\mathfrak{M}(\{0\}) = \mathbb{C}1$.
4) $\mathfrak{M}\Big(\bigcap\limits_{i\in I} \mathcal{V}_i\Big) = \bigcap\limits_{i\in I} \mathfrak{M}(\mathcal{V}_i)$.
5) $\mathfrak{M}\Big(\bigvee\limits_{i\in I} \mathcal{V}_i\Big) = \bigvee\limits_{i\in I} \mathfrak{M}(\mathcal{V}_i)$.
6) $\mathfrak{M}(\mathcal{V})' = \Lambda\mathfrak{M}(\mathrm{i}\mathcal{V}^{\mathrm{perp}})\Lambda$.

The proof of the above theorem is very similar to the proof of Theorem 54 from the bosonic case. The main additional difficulty is the behavior of fermionic fields under the tensor product. They are studied in the following theorem.

Theorem 56. *Let $\mathcal{W}_i$, $i = 1,2$ be two Hilbert spaces and $\mathcal{W} = \mathcal{W}_1 \oplus \mathcal{W}_2$ Let N_i be the number operators in $\Gamma_{\mathrm{a}}(\mathcal{Z}_i)$, $i = 1,2$, $I_i := (-1)^{N_i}$ and $\Lambda_i := (-1)^{N_i(N_i-1)/2}$. We identify the operators on $\Gamma_{\mathrm{a}}(\mathcal{W})$ with those on $\Gamma_{\mathrm{a}}(\mathcal{W}_1) \otimes \Gamma_{\mathrm{a}}(\mathcal{W}_2)$ using U defined in (36). This identification is denoted by $\simeq$. Let $\mathcal{V}_i$, $i = 1,2$ be real closed subspaces of $\mathcal{W}_i$, $i = 1,2$ resp. Then*

$$\mathfrak{M}(\mathcal{V}_1 \oplus \mathcal{V}_2) \simeq \big(\mathfrak{M}(\mathcal{V}_1) \otimes 1 + (-1)^{N_1 \otimes N_2} 1 \otimes \mathfrak{M}(\mathcal{V}_2)(-1)^{N_1 \otimes N_2}\big)''\,, \quad (115)$$

$$\begin{aligned}\mathfrak{M}(\mathcal{V}_1 \oplus \{0\}) &\simeq \mathfrak{M}(\mathcal{V}_1) \otimes 1\,,\\ \mathfrak{M}(\mathcal{W}_1 \oplus \mathcal{V}_2) &\simeq B(\Gamma_{\mathrm{a}}(\mathcal{W}_1)) \otimes \mathfrak{M}(\mathcal{V}_2)\,,\end{aligned} \tag{116}$$

$$\begin{aligned}\Lambda\mathfrak{M}(\mathcal{V}_1 \oplus \mathcal{W}_2)\Lambda &\simeq \Lambda_1\mathfrak{M}(\mathcal{V}_1)\Lambda_1 \otimes B(\Gamma_{\mathrm{a}}(\mathcal{W}_2))\,,\\ \Lambda\mathfrak{M}(\{0\} \oplus \mathcal{V}_2)\Lambda &\simeq\ 1 \otimes \Lambda_2\mathfrak{M}(\mathcal{V}_2)\Lambda_2\,.\end{aligned} \tag{117}$$

Proof. Let $v \in \mathcal{V}_2$. Then, by Theorem 19 2), we have the identification

$$\phi(0, v) \simeq (-1)^{N_1} \otimes \phi(v) = (-1)^{N_1\otimes N_2}\, 1 \otimes \phi(v)\, (-1)^{N_1\otimes N_2}.$$

Therefore, the von Neumann algebra generated by $\phi(0, v)$, $v \in \mathcal{V}_2$, equals $(-1)^{N_1\otimes N_2} 1 \otimes \mathfrak{M}(\mathcal{V}_2)(-1)^{N_1\otimes N_2}$.

Clearly, the von Neumann algebra generated by $\phi(v, 0)$, $v \in \mathcal{V}_1$, equals $\mathfrak{M}(\mathcal{V}_1) \otimes 1$. This implies (115).

Equation (115) implies immediately (116). It also implies

$$\begin{aligned}\mathfrak{M}(\mathcal{V}_1 \oplus \mathcal{W}_2) &\simeq (-1)^{N_1\otimes N_2}\mathfrak{M}(\mathcal{V}_1) \otimes B(\Gamma_{\mathrm{a}}(\mathcal{W}_2))(-1)^{N_1\otimes N_2}\,,\\ \mathfrak{M}(\{0\} \oplus \mathcal{V}_2) &\simeq (-1)^{N_1\otimes N_2}\, 1 \otimes \mathfrak{M}(\mathcal{V}_2)\, (-1)^{N_1\otimes N_2}\,,\end{aligned} \tag{118}$$

from which (117) follows by Theorem 20 1). □

Proof of Theorem 55. Let us first prove 1). Assume that $\mathcal{V}_1$ and $\mathcal{V}_2$ are distinct closed subspaces. It is enough to assume that $\mathcal{V}_2 \not\subset \mathcal{V}_1$. Then we can find $w \in \mathrm{i}\mathcal{V}_1^{\mathrm{perp}} \backslash \mathrm{i}\mathcal{V}_2^{\mathrm{perp}}$. Now $\Lambda\phi(w)\Lambda \in \mathfrak{M}(\mathcal{V}_1)' \backslash \mathfrak{M}(\mathcal{V}_2)'$. Thus $\mathfrak{M}(\mathcal{V}_1)' \neq \mathfrak{M}(\mathcal{V}_2)'$, which yields 1).

Similarly as in the proof of Theorem 54 the only difficult part is a proof of 6).

Assume first that $\mathcal{V}$ satisffies $\mathcal{V} \cap \mathrm{i}\mathcal{V} = \mathcal{V}^{\mathrm{perp}} \cap \mathrm{i}\mathcal{V}^{\mathrm{perp}} = \{0\}$. By Theorem 52, we can write $\mathcal{W} = \mathcal{W}_0 \oplus \mathcal{W}_1$ and $\mathcal{V} = \mathcal{V}_0 \oplus \mathcal{V}_1$ where $\mathcal{V}_0$ is in general position in $\mathcal{W}_0$ and $\mathcal{V}_1^{\mathrm{perp}} = \mathrm{i}\mathcal{V}_1$ inside $\mathcal{W}_1$. By Theorem 53, we can find a complex subspace $\mathcal{Z}$ of $\mathcal{W}_0$, an antilinear involution ϵ on $\mathcal{W}_0$ and a self-adjoint operator $0 \le \chi \le \frac{1}{2}$ such that $\epsilon\mathcal{Z} = \mathcal{Z}^{\perp}$, $\mathrm{Ker}\chi = \mathrm{Ker}(\chi - \frac{1}{2}) = \{0\}$ and

$$\{(1-\chi)^{\frac{1}{2}} z + \epsilon\chi^{\frac{1}{2}} z \;:\; z \in \mathcal{Z}\} \oplus \mathcal{V}_1 = \mathcal{V}\,,$$

$$\{\chi^{\frac{1}{2}} z + \epsilon(1-\chi)^{\frac{1}{2}} z \;:\; z \in \mathcal{Z}\} \oplus \mathcal{V}_1 = \mathrm{i}\mathcal{V}^{\mathrm{perp}}.$$

We can identify $\epsilon\mathcal{Z}$ with $\overline{\mathcal{Z}}$, using ϵ as the conjugation. Then we are precisely in the framework of Theorem 46, which implies that $\mathfrak{M}(\mathcal{V})' = \Lambda\mathfrak{M}(\mathrm{i}\mathcal{V}^{\mathrm{perp}})\Lambda$.

For an arbitrary $\mathcal{V}$, we decompose $\mathcal{W} = \mathcal{W}_+ \oplus \mathcal{W}_0 \oplus \mathcal{W}_1 \oplus \mathcal{W}_-$ and $\mathcal{V} = \mathcal{W}_+ \oplus \mathcal{V}_0 \oplus \mathcal{V}_1 \oplus \{0\}$ as in Theorem 52. Then we can write

$$B(\Gamma_{\mathrm{a}}(\mathcal{W})) \simeq B(\Gamma_{\mathrm{a}}(\mathcal{W}_+)) \otimes B(\Gamma_{\mathrm{a}}(\mathcal{W}_0 \oplus \mathcal{W}_1)) \otimes B(\Gamma_{\mathrm{a}}(\mathcal{W}_-))\,,$$

$$\mathfrak{M}(\mathcal{V}) \simeq B(\Gamma_{\mathrm{a}}(\mathcal{W}_+)) \otimes \mathfrak{M}(\mathcal{V}_0 \oplus \mathcal{V}_1) \otimes 1$$

Let N_{01} be the number operator on $\Gamma_{\mathrm{a}}(\mathcal{W}_0 \oplus \mathcal{W}_1)$ and $\Lambda_{01} := (-1)^{N_{01}(N_{01}-1)/2}$. The commutant of $\mathfrak{M}(\mathcal{V})$ equals

$$\begin{aligned}\mathfrak{M}(\mathcal{V})' &\simeq 1 \otimes \mathfrak{M}(\mathcal{V}_0 \oplus \mathcal{V}_1)' \otimes B(\Gamma_{\mathrm{a}}(\mathcal{W}_-)) ,\\ &= 1 \otimes \Lambda_{01}\mathfrak{M}(\mathrm{i}(\mathcal{V}_0 \oplus \mathcal{V}_1)^{\mathrm{perp}})\Lambda_{01} \otimes B(\Gamma_{\mathrm{a}}(\mathcal{W}_-))\\ &\simeq \Lambda\mathfrak{M}(\mathrm{i}\mathcal{V}^{\mathrm{perp}})\Lambda ,\end{aligned}$$

where in the last step we used Theorem 56. □

13 Pauli-Fierz Systems

In this section we discuss the following setup. We start from a certain physically well motivated quantum system describing a small system interacting with the Bose gas. The Hamiltonian that generates the dynamics is a certain self-adjoint operator, bounded from below, partly expressed in terms of the usual creation and annihillation operators.

Suppose that we want to consider the same system in the thermodynamical limit corresponding to a nonzero density ρ (and the corresponding γ defined by (73)). For instance, we are interested in the density given by the Planck law at inverse temperature β. We can do this as follows: we change the representation of the CCR from the original Fock representation to the Araki-Woods representation at γ. We still assume that the dynamics is formally generated by the same expression.

To make this idea rigorous, it is convenient to use the framework of W^*-dynamical systems. In fact, what we obtain is a family of W^*-dynamical systems $(\mathfrak{M}_\gamma, \tau_\gamma)$ depending on γ, in general non-isomorphic to one another.

Even if we fix γ, then we can consider various unitarily non-equivalent representations of the W^*-dynamical system $(\mathfrak{M}_\gamma, \tau_\gamma)$. In fact, in the literature such systems are considered in at least two different representations.

The first one is what we call the semistandard representation. It was used mostly in the older literature, e.g. by Davies [23]. It is quite simple: the small system is assumed to interact with positive density Araki-Woods fields. In this representation, the dynamics has a unitary implementation given by the unitary group generated by, what we call, the semi-Liouvillean.

The second one is the standard representation. It is commonly used in the more recent literature [11, 27]. One can argue that it is the most natural representation from the point of view of theory of W^*-algebras. In any case, it is a useful tool to study various properties of $(\mathfrak{M}_\gamma, \tau_\gamma)$. On the other hand, it is more complicated than the semistandard representation. The natural implementation of the dynamics in this representation is generated by the standard Liouvillean.

If the bosons are confined in the sense of Subsects. 11.1, then $(\mathfrak{M}_\gamma, \tau_\gamma)$ are for various γ isomorphic. In this case, the algebra $\mathfrak{M}_\gamma$ has also a third useful representation: the irreducible one, which is not available in the general case.

The main goal of this section is to illustrate the above ideas with the so-called Pauli-Fierz systems. We will use the name *a Pauli-Fierz operator* to denote a self-adjoint operator describing bosons interacting with a small quantum system with an interaction linear in fields. We reserve the name *a Pauli-Fierz Hamiltonian* to Pauli-Fierz operators with a positive dispersion relation. This condition guarantees that they are bounded from below. (Note that Pauli-Fierz Liouvilleans and semi-Liouvilleans are in general not bounded from below).

Pauli-Fierz Hamiltonians arise in quantum physics in various contexts and are known under many names (e.g. the spin-boson Hamiltonian). The Hamiltonian of QED in the dipole approximation is an example of such an operator.

Several aspects of Pauli-Fierz operators have been recently studied in mathematical literature, both because of their physical importance and because of their interesting mathematical properties, see [11, 25–27] and references therein.

The plan of this section is as follows. First we fix some notation useful in describing small quantum systems interacting with Bose gas (following mostly [27]). Then we describe a Pauli-Fierz Hamiltonian. It is described by a positive boson 1-particle energy h, small system Hamiltonian K and a coupling operator v. (Essentially the only reason to assume that h is positive is the fact that such 1-particle energies are typical for physical systems). The corresponding W^*-algebraic system is just the algebra of all bounded operators on the Hilbert space with the Heisenberg dynamics generated by the Hamiltonian.

Given the operator γ describing the boson fields (and the corresponding operator ρ describing the boson density, related to γ by (73)) we construct the W^*-dynamical system $(\mathfrak{M}_\gamma, \tau_\gamma)$ – the Pauli-Fierz system corresponding to γ. The system $(\mathfrak{M}_\gamma, \tau_\gamma)$ is described in two representations: the semistandard and the standard one.

Parallel to the general case, we describe the confined case. We show, in particular, that in the confined case the semi-Liouvilleans and Liouvilleans are unitarily equivalent for various densities γ.

The constructions presented in this section are mostly taken from [27]. The only new material is the discussion of the confined case, which, even if straightforward, we believe to be quite instructive.

Remark 4. In all our considerations about Pauli-Fierz systems we restrict ourselves to the W^*-algebraic formalism. It would be tempting to apply the C^*-algebraic approach to describe Pauli-Fierz systems [17]. This approach proposes that a quantum system should be described by a certain C^*-dynamical system (a C^*-algebra with a strongly continuous dynamics). By considering various representations of this C^*-dynamical system one could describe its various thermodynamical behaviors.

Such an approach works usually well in the case of infinitely extended spin or fermionic systems, because in a finite volume typical interactions are bounded [17], and in algebraic local quantum field theory, because of the finite speed of propagation [40]. Unfortunately, for Pauli-Fierz systems the C^*-approach seems to be inappropriate – we do not know of a good choice of a C^*-algebra with the C^*-dynamics generated by a non-trivial Pauli-Fierz Hamiltonian. The problem is related to the unboundedness of bosonic fields that are involved in Pauli-Fierz Hamiltonians.

13.1 Creation and Annihilation Operators in Coupled Systems

Suppose that $\mathcal{W}$ is a Hilbert space. Consider a bosonic system described by the Fock space $\Gamma_{\mathrm{s}}(\mathcal{W})$ interacting with a quantum system described by a Hilbert space $\mathcal{E}$. The composite system is described by the Hilbert space $\mathcal{E} \otimes \Gamma_{\mathrm{s}}(\mathcal{W})$. In this subsection we discuss the formalism that we will use to describe the interaction of such coupled systems.

Let $q \in B(\mathcal{E}, \mathcal{E} \otimes \mathcal{W})$. The annihilation operator $a(q)$ is a densely defined operator on $\mathcal{E} \otimes \Gamma_{\mathrm{s}}(\mathcal{W})$ with the domain equal to the finite particle subspace of $\mathcal{E} \otimes \Gamma_{\mathrm{s}}(\mathcal{W})$. For $\Psi \in \mathcal{E} \otimes \Gamma_{\mathrm{s}}^n(\mathcal{W})$ we set

$$a(q)\Psi := \sqrt{n} q^* \otimes 1_{\mathcal{W}}^{\otimes(n-1)} \Psi \in \mathcal{E} \otimes \Gamma_{\mathrm{s}}^{n-1}(\mathcal{W}) . \tag{119}$$

($\mathcal{E} \otimes \Gamma_{\mathrm{s}}^n(\mathcal{W})$ can be viewed as a subspace of $\mathcal{E} \otimes \mathcal{W}^{\otimes n}$. Moreover, $q^* \otimes 1_{\mathcal{W}}^{\otimes(n-1)}$ is an operator from $\mathcal{E} \otimes \mathcal{W}^{\otimes n}$ to $\mathcal{E} \otimes \mathcal{W}^{\otimes(n-1)}$, which maps $\mathcal{E} \otimes \Gamma_{\mathrm{s}}^n(\mathcal{W})$ into $\mathcal{E} \otimes \Gamma_{\mathrm{s}}^{n-1}(\mathcal{W})$. Therefore, (119) makes sense).

The operator $a(q)$ is closable and we will denote its closure by the same symbol. The creation operator $a^*(q)$ is defined as

$$a^*(q) := a(q)^* .$$

Note that if $q = B \otimes |w)$, for $B \in B(\mathcal{E})$ and $w \in \mathcal{W}$, then

$$a^*(q) = B \otimes a^*(w), \quad a(q) = B^* \otimes a(w) ,$$

where $a^*(w)/a(w)$ are the usual creation/annihilion operators on the Fock space $\Gamma_{\mathrm{s}}(\mathcal{W})$.

13.2 Pauli-Fierz Hamiltonians

Throughout this section we assume that K is a self-adjoint operator on a finite dimensional Hilbert space $\mathcal{K}$, h is a positive operator on a Hilbert space $\mathcal{Z}$ and $v \in B(\mathcal{K}, \mathcal{K} \otimes \mathcal{Z})$. The self-adjoint operator

$$H_{\mathrm{fr}} := K \otimes 1 + 1 \otimes \mathrm{d}\Gamma(h)$$

on $\mathcal{K}\otimes\Gamma_{\rm s}(\mathcal{Z})$ will be called a *free Pauli-Fierz Hamiltonian.* The interaction is described by the self-adjoint operator

$$V = a^*(v) + a(v) .$$

The operator

$$H := H_{\rm fr} + V$$

is called a *full Pauli-Fierz Hamiltonian.*

If

$$h^{-\frac{1}{2}}v \in B(\mathcal{K},\mathcal{K}\otimes\mathcal{Z}) , \tag{120}$$

then H is self-adjoint on $\mathrm{Dom}(H_{\rm fr})$ and bounded from below, see e.g [27].

$\big(B(\mathcal{K}\otimes\Gamma_{\rm s}(\mathcal{Z})), \mathrm{e}^{\mathrm{i}tH}\cdot\mathrm{e}^{-\mathrm{i}tH}\big)$ will be called a *Pauli-Fierz W^*-dynamical system at zero density.*

Remark 5. We will usually drop $1_\mathcal{K}\otimes$ in formulas, so that $h^{-\frac{1}{2}}v$ above should be read $(h^{-\frac{1}{2}}\otimes 1_\mathcal{K})v$.

Remark 6. Self-adjoint operators of the form of a Pauli-Fierz Hamiltonian, but without the requirement that the boson energy is positive, will be called Pauli-Fierz operators.

13.3 More Notation

In order to describe Pauli-Fierz systems at a positive density in a compact and elegant way we need more notation.

Let $\mathcal{K}$ and $\mathcal{Z}$ be Hilbert spaces. Remember that we assume $\mathcal{K}$ to be finite dimensional. First we introduce a certain antilinear map $\star$ from $B(\mathcal{K},\mathcal{K}\otimes\mathcal{Z})$ to $B(\mathcal{K},\mathcal{K}\otimes\overline{\mathcal{Z}})$.

Let $v\in B(\mathcal{K},\mathcal{K}\otimes\mathcal{Z})$. We define $v^\star\in B(\mathcal{K},\mathcal{K}\otimes\overline{\mathcal{Z}})$ such that for $\Phi,\Psi\in\mathcal{K}$ and $w\in\mathcal{Z}$,

$$(\Phi\otimes w\,|\,v\Psi)_{\mathcal{K}\otimes\mathcal{Z}} = (v^\star\Phi|\Psi\otimes\overline{w})_{\mathcal{K}\otimes\overline{\mathcal{Z}}} \cdot$$

It is easy to see that $v^\star$ is uniquely defined. (Note that $\star$ is different from $*$ denoting the Hermitian conjugation).

Remark 7. Given an orthonormal basis $\{w_i\ :\ i\in I\}$ in $\mathcal{Z}$, any $v\in B(\mathcal{K},\mathcal{K}\otimes\mathcal{Z})$ can be decomposed as

$$v = \sum_{i\in I} B_i\otimes|w_i) , \tag{121}$$

where $B_i\in B(\mathcal{K})$, then

$$v^\star := \sum_{i\in I} B_i^*\otimes|\overline{w_i}) .$$

Next we introduce the operation $\check{\otimes}$, which can be called *tensoring in the middle*. Let $\mathcal{H}_1$, $\mathcal{H}_2$ be Hilbert spaces. If $\overline{B} \in B(\overline{\mathcal{K}})$, $A \in B(\mathcal{K}\otimes\mathcal{H}_1, \mathcal{K}\otimes\mathcal{H}_2)$, we define

$$\overline{B}\check{\otimes}A := (\theta^{-1}\otimes 1_{\mathcal{H}_2})\,(\overline{B}\otimes A)\,(\theta\otimes 1_{\mathcal{H}_1}) \in B(\mathcal{K}\otimes\overline{\mathcal{K}}\otimes\mathcal{H}_1, \mathcal{K}\otimes\overline{\mathcal{K}}\otimes\mathcal{H}_2)\,, \quad (122)$$

where $\theta : \mathcal{K}\otimes\overline{\mathcal{K}} \to \overline{\mathcal{K}}\otimes\mathcal{K}$ is defined as $\theta\, \Psi_1\otimes\overline{\Psi}_2 := \overline{\Psi}_2\otimes\Psi_1$.

Remark 8. If $C \in B(\mathcal{K})$, $A \in B(\mathcal{H}_1, \mathcal{H}_2)$, then $\overline{B}\,\check{\otimes}\,(C\otimes A) := C\otimes\overline{B}\otimes A$.

13.4 Pauli-Fierz Systems at a Positive Density

In this subsection we introduce Pauli-Fierz W^*-dynamical systems. They will be the main subject of the remaining part of this section.

Let ρ be a positive operator commuting with h having the interpretation of the *radiation density.* Let γ be the operator related to ρ as in (73). Let $\mathfrak{M}_{\gamma,\mathrm{l}}^{\mathrm{AW}} \subset B(\Gamma_{\mathrm{s}}(\mathcal{Z}\oplus\overline{\mathcal{Z}}))$ be the left Araki-Woods algebra introduced in Subsect. 9.3. The *Pauli-Fierz algebra corresponding to* γ is defined by

$$\mathfrak{M}_\gamma := B(\mathcal{K})\otimes\mathfrak{M}_{\gamma,\mathrm{l}}^{\mathrm{AW}}. \quad (123)$$

The identity map

$$\mathfrak{M}_\gamma \to B(\mathcal{K}\otimes\Gamma_{\mathrm{s}}(\mathcal{Z}\oplus\overline{\mathcal{Z}})) \quad (124)$$

will be called *the semistandard representation of* $\mathfrak{M}_\gamma$. (The bosonic part of (124) is already standard, the part involving $\mathcal{K}$ is not – hence the name).

Proposition 2. *Assume that*

$$(1+\rho)^{1/2}v \in B(\mathcal{K}, \mathcal{K}\otimes\mathcal{Z})\,. \quad (125)$$

Let

$$V_\gamma := a^*\Big((1+\rho)^{\frac{1}{2}}v, \overline{\rho}^{\frac{1}{2}}v^\star\Big) + a\Big((1+\rho)^{\frac{1}{2}}v, \overline{\rho}^{\frac{1}{2}}v^\star\Big).$$

Then the operator V_γ is essentially self-adjoint on the space of finite particle vectors and affiliated to $\mathfrak{M}_\gamma$.

The *free Pauli-Fierz semi-Liouvillean* is the self-adjoint operator on $\mathcal{K}\otimes\Gamma_{\mathrm{s}}(\mathcal{Z}\oplus\overline{\mathcal{Z}})$ defined as

$$L_{\mathrm{fr}}^{\mathrm{semi}} := K\otimes 1 + 1\otimes \mathrm{d}\Gamma(h\oplus(-\overline{h})).$$

The *full Pauli-Fierz semi-Liouvillean corresponding to* γ is

$$L_\gamma^{\mathrm{semi}} := L_{\mathrm{fr}}^{\mathrm{semi}} + V_\gamma. \quad (126)$$

Let us formulate the following assumption:

$$(1+h)(1+\rho)^{1/2}v \in B(\mathcal{K}, \mathcal{K}\otimes\mathcal{Z}). \quad (127)$$

Using Theorem 3.3 of [29] we obtain

Theorem 57. 1)
$$\tau_{\rm fr}^t(A) := \mathrm{e}^{\mathrm{i}tL_{\rm fr}^{\rm semi}} A \mathrm{e}^{-\mathrm{i}tL_{\rm fr}^{\rm semi}}$$
is a W^*-dynamics on $\mathfrak{M}_\gamma$.

2) *Suppose that (127) holds. Then $L_\gamma^{\rm semi}$ is essentially self-adjoint on* $\mathrm{Dom}(L_{\rm fr}^{\rm semi}) \cap \mathrm{Dom}(V_\gamma)$ *and*
$$\tau_\gamma^t(A) := \mathrm{e}^{\mathrm{i}tL_\gamma^{\rm semi}} A \mathrm{e}^{-\mathrm{i}tL_\gamma^{\rm semi}}$$
is a W^*-dynamics on $\mathfrak{M}_\gamma$.

The pair $(\mathfrak{M}_\gamma, \tau_\gamma)$ will be called the *Pauli-Fierz W^*-dynamical system corresponding to γ.*

13.5 Confined Pauli-Fierz Systems – Semistandard Representation

In this subsection we make the assumption $\mathrm{Tr}\gamma < \infty$. As before, we will call it the confined case.

We can use the identity representation for $B(\mathcal{K})$ and the Araki-Woods representation $\theta_{\gamma,\mathrm{l}}$ for $B(\Gamma_{\rm s}(\mathcal{Z}))$. Thus we obtain the faithful representation
$$\pi_\gamma^{\rm semi} : B(\mathcal{K} \otimes \Gamma_{\rm s}(\mathcal{Z})) \to B(\mathcal{K} \otimes \Gamma_{\rm s}(\mathcal{Z} \oplus \overline{\mathcal{Z}})) ,$$
which will be called the semistandard representation of $B(\mathcal{K} \otimes \Gamma_{\rm s}(\mathcal{Z}))$. In other words, $\pi_\gamma^{\rm semi}$ is defined by
$$\pi_\gamma^{\rm semi}(A) = U_\gamma^{\rm semi} A \otimes 1_{\Gamma_{\rm s}(\overline{\mathcal{Z}})} U_\gamma^{\rm semi *} \quad A \in B(\mathcal{K} \otimes \Gamma_{\rm s}(\mathcal{Z})) ,$$
where
$$U_\gamma^{\rm semi} := 1_\mathcal{K} \otimes R_\gamma U ,$$
and U was defined in (36) and R_γ in (110).

Theorem 58.
$$\pi_\gamma^{\rm semi}\left(B(\mathcal{K} \otimes \Gamma_{\rm s}(\mathcal{Z}))\right) = \mathfrak{M}_\gamma ,$$
$$\pi_\gamma^{\rm semi}\left(\mathrm{e}^{\mathrm{i}tH} A \mathrm{e}^{-\mathrm{i}tH}\right) = \tau_\gamma^t\left(\pi_\gamma^{\rm semi}(A)\right), \quad A \in B(\mathcal{K} \otimes \Gamma_{\rm s}(\mathcal{Z})) ,$$
$$L_\gamma^{\rm semi} = U_\gamma^{\rm semi}\left(H \otimes 1_{\Gamma_{\rm s}(\overline{\mathcal{Z}})} - 1_{\mathcal{K} \otimes \Gamma_{\rm s}(\mathcal{Z})} \otimes \mathrm{d}\Gamma(\overline{h})\right) U_\gamma^{\rm semi *}.$$

Let us stress that in the confined case the semi-Liouvilleans $L_\gamma^{\rm semi}$ and the W^*-dynamical systems $(\mathfrak{M}_{\gamma,\mathrm{l}}, \tau_\gamma)$ are unitarily equivalent for different γ.

13.6 Standard Representation of Pauli-Fierz Systems

In this subsection we drop the assumption $\mathrm{Tr}\gamma < \infty$ about the confinement of the bosons and we consider the general case again.

Consider the representation

$$\pi : \mathfrak{M}_\gamma \to B(\mathcal{K} \otimes \overline{\mathcal{K}} \otimes \Gamma_{\mathrm{s}}(\mathcal{Z} \oplus \overline{\mathcal{Z}}))$$

defined by

$$\pi(A) := 1_{\overline{\mathcal{K}}} \check{\otimes} A, \quad A \in \mathfrak{M}_\gamma ,$$

where $\check{\otimes}$ was introduced in (122). Clearly,

$$\pi(\mathfrak{M}_\gamma) = B(\mathcal{K}) \otimes 1_{\overline{\mathcal{K}}} \otimes \mathfrak{M}_{\gamma,\mathrm{l}}^{\mathrm{AW}} .$$

Set $J := J_{\mathcal{K}} \otimes \Gamma(\epsilon)$, where

$$J_{\mathcal{K}} \Psi_1 \otimes \overline{\Psi}_2 := \Psi_2 \otimes \overline{\Psi}_1, \qquad \Psi_1, \Psi_2 \in \mathcal{K} , \tag{128}$$

and ϵ was introduced in (75). Note that

$$J\, B(\mathcal{K}) \otimes 1_{\overline{\mathcal{K}}} \otimes \mathfrak{M}_{\gamma,\mathrm{l}}^{\mathrm{AW}}\, J = 1_{\mathcal{K}} \otimes B(\overline{\mathcal{K}}) \otimes \mathfrak{M}_{\gamma,\mathrm{r}}^{\mathrm{AW}} ,$$

and if $A \in B(\mathcal{K}) \otimes \mathfrak{M}_{\gamma,\mathrm{l}}^{\mathrm{AW}}$, then

$$J\pi(A)J = 1_{\mathcal{K}} \otimes \left(1_{\overline{\mathcal{K}}} \otimes \Gamma(\tau)\, \overline{A}\, 1_{\overline{\mathcal{K}}} \otimes \Gamma(\tau)\right) ,$$

where τ was introduced in (74).

Proposition 3.

$$\left(\pi,\ \mathcal{K} \otimes \overline{\mathcal{K}} \otimes \Gamma_{\mathrm{s}}(\mathcal{Z} \oplus \overline{\mathcal{Z}}),\ J,\ \mathcal{H}_\gamma^+\right)$$

is a standard representation of $\mathfrak{M}_\gamma$, *where*

$$\mathcal{H}_\gamma^+ := \{\pi(A) J \pi(A)\ B \otimes \Omega\ : B \in B_+^2(\mathcal{K}), A \in \mathfrak{M}_\gamma\}^{\mathrm{cl}} .$$

Set

$$L_{\mathrm{fr}} := K \otimes 1 \otimes 1 - 1 \otimes \overline{K} \otimes 1 + 1 \otimes 1 \otimes \mathrm{d}\Gamma(h \oplus (-\overline{h})) .$$

Proposition 4. *Assume (125). Then*

$$\begin{aligned}\pi(V_\gamma) &:= 1_{\overline{\mathcal{K}}} \check{\otimes} V_\gamma \\ &= 1_{\overline{\mathcal{K}}} \check{\otimes} a^*\left((1+\rho)^{\frac{1}{2}} v, \overline{\rho}^{\frac{1}{2}} v^\star\right) + 1_{\overline{\mathcal{K}}} \check{\otimes} a\left((1+\rho)^{\frac{1}{2}} v, \overline{\rho}^{\frac{1}{2}} v^\star\right)\end{aligned}$$

is essentially self-adjoint on finite particle vectors of $\mathcal{K} \otimes \overline{\mathcal{K}} \otimes \Gamma_{\mathrm{s}}(\mathcal{Z} \oplus \overline{\mathcal{Z}})$ *and is affiliated to the* W^**-algebra* $B(\mathcal{K}) \otimes 1_{\overline{\mathcal{K}}} \otimes \mathfrak{M}_{\gamma,\mathrm{l}}^{\mathrm{AW}}$. *Moreover,*

$$\begin{aligned}J\pi(V_\gamma)J &:= 1_{\mathcal{K}} \otimes \left(1_{\overline{\mathcal{K}}} \otimes \Gamma(\tau)\, \overline{V_\gamma}\, 1_{\overline{\mathcal{K}}} \otimes \Gamma(\tau)\right) \\ &= 1_{\mathcal{K}} \otimes a^*\left(\rho^{\frac{1}{2}} \overline{v}^\star, (1+\overline{\rho})^{\frac{1}{2}} \overline{v}\right) + 1_{\mathcal{K}} \otimes a\left(\rho^{\frac{1}{2}} \overline{v}^\star, (1+\overline{\rho})^{\frac{1}{2}} \overline{v}\right) .\end{aligned}$$

Set
$$L_\gamma := L_{\rm fr} + \pi(V_\gamma) - J\pi(V_\gamma)J\,. \tag{129}$$

Theorem 59. *1) $L_{\rm fr}$ is the standard Liouvillean of the free Pauli-Fierz system $(\mathfrak{M}_\gamma, \tau_{\rm fr})$.*
2) Suppose that (127) holds. Then L_γ is essentially self-adjoint on $\mathrm{Dom}(L_{\rm fr}) \cap \mathrm{Dom}(\pi(V_\gamma)) \cap \mathrm{Dom}(J\pi(V_\gamma)J)$ and is the Liouvillean of the Pauli-Fierz system $(\mathfrak{M}_\gamma, \tau_\gamma)$.

13.7 Confined Pauli-Fierz Systems – Standard Representation

Again we make the assumption $\mathrm{Tr}\gamma < \infty$ about the confinement of the bosons.

We can use the standard representation $\pi_{\rm l}$ for $B(\mathcal{K})$ in $B(\mathcal{K}\otimes\overline{\mathcal{K}})$ (in the form of Subsect. 8.6) and the Araki-Woods representation $\theta_{\gamma,{\rm l}}$ for $B(\Gamma_{\rm s}(\mathcal{Z}))$ in $B(\Gamma_{\rm s}(\mathcal{Z}\oplus\overline{\mathcal{Z}}))$. Thus we obtain the representation

$$\pi_{\gamma,{\rm l}} : B(\mathcal{K}\otimes\Gamma_{\rm s}(\mathcal{Z})) \to B(\mathcal{K}\otimes\overline{\mathcal{K}}\otimes\Gamma_{\rm s}(\mathcal{Z}\oplus\overline{\mathcal{Z}}))$$

defined by

$$\pi_{\gamma,{\rm l}}(A_1\otimes A_2) = A_1\otimes 1_{\overline{\mathcal{K}}}\otimes\theta_{\gamma,{\rm l}}(A_2), \quad A_1\in B(\mathcal{K}), \quad A_2\in B(\Gamma_{\rm s}(\mathcal{Z}))\,.$$

Note that

$$\pi_{\gamma,{\rm l}}(A) := 1_{\overline{\mathcal{K}}}\check{\otimes}\pi_\gamma^{\rm semi}(A), \qquad A\in B(\mathcal{K}\otimes\Gamma_{\rm s}(\mathcal{Z}))\,.$$

One can put it in a different way. Introduce the obvious unitary identification

$$\tilde{U} : \mathcal{K}\otimes\Gamma_{\rm s}(\mathcal{Z})\otimes\overline{\mathcal{K}}\otimes\Gamma_{\rm s}(\overline{\mathcal{Z}}) \to \mathcal{K}\otimes\overline{\mathcal{K}}\otimes\Gamma_{\rm s}(\mathcal{Z}\oplus\overline{\mathcal{Z}})\,.$$

Set
$$U_\gamma := 1_{\mathcal{K}\otimes\overline{\mathcal{K}}}\otimes R_\gamma\,\tilde{U}\,.$$

Then
$$\pi_{\gamma,{\rm l}}(A) = U_\gamma\,A\otimes 1_{\overline{\mathcal{K}}\otimes\Gamma_{\rm s}(\overline{\mathcal{Z}})}\,U_\gamma^*, \quad A\in B(\mathcal{K}\otimes\Gamma_{\rm s}(\mathcal{Z}))\,.$$

Theorem 60.

$$\pi_{\gamma,{\rm l}}\left(B(\mathcal{K}\otimes\Gamma_{\rm s}(\mathcal{Z}))\right) = \pi(\mathfrak{M}_\gamma)\,,$$
$$\pi_{\gamma,{\rm l}}\left(\mathrm{e}^{\mathrm{i}tH}A\mathrm{e}^{-\mathrm{i}tH}\right) = \pi\left(\tau_\gamma^t(\pi_\gamma^{\rm semi}(A))\right), \quad A\in B(\mathcal{K}\otimes\Gamma_{\rm s}(\mathcal{Z}))\,,$$
$$L_\gamma = U_\gamma\left(H\otimes 1_{\overline{\mathcal{K}}\otimes\Gamma_{\rm s}(\overline{\mathcal{Z}})} - 1_{\mathcal{K}\otimes\Gamma_{\rm s}(\mathcal{Z})}\otimes\overline{H}\right)U_\gamma^*.$$

Let us stress that in the confined case the Liouvilleans L_γ are unitarily equivalent for different γ.

References

1. Araki, H.: A lattice of von Neumann algebras associated with the quantum theory of free Bose field, Journ. Math. Phys. 4 (1963) 1343–1362.
2. Araki, H.: Type of von Neumann algebra associated with free field, Prog. Theor. Phys. 32 (1964) 956–854.
3. Araki, H.: Relative Hamiltonian for faithful normal states of a von Neumann algebra, Pub. R.I.M.S. Kyoto Univ. **9**, 165 (1973).
4. Araki, H.: On quasi-free states of CAR and Bogolubov automorphism, Publ. RIMS Kyoto Univ. 6 (1970) 385–442.
5. Araki, H.: On quasi-free states of canonical commutation relations II, Publ. RIMS Kyoto Univ. 7 (1971/72) 121–152.
6. Araki, H.: Canonical Anticommutation Relations, Contemp. Math. 62 (1987) 23.
7. Araki, H., Shiraishi, M.: On quasi-free states of canonical commutation relations I, Publ. RIMS Kyoto Univ. 7 (1971/72) 105–120.
8. Araki, H., Woods, E.J.: Representations of the canonical commutation relations describing a nonrelativistic infinite free Bose gas, J. Math. Phys. **4**, 637 (1963).
9. Araki, W., Wyss, W.: Representations of canonical anticommunication relations, Helv. Phys. Acta 37 (1964) 139–159.
10. Araki, H., Yamagami, S.: On quasi-equivalence of quasi-free states of canonical commutation relations, Publ. RIMS, Kyoto Univ. 18, 283–338 (1982).
11. Bach, V., Fröhlich, J., Sigal, I.: Return to equilibrium. J. Math. Phys. **41**, 3985 (2000).
12. Brauer, R., Weyl, H.: Spinors in n dimensions. Amer. Journ. Math. 57 (1935) 425–449.
13. Baez, J.C., Segal, I.E., Zhou, Z.: *Introduction to algebraic and constructive quantum field theory*, Princeton NJ, Princeton University Press 1991.
14. Berezin, F. A. *The method of the Second Quantization*, (Russian) 2nd ed. Nauka 1986.
15. Bogolubov, N.N. Zh. Exp. Teckn. Fiz. 17 (1947) 23.
16. Brattelli, O., Robinson D. W.: *Operator Algebras and Quantum Statistical Mechanics, Volume 1*, Springer-Verlag, Berlin, second edition 1987.
17. Brattelli, O., Robinson D. W.: *Operator Algebras and Quantum Statistical Mechanics, Volume 2*, Springer-Verlag, Berlin, second edition 1996.
18. Cartan, E.: Théorie des spineurs, Actualités Scientifiques et Industrielles No 643 et 701 (1938), Paris, Herman.
19. Clifford, Applications of Grassmann' extensive algebra, Amer. Journ. Math. 1 (1878) 350–358.
20. Connes, A.: Characterization des espaces vectoriels ordonnées sous-jacentes aux algébres de von Neumann, Ann. Inst. Fourier, Grenoble **24**, 121 (1974).
21. Cook, J.: The mathematics of second quantization, Trans. Amer. Math. Soc. 74 (1953) 222–245.
22. van Daele, A.: Quasi-equivalence of quasi-free states on the Weyl algebra, Comm. Math. Phys. 21 (1971) 171–191.
23. Davies, E. B.: Markovian master equations. Commun. Math. Phys. **39** (1974) 91.
24. De Bièvre, S.: *Local states of free Bose fields*. Lect. Notes Phys. **695**, 17–63 (2006).

25. Dereziński, J., Gérard, C.: Asymptotic completeness in quantum field theory. Massive Pauli-Fierz Hamiltonians, Rev. Math. Phys. **11**, 383 (1999).
26. Dereziński, J., Jakšić, V.: Spectral theory of Pauli-Fierz operators, Journ. Func. Anal. **180**, 241 (2001).
27. Dereziński, J., Jakšić, V.: Return to equilibrium for Pauli-Fierz systems, Ann. H. Poincare 4, 739 (2003).
28. Dixmiere, J.: Positions relative de deux varietés linneaires fermées dans un espace de Hilbert, Rev. Sci. 86 (1948) 387.
29. Dereziński, J., Jaksic, V., Pillet, C.A.: Perturbation theory of W^*-dynamics, Liouvilleans and KMS-states, Rev. Math. Phys. 15 (2003) 447–489.
30. Dirac, P.A.M.: The quantum theory of the emission and absorption of radiation. Proc. Royal Soc. London, Series A **114**, 243 (1927).
31. Dirac, P.A.M.: The quantum theory of the electron, Proc. Roy. Soc. London A 117 (1928) 610–624.
32. Eckmann, J. P., Osterwalder, K.: An application of Tomita's theory of modular algebras to duality for free Bose algebras, Journ. Func. Anal. 13 (1973) 1–12.
33. Emch, G.: Algebraic methods instatistical mechanics and quantum field theory, Wiley-Interscience 1972.
34. Fetter, A. L., Walecka, J. D.: Quantum theory of many-particle systems, McGraw-Hill Book Company 1971.
35. Fock, V.: Konfigurationsraum und zweite Quantelung, Z. Phys.: 75 (1932) 622–647.
36. Folland, G.: Harmonic analysis in phase space, Princeton University Press, Princeton, 1989.
37. Friedrichs, K. O. *Mathematical aspects of quantum theory of fields*, New York 1953.
38. Gaarding, L. and Wightman, A. S.: Representations of the commutations and anticommutation relations, Proc. Nat. Acad. Sci. USA 40 (1954) 617–626.
39. Glimm, J., Jaffe, A.: *Quantum Physics. A Functional Integral Point of View*, second edition, Springer-Verlag, New-York, 1987.
40. Haag, R.: *Local quantum physics*, Springer 1992.
41. Haag, R., Hugenholtz, N. M., Winnink, M.: On the equilibrium states in quantum mechanics, Comm. Math. Phys.**5**, 215–236.
42. Haagerup, U.: The standard form of a von Nemann algebra, Math. Scand. **37**, 271 (1975).
43. Halmos, P. R.: Two subspaces, Trans. Amer. Math. Soc. 144 (1969) 381.
44. Jakšić, V., Pillet, C.-A.: Mathematical theory of non-equilibrium quantum statistical mechanics. J. Stat. Phys. **108**, 787 (2002).
45. Jordan, P. and Wigner, E.: Pauli's equivalence prohibition, Zetschr. Phys. 47 (1928) 631.
46. Kato, T.: *Perturbation Theory for Linear Operators*, second edition, Springer-Verlag, Berlin 1976.
47. Lounesto, P.: Clifford algebras and spinors, second edition, Cambridge University Press 2001.
48. Lawson, H. B., Michelson, M.-L.: Spin geometry, Princeton University Press 1989.
49. Lundberg, L.E. Quasi-free "second-quantization", Comm. Math. Phys. 50 (1976) 103–112.
50. Neretin, Y. A.: Category of Symmetries and Infinite-Dimensional Groups, Clarendon Press, Oxford 1996.

51. von Neumann, J.: Die Eindeutigkeit der Schrödingerschen operatoren, Math. Ann. 104 (1931) 570–578.
52. Pauli, W.: Zur Quantenmechanik des magnetischen Elektrons, Z. Physik 43 (1927) 601–623.
53. Pedersen, G. K.: *Analysis Now, revised printing*, Springer 1995.
54. Powers, R.: Representations of uniformly hyperfine algebras and their associated von Neumann rings, Ann. Math. 86 (1967) 138–171.
55. Powers, R. and Stoermer, E.: Free states of the canonical anticommutation relations, Comm. Math. Phys. 16 (1970) 1–33.
56. Reed, M., Simon, B.: *Methods of Modern Mathematical Physics, I. Functional Analysis*, London, Academic Press 1980.
57. Reed, M., Simon, B.: *Methods of Modern Mathematical Physics, IV. Analysis of Operators*, London, Academic Press 1978.
58. Robinson, D.: The ground state of the Bose gas, Comm. Math. Phys. 1 (1965) 159–174.
59. Ruijsenaars, S. N. M.: On Bogoliubov transformations for systems of relativistic charged particles, J. Math. Phys. 18 (1977) 517–526.
60. Ruijsenaars, S. N. M.: On Bogoliubov transformations II. The general case. Ann. Phys. 116 (1978) 105–132.
61. Shale, D.: Linear symmetries of free boson fields, Trans. Amer. Math. Soc. 103 (1962) 149–167.
62. Shale, D. and Stinespring, W.F.: States on the Clifford algebra, Ann. Math. 80 (1964) 365–381.
63. Segal, I. E.: Foundations of the theory of dynamical systems of infinitely many degrees of freedom (I), Mat. Fys. Medd. Danske Vid. Soc. 31, No 12 (1959) 1–39.
64. Segal, I. E.: Mathematical Problems of Relativistic Physics, Amer. Math. Soc. Providence RI 1963.
65. Simon, B.: The $P(\phi)_2$ Euclidean (quantum) field theory, Princeton Univ. Press 1974.
66. Slawny, J.: On factor representations of the C^*-algebra of canonical commutation relations, Comm. Math. Phys. 24 (1971) 151–170.
67. Stratila, S.: *Modular Theory in Operator Algebras*, Abacus Press, Turnbridge Wells 1981.
68. Summers, S. J.: On the Stone – von Neumann Uniqueness Theorem and its ramifications, to appear in "John von Neumann and the Foundations of Quantum Mechanics, eds M. Redei and M. Stoelzner.
69. Takesaki, M.: Theory of Operator Algerbras I, Springer 1979.
70. Takesaki, M.: Theory of Operator Algerbras II, Springer 2003.
71. Trautman, A.: Clifford algebras and their representations, to appear in Encyclopedia of Mathematical Physics, Elsevier.
72. Varilly, J. C. and Gracia-Bondia, The metaplectic representation and boson fields Mod. Phys. Lett. A7 (1992) 659.
73. Varilly, J. C. and Gracia-Bondia, QED in external fields from the spin representation, arXiv:hep-th/9402098 v1.
74. Weil A. Sur certains groupes d'operateurs unitaires, Acta Math. 111 (1964) 143–211.
75. Weyl H. *The Theory of Groups and Quantum Mechanics*, Meuthen, London 1931.

Mathematical Theory of the Wigner-Weisskopf Atom

V. Jakšić[1], E. Kritchevski[1], and C.-A. Pillet[2]

[1] Department of Mathematics and Statistics, McGill University, 805 Sherbrooke Street West Montreal, QC, H3A 2K6, Canada
[2] CPT-CNRS, UMR 6207, Université de Toulon, B.P. 20132, 83957 La Garde Cedex, France

V. Jakšić et al.: *Mathematical Theory of the Wigner-Weisskopf Atom*, Lect. Notes Phys. **695**, 145–215 (2006)
www.springerlink.com

1 Introduction

In these lectures we shall study an "atom", $\mathcal{S}$, described by finitely many energy levels, coupled to a "radiation field", $\mathcal{R}$, described by another set (typically continuum) of energy levels. More precisely, assume that $\mathcal{S}$ and $\mathcal{R}$ are described, respectively, by the Hilbert spaces $\mathfrak{h}_{\mathcal{S}}$, $\mathfrak{h}_{\mathcal{R}}$ and the Hamiltonians $h_{\mathcal{S}}$, $h_{\mathcal{R}}$. Let $\mathfrak{h} = \mathfrak{h}_{\mathcal{S}} \oplus \mathfrak{h}_{\mathcal{R}}$ and $h_0 = h_{\mathcal{S}} \oplus h_{\mathcal{R}}$. If v is a self-adjoint operator on $\mathfrak{h}$ describing the coupling between $\mathcal{S}$ and $\mathcal{R}$, then the Hamiltonian we shall study is $h_\lambda \equiv h_0 + \lambda v$, where $\lambda \in \mathbb{R}$ is a coupling constant.

For reasons of space we shall restrict ourselves here to the case where $\mathcal{S}$ is a single energy level, i.e., we shall assume that $\mathfrak{h}_{\mathcal{S}} \equiv \mathbb{C}$ and that $h_{\mathcal{S}} \equiv \omega$ is the operator of multiplication by a real number ω. The multilevel case will be considered in the continuation of these lecture notes [41]. We will keep $\mathfrak{h}_{\mathcal{R}}$ and $h_{\mathcal{R}}$ general and we will assume that the interaction has the form $v = w + w^*$, where $w : \mathbb{C} \to \mathfrak{h}_{\mathcal{R}}$ is a linear map.

With a slight abuse of notation, in the sequel we will drop $\oplus$ whenever the meaning is clear within the context. Hence, we will write α for $\alpha \oplus 0$, g for $0 \oplus g$, etc. If $w(1) = f$, then $w = (1|\cdot)f$ and $v = (1|\cdot)f + (f|\cdot)1$.

In physics literature, a Hamiltonian of the form

$$h_\lambda = h_0 + \lambda((1|\cdot)f + (f|\cdot)1)\,, \tag{1}$$

with $\lambda \in \mathbb{R}$ is sometimes called the *Wigner-Weisskopf atom* (abbreviated WWA) and we will adopt that name. Operators of the type (1) are also often called *Friedrichs Hamiltonians* [28]. The WWA is a toy model invented to illuminate various aspects of quantum physics; see [2, 4, 5, 9, 13, 14, 17, 24, 28, 29, 34, 46, 49, 53].

Our study of the WWA naturally splits into several parts. Non-perturbative and perturbative spectral analysis are discussed respectively in Sects. 2 and 3. The fermionic second quantization of the WWA is discussed in Sects. 4 and 5.

In Sect. 2 we place no restrictions on $h_{\mathcal{R}}$ and we obtain qualitative information on the spectrum of h_λ which is valid either for all or for Lebesgue a.e. $\lambda \in \mathbb{R}$. Our analysis is based on the spectral theory of rank one perturbations [37, 59]. The theory discussed in this section naturally applies to the cases where $\mathcal{R}$ describes a quasi-periodic or a random structure, or the coupling constant λ is large.

Quantitative information about the WWA can be obtained only in the perturbative regime and under suitable regularity assumptions. In Sect. 3.2 we assume that the spectrum of $h_{\mathcal{R}}$ is purely absolutely continuous, and we study spectral properties of h_λ for small, non-zero λ. The main subject of Sect. 3.2 is the perturbation theory of embedded eigenvalues and related topics (complex resonances, radiative life-time, spectral deformations, weak coupling limit). Although the material covered in this section is very well known, our exposition is not traditional and we hope that the reader will learn something new. The reader may benefit by reading this section in parallel with Complement C_{III} in [13].

The second quantizations of the WWA lead to the simplest non-trivial examples of open systems in quantum statistical mechanics. We shall call the fermionic second quantization of the WWA the *Simple Electronic Black Box* (SEBB) model. The SEBB model in the perturbative regime has been studied in the recent lecture notes [2]. In Sects. 4 and 5 we extend the analysis and results of [2] to the non-perturbative regime. For additional information about the Electronic Black Box models we refer the reader to [3].

Assume that $\mathfrak{h}_{\mathcal{R}}$ is a *real* Hilbert space and consider the WWA (1) over the real Hilbert space $\mathbb{R} \oplus \mathfrak{h}_{\mathcal{R}}$. The bosonic second quantization of the wave equation $\partial_t^2 \psi_t + h_\lambda \psi_t = 0$ (see Sect. 6.3 in [10] and the lectures [18,19] in this volume) leads to the so called *FC (fully coupled) quantum oscillator model.* This model has been extensively discussed in the literature. The well-known references in the mathematics literature are [5, 14, 30]. For references in the physics literature the reader may consult [7, 47]. One may use the results of these lecture notes to completely describe spectral theory, scattering theory, and statistical mechanics of the FC quantum oscillator model. For reasons of space we shall not discuss this topic here (see [41]).

These lecture notes are on a somewhat higher technical level than the recent lecture notes of the first and the third author [2, 37, 51]. The first two sections can be read as a continuation (i.e. the final section) of the lecture notes [37]. In these two sections we have assumed that the reader is familiar with elementary aspects of spectral theory and harmonic analysis discussed in [37]. Alternatively, all the prerequisites can be found in [42, 44, 54–58]. In Sect. 2 we have assumed that the reader is familiar with basic results of the rank one perturbation theory [37, 59]. In Sects. 4 and 5 we have assumed that the reader is familiar with basic notions of quantum statistical mechanics [8–10, 32]. The reader with no previous exposure to open quantum systems would benefit by reading the last two sections in parallel with [2].

The notation used in these notes is standard except that we denote the spectrum of a self-adjoint operator A by $\mathrm{sp}(A)$. The set of eigenvalues, the absolutely continuous, the pure point and the singular continuous spectrum of A are denoted respectively by $\mathrm{sp}_{\mathrm{p}}(A)$, $\mathrm{sp}_{\mathrm{ac}}(A)$, $\mathrm{sp}_{\mathrm{pp}}(A)$, and $\mathrm{sp}_{\mathrm{sc}}(A)$. The singular spectrum of A is $\mathrm{sp}_{\mathrm{sing}}(A) = \mathrm{sp}_{\mathrm{pp}}(A) \cup \mathrm{sp}_{\mathrm{sc}}(A)$. The spectral subspaces associated to the absolutely continuous, the pure point, and the singular continuous spectrum of A are denoted by $\mathfrak{h}_{\mathrm{ac}}(A)$, $\mathfrak{h}_{\mathrm{pp}}(A)$, and $\mathfrak{h}_{\mathrm{sc}}(A)$. The projections on these spectral subspaces are denoted by $\mathbf{1}_{\mathrm{ac}}(A)$, $\mathbf{1}_{\mathrm{pp}}(A)$, and $\mathbf{1}_{\mathrm{sc}}(A)$.

Acknowledgment

These notes are based on the lectures the first author gave in the Summer School "Large Coulomb Systems – QED", held in Nordfjordeid, August 11 – 18 2003. V.J. is grateful to Jan Dereziński and Heinz Siedentop for the invitation to speak and for their hospitality. The research of V.J. was partly supported by NSERC. The research of E.K. was supported by an FCAR scholarship. We are grateful to S. De Bièvre and J. Dereziński for enlightening remarks on an earlier version of these lecture notes.

2 Non-Perturbative Theory

Let ν be a positive Borel measure on $\mathbb{R}$. We denote by ν_{ac}, ν_{pp}, and ν_{sc} the absolutely continuous, the pure point and the singular continuous part of ν w.r.t. the Lebesgue measure. The singular part of ν is $\nu_{\mathrm{sing}} = \nu_{\mathrm{pp}} + \nu_{\mathrm{sc}}$. We adopt the definition of a complex Borel measure given in [37,58]. In particular, any complex Borel measure on $\mathbb{R}$ is finite.

Let ν be a complex Borel measure or a positive measure such that

$$\int_{\mathbb{R}} \frac{\mathrm{d}\nu(t)}{1+|t|} < \infty\,.$$

The Borel transform of ν is the analytic function

$$F_\nu(z) \equiv \int_{\mathbb{R}} \frac{\mathrm{d}\nu(t)}{t-z}, \qquad z \in \mathbb{C}\setminus\mathbb{R}\,.$$

Let ν be a complex Borel measure or a positive measure such that

$$\int_{\mathbb{R}} \frac{\mathrm{d}\nu(t)}{1+t^2} < \infty\,. \tag{2}$$

The Poisson transform of ν is the harmonic function

$$P_\nu(x,y) \equiv y\int_{\mathbb{R}} \frac{\mathrm{d}\nu(t)}{(x-t)^2+y^2}, \qquad x+\mathrm{i}y \in \mathbb{C}_+\,,$$

where $\mathbb{C}_\pm \equiv \{z \in \mathbb{C} \mid \pm\operatorname{Im} z > 0\}$.

The Borel transform of a positive Borel measure is a Herglotz function, i.e., an analytic function on $\mathbb{C}_+$ with positive imaginary part. In this case

$$P_\nu(x,y) = \operatorname{Im} F_\nu(x+\mathrm{i}y)\,,$$

is a positive harmonic function. The G-function of ν is defined by

$$G_\nu(x) \equiv \int_{\mathbb{R}} \frac{\mathrm{d}\nu(t)}{(x-t)^2} = \lim_{y\downarrow 0} \frac{P_\nu(x,y)}{y}, \qquad x \in \mathbb{R}\,.$$

We remark that G_ν is an everywhere defined function on $\mathbb{R}$ with values in $[0,\infty]$. Note also that if $G_\nu(x)<\infty$, then $\lim_{y\downarrow 0}\operatorname{Im} F_\nu(x+\mathrm{i}y)=0$.

If $h(z)$ is analytic in the half-plane $\mathbb{C}_\pm$, we set

$$h(x\pm \mathrm{i}0)\equiv \lim_{y\downarrow 0} h(x\pm \mathrm{i}y)\;,$$

whenever the limit exist. In these lecture notes we will use a number of standard results concerning the boundary values $F_\nu(x\pm\mathrm{i}0)$. The proofs of these results can be found in [37] or in any book on harmonic analysis. We note in particular that $F_\nu(x\pm\mathrm{i}0)$ exist and is finite for Lebesgue a.e. $x\in\mathbb{R}$. If ν is real-valued and non-vanishing, then for any $a\in\mathbb{C}$ the sets $\{x\in\mathbb{R}\,|\,F_\nu(x\pm\mathrm{i}0)=a\}$ have zero Lebesgue measure.

Let ν be a positive Borel measure. For later reference, we describe some elementary properties of its Borel transform. First, the Cauchy-Schwartz inequality yields that for $y>0$

$$\nu(\mathbb{R})\operatorname{Im} F_\nu(x+\mathrm{i}y)\geq y\,|F_\nu(x+\mathrm{i}y)|^2\;. \tag{3}$$

The dominated convergence theorem yields

$$\lim_{y\to\infty} y\operatorname{Im} F_\nu(\mathrm{i}y)=\lim_{y\to\infty} y\,|F_\nu(\mathrm{i}y)|=\nu(\mathbb{R})\;. \tag{4}$$

Assume in addition that $\nu(\mathbb{R})=1$. The monotone convergence theorem yields

$$\begin{aligned}
&\lim_{y\to\infty} y^2\left(y\operatorname{Im} F_\nu(\mathrm{i}y)-y^2\,|F_\nu(\mathrm{i}y)|^2\right)\\
&=\lim_{y\to\infty}\frac{y^4}{2}\int_{\mathbb{R}\times\mathbb{R}}\left(\frac{1}{t^2+y^2}+\frac{1}{s^2+y^2}-\frac{2}{(t-\mathrm{i}y)(s+\mathrm{i}y)}\right)\mathrm{d}\nu(t)\,\mathrm{d}\nu(s)\\
&=\lim_{y\to\infty}\frac{1}{2}\int_{\mathbb{R}\times\mathbb{R}}\frac{y^2}{t^2+y^2}\frac{y^2}{s^2+y^2}(t-s)^2\mathrm{d}\nu(t)\,\mathrm{d}\nu(s)\\
&=\frac{1}{2}\int_{\mathbb{R}\times\mathbb{R}}(t-s)^2\mathrm{d}\nu(t)\,\mathrm{d}\nu(s).
\end{aligned}$$

If ν has finite second moment, $\int_{\mathbb{R}} t^2\mathrm{d}\nu(t)<\infty$, then

$$\frac{1}{2}\int_{\mathbb{R}\times\mathbb{R}}(t-s)^2\mathrm{d}\nu(t)\,\mathrm{d}\nu(s)=\int_{\mathbb{R}} t^2\mathrm{d}\nu(t)-\left(\int_{\mathbb{R}} t\mathrm{d}\nu(t)\right)^2\;. \tag{5}$$

If $\int_{\mathbb{R}} t^2\mathrm{d}\nu(t)=\infty$, then it is easy to see that the both sides in (5) are also infinite. Combining this with (4) we obtain

$$\lim_{y\to\infty}\frac{y\operatorname{Im} F_\nu(\mathrm{i}y)-y^2\,|F_\nu(\mathrm{i}y)|^2}{|F_\nu(\mathrm{i}y)|^2}=\int_{\mathbb{R}} t^2\mathrm{d}\nu(t)-\left(\int_{\mathbb{R}} t\mathrm{d}\nu(t)\right)^2\;, \tag{6}$$

where the right hand side is defined to be ∞ whenever $\int_{\mathbb{R}} t^2\mathrm{d}\nu(t)=\infty$.

In the sequel $|B|$ denotes the Lebesgue measure of a Borel set B and δ_y the delta-measure at $y\in\mathbb{R}$.

2.1 Basic Facts

Let $\mathfrak{h}_{\mathcal{R},f} \subset \mathfrak{h}_{\mathcal{R}}$ be the cyclic space generated by $h_{\mathcal{R}}$ and f. We recall that $\mathfrak{h}_{\mathcal{R},f}$ is the closure of the linear span of the set of vectors $\{(h_{\mathcal{R}}-z)^{-1}f \mid z \in \mathbb{C}\setminus\mathbb{R}\}$. Since $(\mathbb{C} \oplus \mathfrak{h}_{\mathcal{R},f})^{\perp}$ is invariant under h_λ for all λ and

$$h_\lambda|_{(\mathbb{C}\oplus\mathfrak{h}_{\mathcal{R},f})^{\perp}} = h_{\mathcal{R}}|_{(\mathbb{C}\oplus\mathfrak{h}_{\mathcal{R},f})^{\perp}} ,$$

in this section without loss of generality we may assume that $\mathfrak{h}_{\mathcal{R},f} = \mathfrak{h}_{\mathcal{R}}$, namely that f is a cyclic vector for $h_{\mathcal{R}}$. By the spectral theorem, w.l.o.g. we may assume that

$$\mathfrak{h}_{\mathcal{R}} = L^2(\mathbb{R}, \mathrm{d}\mu_{\mathcal{R}}) ,$$

and that $h_{\mathcal{R}} \equiv x$ is the operator of multiplication by the variable x. We will write

$$F_{\mathcal{R}}(z) \equiv (f|(h_{\mathcal{R}} - z)^{-1}f) .$$

Note that $F_{\mathcal{R}} = F_{f^2\mu_{\mathcal{R}}}$. Similarly, we denote $P_{\mathcal{R}}(x,y) = \operatorname{Im} F_{\mathcal{R}}(x+\mathrm{i}y)$, etc.

As we shall see, in the non-perturbative theory of the WWA it is very natural to consider the Hamiltonian (1) as an operator-valued function of two real parameters λ and ω. Hence, in this section we will write

$$h_{\lambda,\omega} \equiv h_0 + \lambda v = \omega \oplus x + \lambda\left((f|\cdot)1 + (1|\cdot)f\right) .$$

We start with some basic formulas. The relation

$$A^{-1} - B^{-1} = A^{-1}(B-A)B^{-1} ,$$

yields that

$$\begin{aligned} (h_{\lambda,\omega} - z)^{-1}1 &= (\omega - z)^{-1}1 - \lambda(\omega - z)^{-1}(h_{\lambda,\omega} - z)^{-1}f , \\ (h_{\lambda,\omega} - z)^{-1}f &= (h_{\mathcal{R}} - z)^{-1}f - \lambda(f|(h_{\mathcal{R}} - z)^{-1}f)(h_{\lambda,\omega} - z)^{-1}1. \end{aligned} \tag{7}$$

It follows that the cyclic subspace generated by $h_{\lambda,\omega}$ and the vectors $1, f$, is independent of λ and equal to $\mathfrak{h}$, and that for $\lambda \neq 0$, 1 is a cyclic vector for $h_{\lambda,\omega}$. We denote by $\mu^{\lambda,\omega}$ the spectral measure for $h_{\lambda,\omega}$ and 1. The measure $\mu^{\lambda,\omega}$ contains full spectral information about $h_{\lambda,\omega}$ for $\lambda \neq 0$. We also denote by $F_{\lambda,\omega}$ and $G_{\lambda,\omega}$ the Borel transform and the G-function of $\mu^{\lambda,\omega}$. The formulas (7) yield

$$F_{\lambda,\omega}(z) = \frac{1}{\omega - z - \lambda^2 F_{\mathcal{R}}(z)} . \tag{8}$$

Since $F_{\lambda,\omega} = F_{-\lambda,\omega}$, the operators $h_{\lambda,\omega}$ and $h_{-\lambda,\omega}$ are unitarily equivalent.

According to the decomposition $\mathfrak{h} = \mathfrak{h}_{\mathcal{S}} \oplus \mathfrak{h}_{\mathcal{R}}$ we can write the resolvent $r_{\lambda,\omega}(z) \equiv (h_{\lambda,\omega} - z)^{-1}$ in matrix form

$$r_{\lambda,\omega}(z) = \begin{bmatrix} r_{\lambda,\omega}^{\mathcal{SS}}(z) & r_{\lambda,\omega}^{\mathcal{SR}}(z) \\ r_{\lambda,\omega}^{\mathcal{RS}}(z) & r_{\lambda,\omega}^{\mathcal{RR}}(z) \end{bmatrix} .$$

A simple calculation leads to the following formulas for its matrix elements

$$\begin{aligned}
r_{\lambda,\omega}^{\mathcal{SS}}(z) &= F_{\lambda,\omega}(z), \\
r_{\lambda,\omega}^{\mathcal{SR}}(z) &= -\lambda F_{\lambda,\omega}(z)1(f|(h_{\mathcal{R}}-z)^{-1}\,\cdot\,), \\
r_{\lambda,\omega}^{\mathcal{RS}}(z) &= -\lambda F_{\lambda,\omega}(z)(h_{\mathcal{R}}-z)^{-1}f(1|\,\cdot\,), \\
r_{\lambda,\omega}^{\mathcal{RR}}(z) &= (h_{\mathcal{R}}-z)^{-1}+\lambda^2 F_{\lambda,\omega}(z)(h_{\mathcal{R}}-z)^{-1}f(f|(h_{\mathcal{R}}-z)^{-1}\,\cdot\,).
\end{aligned} \tag{9}$$

Note that for $\lambda \neq 0$,

$$F_{\lambda,\omega}(z) = \frac{F_{\lambda,0}(z)}{1+\omega F_{\lambda,0}(z)} .$$

This formula should not come as a surprise. For fixed $\lambda \neq 0$,

$$h_{\lambda,\omega} = h_{\lambda,0} + \omega(1|\cdot)1 ,$$

and since 1 is a cyclic vector for $h_{\lambda,\omega}$, we are in the usual framework of the rank one perturbation theory with ω as the perturbation parameter! This observation will allow us to naturally embed the spectral theory of $h_{\lambda,\omega}$ into the spectral theory of rank one perturbations.

By taking the imaginary part of Relation (8) we can relate the G-functions of $\mu_{\mathcal{R}}$ and $\mu^{\lambda,\omega}$ as

$$G_{\lambda,\omega}(x) = \frac{1+\lambda^2 G_{\mathcal{R}}(x)}{|\omega - x - \lambda^2 F_{\mathcal{R}}(x+\mathrm{i}0)|^2} , \tag{10}$$

whenever the boundary value $F_{\mathcal{R}}(x+\mathrm{i}0)$ exists and the numerator and denominator of the right hand side are not both infinite.

It is important to note that, subject to a natural restriction, every rank one spectral problem can be put into the form $h_{\lambda,\omega}$ for a fixed $\lambda \neq 0$.

Proposition 1. *Let ν be a Borel probability measure on $\mathbb{R}$, $f(x) = 1$ for all $x \in \mathbb{R}$, and $\lambda \neq 0$. Then the following statements are equivalent:*

1. *There exists a Borel probability measure $\mu_{\mathcal{R}}$ on $\mathbb{R}$ such that the corresponding $\mu^{\lambda,0}$ is equal to ν.*
2. *$\int_{\mathbb{R}} t\mathrm{d}\nu(t) = 0$ and $\int_{\mathbb{R}} t^2\mathrm{d}\nu(t) = \lambda^2$.*

Proof. (1) $\Rightarrow$ (2) Assume that $\mu_{\mathcal{R}}$ exists. Then $h_{\lambda,0}1 = \lambda f$ and hence

$$\int_{\mathbb{R}} t\mathrm{d}\nu(t) = (1|h_{\lambda,0}1) = 0 ,$$

and

$$\int_{\mathbb{R}} t^2\mathrm{d}\nu(t) = \|h_{\lambda,0}1\|^2 = \lambda^2 .$$

(2) $\Rightarrow$ (1) We need to find a probability measure $\mu_{\mathcal{R}}$ such that

$$F_{\mathcal{R}}(z) = \lambda^{-2}\left(-z - \frac{1}{F_\nu(z)}\right) , \tag{11}$$

for all $z \in \mathbb{C}_+$. Set

$$H_\nu(z) \equiv -z - \frac{1}{F_\nu(z)} .$$

Equation (3) yields that $\mathbb{C}_+ \ni z \mapsto \lambda^{-2} \operatorname{Im} H_\nu(z)$ is a non-negative harmonic function. Hence, by a well-known result in harmonic analysis (see e.g. [37,44]), there exists a Borel measure $\mu_{\mathcal{R}}$ which satisfies (2) and a constant $C \geq 0$ such that

$$\lambda^{-2} \operatorname{Im} H_\nu(x + \mathrm{i}y) = P_{\mathcal{R}}(x, y) + Cy , \tag{12}$$

for all $x + \mathrm{i}y \in \mathbb{C}_+$. The dominated convergence theorem and (2) yield that

$$\lim_{y\to\infty} \frac{P_{\mathcal{R}}(0, y)}{y} = \lim_{y\to\infty} \int_{\mathbb{R}} \frac{\mathrm{d}\mu_{\mathcal{R}}(t)}{t^2 + y^2} = 0 .$$

Note that

$$y \operatorname{Im} H_\nu(\mathrm{i}y) = \frac{y \operatorname{Im} F_\nu(\mathrm{i}y) - y^2 \, |F_\nu(\mathrm{i}y)|^2}{|F_\nu(\mathrm{i}y)|^2} , \tag{13}$$

and so (6) yields

$$\lim_{y\to\infty} \frac{\operatorname{Im} H_\nu(\mathrm{i}y)}{y} = 0 .$$

This fact and (12) yield that $C = 0$ and that

$$F_{\mathcal{R}}(z) = \lambda^{-2} H_\nu(z) + C_1 , \tag{14}$$

where C_1 is a real constant. From (4), (13) and (6) we get

$$\begin{aligned}
\mu_{\mathcal{R}}(\mathbb{R}) &= \lim_{y\to\infty} y \operatorname{Im} F_{\mathcal{R}}(\mathrm{i}y) \\
&= \lambda^{-2} \lim_{y\to\infty} y \operatorname{Im} H_\nu(\mathrm{i}y) \\
&= \lambda^{-2}\left(\int_{\mathbb{R}} t^2 \mathrm{d}\nu(t) - \left(\int_{\mathbb{R}} t\mathrm{d}\nu(t)\right)^2\right) = 1 ,
\end{aligned}$$

and so $\mu_{\mathcal{R}}$ is probability measure. Since

$$\operatorname{Re} H_\nu(\mathrm{i}y) = -\frac{\operatorname{Re} F_\nu(\mathrm{i}y)}{|F_\nu(\mathrm{i}y)|^2},$$

Equation (14), (4) and the dominated convergence theorem yield that

$$\begin{aligned}
\lambda^2 C_1 &= -\lim_{y\to\infty} \operatorname{Re} H_\nu(\mathrm{i}y) \\
&= \lim_{y\to\infty} y^2 \operatorname{Re} F_\nu(\mathrm{i}y) \\
&= \lim_{y\to\infty} \int_{\mathbb{R}} \frac{ty^2}{t^2 + y^2} \mathrm{d}\nu(t) \\
&= \int_{\mathbb{R}} t\mathrm{d}\nu(t) = 0 .
\end{aligned}$$

Hence, $C_1 = 0$ and (11) holds. □

2.2 Aronszajn-Donoghue Theorem

For $\lambda \neq 0$ define

$$\begin{aligned} T_{\lambda,\omega} &\equiv \{x \in \mathbb{R} \mid G_{\mathcal{R}}(x) < \infty,\ x - \omega + \lambda^2 F_{\mathcal{R}}(x + \mathrm{i}0) = 0\}\,, \\ S_{\lambda,\omega} &\equiv \{x \in \mathbb{R} \mid G_{\mathcal{R}}(x) = \infty,\ x - \omega + \lambda^2 F_{\mathcal{R}}(x + \mathrm{i}0) = 0\}\,, \\ L &\equiv \{x \in \mathbb{R} \mid \operatorname{Im} F_{\mathcal{R}}(x + \mathrm{i}0) > 0\}\,. \end{aligned} \tag{15}$$

Since the analytic function $\mathbb{C}_+ \ni z \mapsto z - \omega + \lambda^2 F_{\mathcal{R}}(z)$ is non-constant and has a positive imaginary part, by a well known result in harmonic analysis $|T_{\lambda,\omega}| = |S_{\lambda,\omega}| = 0$. Equation (8) implies that, for $\omega \neq 0$, $x - \omega + \lambda^2 F_{\mathcal{R}}(x + \mathrm{i}0) = 0$ is equivalent to $F_{\lambda,0}(x + \mathrm{i}0) = -\omega^{-1}$. Moreover, if one of these conditions is satisfied, then (10) yields

$$\omega^2 G_{\lambda,0}(x) = 1 + \lambda^2 G_{\mathcal{R}}(x)\,.$$

Therefore, if $\omega \neq 0$, then

$$\begin{aligned} T_{\lambda,\omega} &= \{x \in \mathbb{R} \mid G_{\lambda,0}(x) < \infty,\ F_{\lambda,0}(x + \mathrm{i}0) = -\omega^{-1}\}\,, \\ S_{\lambda,\omega} &= \{x \in \mathbb{R} \mid G_{\lambda,0}(x) = \infty,\ F_{\lambda,0}(x + \mathrm{i}0) = -\omega^{-1}\}\,. \end{aligned}$$

The well-known Aronszajn-Donoghue theorem in spectral theory of rank one perturbations (see [37, 59]) translates to the following result concerning the WWA.

Theorem 1. *1. $T_{\lambda,\omega}$ is the set of eigenvalues of $h_{\lambda,\omega}$. Moreover,*

$$\mu_{\mathrm{pp}}^{\lambda,\omega} = \sum_{x \in T_{\lambda,\omega}} \frac{1}{1 + \lambda^2 G_{\mathcal{R}}(x)}\, \delta_x\,. \tag{16}$$

If $\omega \neq 0$, then also

$$\mu_{\mathrm{pp}}^{\lambda,\omega} = \sum_{x \in T_{\lambda,\omega}} \frac{1}{\omega^2\, G_{\lambda,0}(x)}\, \delta_x\,.$$

2. ω is not an eigenvalue of $h_{\lambda,\omega}$ for all $\lambda \neq 0$.
3. $\mu_{\mathrm{sc}}^{\lambda,\omega}$ is concentrated on $S_{\lambda,\omega}$.
4. For all λ, ω, the set L is an essential support of the absolutely continuous spectrum of $h_{\lambda,\omega}$. Moreover $\mathrm{sp}_{\mathrm{ac}}(h_{\lambda,\omega}) = \mathrm{sp}_{\mathrm{ac}}(h_{\mathcal{R}})$ and

$$\mathrm{d}\mu_{\mathrm{ac}}^{\lambda,\omega}(x) = \frac{1}{\pi} \operatorname{Im} F_{\lambda,\omega}(x + \mathrm{i}0)\, \mathrm{d}x\,.$$

5. For a given ω, $\{\mu_{\mathrm{sing}}^{\lambda,\omega} \mid \lambda > 0\}$ is a family of mutually singular measures.
6. For a given $\lambda \neq 0$, $\{\mu_{\mathrm{sing}}^{\lambda,\omega} \mid \omega \neq 0\}$ is a family of mutually singular measures.

2.3 The Spectral Theorem

In this subsection $\lambda \neq 0$ and ω are given real numbers. By the spectral theorem, there exists a unique unitary operator

$$U^{\lambda,\omega} : \mathfrak{h} \to L^2(\mathbb{R}, \mathrm{d}\mu^{\lambda,\omega}) , \tag{17}$$

such that $U^{\lambda,\omega} h_{\lambda,\omega} (U^{\lambda,\omega})^{-1}$ is the operator of multiplication by x on the Hilbert space $L^2(\mathbb{R}, \mathrm{d}\mu^{\lambda,\omega})$ and $U^{\lambda,\omega} 1 = \mathbb{1}$, where $\mathbb{1}(x) = 1$ for all $x \in \mathbb{R}$. Moreover,

$$U^{\lambda,\omega} = U^{\lambda,\omega}_{\mathrm{ac}} \oplus U^{\lambda,\omega}_{\mathrm{pp}} \oplus U^{\lambda,\omega}_{\mathrm{sc}} ,$$

where

$$U^{\lambda,\omega}_{\mathrm{ac}} : \mathfrak{h}_{\mathrm{ac}}(h_{\lambda,\omega}) \to L^2(\mathbb{R}, \mathrm{d}\mu^{\lambda,\omega}_{\mathrm{ac}}) ,$$
$$U^{\lambda,\omega}_{\mathrm{pp}} : \mathfrak{h}_{\mathrm{pp}}(h_{\lambda,\omega}) \to L^2(\mathbb{R}, \mathrm{d}\mu^{\lambda,\omega}_{\mathrm{pp}}) ,$$
$$U^{\lambda,\omega}_{\mathrm{sc}} : \mathfrak{h}_{\mathrm{sc}}(h_{\lambda,\omega}) \to L^2(\mathbb{R}, \mathrm{d}\mu^{\lambda,\omega}_{\mathrm{sc}}) ,$$

are unitary. In this subsection we will describe these unitary operators. We shall make repeated use of the following fact. Let μ be a positive Borel measure on $\mathbb{R}$. For any complex Borel measure ν on $\mathbb{R}$ denote by $\nu = \nu_{\mathrm{ac}} + \nu_{\mathrm{sing}}$ the Lebesgue decomposition of ν into absolutely continuous and singular parts w.r.t. μ. The Radon-Nikodym derivative of ν_{ac} w.r.t. μ is given by

$$\lim_{y \downarrow 0} \frac{P_\nu(x,y)}{P_\mu(x,y)} = \frac{\mathrm{d}\nu_{\mathrm{ac}}}{\mathrm{d}\mu}(x) ,$$

for μ-almost every x (see [37]). In particular, if μ is Lebesgue measure, then

$$\lim_{y \downarrow 0} P_\nu(x,y) = \pi \frac{\mathrm{d}\nu_{\mathrm{ac}}}{\mathrm{d}x}(x) , \tag{18}$$

for Lebesgue a.e. x. By (8),

$$\operatorname{Im} F_{\lambda,\omega}(x + \mathrm{i}0) = \lambda^2 \, |F_{\lambda,\omega}(x + \mathrm{i}0)|^2 \operatorname{Im} F_{\mathcal{R}}(x + \mathrm{i}0) , \tag{19}$$

and so (18) yields that

$$\frac{\mathrm{d}\mu^{\lambda,\omega}_{\mathrm{ac}}}{\mathrm{d}x} = \lambda^2 |F_{\lambda,\omega}(x + \mathrm{i}0)|^2 |f(x)|^2 \frac{\mathrm{d}\mu_{\mathcal{R},\mathrm{ac}}}{\mathrm{d}x} . \tag{20}$$

In particular, since $F_{\lambda,\omega}(x + \mathrm{i}0) \neq 0$ for Lebesgue a.e. x and $f(x) \neq 0$ for $\mu_{\mathcal{R}}$-a.e. x, $\mu^{\lambda,\omega}_{\mathrm{ac}}$ and $\mu_{\mathcal{R},\mathrm{ac}}$ are equivalent measures.

Let $\phi = \alpha \oplus \varphi \in \mathfrak{h}$ and

$$M(z) \equiv \frac{1}{2\mathrm{i}} \left[(1|(h_{\lambda,\omega} - z)^{-1}\phi) - (1|(h_{\lambda,\omega} - \overline{z})^{-1}\phi) \right], \qquad z \in \mathbb{C}_+ .$$

The formulas (7) and (9) yield that

$$(1|(h_{\lambda,\omega} - z)^{-1}\phi) = F_{\lambda,\omega}(z)\Big(\alpha - \lambda(f|(h_{\mathcal{R}} - z)^{-1}\varphi)\Big), \tag{21}$$

and so

$$\begin{aligned} M(z) &= \operatorname{Im} F_{\lambda,\omega}(z)\Big(\alpha - \lambda(f|(h_{\mathcal{R}} - z)^{-1}\varphi)\Big) \\ &\quad - \lambda F_{\lambda,\omega}(\overline{z})\Big(y\,(f|((h_{\mathcal{R}} - x)^2 + y^2)^{-1}\varphi)\Big) \\ &= \operatorname{Im} F_{\lambda,\omega}(z)\Big(\alpha - \lambda(f|(h_{\mathcal{R}} - z)^{-1}\varphi)\Big) \\ &\quad - \lambda F_{\lambda,\omega}(\overline{z})\, y \int_{\mathbb{R}} \frac{\overline{f(t)}\varphi(t)}{(t-x)^2 + y^2}\, \mathrm{d}\mu_{\mathcal{R}}(t). \end{aligned}$$

This relation and (18) yield that for $\mu_{\mathcal{R},\mathrm{ac}}$-a.e. x,

$$\begin{aligned} M(x + \mathrm{i}0) &= \operatorname{Im} F_{\lambda,\omega}(x + \mathrm{i}0)\Big(\alpha - \lambda(f|(h_{\mathcal{R}} - x - \mathrm{i}0)^{-1}\varphi)\Big) \\ &\quad - \lambda F_{\lambda,\omega}(x - \mathrm{i}0)\overline{f(x)}\varphi(x)\, \pi \frac{\mathrm{d}\mu_{\mathcal{R}\,\mathrm{ac}}}{\mathrm{d}x}(x). \end{aligned} \tag{22}$$

On the other hand, computing $M(z)$ in the spectral representation (17) we get

$$M(z) = y \int_{\mathbb{R}} \frac{(U^{\lambda,\omega}\phi)(t)}{(t-x)^2 + y^2}\, \mathrm{d}\mu^{\lambda,\omega}(t).$$

This relation and (18) yield that for Lebesgue a.e. x,

$$M(x + \mathrm{i}0) = (U^{\lambda,\omega}_{\mathrm{ac}}\phi)(x)\, \pi \frac{\mathrm{d}\mu^{\lambda,\omega}_{\mathrm{ac}}}{\mathrm{d}x}(x).$$

Since $\mu_{\mathcal{R},\mathrm{ac}}$ and $\mu^{\lambda,\omega}_{\mathrm{ac}}$ are equivalent measures, comparison with the expression (22) and use of (8) yield

Proposition 2. *Let $\phi = \alpha \oplus \varphi \in \mathfrak{h}$. Then*

$$(U^{\lambda,\omega}_{\mathrm{ac}}\phi)(x) = \alpha - \lambda(f|(h_{\mathcal{R}} - x - \mathrm{i}0)^{-1}\varphi) - \frac{\varphi(x)}{\lambda F_{\lambda,\omega}(x + \mathrm{i}0) f(x)}.$$

We now turn to the pure point part $U^{\lambda,\omega}_{\mathrm{pp}}$. Recall that $T_{\lambda,\omega}$ is the set of eigenvalues of $h_{\lambda,\omega}$. Using the spectral representation (17), it is easy to prove that for $x \in T_{\lambda,\omega}$

$$\lim_{y\downarrow 0} \frac{(1|(h_{\lambda,\omega} - x - \mathrm{i}y)^{-1}\phi)}{(1|(h_{\lambda,\omega} - x - \mathrm{i}y)^{-1}1)} = \lim_{y\downarrow 0} \frac{F_{(U^{\lambda,\omega}\phi)\mu^{\lambda,\omega}}(x + \mathrm{i}y)}{F_{\lambda,\omega}(x + \mathrm{i}y)} = (U^{\lambda,\omega}\phi)(x). \tag{23}$$

The relations (21) and (23) yield that for $x \in T_{\lambda,\omega}$ the limit

$$H_{\varphi}(x + \mathrm{i}0) \equiv \lim_{y\downarrow 0} (f|(h_{\mathcal{R}} - x - \mathrm{i}y)^{-1}\varphi), \tag{24}$$

exists and that $(U^{\lambda,\omega}\phi)(x) = \alpha - \lambda H_{\varphi}(x + \mathrm{i}0)$. Hence, we have:

Proposition 3. *Let $\phi = \alpha \oplus \varphi \in \mathfrak{h}$. Then for $x \in T_{\lambda,\omega}$,*

$$(U_{\rm pp}^{\lambda,\omega}\phi)(x) = \alpha - \lambda H_\varphi(x + {\rm i}0) .$$

The assumption $x \in T_{\lambda,\omega}$ makes the proof of (23) easy. However, this formula holds in a much stronger form. It is a deep result of Poltoratskii [52] (see also [37, 38]) that

$$\lim_{y \downarrow 0} \frac{(1|(h_{\lambda,\omega} - x - {\rm i}y)^{-1}\phi)}{(1|(h_{\lambda,\omega} - x - {\rm i}y)^{-1}1)} = (U^{\lambda,\omega}\phi)(x) \qquad \text{for } \mu_{\rm sing}^{\lambda,\omega} - \text{a.e. } x . \tag{25}$$

Hence, the limit (24) exists and is finite for $\mu_{\rm sing}^{\lambda,\omega}$-a.e. x. Thus, we have:

Proposition 4. *Let $\phi = \alpha \oplus \varphi \in \mathfrak{h}$. Then,*

$$(U_{\rm sing}^{\lambda,\omega}\phi)(x) = \alpha - \lambda H_\varphi(x + {\rm i}0) ,$$

where $U_{\rm sing}^{\lambda,\omega} = U_{\rm pp}^{\lambda,\omega} \oplus U_{\rm sc}^{\lambda,\omega}$.

We finish this subsection with the following remark. There are many unitaries

$$W : \mathfrak{h} \to L^2(\mathbb{R}, {\rm d}\mu^{\lambda,\omega}) ,$$

such that $Wh_{\lambda,\omega}W^{-1}$ is the operator of multiplication by x on the Hilbert space $L^2(\mathbb{R}, {\rm d}\mu^{\lambda,\omega})$. Such unitaries are completely determined by their action on the vector 1 and can be classified as follows. The operator

$$U^{\lambda,\omega}W^{-1} : L^2(\mathbb{R}, {\rm d}\mu^{\lambda,\omega}) \to L^2(\mathbb{R}, {\rm d}\mu^{\lambda,\omega}) ,$$

is a unitary which commutes with the operator of multiplication by x. Hence, there exists $\theta \in L^\infty(\mathbb{R}, {\rm d}\mu^{\lambda,\omega})$ such that $|\theta| = 1$ and

$$W = \theta U^{\lambda,\omega} .$$

We summarize:

Proposition 5. *Let $W : \mathfrak{h} \to L^2(\mathbb{R}, {\rm d}\mu^{\lambda,\omega})$ be a unitary operator. Then the following statements are equivalent:*

1. *$Wh_{\lambda,\omega}W^{-1}$ is the operator of multiplication by x on the Hilbert space $L^2(\mathbb{R}, {\rm d}\mu^{\lambda,\omega})$.*
2. *There exists $\theta \in L^\infty(\mathbb{R}, {\rm d}\mu^{\lambda,\omega})$ satisfying $|\theta| = 1$ such that*

$$(W\phi)(x) = \theta(x)(U^{\lambda,\omega}\phi)(x) .$$

2.4 Scattering Theory

Recall that $h_{\mathcal{R}}$ is the operator of multiplication by the variable x on the space $L^2(\mathbb{R}, \mathrm{d}\mu_{\mathcal{R}})$. $U^{\lambda,\omega} h_{\lambda,\omega} (U^{\lambda,\omega})^{-1}$ is the operator of multiplication by x on the space $L^2(\mathbb{R}, \mathrm{d}\mu^{\lambda,\omega})$. Set

$$h_{\mathcal{R},\mathrm{ac}} \equiv h_{\mathcal{R}}|_{\mathfrak{h}_{\mathrm{ac}}(h_{\mathcal{R}})}, \qquad h_{\lambda,\omega,\mathrm{ac}} \equiv h_{\lambda,\omega}|_{\mathfrak{h}_{\mathrm{ac}}(h_{\lambda,\omega})} .$$

Since $\mathfrak{h}_{\mathrm{ac}}(h_{\mathcal{R}}) = L^2(\mathbb{R}, \mathrm{d}\mu_{\mathcal{R},\mathrm{ac}})$,

$$\mathfrak{h}_{\mathrm{ac}}(h_{\lambda,\omega}) = (U^{\lambda,\omega}_{\mathrm{ac}})^{-1} L^2(\mathbb{R}, \mathrm{d}\mu^{\lambda,\omega}_{\mathrm{ac}}) ,$$

and the measures $\mu_{\mathcal{R},\mathrm{ac}}$ and $\mu^{\lambda,\omega}_{\mathrm{ac}}$ are equivalent, the operators $h_{\mathcal{R},\mathrm{ac}}$ and $h_{\lambda,\omega,\mathrm{ac}}$ are unitarily equivalent. Using (20) and the chain rule one easily checks that the operator

$$(W^{\lambda,\omega}\phi)(x) = \sqrt{\frac{\mathrm{d}\mu^{\lambda,\omega}_{\mathrm{ac}}}{\mathrm{d}\mu_{\mathcal{R},\mathrm{ac}}}(x)}\,(U^{\lambda,\omega}_{\mathrm{ac}}\phi)(x) = |\lambda F_{\lambda,\omega}(x+\mathrm{i}0) f(x)| (U^{\lambda,\omega}_{\mathrm{ac}}\phi)(x) ,$$

is an explicit unitary which takes $\mathfrak{h}_{\mathrm{ac}}(h_{\lambda,\omega})$ onto $\mathfrak{h}_{\mathrm{ac}}(h_{\mathcal{R}})$ and satisfies

$$W^{\lambda,\omega} h_{\lambda,\omega,\mathrm{ac}} = h_{\mathcal{R},\mathrm{ac}} W^{\lambda,\omega} .$$

Moreover, we have:

Proposition 6. *Let $W : \mathfrak{h}_{\mathrm{ac}}(h_{\lambda,\omega}) \to \mathfrak{h}_{\mathrm{ac}}(h_{\mathcal{R}})$ be a unitary operator. Then the following statements are equivalent:*

1. *W intertwines $h_{\lambda,\omega,\mathrm{ac}}$ and $h_{\mathcal{R},\mathrm{ac}}$, i.e.,*

$$W h_{\lambda,\omega,\mathrm{ac}} = h_{\mathcal{R},\mathrm{ac}} W . \tag{26}$$

2. *There exists $\theta \in L^\infty(\mathbb{R}, \mathrm{d}\mu_{\mathcal{R},\mathrm{ac}})$ satisfying $|\theta| = 1$ such that*

$$(W\phi)(x) = \theta(x)(W^{\lambda,\omega}\phi)(x) .$$

In this subsection we describe a particular pair of unitaries, called wave operators, which satisfy (26).

Theorem 2. *1. The strong limits*

$$U^{\pm}_{\lambda,\omega} \equiv \mathrm{s}-\lim_{t\to\pm\infty} \mathrm{e}^{\mathrm{i}th_{\lambda,\omega}} \mathrm{e}^{-\mathrm{i}th_0} \mathbf{1}_{\mathrm{ac}}(h_0) , \tag{27}$$

exist and $\mathrm{Ran}\, U^{\pm}_{\lambda,\omega} = \mathfrak{h}_{\mathrm{ac}}(h_{\lambda,\omega})$.

2. The strong limits

$$\Omega^{\pm}_{\lambda,\omega} \equiv \mathrm{s}-\lim_{t\to\pm\infty} \mathrm{e}^{\mathrm{i}th_0} \mathrm{e}^{-\mathrm{i}th_{\lambda,\omega}} \mathbf{1}_{\mathrm{ac}}(h_{\lambda,\omega}) , \tag{28}$$

exist and $\mathrm{Ran}\, \Omega^{\pm}_{\lambda,\omega} = \mathfrak{h}_{\mathrm{ac}}(h_0)$.

3. *The maps* $U^{\pm}_{\lambda,\omega} : \mathfrak{h}_{\mathrm{ac}}(h_0) \to \mathfrak{h}_{\mathrm{ac}}(h_{\lambda,\omega})$ *and* $\Omega^{\pm}_{\lambda,\omega} : \mathfrak{h}_{\mathrm{ac}}(h_{\lambda,\omega}) \to \mathfrak{h}_{\mathrm{ac}}(h_0)$ *are unitary.* $U^{\pm}_{\lambda,\omega}\Omega^{\pm}_{\lambda,\omega} = \mathbf{1}_{\mathrm{ac}}(h_{\lambda,\omega})$ *and* $\Omega^{\pm}_{\lambda,\omega}U^{\pm}_{\lambda,\omega} = \mathbf{1}_{\mathrm{ac}}(h_0)$. *Moreover,* $\Omega^{\pm}_{\lambda,\omega}$ *satisfies the intertwining relation (26).*
4. *The S-matrix* $S \equiv \Omega^{+}_{\lambda,\omega}U^{-}_{\lambda,\omega}$ *is unitary on* $\mathfrak{h}_{\mathrm{ac}}(h_0)$ *and commutes with* $h_{0,\mathrm{ac}}$.

This theorem is a basic result in scattering theory. The detailed proof can be found in [42, 56].

The wave operators and the S-matrix can be described as follows.

Proposition 7. *Let* $\phi = \alpha \oplus \varphi \in \mathfrak{h}$. *Then*

$$(\Omega^{\pm}_{\lambda,\omega}\phi)(x) = \varphi(x) - \lambda f(x) F_{\lambda,\omega}(x \pm \mathrm{i}0)(\alpha - \lambda(f|(h_{\mathcal{R}} - x \mp \mathrm{i}0)^{-1}\varphi)) \,. \quad (29)$$

Moreover, for any $\psi \in \mathfrak{h}_{\mathrm{ac}}(h_0)$ *one has* $(S\psi)(x) = S(x)\psi(x)$ *with*

$$S(x) = 1 + 2\pi\mathrm{i}\lambda^2 F_{\lambda,\omega}(x + \mathrm{i}0)|f(x)|^2 \frac{\mathrm{d}\mu_{\mathcal{R},\mathrm{ac}}}{\mathrm{d}x}(x) \,. \quad (30)$$

Remark. The assumption that f is a cyclic vector for $h_{\mathcal{R}}$ is not needed in Theorem 2 and Proposition 7.

Proof. We will compute $\Omega^{+}_{\lambda,\omega}$. The case of $\Omega^{-}_{\lambda,\omega}$ is completely similar. Let $\psi \in \mathfrak{h}_{\mathrm{ac}}(h_0) = \mathfrak{h}_{\mathrm{ac}}(h_{\mathcal{R}})$. We start with the identity

$$(\psi|\mathrm{e}^{\mathrm{i}th_0}\mathrm{e}^{-\mathrm{i}th_{\lambda,\omega}}\phi) = (\psi|\phi) - \mathrm{i}\lambda\int_0^t (\psi|\mathrm{e}^{\mathrm{i}sh_0}f)(1|\mathrm{e}^{-\mathrm{i}sh_{\lambda,\omega}}\phi)\,\mathrm{d}s \,. \quad (31)$$

Note that $(\psi|\phi) = (\psi|\varphi)$, $(\psi|\mathrm{e}^{\mathrm{i}sh_0}f) = (\psi|\mathrm{e}^{\mathrm{i}sh_{\mathcal{R}}}f)$, and that

$$\begin{aligned}
\lim_{t\to\infty}(\psi|\mathrm{e}^{\mathrm{i}th_0}\mathrm{e}^{-\mathrm{i}th_{\lambda,\omega}}\phi) &= \lim_{t\to\infty}(\mathrm{e}^{\mathrm{i}th_{\lambda,\omega}}\mathrm{e}^{-\mathrm{i}th_0}\psi|\phi) \\
&= (U^{+}_{\lambda,\omega}\psi|\phi) \\
&= (\psi|\Omega^{+}_{\lambda,\omega}\phi).
\end{aligned}$$

Hence, by the Abel theorem,

$$(\psi|\Omega^{+}_{\lambda,\omega}\phi) = (\psi|\varphi) - \lim_{y\downarrow 0}\mathrm{i}\lambda L(y) \,, \quad (32)$$

where

$$L(y) = \int_0^\infty \mathrm{e}^{-ys}(\psi|\mathrm{e}^{\mathrm{i}sh_0}f)(1|\mathrm{e}^{-\mathrm{i}sh_{\lambda,\omega}}\phi)\,\mathrm{d}s \,.$$

Now,

$$
\begin{aligned}
L(y) &= \int_0^\infty \mathrm{e}^{-ys}(\psi|\mathrm{e}^{\mathrm{i}sh_0}f)(1|\mathrm{e}^{-\mathrm{i}sh_{\lambda,\omega}}\phi)\,\mathrm{d}s \\
&= \int_{\mathbb{R}} \overline{\psi(x)}f(x)\left[\int_0^\infty (1|\mathrm{e}^{\mathrm{i}s(x+\mathrm{i}y-h_{\lambda,\omega})}\phi)\mathrm{d}s\right]\mathrm{d}\mu_{\mathcal{R},\mathrm{ac}}(x) \\
&= -\mathrm{i}\int_{\mathbb{R}} \overline{\psi(x)}f(x)(1|(h_{\lambda,\omega}-x-\mathrm{i}y)^{-1}\phi)\,\mathrm{d}\mu_{\mathcal{R},\mathrm{ac}}(x) \\
&= -\mathrm{i}\int_{\mathbb{R}} \overline{\psi(x)}f(x)g_y(x)\,\mathrm{d}\mu_{\mathcal{R},\mathrm{ac}}(x)\;,
\end{aligned}
\tag{33}
$$

where

$$
g_y(x) \equiv (1|(h_{\lambda,\omega}-x-\mathrm{i}y)^{-1}\phi)\;.
$$

Recall that for Lebesgue a.e. x,

$$
g_y(x) \to g(x) \equiv (1|(h_{\lambda,\omega}-x-\mathrm{i}0)^{-1}\phi)\;, \tag{34}
$$

as $y \downarrow 0$. By the Egoroff theorem (see e.g. Problem 16 in Chap. 3 of [58], or any book on measure theory), for any $n > 0$ there exists a measurable set $R_n \subset \mathbb{R}$ such that $|\mathbb{R} \setminus R_n| < 1/n$ and $g_y \to g$ uniformly on R_n. The set

$$
\bigcup_{n>0}\{\psi \in L^2(\mathbb{R}, \mathrm{d}\mu_{\mathcal{R},\mathrm{ac}})\,|\,\mathrm{supp}\,\psi \subset R_n\}\;,
$$

is clearly dense in $\mathfrak{h}_{\mathrm{ac}}(h_{\mathcal{R}})$. For any ψ in this set the uniform convergence $g_y \to g$ on $\mathrm{supp}\,\psi$ implies that there exists a constant C_ψ such that

$$
|\overline{\psi}f(g_y-g)| \le C_\psi|\overline{\psi}f| \in L^1(\mathbb{R}, \mathrm{d}\mu_{\mathcal{R},\mathrm{ac}})\;.
$$

This estimate and the dominated convergence theorem yield that

$$
\lim_{y\downarrow 0}\int_{\mathbb{R}} \overline{\psi}\, f(g_y-g)\mathrm{d}\mu_{\mathcal{R},\mathrm{ac}} = 0\;.
$$

On the other hand, (32) and (33) yield that the limit

$$
\lim_{y\downarrow 0}\int_{\mathbb{R}} \overline{\psi}\, f g_y \mathrm{d}\mu_{\mathcal{R}}\;,
$$

exists, and so the relation

$$
(\psi|\Omega^+_{\lambda,\omega}\phi) = (\psi|\varphi) - \lambda\int_{\mathbb{R}} \overline{\psi(x)}f(x)(1|(h_{\lambda,\omega}-x-\mathrm{i}0)^{-1}\phi)\mathrm{d}\mu_{\mathcal{R},\mathrm{ac}}(x)\;,
$$

holds for a dense set of vectors ψ. Hence,

$$
(\Omega^+_{\lambda,\omega}\phi)(x) = \varphi(x) - \lambda f(x)(1|(h_{\lambda,\omega}-x-\mathrm{i}0)^{-1}\phi)\;,
$$

and the formula (21) completes the proof.

To compute the S-matrix, note that by Proposition 6, $\Omega^{\pm}_{\lambda,\omega} = \theta_{\pm} W^{\lambda,\omega}$, where

$$\theta_{\pm}(x) = \frac{(\Omega^{\pm}_{\lambda,\omega} 1)(x)}{(W^{\lambda,\omega}1)(x)} = -\frac{\lambda F_{\lambda,\omega}(x \pm \mathrm{i}0) f(x)}{|\lambda F_{\lambda,\omega}(x + \mathrm{i}0) f(x)|} .$$

Since

$$S = \Omega^{+}_{\lambda,\omega} U^{-}_{\lambda,\omega} = \Omega^{+}_{\lambda,\omega} (\Omega^{-}_{\lambda,\omega})^{*} = \theta_{+} W^{\lambda,\omega} (W^{\lambda,\omega})^{*} \overline{\theta_{-}} = \theta_{+} \overline{\theta_{-}} ,$$

we see that $(S\psi)(x) = S(x)\psi(x)$, where

$$S(x) = \theta_{+}(x)\overline{\theta_{-}(x)} = \frac{F_{\lambda,\omega}(x + \mathrm{i}0)}{F_{\lambda,\omega}(x - \mathrm{i}0)} = \frac{\omega - x - \lambda^2 F_{\mathcal{R}}(x - \mathrm{i}0)}{\omega - x - \lambda^2 F_{\mathcal{R}}(x + \mathrm{i}0)} .$$

Hence,

$$\begin{aligned} S(x) &= 1 + 2\mathrm{i}\lambda^2 F_{\lambda,\omega}(x + \mathrm{i}0) \mathrm{Im}\, F_{\mathcal{R}}(x + \mathrm{i}0) \\ &= 1 + 2\pi \mathrm{i}\lambda^2 F_{\lambda,\omega}(x + \mathrm{i}0) |f(x)|^2 \frac{\mathrm{d}\mu_{\mathcal{R},\mathrm{ac}}}{\mathrm{d}x}(x) . \end{aligned}$$

□

2.5 Spectral Averaging

We will freely use the standard measurability results concerning the measure-valued function $(\lambda, \omega) \mapsto \mu^{\lambda,\omega}$. The reader not familiar with these facts may consult [11,12,37].

Let $\lambda \neq 0$ and

$$\overline{\mu}^{\lambda}(B) = \int_{\mathbb{R}} \mu^{\lambda,\omega}(B)\, \mathrm{d}\omega ,$$

where $B \subset \mathbb{R}$ is a Borel set. Obviously, $\overline{\mu}^{\lambda}$ is a Borel measure on $\mathbb{R}$. The following (somewhat surprising) result is often called *spectral averaging.*

Proposition 8. *The measure $\overline{\mu}^{\lambda}$ is equal to the Lebesgue measure and for all $g \in L^1(\mathbb{R}, \mathrm{d}x)$,*

$$\int_{\mathbb{R}} g(x)\mathrm{d}x = \int_{\mathbb{R}} \left[\int_{\mathbb{R}} g(x) \mathrm{d}\mu^{\lambda,\omega}(x) \right] \mathrm{d}\omega .$$

The proof of this proposition is elementary and can be found in [37,59].

One can also average with respect to both parameters. It follows from Proposition 8 that the averaged measure

$$\overline{\mu}(B) = \frac{1}{\pi} \int_{\mathbb{R}^2} \frac{\mu^{\lambda,\omega}(B)}{1 + \lambda^2}\, \mathrm{d}\lambda \mathrm{d}\omega ,$$

is also equal to the Lebesgue measure.

2.6 Simon-Wolff Theorems

Recall that $x + \lambda^2 F_{\mathcal{R}}(x + \mathrm{i}0)$ and $F_{\lambda,0}(x + \mathrm{i}0)$ are finite and non-vanishing for Lebesgue a.e. x. For $\lambda \neq 0$, (10) gives that for Lebesgue a.e x,

$$G_{\lambda,0}(x) = \frac{1 + \lambda^2 G_{\mathcal{R}}(x)}{|x + \lambda^2 F_{\mathcal{R}}(x + \mathrm{i}0)|^2} = |F_{\lambda,0}(x + \mathrm{i}0)|^2(1 + \lambda^2 G_{\mathcal{R}}(x)) .$$

These observations yield:

Lemma 1. *Let $B \subset \mathbb{R}$ be a Borel set and $\lambda \neq 0$. Then $G_{\mathcal{R}}(x) < \infty$ for Lebesgue a.e. $x \in B$ iff $G_{\lambda,0}(x) < \infty$ for Lebesgue a.e. $x \in B$.*

This lemma and the Simon-Wolff theorems in rank one perturbation theory (see [37, 59, 61]) yield:

Theorem 3. *Let $B \subset \mathbb{R}$ be a Borel set. Then the following statements are equivalent:*

1. *$G_{\mathcal{R}}(x) < \infty$ for Lebesgue a.e. $x \in B$.*
2. *For all $\lambda \neq 0$, $\mu_{\mathrm{cont}}^{\lambda,\omega}(B) = 0$ for Lebesgue a.e. $\omega \in \mathbb{R}$. In particular, $\mu_{\mathrm{cont}}^{\lambda,\omega}(B) = 0$ for Lebesgue a.e. $(\lambda, \omega) \in \mathbb{R}^2$.*

Theorem 4. *Let $B \subset \mathbb{R}$ be a Borel set. Then the following statements are equivalent:*

1. *$\mathrm{Im}\, F_{\mathcal{R}}(x + \mathrm{i}0) = 0$ and $G_{\mathcal{R}}(x) = \infty$ for Lebesgue a.e. $x \in B$.*
2. *For all $\lambda \neq 0$, $\mu_{\mathrm{ac}}^{\lambda,\omega}(B) + \mu_{\mathrm{pp}}^{\lambda,\omega}(B) = 0$ for Lebesgue a.e. $\omega \in \mathbb{R}$. In particular, $\mu_{\mathrm{ac}}^{\lambda,\omega}(B) + \mu_{\mathrm{pp}}^{\lambda,\omega}(B) = 0$ for Lebesgue a.e. $(\lambda, \omega) \in \mathbb{R}^2$.*

Theorem 5. *Let $B \subset \mathbb{R}$ be a Borel set. Then the following statements are equivalent:*

1. *$\mathrm{Im}\, F_{\mathcal{R}}(x + \mathrm{i}0) > 0$ for Lebesgue a.e. $x \in B$.*
2. *For all $\lambda \neq 0$, $\mu_{\mathrm{sing}}^{\lambda,\omega}(B) = 0$ for Lebesgue a.e. $\omega \in \mathbb{R}$. In particular, $\mu_{\mathrm{sing}}^{\lambda,\omega}(B) = 0$ for Lebesgue a.e. $(\lambda, \omega) \in \mathbb{R}^2$.*

Note that while the Simon-Wolff theorems hold for a fixed λ and for a.e. ω, we cannot claim that they hold for a fixed ω and for a.e. λ – from Fubini's theorem we can deduce only that for a.e. ω the results hold for a.e. λ. This is somewhat annoying since in many applications for physical reasons it is natural to fix ω and vary λ. The next subsection deals with this issue.

2.7 Fixing ω

The results discussed in this subsection are not an immediate consequence of the standard results of rank one perturbation theory and for this reason we will provide complete proofs.

In this subsection ω is a fixed real number. Let

$$\overline{\mu}^{\omega}(B) = \int_{\mathbb{R}} \mu^{\lambda,\omega}(B)\mathrm{d}\lambda \,,$$

where $B \subset \mathbb{R}$ is a Borel set. Obviously, $\overline{\mu}^{\omega}$ is a positive Borel measure on $\mathbb{R}$ and for all Borel measurable $g \geq 0$,

$$\int_{\mathbb{R}} g(t)\mathrm{d}\overline{\mu}^{\omega}(t) = \int_{\mathbb{R}} \left[\int_{\mathbb{R}} g(t)\mathrm{d}\mu^{\lambda,\omega}(t) \right] \mathrm{d}\lambda \,,$$

where both sides are allowed to be infinite.

We will study the measure $\overline{\mu}^{\omega}$ by examining the boundary behavior of its Poisson transform $P_\omega(x,y)$ as $y \downarrow 0$. In this subsection we set

$$l(z) \equiv (\omega - z)F_{\mathcal{R}}(z) \,.$$

Lemma 2. *For* $z \in \mathbb{C}_+$,

$$P_\omega(z) = \frac{\pi}{\sqrt{2}} \frac{\sqrt{|l(z)| + \mathrm{Re}\, l(z)}}{|l(z)|} \,.$$

Proof. We start with

$$\begin{aligned} P_\omega(x,y) &= \int_{\mathbb{R}} \left[\int_{\mathbb{R}} \frac{y}{(t-x)^2 + y^2} \mathrm{d}\mu^{\lambda,\omega}(t) \right] \mathrm{d}\lambda \\ &= \mathrm{Im} \int_{\mathbb{R}} F_{\lambda,\omega}(x + \mathrm{i}y)\, \mathrm{d}\lambda \,. \end{aligned}$$

Equation (8) and a simple residue calculation yield

$$\int_{\mathbb{R}} F_{\lambda,\omega}(x+\mathrm{i}y)\mathrm{d}\lambda = \frac{-\pi \mathrm{i}}{F_{\mathcal{R}}(z)\sqrt{\frac{\omega - z}{F_{\mathcal{R}}(z)}}} \,,$$

where the branch of the square root is chosen to be in $\mathbb{C}_+$. An elementary calculation shows that

$$P_\omega(x,y) = \mathrm{Im}\, \frac{\mathrm{i}\pi}{\sqrt{l(x+\mathrm{i}y)}} \,,$$

where the branch of the square root is chosen to have positive real part, explicitly

$$\sqrt{w} \equiv \frac{1}{\sqrt{2}} \left(\sqrt{|w| + \mathrm{Re}\, w} + \mathrm{i}\, \mathrm{sign}(\mathrm{Im}\, w)\, \sqrt{|w| - \mathrm{Re}\, w} \right) \,. \qquad (35)$$

This yields the statement. □

Theorem 6. *The measure $\overline{\mu}^\omega$ is absolutely continuous with respect to Lebesgue measure and*

$$\frac{\mathrm{d}\overline{\mu}^\omega}{\mathrm{d}x}(x) = \frac{\sqrt{|l(x+\mathrm{i}0)| + \mathrm{Re}\, l(x+\mathrm{i}0)}}{\sqrt{2}\,|l(x+\mathrm{i}0)|} . \tag{36}$$

The set

$$\mathcal{E} \equiv \{x \mid \mathrm{Im}\, F_{\mathcal{R}}(x+\mathrm{i}0) > 0\} \cup \{x \mid (\omega - x)F_{\mathcal{R}}(x+\mathrm{i}0) > 0\} ,$$

is an essential support for $\overline{\mu}^\omega$ and $\mu^{\lambda,\omega}$ is concentrated on $\mathcal{E}$ for all $\lambda \neq 0$.

Proof. By Theorem 1, ω is not an eigenvalue of $h_{\lambda,\omega}$ for $\lambda \neq 0$. This implies that $\overline{\mu}^\omega(\{\omega\}) = 0$. By the theorem of de la Vallée Poussin (for detailed proof see e.g. [37]), $\overline{\mu}^\omega_{\mathrm{sing}}$ is concentrated on the set

$$\{x \mid x \neq \omega \text{ and } \lim_{y \downarrow 0} P_\omega(x+\mathrm{i}y) = \infty\} .$$

By Lemma 2, this set is contained in

$$\mathcal{S} \equiv \{x \mid \lim_{y\downarrow 0} F_{\mathcal{R}}(x+\mathrm{i}y) = 0\} .$$

Since $\mathcal{S} \cap S_{\lambda,\omega} \subset \{\omega\}$, Theorem 1 implies that $\mu^{\lambda,\omega}_{\mathrm{sing}}(\mathcal{S}) = 0$ for all $\lambda \neq 0$. Since $|\mathcal{S}| = 0$, $\mu^{\lambda,\omega}_{\mathrm{ac}}(\mathcal{S}) = 0$ for all λ. We conclude that $\mu^{\lambda,\omega}(\mathcal{S}) = 0$ for all $\lambda \neq 0$, and so

$$\overline{\mu}^\omega(\mathcal{S}) = \int_{\mathbb{R}} \mu^{\lambda,\omega}(\mathcal{S})\,\mathrm{d}\lambda = 0 .$$

Hence, $\overline{\mu}^\omega_{\mathrm{sing}} = 0$. From Theorem 1 we now get

$$\mathrm{d}\overline{\mu}^\omega(x) = \mathrm{d}\overline{\mu}^\omega_{\mathrm{ac}}(x) = \frac{1}{\pi}\,\mathrm{Im}\, F_\omega(x+\mathrm{i}0)\,\mathrm{d}x ,$$

and (36) follows from Lemma 2. The remaining statements are obvious. □

We are now ready to state and prove the Simon-Wolff theorems for fixed ω.

Theorem 7. *Let $B \subset \mathbb{R}$ be a Borel set. Consider the following statements:*

1. *$G_{\mathcal{R}}(x) < \infty$ for Lebesgue a.e. $x \in B$.*
2. *$\mu^{\lambda,\omega}_{\mathrm{cont}}(B) = 0$ for Lebesgue a.e. $\lambda \in \mathbb{R}$.*

Then (1) $\Rightarrow$ (2). If $B \subset \mathcal{E}$, then also (2) $\Rightarrow$ (1).

Proof. Let $A \equiv \{x \in B \mid G_{\mathcal{R}}(x) = \infty\} \cap \mathcal{E}$.
(1)$\Rightarrow$(2) By assumption, A has zero Lebesgue measure. Theorem 6 yields that $\overline{\mu}^\omega(A) = 0$. Since $G_{\mathcal{R}}(x) < \infty$ for Lebesgue a.e. $x \in B$, $\mathrm{Im}\, F_{\mathcal{R}}(x+\mathrm{i}0) = 0$ for Lebesgue a.e. $x \in B$. Hence, for all λ, $\mathrm{Im}\, F_{\lambda,\omega}(x+\mathrm{i}0) = 0$ for Lebesgue a.e.

$x \in B$. By Theorem 1, $\mu_{\mathrm{ac}}^{\lambda,\omega}(B) = 0$ and the measure $\mu_{\mathrm{sc}}^{\lambda,\omega}|_B$ is concentrated on the set A for all $\lambda \neq 0$. Then,

$$\int_{\mathbb{R}} \mu_{\mathrm{sc}}^{\lambda,\omega}(B)\, \mathrm{d}\lambda = \int_{\mathbb{R}} \mu_{\mathrm{sc}}^{\lambda,\omega}(A)\, \mathrm{d}\lambda \leq \int_{\mathbb{R}} \mu^{\lambda,\omega}(A)\, \mathrm{d}\lambda = \overline{\mu}^{\,\omega}(A) = 0 \, .$$

Hence, $\mu_{\mathrm{sc}}^{\lambda,\omega}(B) = 0$ for Lebesgue a.e λ.
(2)⇒(1) Assume that the set A has positive Lebesgue measure. By Theorem 1, $\mu_{\mathrm{pp}}^{\lambda,\omega}(A) = 0$ for all $\lambda \neq 0$, and

$$\int_{\mathbb{R}} \mu_{\mathrm{cont}}^{\lambda,\omega}(A)\, \mathrm{d}\lambda = \int_{\mathbb{R}} \mu^{\lambda,\omega}(A)\, \mathrm{d}\lambda = \overline{\mu}^{\omega}(A) > 0 \, .$$

Hence, for a set of λ's of positive Lebesgue measure, $\mu_{\mathrm{cont}}^{\lambda,\omega}(B) > 0$. □

Theorem 8. *Let $B \subset \mathbb{R}$ be a Borel set. Consider the following statements:*

1. $\mathrm{Im}\, F_{\mathcal{R}}(x + \mathrm{i}0) = 0$ *and* $G_{\mathcal{R}}(x) = \infty$ *for Lebesgue a.e.* $x \in B$.
2. $\mu_{\mathrm{ac}}^{\lambda,\omega}(B) + \mu_{\mathrm{pp}}^{\lambda,\omega}(B) = 0$ *for Lebesgue a.e.* $\lambda \in \mathbb{R}$.

Then (1) ⇒ (2). *If* $B \subset \mathcal{E}$, *then also* (2) ⇒ (1).

Proof. Let $A \equiv \{x \in B \,|\, G_{\mathcal{R}}(x) < \infty\} \cap \mathcal{E}$.
(1)⇒(2) Since $\mathrm{Im}\, F_{\mathcal{R}}(x + \mathrm{i}0) = 0$ for Lebesgue a.e. $x \in B$, Theorem 1 implies that $\mu_{\mathrm{ac}}^{\lambda,\omega}(B) = 0$ for all λ. By Theorems 1 and 6, for $\lambda \neq 0$, $\mu_{\mathrm{pp}}^{\lambda,\omega}|_B$ is concentrated on the set A. Since A has Lebesgue measure zero,

$$\int_{\mathbb{R}} \mu_{\mathrm{pp}}^{\lambda,\omega}(A)\, \mathrm{d}\lambda \leq \overline{\mu}^{\,\omega}(A) = 0 \, ,$$

and so $\mu_{\mathrm{pp}}^{\lambda,\omega}(B) = 0$ for Lebesgue a.e. λ.
(2)⇒(1) If $\mathrm{Im}\, F_{\mathcal{R}}(x+\mathrm{i}0) > 0$ for a set of $x \in B$ of positive Lebesgue measure, then, by Theorem 1, $\mu_{\mathrm{ac}}^{\lambda,\omega}(B) > 0$ for all λ. Assume that $\mathrm{Im}\, F_{\mathcal{R}}(x+\mathrm{i}0) = 0$ for Lebesgue a.e. $x \in B$ and that A has positive Lebesgue measure. By Theorem 1, $\mu_{\mathrm{cont}}^{\lambda,\omega}(A) = 0$ for all $\lambda \neq 0$ and since $A \subset \mathcal{E}$, Theorem 6 implies

$$\int_{\mathbb{R}} \mu_{\mathrm{pp}}^{\lambda,\omega}(A)\, \mathrm{d}\lambda = \int_{\mathbb{R}} \mu^{\lambda,\omega}(A)\, \mathrm{d}\lambda = \overline{\mu}^{\omega}(A) > 0 \, .$$

Thus, we must have that $\mu_{\mathrm{pp}}^{\lambda,\omega}(B) > 0$ for a set of λ's of positive Lebesgue measure. □

Theorem 9. *Let $B \subset \mathbb{R}$ be a Borel set. Consider the following statements:*

1. $\mathrm{Im}\, F_{\mathcal{R}}(x + \mathrm{i}0) > 0$ *for Lebesgue a.e.* $x \in B$.
2. $\mu_{\mathrm{sing}}^{\lambda,\omega}(B) = 0$ *for Lebesgue a.e.* $\lambda \in \mathbb{R}$.

Then (1) ⇒ (2). *If* $B \subset \mathcal{E}$, *then also* (2) ⇒ (1).

Proof. (1)⇒(2) By Theorem 1, for $\lambda \neq 0$ the measure $\mu_{\text{sing}}^{\lambda,\omega}|_B$ is concentrated on the set $A \equiv \{x \in B \,|\, \operatorname{Im} F_{\mathcal{R}}(x + \mathrm{i}0) = 0\} \cap \mathcal{E}$. By assumption, A has Lebesgue measure zero and

$$\int_{\mathbb{R}} \mu_{\text{sing}}^{\lambda,\omega}(A)\,\mathrm{d}\lambda \le \int_{\mathbb{R}} \mu^{\lambda,\omega}(A)\,\mathrm{d}\lambda = \overline{\mu}^{\omega}(A) = 0 \,.$$

Hence, for Lebesgue a.e. $\lambda \in \mathbb{R}$, $\mu_{\text{sing}}^{\lambda}(B) = 0$.
(2)⇒(1) Assume that $B \subset \mathcal{E}$ and that the set

$$A \equiv \{x \in B \,|\, \operatorname{Im} F_{\mathcal{R}}(x + \mathrm{i}0) = 0\} \,,$$

has positive Lebesgue measure. By Theorem 1, $\mu_{\text{ac}}^{\lambda,\omega}(A) = 0$ for all λ, and

$$\int_{\mathbb{R}} \mu_{\text{sing}}^{\lambda,\omega}(A)\,\mathrm{d}\lambda = \int_{\mathbb{R}} \mu^{\lambda,\omega}(A)\,\mathrm{d}\lambda = \overline{\mu}^{\omega}(A) > 0 \,.$$

Hence, for a set of λ's of positive Lebesgue measure, $\mu_{\text{sing}}^{\lambda,\omega}(B) > 0$. □

2.8 Examples

In all examples in this subsection $\mathfrak{h}_{\mathcal{R}} = L^2([a,b], \mathrm{d}\mu_{\mathcal{R}})$ and $h_{\mathcal{R}}$ is the operator of multiplication by x. In Examples 1–9 $[a,b] = [0,1]$. In Examples 1 and 2 we do not assume that f is a cyclic vector for $h_{\mathcal{R}}$.

Example 1. In this example we deal with the spectrum outside $]0,1[$. Let

$$\Lambda_0 = \int_0^1 \frac{|f(x)|^2}{x}\,\mathrm{d}\mu_{\mathcal{R}}(x), \qquad \Lambda_1 = \int_0^1 \frac{|f(x)|^2}{x-1}\,\mathrm{d}\mu_{\mathcal{R}}(x) \,.$$

Obviously, $\Lambda_0 \in]0,\infty]$ and $\Lambda_1 \in [-\infty,0[$. If $\lambda^2 > \omega/\Lambda_0$, then $h_{\lambda,\omega}$ has a unique eigenvalue $e < 0$ which satisfies

$$\omega - e - \lambda^2 \int_0^1 \frac{|f(x)|^2}{x-e}\,\mathrm{d}\mu_{\mathcal{R}}(x) = 0 \,. \tag{37}$$

If $\lambda^2 < \omega/\Lambda_0$, then $h_{\lambda,\omega}$ has no eigenvalue in $]-\infty,0[$. 0 is an eigenvalue of $h_{\lambda,\omega}$ iff $\lambda^2 = \omega/\Lambda_0$ and $\int_0^1 |f(x)|^2 x^{-2}\mathrm{d}\mu_{\mathcal{R}}(x) < \infty$. Similarly, if

$$(\omega - 1)/\Lambda_1 < \lambda^2 \,,$$

then $h_{\lambda,\omega}$ has a unique eigenvalue $e > 1$ which satisfies (37), and if

$$(\omega - 1)/\Lambda_1 > \lambda^2 \,,$$

then $h_{\lambda,\omega}$ has no eigenvalue in $]1,\infty[$. 1 is an eigenvalue of $h_{\lambda,\omega}$ iff

$$(\omega - 1)/\Lambda_1 = \lambda^2 ,$$

and $\int_0^1 |f(x)|^2 (x-1)^{-2} \mathrm{d}\mu_{\mathcal{R}}(x) < \infty$.

Example 2. Let $\mathrm{d}\mu_{\mathcal{R}}(x) \equiv \mathrm{d}x|_{[0,1]}$, let f be a continuous function on $]0,1[$, and let

$$\mathcal{S} = \{x \in]0,1[\,|\, f(x) \neq 0\} .$$

The set $\mathcal{S}$ is open in $]0,1[$, and the cyclic space generated by $h_{\mathcal{R}}$ and f is $L^2(\mathcal{S}, \mathrm{d}x)$. The spectrum of

$$h_{\lambda,\omega}|_{(\mathbb{C} \oplus L^2(\mathcal{S},\mathrm{d}x))^\perp} ,$$

is purely absolutely continuous and equal to $[0,1] \setminus \mathcal{S}$. Since for $x \in \mathcal{S}$, $\lim_{y \downarrow 0} \mathrm{Im}\, F_{\mathcal{R}}(x + \mathrm{i}y) = \pi |f(x)|^2 > 0$, the spectrum of $h_{\lambda,\omega}$ in $\mathcal{S}$ is purely absolutely continuous for all $\lambda \neq 0$. Hence, if

$$\mathcal{S} = \bigcup_n]a_n, b_n[,$$

is the decomposition of $\mathcal{S}$ into connected components, then the singular spectrum of $h_{\lambda,\omega}$ inside $[0,1]$ is concentrated on the set $\cup_n \{a_n, b_n\}$. In particular, $h_{\lambda,\omega}$ has no singular continuous spectrum. A point $e \in \cup_n \{a_n, b_n\}$ is an eigenvalue of $h_{\lambda,\omega}$ iff

$$\int_0^1 \frac{|f(x)|^2}{(x-e)^2} \mathrm{d}x < \infty \qquad \text{and} \qquad \omega - e - \lambda^2 \int_0^1 \frac{|f(x)|^2}{x-e} \mathrm{d}x = 0 . \tag{38}$$

Given ω, for each e for which the first condition holds there are precisely two λ's such that e is an eigenvalue of $h_{\lambda,\omega}$. Hence, given ω, the set of λ's for which $h_{\lambda,\omega}$ has eigenvalues in $]0,1[$ is countable. Similarly, given λ, the set of ω's for which $h_{\lambda,\omega}$ has eigenvalues in $]0,1[$ is countable.

Let

$$Z \equiv \{x \in [0,1] \,|\, f(x) = 0\} ,$$

and $\mathfrak{g} \equiv \sup_{x \in Z} G_{\mathcal{R}}(x)$. By (16), the number of eigenvalues of $h_{\lambda,\omega}$ is bounded by $1 + \lambda^2 \mathfrak{g}$. Hence, if $\mathfrak{g} < \infty$, then $h_{\lambda,\omega}$ can have at most finitely many eigenvalues. If, for example,

$$|f(x) - f(y)| \leq C|x-y|^\delta ,$$

for all $x, y \in [0,1]$ and some $\delta > 1/2$, then

$$\begin{aligned} \mathfrak{g} &= \sup_{x \in Z} \int_0^1 \frac{|f(t)|^2}{(t-x)^2} \mathrm{d}t = \sup_{x \in Z} \int_0^1 \frac{|f(t) - f(x)|^2}{(t-x)^2} \mathrm{d}t \\ &\leq \sup_{x \in Z} \int_0^1 \frac{C}{(t-x)^{2(1-\delta)}} \mathrm{d}t < \infty , \end{aligned}$$

and $h_{\lambda,\omega}$ has at most finitely many eigenvalues. On the other hand, given $\lambda \neq 0$, ω, and a finite sequence $E \equiv \{e_1, \ldots, e_n\} \in]0,1[$, one can construct a C^∞ function f with bounded derivatives such that E is precisely the set of eigenvalues of $h_{\lambda,\omega}$ in $]0,1[$.

More generally, let $E \equiv \{e_n\} \subset]0,1[$ be a discrete set. (By discrete we mean that for all n, $\inf_{j\neq n} |e_n - e_j| > 0$ – the accumulation points of E are not in E). Let $\lambda \neq 0$ and ω be given and assume that ω is not an accumulation point of E. Then there is a C^∞ function f such that E is precisely the set of eigenvalues of $h_{\lambda,\omega}$ in $]0,1[$. Of course, in this case $f'(x)$ cannot be bounded. The construction of a such f is somewhat lengthy and can be found in [45].

In the remaining examples we assume $f = \mathbb{1}$. The next two examples are based on [36].

Example 3. Let $\mu_\mathcal{R}$ be a pure point measure with atoms $\mu_\mathcal{R}(x_n) = a_n$. Then

$$G_\mathcal{R}(x) = \sum_{n=1}^{\infty} \frac{a_n}{(x - x_n)^2} \, .$$

If $\sum_n \sqrt{a_n} < \infty$, then $G_\mathcal{R}(x) < \infty$ for Lebesgue a.e. $x \in [0,1]$ (see Theorem 3.1 in [36]). Hence, by Simon-Wolff theorems 3 and 7, for a fixed $\lambda \neq 0$ and Lebesgue a.e. ω, and for a fixed ω and Lebesgue a.e. λ, $h_{\lambda,\omega}$ has only a pure point spectrum. On the other hand, for a fixed $\lambda \neq 0$, there is a dense G_δ set of $\omega \in \mathbb{R}$ such that the spectrum of $h_{\lambda,\omega}$. on $]0,1[$ is purely singular continuous [25, 31].

Example 4 (continuation). Assume that $x_n = x_n(w)$ are independent random variables uniformly distributed on $[0,1]$. We keep the a_n's deterministic and assume that $\sum \sqrt{a}_n = \infty$. Then, for a.e. w, $G_{\mathcal{R},w}(x) = \infty$ for Lebesgue a.e. $x \in [0,1]$ (see Theorem 3.2 in [36]). Hence, by Simon-Wolff theorems 4 and 8, for a fixed $\lambda \neq 0$ and Lebesgue a.e. ω, and for a fixed ω and Lebesgue a.e. λ, the spectrum of $h_{\lambda,\omega}(w)$ on $[0,1]$ is singular continuous with probability 1.

Example 5. Let ν be a probability measure on $[0,1]$ and

$$\mathrm{d}\mu_\mathcal{R}(x) = \frac{1}{2}\left(\mathrm{d}x|_{[0,1]} + \mathrm{d}\nu(x)\right) \, .$$

Since for all $x \in]0,1[$,

$$\liminf_{y \downarrow 0} \operatorname{Im} F_\mathcal{R}(x + \mathrm{i}y) \geq \frac{\pi}{2} \, ,$$

the operator $h_{\lambda,\omega}$ has purely absolutely continuous spectrum on $[0,1]$ for all $\lambda \neq 0$. In particular, the singular spectrum of h_0 associated to ν_{sing} disappears under the influence of the perturbation for all $\lambda \neq 0$.

Example 6. This example is due to Simon-Wolff [61]. Let

$$\mu_n = 2^{-n}\sum_{j=1}^{2^n}\delta_{j2^{-n}} ,$$

and $\mu_{\mathcal{R}} = \sum_n a_n\mu_n$, where $a_n > 0$, $\sum_n a_n = 1$ and $\sum_n 2^n a_n = \infty$. The spectrum of $h_{0,\omega}$ is pure point and equal to $[0,1]\cup\{\omega\}$. For any $x\in[0,1]$ there is j_x such that $|j_x/2^n - x| \le 2^{-n}$. Hence, for all n,

$$\int_{\mathbb{R}}\frac{\mathrm{d}\mu_n(t)}{(t-x)^2} \ge 2^n ,$$

and $G_{\mathcal{R}}(x) = \infty$ for all $x\in[0,1]$. We conclude that for all $\lambda\neq 0$ and all ω the spectrum of $h_{\lambda,\omega}$ on $[0,1]$ is purely singular continuous.

Example 7. Let μ_C be the standard Cantor measure (see [54]). Set

$$\nu_{j,n}(A) \equiv \mu_C(A + j2^{-n}) ,$$

and

$$\mu_{\mathcal{R}} \equiv c\,\chi_{[0,1]}\sum_{n=1}^{\infty} n^{-2}\sum_{j=1}^{2^n}\nu_{j,n} ,$$

where c is the normalization constant. Then $G_{\mathcal{R}}(x) = \infty$ for all $x\in[0,1]$ (see Example 5 in Sect. II.5 of [60]), and the spectrum of $h_{\lambda,\omega}$ on $[0,1]$ is purely singular continuous for all λ,ω.

Example 8. The following example is due to del Rio and Simon (Example 7 in Sect. II.5 of [60]). Let $\{r_n\}$ be the set of rationals in $]0,1/2[$, $a_n = \min(3^{-n-1}, r_n, 1/2 - r_n)$,

$$I_n =]r_n - a_n, r_n + a_n[\ \cup\ ]1 - r_n - a_n, 1 - r_n + a_n[,$$

and $S = \cup_n I_n$. The set S is dense in $[0,1]$ and $|S| \le 2/3$. Let $\mathrm{d}\mu_{\mathcal{R}} = |S|^{-1}\chi_S\mathrm{d}x$. The spectrum of $h_{\mathcal{R}}$ is purely absolutely continuous and equal to $[0,1]$. The set S is the essential support of this absolutely continuous spectrum. Clearly, for all λ,ω, $\mathrm{sp}_{\mathrm{ac}}(h_{\lambda,\omega}) = [0,1]$. By Theorem 5, for any fixed $\lambda\neq 0$, $h_{\lambda,\omega}$ will have some singular spectrum in $[0,1]\setminus S$ for a set of ω's of positive Lebesgue measure. It is not difficult to show that $G_{\mathcal{R}}(x) < \infty$ for Lebesgue a.e. $x\in[0,1]\setminus S$ (see [60]). Hence, for a fixed λ, $h_{\lambda,\omega}$ will have no singular continuous spectrum for Lebesgue a.e. ω but it has some point spectrum in $[0,1]\setminus S$ for a set of ω's of positive Lebesgue measure.
For a given ω, $h_{\lambda,\omega}$ has no singular continuous spectrum for Lebesgue a.e. λ. Note that for Lebesque a.e. $x\in\mathbb{R}\setminus S$, $F_{\mathcal{R}}(x+\mathrm{i}0) = \mathrm{Re}\,F_{\mathcal{R}}(x+\mathrm{i}0)\neq 0$. Since the set S is symmetric with respect to the point $1/2$, we have that for all $z\in\mathbb{C}_{\pm}$, $\mathrm{Re}\,F_{\mathcal{R}}(z) = -\mathrm{Re}\,F_{\mathcal{R}}(-z+1/2)$. Hence,

$$\mathrm{Re}\, F_{\mathcal{R}}(x) = -\mathrm{Re}\, F_{\mathcal{R}}(-x+1/2)\ , \tag{39}$$

and if $|\omega| \geq 1$, then the set

$$\{x \in [0,1] \setminus S \mid (\omega - x)F_{\mathcal{R}}(x) > 0\}\ , \tag{40}$$

has positive Lebesgue measure. Theorem 9 yields that for a given $\omega \notin]0,1[$, $h_{\lambda,\omega}$ will have some point spectrum in $[0,1] \setminus S$ for a set of λ's of positive Lebesgue measure. If $\omega \in]0,1[$, the situation is more complex and depends on the choice of enumeration of the rationals. The enumeration can be always chosen in such a way that for all $0 < \epsilon < 1$, $|S \cap [0,\epsilon]| < \epsilon$. In this case for any given ω the set (40) has positive Lebesgue measure and $h_{\lambda,\omega}$ will have some singular continuous spectrum in $[0,1] \setminus S$ for a set of λ's of positive Lebesgue measure.

Example 9. This example is also due to del Rio and Simon (Example 8 in Sect. II.5 of [60]). Let

$$S_n = \bigcup_{j=1}^{2^n-1} \left] \frac{j}{2^n} - \frac{1}{4n^2 2^n}, \frac{j}{2^n} + \frac{1}{4n^2 2^n} \right[\ ,$$

and $S = \cup_n S_n$. The set S is dense in $[0,1]$ and $|S| < 1$. Let $\mathrm{d}\mu_{\mathcal{R}} = |S|^{-1}\chi_S\, \mathrm{d}x$. Then the absolutely continuous spectrum of $h_{\lambda,\omega}$ is equal to $[0,1]$ for all λ, ω. One easily shows that $G_{\mathcal{R}}(x) = \infty$ on $[0,1]$ (see [60]). Hence, for a fixed λ, $h_{\lambda,\omega}$ will have no point spectrum on $[0,1]$ for Lebesgue a.e. ω but it has some singular continuous spectrum in $[0,1] \setminus S$ for a set of ω's of positive Lebesgue measure.

For a given ω, $h_{\lambda,\omega}$ will have no point spectrum inside $[0,1]$ for Lebesgue a.e. λ. The set S is symmetric with respect to $1/2$ and (39) holds. Since for any $0 < \epsilon < 1$, $|S \cap [0,\epsilon]| < \epsilon$, for any given ω the set

$$\{x \in [0,1] \setminus S \mid (\omega - x)F_{\mathcal{R}}(x) > 0\}\ ,$$

has positive Lebesgue measure. Hence, Theorem 9 yields that for a given ω, $h_{\lambda,\omega}$ will have some singular continuous spectrum in $[0,1] \setminus S$ for a set of λ's of positive Lebesgue measure.

Example 10. Proposition 1 and a theorem of del Rio and Simon [DS] yield that there exist a bounded interval $[a,b]$, a Borel probability measure $\mu_{\mathcal{R}}$ on $[a,b]$ and $\lambda_0 > 0$ such that:

1. $\mathrm{sp}_{\mathrm{ac}}(h_{\lambda,\omega}) = [a,b]$ for all λ, ω.
2. for a set of ω's of positive Lebesgue measure, $h_{\lambda_0,\omega}$ has embedded point spectrum in $[a,b]$.
3. for a set of ω's of positive Lebesgue measure, $h_{\lambda_0,\omega}$ has embedded singular continuous spectrum in $[a,b]$.

Example 11. Proposition 1 and a theorem of del Rio-Fuentes-Poltoratskii [22] yield that there exist a bounded interval $[a, b]$, a Borel probability measure $\mu_{\mathcal{R}}$ on $[a, b]$ and $\lambda_0 > 0$ such that:

1. $\mathrm{sp}_{\mathrm{ac}}(h_{\lambda,\omega}) = [a, b]$ for all λ, ω.
2. for all $\omega \in [0, 1]$, the spectrum of $h_{\lambda_0,\omega}$ is purely absolutely continuous.
3. for all $\omega \notin [0, 1]$, $[a, b] \subset \mathrm{sp}_{\mathrm{sing}}(h_{\lambda_0,\omega})$.

2.9 Digression: the Semi-Circle Law

In the proof of Proposition 1 we have solved the equation (11) for $\mu_{\mathcal{R}}$. In this subsection we will find the fixed point of the (11). More precisely, we will find a finite Borel measure ν whose Borel transform satisfies the functional equation

$$H(z) = \frac{1}{-z - \lambda^2 H(z)} ,$$

or, equivalently

$$\lambda^2 H(z)^2 + zH(z) + 1 = 0 . \tag{41}$$

The unique analytic solution of this equation is

$$H(z) = \frac{\sqrt{z^2 - 4\lambda^2} - z}{2\lambda^2} ,$$

a two-valued function which can be made single valued by cutting the complex plane along the line segment $[-2|\lambda|, 2|\lambda|]$. Only one branch has the Herglotz property $H(\mathbb{C}_+) \subset \mathbb{C}_+$. This branch is explicitly given by

$$H(z) = \frac{1}{|\lambda|}\frac{\xi - 1}{\xi + 1}, \qquad \xi \equiv \sqrt{\frac{z - 2|\lambda|}{z + 2|\lambda|}} ,$$

where the branch of the square root is determined by $\mathrm{Re}\,\xi > 0$ (the so-called principal branch). In particular, $H(x + \mathrm{i}y) \sim \mathrm{i}y^{-1}$ as $y \to +\infty$, and by a well known result in harmonic analysis (see e.g. [37]) there exists a unique Borel probability measure ν such that $F_\nu(z) = H(z)$ for $z \in \mathbb{C}_+$. For all $x \in \mathbb{R}$,

$$\lim_{y \downarrow 0} \mathrm{Im}\, F_\nu(x + \mathrm{i}y) = s_\lambda(x) ,$$

where

$$s_\lambda(x) = \begin{cases} \dfrac{\sqrt{4\lambda^2 - x^2}}{2\lambda^2} & \text{if } |x| \le 2|\lambda| , \\ 0 & \text{if } |x| > 2|\lambda| . \end{cases}$$

We deduce that the measure ν is absolutely continuous w.r.t. Lebesgue measure and that

$$\mathrm{d}\nu(x) = \pi^{-1} s_\lambda(x)\mathrm{d}x .$$

Of course, ν is the celebrated Wigner semi-circle law which naturally arises in the study of the eigenvalue distribution of certain random matrices, see e.g. [48]. The result of this computation will be used in several places in the remaining part of our lectures.

3 The Perturbative Theory

3.1 The Radiating Wigner-Weisskopf Atom

In this section we consider a specific class of WWA models which satisfy the following set of assumptions.

Assumption (A1) $\mathfrak{h}_{\mathcal{R}} = L^2(X, \mathrm{d}x; \mathfrak{K})$, where $X = (e_-, e_+) \subset \mathbb{R}$ is an open (possibly infinite) interval and $\mathfrak{K}$ is a separable Hilbert space. The Hamiltonian $h_{\mathcal{R}} \equiv x$ is the operator of multiplication by x.

Note that the spectrum of $h_{\mathcal{R}}$ is purely absolutely continuous and equal to $\overline{X}$. For notational simplicity in this section we do not assume that f is a cyclic vector for $h_{\mathcal{R}}$. This assumption is irrelevant for our purposes: since the cyclic space $\mathfrak{h}_1$ generated by h_λ and 1 is independent of λ for $\lambda \neq 0$, so is $\mathfrak{h}_1^\perp \subset \mathfrak{h}_{\mathcal{R}}$ and $h_\lambda|_{\mathfrak{h}_1^\perp} = h_{\mathcal{R}}|_{\mathfrak{h}_1^\perp}$ has purely absolutely continuous spectrum.

Assumption (A2) The function

$$g(t) = \int_X \mathrm{e}^{-\mathrm{i}tx} \|f(x)\|_{\mathfrak{K}}^2 \, \mathrm{d}x \, ,$$

is in $L^1(\mathbb{R}, \mathrm{d}t)$.

This assumption implies that $x \mapsto \|f(x)\|_{\mathfrak{K}}$ is a bounded continuous function on $\overline{X}$. Note also that for $\operatorname{Im} z > 0$,

$$F_{\mathcal{R}}(z) = \int_X \frac{\|f(x)\|_{\mathfrak{K}}^2}{x - z} \, \mathrm{d}x = \mathrm{i} \int_0^\infty \mathrm{e}^{\mathrm{i}zs} g(s) \, \mathrm{d}s \, .$$

Hence, $F_{\mathcal{R}}(z)$ is bounded and continuous on the closed half-plane $\overline{\mathbb{C}_+}$. In particular, the function $F_{\mathcal{R}}(x + \mathrm{i}0)$ is bounded and continuous on $\mathbb{R}$. If in addition $t^n g(t) \in L^1(\mathbb{R}, \mathrm{d}t)$ for some positive integer n, then $\|f(x)\|_{\mathfrak{K}}^2$ and $F_{\mathcal{R}}(x + \mathrm{i}0)$ are n-times continuously differentiable with bounded derivatives.

Assumption (A3) $\omega \in X$ and $\|f(\omega)\|_{\mathfrak{K}} > 0$.

This assumption implies that the eigenvalue ω of h_0 is embedded in its absolutely continuous spectrum.

Until the end of this section we will assume that Assumptions (A1)–(A3) hold. We will call the WWA which satisfies (A1)–(A3) the Radiating Wigner-Weisskopf Atom (abbreviated RWWA).

In contrast to the previous section, until the end of the paper we will keep ω fixed and consider *only* λ as the perturbation parameter. In the sequel we drop the subscript ω and write F_λ for $F_{\lambda,\omega}$, etc.

Since $\|f(x)\|_{\mathfrak{K}}$ is a continuous function of x, the argument of Example 2 in Subsect. 2.8 yields that h_λ has no singular continuous spectrum for all λ. However, h_λ may have eigenvalues (and, if $X \neq \mathbb{R}$, it will certainly have them for λ large enough). For λ small, however, the spectrum of h_λ is purely absolutely continuous.

Proposition 9. *There exists $\Lambda > 0$ such that for $0 < |\lambda| < \Lambda$ the spectrum of h_λ is purely absolutely continuous and equal to $\overline{X}$.*

Proof. By Theorem 1, the singular spectrum of h_λ is concentrated on the set

$$S = \{x \in \mathbb{R} \,|\, \omega - x - \lambda^2 F_{\mathcal{R}}(x + \mathrm{i}0) = 0\} \,.$$

Since $\mathrm{Im}\, F_{\mathcal{R}}(\omega + \mathrm{i}0) = \pi\|f(\omega)\|^2_{\mathfrak{K}} > 0$, there is $\epsilon > 0$ such that

$$\mathrm{Im}\, F_{\mathcal{R}}(x + \mathrm{i}0) > 0 \,,$$

for $|x - \omega| < \epsilon$. Let $m \equiv \max_{x\in\mathbb{R}} |F_{\mathcal{R}}(x + \mathrm{i}0)|$ and $\Lambda \equiv (\epsilon/m)^{1/2}$. Then, for $|\lambda| < \Lambda$ and $x \notin \,]\omega - \epsilon, \omega + \epsilon[$, one has $|\omega - x| > \lambda^2 |F_{\mathcal{R}}(x + \mathrm{i}0)|$. Hence, S is empty for $0 < |\lambda| < \Lambda$, and the spectrum of $h_\lambda|_{\mathfrak{h}_1}$ is purely absolutely continuous. □

We finish this subsection with two examples.

Example 1. Assume that $\mathfrak{h}_{\mathcal{R}} = L^2(\mathbb{R}^d, \mathrm{d}^d x)$ and let $h_{\mathcal{R}} = -\Delta$, where Δ is the usual Laplacian in $\mathbb{R}^d$. The Fourier transform

$$\tilde{\varphi}(k) = \frac{1}{(2\pi)^{d/2}} \int_{\mathbb{R}^d} \mathrm{e}^{-\mathrm{i}k\cdot x} \varphi(x) \,\mathrm{d}x \,,$$

maps unitarily $L^2(\mathbb{R}^d, \mathrm{d}^d x)$ onto $L^2(\mathbb{R}^d, \mathrm{d}^d k)$ and the Hamiltonian $h_{\mathcal{R}}$ becomes multiplication by $|k|^2$. By passing to polar coordinates with $r = |k|$ we identify $L^2(\mathbb{R}^d, \mathrm{d}^d k)$ with $L^2(\mathbb{R}_+, r^{d-1}\mathrm{d}r; \mathfrak{K})$, where $\mathfrak{K} = L^2(S^{d-1}, \mathrm{d}\sigma)$, S^{d-1} is the unit sphere in $\mathbb{R}^d$, and $\mathrm{d}\sigma$ is its surface measure. The operator $h_{\mathcal{R}}$ becomes multiplication by r^2. Finally, the map

$$\varphi^\#(x) = 2^{-1/2} x^{\frac{d-2}{4}} \tilde{\varphi}(\sqrt{x}) \,,$$

maps $L^2(\mathbb{R}^d, \mathrm{d}^d x)$ unitarily onto $L^2(X, \mathrm{d}x; \mathfrak{K})$ with $X = (0, \infty)$, and

$$(h_{\mathcal{R}}\varphi)^\#(x) = x\varphi^\#(x) \,.$$

This representation of $\mathfrak{h}_{\mathcal{R}}$ and $h_{\mathcal{R}}$ (sometimes called the *spectral* or the *energy* representation) clearly satisfies (A1).

The function $f^{\#}$ satisfies (A2) iff the function $g(t) = (f|\mathrm{e}^{-ith_{\mathcal{R}}}f)$ is in $L^1(\mathbb{R}, \mathrm{d}t)$. If $f \in L^2(\mathbb{R}^d, \mathrm{d}^d x)$ is compactly supported, then $g(t) = O(t^{-d/2})$, and so if $d \geq 3$, then (A2) holds for all compactly supported f. If $d = 1, 2$, then (A2) holds if f is in the domain of $|x|^2$ and its Fourier transform vanishes in a neighborhood of the origin. The proofs of these facts are simple and can be found in [9], Example 5.4.9.

Example 2. Let $\mathfrak{h}_{\mathcal{R}} = \ell^2(\mathbb{Z}_+)$, where $\mathbb{Z}_+ = \{1, 2, \cdots\}$, and let

$$h_{\mathcal{R}} = \frac{1}{2} \sum_{n \in \mathbb{Z}_+} \Big((\delta_n | \cdot)\delta_{n+1} + (\delta_{n+1} | \cdot)\delta_n \Big) ,$$

where δ_n is the Kronecker delta function at $n \in \mathbb{Z}_+$. $h_{\mathcal{R}}$ is the usual discrete Laplacian on $\ell^2(\mathbb{Z}_+)$ with Dirichlet boundary condition. The Fourier-sine transform

$$\tilde{\varphi}(k) \equiv \sqrt{\frac{2}{\pi}} \sum_{n \in \mathbb{Z}_+} \varphi(n) \sin(kn) ,$$

maps $\ell^2(\mathbb{Z}_+)$ unitarily onto $L^2([0, \pi], \mathrm{d}k)$ and the Hamiltonian $h_{\mathcal{R}}$ becomes multiplication by $\cos k$. Finally, the map

$$\varphi^{\#}(x) = (1 - x^2)^{-1/4} \tilde{\varphi}(\arccos x) ,$$

maps $\ell^2(\mathbb{Z}_+)$ unitarily onto $L^2(X, \mathrm{d}x)$, where $X = (-1, 1)$ and

$$(h_{\mathcal{R}}\varphi)^{\#}(x) = x\varphi^{\#}(x) .$$

If f has bounded support in $\mathbb{Z}_+$, then $|f^{\#}(x)|^2 = (1 - x^2)^{1/2} P_f(x)$, where $P_f(x)$ is a polynomial in x. A simple stationary phase argument yields that $g(t) = O(|t|^{-3/2})$ and Assumption (A2) holds.

3.2 Perturbation Theory of Embedded Eigenvalue

Until the end of this section Λ is the constant in Proposition 9.

Note that the operator $h_0 = \omega \oplus x$ has the eigenvalue ω embedded in the absolutely continuous spectrum of x. On the other hand, for $0 < |\lambda| < \Lambda$ the operator h_λ has no eigenvalue – the embedded eigenvalue has "dissolved" in the absolutely continuous spectrum under the influence of the perturbation. In this subsection we will analyze this phenomenon. At its heart are the concepts of *resonance* and *life-time* of an embedded eigenvalue which are of profound physical importance.

We set $D(w, r) \equiv \{z \in \mathbb{C} \,|\, |z - w| < r\}$. In addition to (A1)–(A3) we will need the following assumption.

Assumption (A4) There exists $\rho > 0$ such that the function

$$\mathbb{C}_+ \ni z \to F_{\mathcal{R}}(z) ,$$

has an analytic continuation across the interval $]\omega - \rho, \omega + \rho[$ to the region $\mathbb{C}_+ \cup D(\omega, \rho)$. We denote the extended function by $F_{\mathcal{R}}^+(z)$.

It is important to note that $F_{\mathcal{R}}^+(z)$ is *different* from $F_{\mathcal{R}}(z)$ for $\mathrm{Im}\, z < 0$. This is obvious from the fact that

$$\mathrm{Im}\, F_{\mathcal{R}}(x + \mathrm{i}0) - \mathrm{Im}\, F_{\mathcal{R}}(x - \mathrm{i}0) = 2\pi \|f(x)\|_{\mathfrak{K}}^2 > 0 ,$$

near ω. In particular, if (A4) holds, then ρ must be such that $]\omega - \rho, \omega + \rho[\subset X$.

Until the end of this subsection we will assume that Assumptions (A1)–(A4) hold.

Theorem 10. *1. The function $F_\lambda(z) = (1|(h_\lambda - z)^{-1}1)$ has a meromorphic continuation from $\mathbb{C}_+$ to the region $\mathbb{C}_+ \cup D(\omega, \rho)$. We denote this continuation by $F_\lambda^+(z)$.*

2. Let $0 < \rho' < \rho$ be given. Then there is $\Lambda' > 0$ such that for $|\lambda| < \Lambda'$ the only singularity of $F_\lambda^+(z)$ in $D(\omega, \rho')$ is a simple pole at $\omega(\lambda)$. The function $\lambda \mapsto \omega(\lambda)$ is analytic for $|\lambda| < \Lambda'$ and

$$\omega(\lambda) = \omega + a_2 \lambda^2 + O(\lambda^4) ,$$

where $a_2 \equiv -F_{\mathcal{R}}(\omega + \mathrm{i}0)$. In particular, $\mathrm{Im}\, a_2 = -\pi \|f(\omega)\|_{\mathfrak{K}}^2 < 0$.

Proof. Part (1) is simple – Assumption A4 and (8) yield that

$$F_\lambda^+(z) = \frac{1}{\omega - z - \lambda^2 F_{\mathcal{R}}^+(z)} ,$$

is the mermorphic continuation of $\mathbb{C}_+ \ni z \mapsto F_\lambda(z)$ to $\mathbb{C}_+ \cup D(\omega, \rho)$.

For a given ρ', choose $\Lambda' > 0$ such that

$$\rho' > |\Lambda'|^2 \sup_{|z|=\rho'} |F_{\mathcal{R}}^+(z)| .$$

By Rouché's theorem, there is an $\epsilon > 0$ such that for $|\lambda| < \Lambda'$ the function $\omega - z - \lambda^2 F_{\mathcal{R}}^+(z)$ has a unique simple zero $\omega(\lambda)$ inside $D(\omega, \rho' + \epsilon)$ such that $|\omega(\lambda) - \omega| < \rho' - \epsilon$. This yields that $F_\lambda^+(z)$ is analytic in $\mathbb{C}_+ \cup D(\omega, \rho' + \epsilon)$ except for a simple pole at $\omega(\lambda)$. The function

$$P(\lambda) \equiv \oint_{|\omega - z| = \rho'} z F_\lambda^+(z) \mathrm{d}z = \sum_{n=0}^{\infty} \lambda^{2n} \oint_{|\omega - z| = \rho'} z \left(\frac{F_{\mathcal{R}}^+(z)}{\omega - z} \right)^n \frac{\mathrm{d}z}{\omega - z} ,$$

is analytic for $|\lambda| < \Lambda'$. Similarly, the function

$$Q(\lambda) \equiv \oint_{|\omega - z| = \rho'} F_\lambda^+(z)\mathrm{d}z = \sum_{n=0}^{\infty} \lambda^{2n} \oint_{|\omega - z| = \rho'} \left(\frac{F_{\mathcal{R}}^+(z)}{\omega - z} \right)^n \frac{\mathrm{d}z}{\omega - z} , \tag{42}$$

is analytic and non-zero for $|\lambda| < \Lambda'$. Since

$$\omega(\lambda) = \frac{P(\lambda)}{Q(\lambda)} ,$$

we see that $\omega(\lambda)$ is analytic for $|\lambda| < \Lambda$ with the power series expansion

$$\omega(\lambda) = \sum_{n=0}^{\infty} \lambda^{2n} a_{2n} .$$

Obviously, $a_0 = \omega$ and

$$a_2 = -\frac{1}{2\pi \mathrm{i}} \oint_{|\omega - z| = \rho'} \frac{F_{\mathcal{R}}^+(z)}{z - \omega} \mathrm{d}z = -F_{\mathcal{R}}^+(\omega) = -F_{\mathcal{R}}(\omega + \mathrm{i}0) .$$

The same formula can be obtained by implicit differentiation of

$$\omega - \omega(\lambda) - \lambda^2 F_{\mathcal{R}}^+(\omega(\lambda)) = 0 ,$$

at $\lambda = 0$. □

Theorem 10 explains the mechanism of "dissolving" of the embedded eigenvalue ω. The embedded eigenvalue ω has moved from the real axis to a point $\omega(\lambda)$ on the second (improperly called "unphysical") Riemann sheet of the function $F_\lambda(z)$. There it remains the singularity of the analytically continued resolvent matrix element $(1|(h_\lambda - z)^{-1}1)$, see Fig. 1.

We now turn to the physically important concept of life-time of the embedded eigenvalue.

Theorem 11. *There exists $\Lambda'' > 0$ such that for $|\lambda| < \Lambda''$ and all $t \geq 0$*

$$(1|\mathrm{e}^{-\mathrm{i}th_\lambda}1) = \mathrm{e}^{-\mathrm{i}t\omega(\lambda)} + O(\lambda^2) .$$

Proof. By Theorem 9 the spectrum of h_λ is purely absolutely continuous for $0 < |\lambda| < \Lambda$. Hence, by Theorem 1,

$$\mathrm{d}\mu^\lambda(x) = \mathrm{d}\mu^\lambda_{\mathrm{ac}}(x) = \frac{1}{\pi} \operatorname{Im} F_\lambda(x + \mathrm{i}0)\,\mathrm{d}x = \frac{1}{\pi} \operatorname{Im} F_\lambda^+(x)\,\mathrm{d}x .$$

Let Λ' and ρ' be the constants in Theorem 10, $\Lambda'' \equiv \min(\Lambda', \Lambda)$, and suppose that $0 < |\lambda| < \Lambda''$. We split the integral representation

$$(1|\mathrm{e}^{-\mathrm{i}th_\lambda}1) = \int_X \mathrm{e}^{-\mathrm{i}tx}\mathrm{d}\mu^\lambda(x) , \tag{43}$$

into three parts as

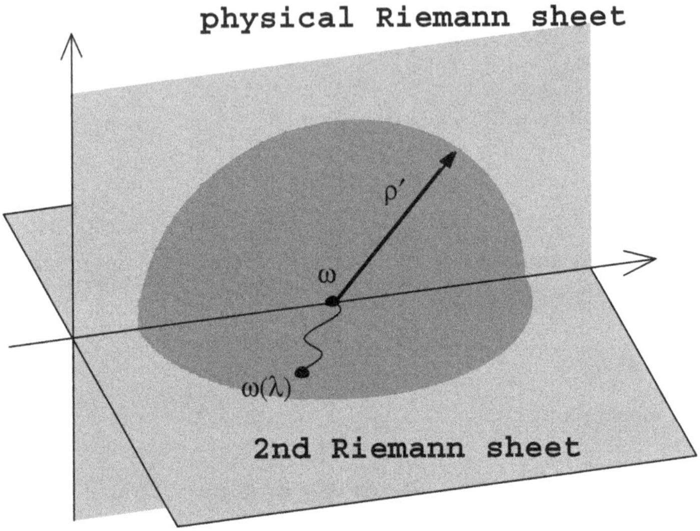

Fig. 1. The resonance pole $\omega(\lambda)$

$$\int_{e_-}^{\omega-\rho\prime} + \int_{\omega-\rho'}^{\omega+\rho'} + \int_{\omega+\rho'}^{e_+} .$$

Equation (8) yields

$$\operatorname{Im} F_\lambda^+(x) = \lambda^2 \frac{\operatorname{Im} F_{\mathcal{R}}^+(x)}{|\omega - x - \lambda^2 F_{\mathcal{R}}^+(x)|^2} ,$$

and so the first term and the third term can be estimated as $O(\lambda^2)$. The second term can be written as

$$I(t) \equiv \frac{1}{2\pi \mathrm{i}} \int_{\omega-\rho'}^{\omega+\rho'} \mathrm{e}^{-\mathrm{i}tx} \left(F_\lambda^+(x) - \overline{F_\lambda^+(x)} \right) \mathrm{d}x .$$

The function $z \mapsto \overline{F_\lambda^+(\overline{z})}$ is meromorphic in an open set containing $D(\omega, \rho)$ with only singularity at $\overline{\omega(\lambda)}$. We thus have

$$I(t) = -R(\lambda)\, \mathrm{e}^{-\mathrm{i}t\omega(\lambda)} + \int_\gamma \mathrm{e}^{-\mathrm{i}tz} \left(F_\lambda^+(z) - \overline{F_\lambda^+(\overline{z})} \right) \mathrm{d}z ,$$

where the half-circle $\gamma = \{z \,|\, |z - \omega| = \rho', \operatorname{Im} z \leq 0\}$ is positively oriented and

$$R(\lambda) = \operatorname{Res}_{z=\omega(\lambda)} F_\lambda^+(z) .$$

By (42), $R(\lambda) = Q(\lambda)/2\pi\mathrm{i}$ is analytic for $|\lambda| < \Lambda''$ and

$$R(\lambda) = -1 + O(\lambda^2) .$$

Equation (8) yields that for $z \in \gamma$

$$F_\lambda^+(z) = \frac{1}{\omega - z} + O(\lambda^2) .$$

Since ω is real, this estimate yields

$$F_\lambda^+(z) - \overline{F_\lambda^+(\overline{z})} = O(\lambda^2) .$$

Combining the estimates we derive the statement. □

If a quantum mechanical system, described by the Hilbert space $\mathfrak{h}$ and the Hamiltonian h_λ, is initially in a pure state described by the vector 1, then

$$P(t) = |(1|\mathrm{e}^{-ith_\lambda}1)|^2 ,$$

is the probability that the system will be in the same state at time t. Since the spectrum of h_λ is purely absolutely continuous, by the Riemann-Lebesgue lemma $\lim_{t\to\infty} P(t) = 0$. On physical grounds one expects more, namely an approximate relation

$$P(t) \sim \mathrm{e}^{-t\Gamma(\lambda)} , \tag{44}$$

where $\Gamma(\lambda)$ is the so-called radiative life-time of the state 1. The strict exponential decay $P(t) = O(\mathrm{e}^{-at})$ is possible only if $X = \mathbb{R}$. Since in a typical physical situation $X \neq \mathbb{R}$, the relation (44) is expected to hold on an intermediate time scale (for times which are not "too long" or "too short"). Theorem 11 is a mathematically rigorous version of these heuristic claims and $\Gamma(\lambda) = -2\,\mathrm{Im}\,\omega(\lambda)$. The computation of the radiative life-time is of paramount importance in quantum mechanics and the reader may consult standard references [13, 34, 49] for additional information.

3.3 Complex Deformations

In this subsection we will discuss Assumption (A4) and the perturbation theory of the embedded eigenvalue in some specific situations.

Example 1. In this example we consider the case $X =]0, \infty[$.

Let $0 < \delta < \pi/2$ and $\mathcal{A}(\delta) = \{z \in \mathbb{C} \,|\, \mathrm{Re}\, z > 0, |\mathrm{Arg}\, z| < \delta\}$. We denote by $H_\mathrm{d}^2(\delta)$ the class of all functions $f : X \to \mathfrak{K}$ which have an analytic continuation to the sector $\mathcal{A}(\delta)$ such that

$$\|f\|_\delta^2 = \sup_{|\theta|<\delta} \int_X \|f(\mathrm{e}^{\mathrm{i}\theta}x)\|_\mathfrak{K}^2 \mathrm{d}x < \infty .$$

The class $H_\mathrm{d}^2(\delta)$ is a Hilbert space. The functions in $H_\mathrm{d}^2(\delta)$ are sometimes called *dilation analytic.*

Proposition 10. *Assume that* $f \in H_\mathrm{d}^2(\delta)$. *Then Assumption* (A4) *holds in the following stronger form:*

1. *The function* $F_{\mathcal{R}}(z)$ *has an analytic continuation to the region* $\mathbb{C}_+ \cup \mathcal{A}(\delta)$. *We denote the extended function by* $F_{\mathcal{R}}^+(z)$.
2. *For* $0 < \delta' < \delta$ *and* $\epsilon > 0$ *one has*

$$\sup_{|z|>\epsilon, z\in\mathcal{A}(\delta')} |F_{\mathcal{R}}^+(z)| < \infty .$$

Proof. The proposition follows from the representation

$$F_{\mathcal{R}}(z) = \int_X \frac{\|f(x)\|_{\mathfrak{K}}^2}{x-z}\,\mathrm{d}x = \mathrm{e}^{\mathrm{i}\theta} \int_X \frac{(f(\mathrm{e}^{-\mathrm{i}\theta}x)|f(\mathrm{e}^{\mathrm{i}\theta}x))_{\mathfrak{K}}}{\mathrm{e}^{\mathrm{i}\theta}x - z}\,\mathrm{d}x , \tag{45}$$

which holds for $\mathrm{Im}\, z > 0$ and $-\delta < \theta \le 0$. This representation can be proven as follows.

Let $\gamma(\theta)$ be the half-line $\mathrm{e}^{\mathrm{i}\theta}\mathbb{R}_+$. We wish to prove that for $\mathrm{Im}\, z > 0$

$$\int_X \frac{\|f(x)\|_{\mathfrak{K}}^2}{x-z}\,\mathrm{d}x = \int_{\gamma(\theta)} \frac{(f(\overline{w})|f(w))_{\mathfrak{K}}}{w-z}\,\mathrm{d}w .$$

To justify the interchange of the line of integration, it suffices to show that

$$\lim_{n\to\infty} r_n \int_\theta^0 \frac{|(f(r_n\mathrm{e}^{-\mathrm{i}\varphi})|f(r_n\mathrm{e}^{\mathrm{i}\varphi}))_{\mathfrak{K}}|}{|r_n\mathrm{e}^{\mathrm{i}\varphi} - z|}\,\mathrm{d}\varphi = 0 ,$$

along some sequence $r_n \to \infty$. This fact follows from the estimate

$$\int_X \left[\int_\theta^0 \frac{x|(f(\mathrm{e}^{-\mathrm{i}\varphi}x)|f(\mathrm{e}^{\mathrm{i}\varphi}x))_{\mathfrak{K}}|}{|\mathrm{e}^{\mathrm{i}\varphi}x - z|}\,\mathrm{d}\varphi\right]\mathrm{d}x \le C_z \|f\|_\delta^2 .$$

□

Until the end of this example we assume that $f \in H_{\mathrm{d}}^2(\delta)$ and that Assumption (A2) holds (this is the case, for example, if $f' \in H_{\mathrm{d}}^2(\delta)$ and $f(0) = 0$). Then, Theorems 10 and 11 hold in the following stronger forms.

Theorem 12. *1. The function*

$$F_\lambda(z) = (1|(h_\lambda - z)^{-1}1) ,$$

has a meromorphic continuation from $\mathbb{C}_+$ *to the region* $\mathbb{C}_+ \cup \mathcal{A}(\delta)$. *We denote this continuation by* $F_\lambda^+(z)$.

2. *Let* $0 < \delta' < \delta$ *be given. Then there is* $\Lambda' > 0$ *such that for* $|\lambda| < \Lambda'$ *the only singularity of* $F_\lambda^+(z)$ *in* $\mathcal{A}(\delta')$ *is a simple pole at* $\omega(\lambda)$. *The function* $\lambda \mapsto \omega(\lambda)$ *is analytic for* $|\lambda| < \Lambda'$ *and*

$$\omega(\lambda) = \omega + \lambda^2 a_2 + O(\lambda^4) ,$$

where $a_2 = -F_{\mathcal{R}}(\omega + \mathrm{i}0)$. *In particular,* $\mathrm{Im}\, a_2 = -\pi\|f(\omega)\|_{\mathfrak{K}}^2 < 0$.

Theorem 13. *There exists $\Lambda'' > 0$ such that for $|\lambda| < \Lambda''$ and all $t \geq 0$,*

$$(1|\mathrm{e}^{-\mathrm{i}th_\lambda}1) = \mathrm{e}^{-\mathrm{i}t\omega(\lambda)} + O(\lambda^2 t^{-1})\ .$$

The proof of Theorem 13 starts with the identity

$$(1|\mathrm{e}^{-\mathrm{i}th_\lambda}1) = \lambda^2 \int_X \mathrm{e}^{-\mathrm{i}tx} \|f(x)\|_{\mathfrak{K}}^2 \, |F_\lambda^+(x)|^2 \, \mathrm{d}x\ .$$

Given $0 < \delta' < \delta$ one can find Λ'' such that for $|\lambda| < \Lambda''$

$$(1|\mathrm{e}^{-\mathrm{i}th_\lambda}1) = \mathrm{e}^{-\mathrm{i}t\omega(\lambda)} + \lambda^2 \int_{\mathrm{e}^{-\mathrm{i}\delta'}\mathbb{R}_+} \mathrm{e}^{-\mathrm{i}tw} (f(\overline{w})|f(w))_{\mathfrak{K}} \, \overline{F_\lambda^+(\overline{w})} F_\lambda^+(w) \, \mathrm{d}w\ , \tag{46}$$

and the integral on the right is easily estimated by $O(t^{-1})$. We leave the details of the proof as an exercise for the reader.

Example 2. We will use the structure of the previous example to illustrate the complex deformation method in study of resonances. In this example we assume that $f \in H_{\mathrm{d}}^2(\delta)$.

We define a group $\{u(\theta) \,|\, \theta \in \mathbb{R}\}$ of unitaries on $\mathfrak{h}$ by

$$u(\theta) : \alpha \oplus f(x) \mapsto \alpha \oplus \mathrm{e}^{\theta/2} f(e^\theta x)\ .$$

Note that $h_{\mathcal{R}}(\theta) \equiv u(-\theta) h_{\mathcal{R}} u(\theta)$ is the operator of multiplication by $\mathrm{e}^{-\theta}x$. Set $h_0(\theta) = \omega \oplus h_{\mathcal{R}}(\theta)$, $f_\theta(x) = u(-\theta) f(x) u(\theta) = f(\mathrm{e}^{-\theta}x)$, and

$$h_\lambda(\theta) = h_0(\theta) + \lambda\left((1|\cdot) f_\theta + (f_\theta|\cdot) 1\right)\ .$$

Clearly, $h_\lambda(\theta) = u(-\theta) h_\lambda u(\theta)$.

We set $S(\delta) \equiv \{z \,|\, |\mathrm{Im}\, z| < \delta\}$ and note that the operator $h_0(\theta)$ and the function f_θ are defined for all $\theta \in S(\delta)$. We define $h_\lambda(\theta)$ for $\lambda \in \mathbb{C}$ and $\theta \in S(\delta)$ by

$$h_\lambda(\theta) = h_0(\theta) + \lambda\left((1|\cdot) f_\theta + (f_{\overline{\theta}}|\cdot) 1\right)\ .$$

The operators $h_\lambda(\theta)$ are called dilated Hamiltonians. The basic properties of this family of operators are:

1. $\mathrm{Dom}\,(h_\lambda(\theta))$ is independent of λ and θ and equal to $\mathrm{Dom}\,(h_0)$.
2. For all $\phi \in \mathrm{Dom}\,(h_0)$ the function $\mathbb{C} \times S(\delta) \ni (\lambda, \theta) \mapsto h_\lambda(\theta)\phi$ is analytic.
3. If $\mathrm{Im}\,\theta = \mathrm{Im}\,\theta'$, then the operators $h_\lambda(\theta)$ and $h_\lambda(\theta')$ are unitarily equivalent, namely

$$h_0(\theta') = u(-(\theta' - \theta)) h_0(\theta) u(\theta' - \theta)\ .$$

4. $\mathrm{sp}_{\mathrm{ess}}(h_0(\theta)) = \mathrm{e}^{-\theta}\mathbb{R}_+$ and $\mathrm{sp}_{\mathrm{disc}}(h_0(\theta)) = \{\omega\}$, see Fig. 2.

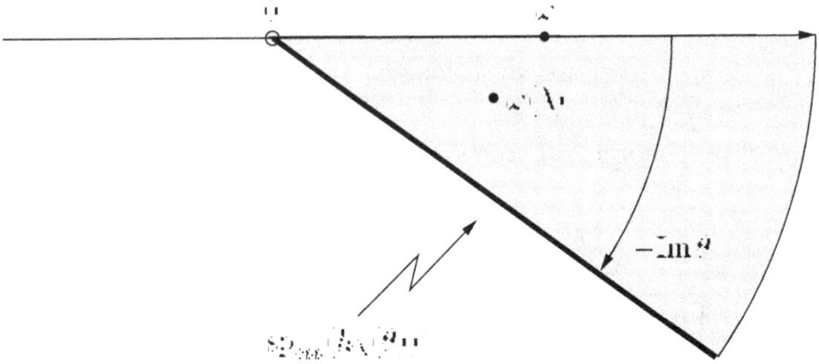

Fig. 2. The spectrum of the dilated Hamiltonian $h_\lambda(\theta)$

The important aspect of (4) is that while ω is an embedded eigenvalue of h_0, it is an isolated eigenvalue of $h_0(\theta)$ as soon as $\operatorname{Im}\theta < 0$. Hence, if $\operatorname{Im}\theta < 0$, then regular perturbation theory can be applied to the isolated eigenvalue ω. Clearly, for all λ, $\mathrm{sp}_{\mathrm{ess}}(h_\lambda(\theta)) = \mathrm{sp}(h_0(\theta))$ and one easily shows that for λ small enough $\mathrm{sp}_{\mathrm{disc}}(h_\lambda)(\theta) = \{\tilde{\omega}(\lambda)\}$ (see Fig. 2). Moreover, if $0 < \rho < \min\{\omega, \omega\tan\theta\}$, then for sufficiently small λ,

$$\tilde{\omega}(\lambda) = \frac{\oint_{|z-\omega|=\rho} z(1|(h_\lambda(\theta)-z)^{-1}1)\,\mathrm{d}z}{\oint_{|z-\omega|=\rho} (1|(h_\lambda(\theta)-z)^{-1}1)\,\mathrm{d}z} .$$

The reader should not be surprised that the eigenvalue $\tilde{\omega}(\lambda)$ is precisely the pole $\omega(\lambda)$ of $F_\lambda^+(z)$ discussed in Theorem 10 (in particular, $\tilde{\omega}(\lambda)$ is independent of θ). To clarify this connection, note that $u(\theta)1 = 1$. Thus, for real θ and $\operatorname{Im} z > 0$,

$$F_\lambda(z) = (1|(h_\lambda - z)^{-1}1) = (1|(h_\lambda(\theta) - z)^{-1}1) .$$

On the other hand, the function $\mathbb{R} \ni \theta \mapsto (1|h_\lambda(\theta) - z)^{-1}1)$ has an analytic continuation to the strip $-\delta < \operatorname{Im}\theta < \operatorname{Im} z$. This analytic continuation is a constant function, and so

$$F_\lambda^+(z) = (1|(h_\lambda(\theta) - z)^{-1}1) ,$$

for $-\delta < \operatorname{Im}\theta < 0$ and $z \in \mathbb{C}_+ \cup \mathcal{A}(|\operatorname{Im}\theta|)$. This yields that $\omega(\lambda) = \tilde{\omega}(\lambda)$.

The above set of ideas plays a very important role in mathematical physics. For additional information and historical perspective we refer the reader to [1,6,11,20,57,60].

Example 3. In this example we consider the case $X = \mathbb{R}$.

Let $\delta > 0$. We denote by $H_{\mathrm{t}}^2(\delta)$ the class of all functions $f : X \to \mathfrak{K}$ which have an analytic continuation to the strip $S(\delta)$ such that

$$\|f\|_\delta^2 \equiv \sup_{|\theta|<\delta} \int_X \|f(x+\mathrm{i}\theta)\|_{\mathfrak{K}}^2 \,\mathrm{d}x < \infty \,.$$

The class $H_{\mathrm{t}}^2(\delta)$ is a Hilbert space. The functions in $H_{\mathrm{t}}^2(\delta)$ are sometimes called *translation analytic*.

Proposition 11. *Assume that* $f \in H_{\mathrm{t}}^2(\delta)$*. Then the function* $F_{\mathcal{R}}(z)$ *has an analytic continuation to the half-plane* $\{z \in \mathbb{C} \,|\, \mathrm{Im}\, z > -\delta\}$.

The proposition follows from the relation

$$F_{\mathcal{R}}(z) = \int_X \frac{\|f(x)\|_{\mathfrak{K}}^2}{x-z} \,\mathrm{d}x = \int_X \frac{(f(x-\mathrm{i}\theta)|f(x+\mathrm{i}\theta))_{\mathfrak{K}}}{x+\mathrm{i}\theta-z} \,\mathrm{d}x \,, \tag{47}$$

which holds for $\mathrm{Im}\, z > 0$ and $-\delta < \theta \leq 0$. The proof of (47) is similar to the proof of (45).

Until the end of this example we will assume that $f \in H_{\mathrm{t}}^2(\delta)$. A change of the line of integration yields that the function

$$g(t) = \int_{\mathbb{R}} \mathrm{e}^{-\mathrm{i}tx} \|f(x)\|_{\mathfrak{K}}^2 \,\mathrm{d}x \,,$$

satisfies the estimate $|g(t)| \leq \mathrm{e}^{-\delta|t|}\|f\|_\delta^2$, and so Assumption (A2) holds. Moreover, Theorems 10 and 11 hold in the following stronger forms.

Theorem 14. *1. The function*

$$F_\lambda(z) = (1|(h_\lambda - z)^{-1}1) \,,$$

has a meromorphic continuation from $\mathbb{C}_+$ *to the half-plane*

$$\{z \in \mathbb{C} \,|\, \mathrm{Im}\, z > -\delta\} \,.$$

We denote this continuation by $F_\lambda^+(z)$.

2. Let $0 < \delta' < \delta$ *be given. Then there is* $\Lambda' > 0$ *such that for* $|\lambda| < \Lambda'$ *the only singularity of* $F_\lambda^+(z)$ *in* $\{z \in \mathbb{C} \,|\, \mathrm{Im}\, z > -\delta'\}$ *is a simple pole at* $\omega(\lambda)$*.* $\omega(\lambda)$ *is analytic for* $|\lambda| < \Lambda'$ *and*

$$\omega(\lambda) = \omega + \lambda^2 a_2 + O(\lambda^4) \,,$$

where $a_2 = -F_{\mathcal{R}}(\omega + \mathrm{i}0)$*. In particular,* $\mathrm{Im}\, a_2 = -\pi\|f(\omega)\|_{\mathfrak{K}}^2 < 0$.

Theorem 15. *Let* $0 < \delta' < \delta$ *be given. Then there exists* $\Lambda'' > 0$ *such that for* $|\lambda| < \Lambda''$ *and all* $t \geq 0$

$$(1|\mathrm{e}^{-\mathrm{i}th_\lambda}1) = \mathrm{e}^{-\mathrm{i}t\omega(\lambda)} + O(\lambda^2\mathrm{e}^{-\delta' t}) \,.$$

In this example the survival probability has strict exponential decay.

We would like to mention two well-known models in mathematical physics for which analogs of Theorems 14 and 15 holds. The first model is the Stark Hamiltonian which describes charged quantum particle moving under the influence of a constant electric field [35]. The second model is the spin-boson system at positive temperature [39, 40].

In the translation analytic case, one can repeat the discussion of the previous example with the analytic family of operators

$$h_\lambda(\theta) = \omega \oplus (x+\theta) + \lambda\left((1|\cdot)f_\theta + (f_{\overline{\theta}}|\cdot)1\right) ,$$

where $f_\theta(x) \equiv f(x+\theta)$ (see Fig. 3). Note that in this case

$$\mathrm{sp}_{\mathrm{ess}}(h_\lambda(\theta)) = \mathrm{sp}_{\mathrm{ess}}(h_0(\theta)) = \mathbb{R} + \mathrm{i}\,\mathrm{Im}\,\theta .$$

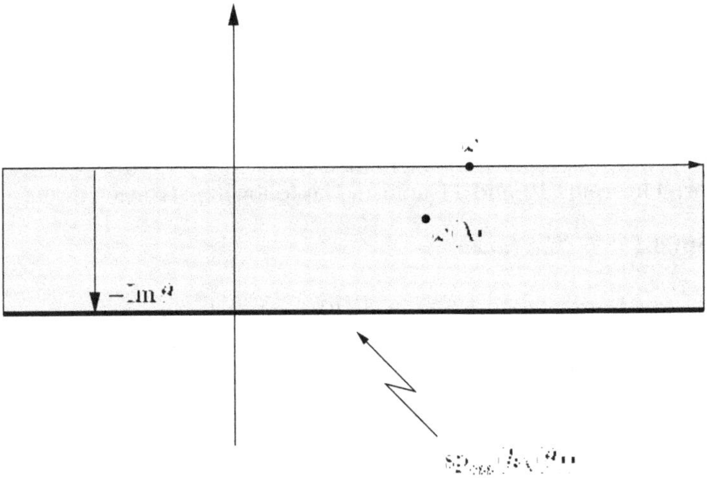

Fig. 3. The spectrum of the translated Hamiltonian $h_\lambda(\theta)$

Example 4. Let us consider the model described in Example 2 of Subsect. 3.1 where $f \in \ell^2(\mathbb{Z}_+)$ has bounded support. In this case $X =]-1,1[$ and

$$F_{\mathcal{R}}(z) = \int_{-1}^{1} \frac{\sqrt{1-x^2}}{x-z} P_f(x)\,\mathrm{d}x , \tag{48}$$

where $P_f(x)$ is a polynomial in x. Since the integrand is analytic in the cut plane $\mathbb{C} \setminus \{x \in \mathbb{R}\,|\,|x| \geq 1\}$, we can deform the path of integration to any curve γ joining -1 to 1 and lying entirely in the lower half-plane. This shows that the function $F_{\mathcal{R}}(z)$ has an analytic continuation from $\mathbb{C}_+$ to the entire cut plane $\mathbb{C} \setminus \{x \in \mathbb{R}\,|\,|x| \geq 1\}$. Assumption (A4) holds in this case.

3.4 Weak Coupling Limit

The first computation of the radiative life-time in quantum mechanics goes back to the seminal papers of Dirac [23] and Wigner and Weisskopf [64]. Consider the survival probability $P(t)$ and assume that $P(t) \sim \mathrm{e}^{-t\Gamma(\lambda)}$ where $\Gamma(\lambda) = \lambda^2\Gamma_2 + O(\lambda^3)$ for λ small. To compute the first non-trivial coefficient Γ_2, Dirac devised a computational scheme called time-dependent perturbation theory. Dirac's formula for Γ_2 was called *Golden Rule* in Fermi's lectures [27], and since then this formula is known as *Fermi's Golden Rule.*

One possible mathematically rigorous approach to time-dependent perturbation theory is the so-called weak coupling (or van Hove) limit. The idea is to study $P(t/\lambda^2)$ as $\lambda \to 0$. Under very general conditions one can prove that

$$\lim_{\lambda\to 0} P(t/\lambda^2) = \mathrm{e}^{-t\Gamma_2} ,$$

and that Γ_2 is given by Dirac's formula (see [15, 16]).

In this section we will discuss the weak coupling limit for the RWWA. We will prove:

Theorem 16. *Suppose that Assumptions* (A1)–(A3) *hold. Then*

$$\lim_{\lambda\to 0} \left| (1|\mathrm{e}^{-ith_\lambda/\lambda^2} 1) - \mathrm{e}^{-it\omega/\lambda^2} \mathrm{e}^{itF_{\mathcal{R}}(\omega+\mathrm{i}0)} \right| = 0 ,$$

for any $t \geq 0$. *In particular,*

$$\lim_{\lambda\to 0} |(1|\mathrm{e}^{-ith_\lambda/\lambda^2} 1)|^2 = \mathrm{e}^{-2\pi\|f(\omega)\|^2_{\mathfrak{K}} t} .$$

Remark. If in addition Assumption (A4) holds, then Theorem 16 is an immediate consequence of Theorem 11. The point is that the leading contribution to the life-time can be rigorously derived under much weaker regularity assumptions.

Lemma 3. *Suppose that Assumptions* (A1)–(A3) *hold. Let* u *be a bounded continuous function on* $\overline{X}$. *Then*

$$\lim_{\lambda\to 0} \left| \lambda^2 \int_X \mathrm{e}^{-itx/\lambda^2} u(x) |F_\lambda(x+\mathrm{i}0)|^2 \,\mathrm{d}x - \frac{u(\omega)}{\|f(\omega)\|^2_{\mathfrak{K}}} \mathrm{e}^{-it(\omega/\lambda^2 - F_{\mathcal{R}}(\omega+\mathrm{i}0))} \right| = 0 ,$$

for any $t \geq 0$.

Proof. We set $l_\omega(x) \equiv |\omega - x - \lambda^2 F_{\mathcal{R}}(\omega+\mathrm{i}0)|^{-2}$ and

$$I_\lambda(t) \equiv \lambda^2 \int_X \mathrm{e}^{-itx/\lambda^2} u(x) |F_\lambda(x+\mathrm{i}0)|^2 \,\mathrm{d}x .$$

We write $u(x)|F_\lambda(x+\mathrm{i}0)|^2$ as

$$u(\omega)l_\omega(x) + (u(x) - u(\omega))l_\omega(x) + u(x)\left(|F_\lambda(x+\mathrm{i}0)|^2 - l_\omega(x)\right) ,$$

and decompose $I_\lambda(t)$ into three corresponding pieces $I_{k,\lambda}(t)$. The first piece is

$$I_{1,\lambda}(t) = \lambda^2 u(\omega) \int_{e_-}^{e_+} \frac{\mathrm{e}^{-\mathrm{i}tx/\lambda^2}}{(\omega - x - \lambda^2 \mathrm{Re}\, F_{\mathcal{R}}(\omega+\mathrm{i}0))^2 + (\lambda^2 \mathrm{Im}\, F_{\mathcal{R}}(\omega+\mathrm{i}0))^2}\, \mathrm{d}x .$$

The change of variable

$$y = \frac{x - \omega + \lambda^2 \mathrm{Re}\, F_{\mathcal{R}}(\omega+\mathrm{i}0)}{\lambda^2 \mathrm{Im}\, F_{\mathcal{R}}(\omega+\mathrm{i}0)} ,$$

and the relation $\mathrm{Im}\, F_{\mathcal{R}}(\omega+\mathrm{i}0) = \pi\|f(\omega)\|^2_{\mathfrak{K}}$ yield that

$$I_{1,\lambda}(t) = \mathrm{e}^{-\mathrm{i}t(\omega/\lambda^2 - \mathrm{Re}\, F_{\mathcal{R}}(\omega+\mathrm{i}0))} \frac{u(\omega)}{\|f(\omega)\|^2_{\mathfrak{K}}} \frac{1}{\pi} \int_{e_-(\lambda)}^{e_+(\lambda)} \frac{\mathrm{e}^{-\mathrm{i}t\mathrm{Im}\, F_{\mathcal{R}}(\omega+\mathrm{i}0)y}}{y^2+1}\, \mathrm{d}y ,$$

where

$$e_\pm(\lambda) \equiv \lambda^{-2} \frac{e_\pm - \omega}{\pi\|f(\omega)\|^2_{\mathfrak{K}}} + \frac{\mathrm{Re}\, F_{\mathcal{R}}(\omega+\mathrm{i}0)}{\pi\|f(\omega)\|^2_{\mathfrak{K}}} \to \pm\infty ,$$

as $\lambda \to 0$. From the formula

$$\frac{1}{\pi}\int_{-\infty}^{\infty} \frac{\mathrm{e}^{-\mathrm{i}t\mathrm{Im}\, F_{\mathcal{R}}(\omega+\mathrm{i}0)y}}{y^2+1}\, \mathrm{d}y = \mathrm{e}^{-t\mathrm{Im}\, F_{\mathcal{R}}(\omega+\mathrm{i}0)} ,$$

we obtain that

$$I_{1,\lambda}(t) = \frac{u(\omega)}{\|f(\omega)\|^2_{\mathfrak{K}}} \mathrm{e}^{-\mathrm{i}t(\omega/\lambda^2 - F_{\mathcal{R}}(\omega+\mathrm{i}0))} (1 + o(1)) , \tag{49}$$

as $\lambda \to 0$.

Using the boundedness and continuity properties of u and l_ω, one easily shows that the second and the third piece can be estimated as

$$|I_{2,\lambda}(t)| \le \lambda^2 \int_X |u(x) - u(\omega)| l_\omega(x)\, \mathrm{d}x,$$

$$|I_{3,\lambda}(t)| \le \lambda^2 \int_X |u(x)| \left||F_\lambda(x+\mathrm{i}0)|^2 - l_\omega(x)\right| \mathrm{d}x .$$

Hence, they vanish as $\lambda \to 0$, and the result follows from (49). □**Proof of Theorem 16.** Let Λ be as in Proposition 9. Recall that for $0 < |\lambda| < \Lambda$ the spectrum of h_λ is purely absolutely continuous. Hence, for λ small,

$$\begin{aligned}(1|\mathrm{e}^{-\mathrm{i}th_\lambda/\lambda^2}1) &= \frac{1}{\pi}\int_X \mathrm{e}^{-\mathrm{i}tx/\lambda^2} \mathrm{Im}\, F_\lambda(x+\mathrm{i}0)\mathrm{d}x \\ &= \frac{1}{\pi}\int_X \mathrm{e}^{-\mathrm{i}tx/\lambda^2} |F_\lambda(x+\mathrm{i}0)|^2 \mathrm{Im}\, F_{\mathcal{R}}(x+\mathrm{i}0)\mathrm{d}x \\ &= \lambda^2 \int_X \mathrm{e}^{-\mathrm{i}tx/\lambda^2} \|f(x)\|^2_{\mathfrak{K}} |F_\lambda(x+\mathrm{i}0)|^2 \mathrm{d}x ,\end{aligned}$$

where we used (19). This formula and Lemma 3 yield Theorem 16. □

The next result we wish to discuss concerns the weak coupling limit for the form of the emitted wave. Let $p_{\mathcal{R}}$ be the orthogonal projection on the subspace $\mathfrak{h}_{\mathcal{R}}$ of $\mathfrak{h}$.

Theorem 17. *For any $g \in C_0(\mathbb{R})$,*

$$\lim_{\lambda \to 0} (p_{\mathcal{R}} \mathrm{e}^{-ith_\lambda/\lambda^2} 1 | g(h_{\mathcal{R}}) p_{\mathcal{R}} \mathrm{e}^{-ith_\lambda/\lambda^2} 1) = g(\omega) \left(1 - \mathrm{e}^{-2\pi \| f(\omega) \|_{\mathfrak{K}}^2 t}\right) . \quad (50)$$

Proof. Using the decomposition

$$\begin{aligned} p_{\mathcal{R}} g(h_{\mathcal{R}}) p_{\mathcal{R}} &= (p_{\mathcal{R}} g(h_{\mathcal{R}}) p_{\mathcal{R}} - g(h_0)) + (g(h_0) - g(h_\lambda)) + g(h_\lambda) \\ &= -g(\omega)(1| \cdot)1 + (g(h_0) - g(h_\lambda)) + g(h_\lambda) , \end{aligned}$$

we can rewrite $(p_{\mathcal{R}} \mathrm{e}^{-ith_\lambda/\lambda^2} 1 | g(h_{\mathcal{R}}) p_{\mathcal{R}} \mathrm{e}^{-ith_\lambda/\lambda^2} 1)$ as a sum of three pieces. The first piece is equal to

$$- g(\omega) |(1| \mathrm{e}^{-ith_\lambda/\lambda^2} 1)|^2 = -g(\omega) e^{-2\pi \| f(\omega) \|_{\mathfrak{K}}^2 t} . \quad (51)$$

Since $\lambda \mapsto h_\lambda$ is continuous in the norm resolvent sense, we have

$$\lim_{\lambda \to 0} \| g(h_\lambda) - g(h_0) \| = 0 ,$$

and the second piece can be estimated

$$(\mathrm{e}^{-ith_\lambda/\lambda^2} 1 | (g(h_0) - g(h_\lambda)) \mathrm{e}^{-ith_\lambda/\lambda^2} 1) = o(1) , \quad (52)$$

as $\lambda \to 0$. The third piece satisfies

$$\begin{aligned} (\mathrm{e}^{-ith_\lambda/\lambda^2} 1 | g(h_\lambda) \mathrm{e}^{-ith_\lambda/\lambda^2} 1) &= (1 | g(h_\lambda) 1) \\ &= (1 | g(h_0) 1) + (1 | (g(h_\lambda) - g(h_0)) 1) \quad (53) \\ &= g(\omega) + o(1) , \end{aligned}$$

as $\lambda \to 0$. Equation (51), (52) and (53) yield the statement. □

Needless to say, Theorems 16 and 17 can be also derived from the general theory of weak coupling limit developed in [15,16]. For additional information about the weak coupling limit we refer the reader to [15, 16, 21, 29, 33, 62].

3.5 Examples

In this subsection we describe the meromorphic continuation of

$$F_\lambda(z) = (1 | (h_\lambda - z)^{-1} 1) ,$$

across $\mathrm{sp_{ac}}(h_\lambda)$ in some specific examples which allow for explicit computations. Since $F_\lambda(z) = F_{-\lambda}(z)$, we need to consider only $\lambda \geq 0$.

Example 1. Let $X =]0, \infty[$ and

$$f(x) \equiv \pi^{-1/2}(2x)^{1/4}(1+x^2)^{-1/2} .$$

Note that $f \in H^2_{\mathrm{d}}(\delta)$ for $0 < \delta < \pi/2$ and so f is dilation analytic. In this specific example one can evaluate $F_{\mathcal{R}}(z)$ directly and describe the entire Riemann surface of $F_\lambda(z)$, thus going far beyond the results of Theorem 12.

For $z \in \mathbb{C} \setminus [0, \infty)$ we set $w \equiv \sqrt{-z}$, where the branch is chosen so that $\mathrm{Re}\, w > 0$. Then $\mathrm{i}w \in \mathbb{C}_+$ and the integral

$$F_{\mathcal{R}}(z) = \frac{1}{\pi} \int_0^\infty \frac{\sqrt{2t}}{1+t^2} \frac{\mathrm{d}t}{t-z} = \frac{\sqrt{2}}{\pi} \int_{-\infty}^\infty \frac{t^2}{1+t^4} \frac{\mathrm{d}t}{t^2+w^2} ,$$

is easily evaluated by closing the integration path in the upper half-plane and using the residue method. We get

$$F_{\mathcal{R}}(z) = \frac{1}{w^2 + \sqrt{2}w + 1} .$$

Thus $F_{\mathcal{R}}$ is a meromorphic function of w with two simple poles at $w = \mathrm{e}^{\pm 3\mathrm{i}\pi/4}$. It follows that $F_{\mathcal{R}}(z)$ is meromorphic on the two-sheeted Riemann surface of $\sqrt{-z}$. On the first (physical) sheet, where $\mathrm{Re}\, w > 0$, it is of course analytic. On the second sheet, where $\mathrm{Re}\, w < 0$, it has two simple poles at $z = \pm\mathrm{i}$.

In term of the uniformizing variable w, we have

$$F_\lambda(z) = \frac{w^2 + \sqrt{2}w + 1}{(w^2+\omega)(w^2+\sqrt{2}w+1) - \lambda^2} .$$

For $\lambda > 0$, this meromorphic function has 4 poles. These poles are analytic functions of λ except at the collision points. For λ small, the poles form two conjugate pairs, one near $\pm\mathrm{i}\sqrt{\omega}$, the other near $e^{\pm 3\mathrm{i}\pi/4}$. Both pairs are on the second sheet. For λ large, a pair of conjugated poles goes to infinity along the asymptote $\mathrm{Re}\, w = -\sqrt{2}/4$. A pair of real poles goes to $\pm\infty$. In particular, one of them enters the first sheet at $\lambda = \sqrt{\omega}$ and h_λ has one negative eigenvalue for $\lambda > \sqrt{\omega}$. Since

$$G_{\mathcal{R}}(x) = \frac{1}{\pi} \int_0^\infty \frac{\sqrt{2}}{1+t^2} \frac{\mathrm{d}t}{(t-x)^2} ,$$

is finite for $x < 0$ and infinite for $x \geq 0$, 0 is not an eigenvalue of h_λ for $\lambda = \sqrt{\omega}$, but a zero energy resonance. Note that the image of the asymptote $\mathrm{Re}\, w = -\sqrt{2}/4$ on the second sheet is the parabola $\{z = x + \mathrm{i}y \,|\, x = 2y^2 - 1/8\}$. Thus, as $\lambda \to \infty$, the poles of $F_\lambda(z)$ move away from the spectrum. This means that there are no resonances in the large coupling limit.

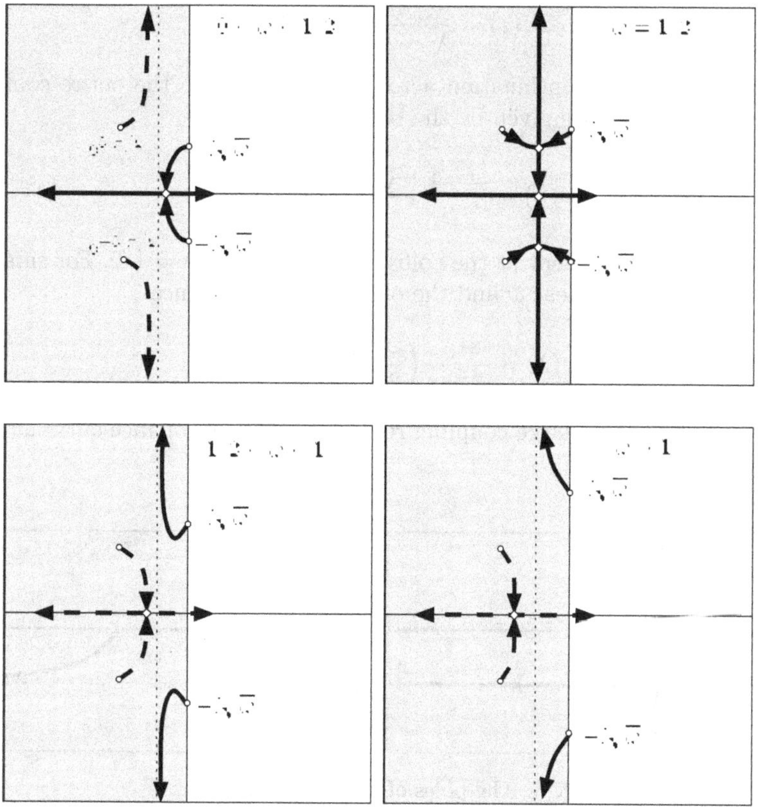

Fig. 4. Trajectories of the poles of $F_\lambda(z)$ in w-space for various values of ω in Example 1. Notice the simultaneous collision of the two pairs of conjugate poles when $\omega = \lambda = 1/2$. The second Riemann sheet is shaded

The qualitative trajectories of the poles (as functions of λ for fixed values of ω) are plotted in Fig. 4.

Example 2. Let $X = \mathbb{R}$ and

$$f(x) \equiv \pi^{-1/2}(1 + x^2)^{-1/2} .$$

Since $f \in H^2_{\mathrm{t}}(\delta)$ for $0 < \delta < 1$, the function f is translation analytic. Here again we can compute explicitly $F_{\mathcal{R}}(z)$. For $z \in \mathbb{C}_+$, a simple residue calculation leads to

$$F_{\mathcal{R}}(z) = \frac{1}{\pi} \int_{-\infty}^{\infty} \frac{1}{1+t^2} \frac{\mathrm{d}t}{t - z} = -\frac{1}{z + \mathrm{i}} .$$

Hence,

$$F_\lambda(z) = \frac{z+\mathrm{i}}{\lambda^2 - (z+\mathrm{i})(z-\omega)},$$

has a meromorphic continuation across the real axis to the entire complex plane. It has two poles given by the two branches of

$$\omega(\lambda) = \frac{\omega - \mathrm{i} + \sqrt{(\omega+\mathrm{i})^2 + 4\lambda^2}}{2},$$

which are analytic except at the collision point $\omega = 0$, $\lambda = 1/2$. For small λ, one of these poles is near ω and the other is near $-\mathrm{i}$. Since

$$\omega(\lambda) = -\frac{\mathrm{i}}{2} + \left(\frac{\omega}{2} \pm \lambda\right) + O(1/\lambda),$$

as $\lambda \to \infty$, h_λ has no large coupling resonances. The resonance curve $\omega(\lambda)$ is plotted in Fig. 5.

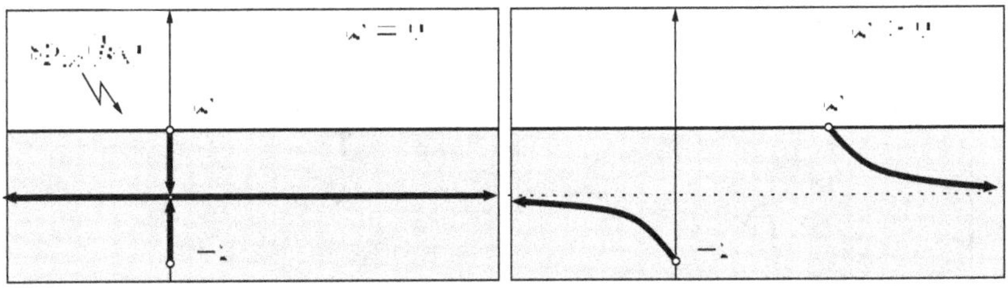

Fig. 5. The poles of $F_\lambda(z)$ for Example 2

Clearly, $\mathrm{sp}(h_\lambda) = \mathbb{R}$ for all ω and λ. Note that for all $x \in \mathbb{R}$, $G_{\mathcal{R}}(x) = \infty$ and

$$\operatorname{Im} F_\lambda(x + \mathrm{i}0) = \frac{\lambda^2}{(x-\omega)^2 + (\lambda^2 - x(x-\omega))^2}.$$

Hence, the operator h_λ has purely absolutely continuous spectrum for all ω and all $\lambda \neq 0$.

Example 3. Let $X =]-1, 1[$ and

$$f(x) \equiv \sqrt{\frac{2}{\pi}}(1-x^2)^{1/4}.$$

(Recall Example 2 in Subsect. 3.1 and Example 4 in Subsect. 3.3 – $\mathfrak{h}_{\mathcal{R}}$ and $h_{\mathcal{R}}$ are $\ell^2(\mathbb{Z}_+)$ and the discrete Laplacian in the energy representation and $f = \delta_1^\#$.) In Subsect. 2.9 we have shown that for $z \in \mathbb{C} \setminus [-1, 1]$,

$$F_{\mathcal{R}}(z) = \frac{2}{\pi}\int_{-1}^{1} \frac{\sqrt{1-t^2}}{t-z}\mathrm{d}t = 2\frac{\xi - 1}{\xi + 1}, \tag{54}$$

where

$$\xi = \sqrt{\frac{z-1}{z+1}} \, .$$

The principal branch of the square root $\mathrm{Re}\,\xi > 0$ corresponds to the first (physical) sheet of the Riemann surface R of $F_{\mathcal{R}}(z)$. The branch $\mathrm{Re}\,\xi < 0$ corresponds to the second sheet of R. In particular,

$$F_{\mathcal{R}}(x + \mathrm{i}0) = 2(-x + \mathrm{i}\sqrt{1-x^2}) \, .$$

To discuss the analytic structure of the Borel transform $F_\lambda(z)$, it is convenient to introduce the uniformizing variable

$$w \equiv -\frac{2}{F_{\mathcal{R}}(z)} = \frac{1+\xi}{1-\xi} \, ,$$

which maps the Riemann surface R to $\mathbb{C} \setminus \{0\}$. Note that the first sheet of R is mapped on the exterior of the unit disk and that the second sheet is mapped on the punctured disk $\{z \in \mathbb{C} \,|\, 0 < |z| < 1\}$ (see Fig. 6). The inverse of this map is

$$z = \frac{1}{2}\left(w + \frac{1}{w}\right) \, .$$

For $z \in \mathbb{C} \setminus [-1, 1]$ the function $F_\lambda(z)$ is given by

$$F_\lambda(z) = \frac{-2w}{w^2 - 2\omega w + 1 - 4\lambda^2} \, ,$$

and thus has a meromorphic continuation to the entire Riemann surface R. The resonance poles in the w-plane are computed by solving

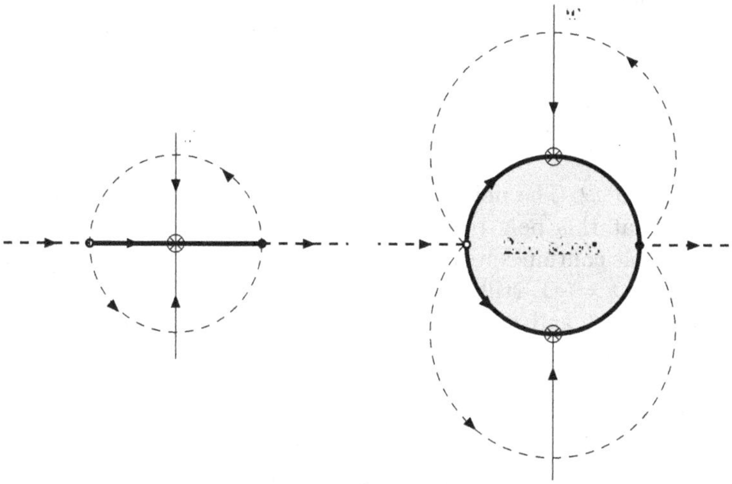

Fig. 6. Mapping the cut plane $\mathbb{C} \setminus [-1, 1]$ to the exterior of the unit disk

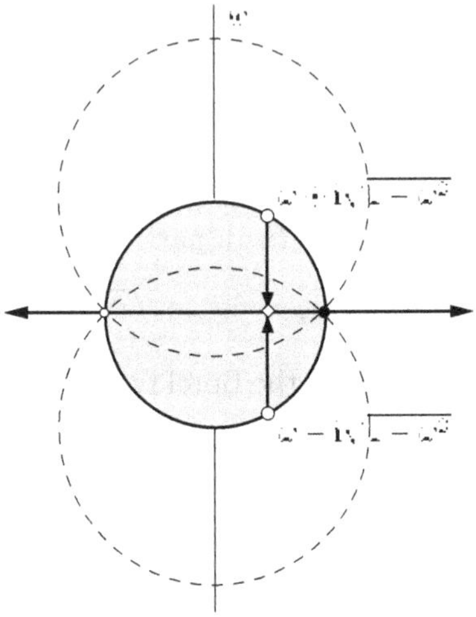

Fig. 7. The trajectories of the poles of F_λ in the w-plane. The second sheet is shaded

$$w^2 - 2\omega w + 1 - 4\lambda^2 = 0 \,,$$

and are given by the two-valued analytic function

$$w = \omega + \sqrt{4\lambda^2 + \omega^2 - 1} \,.$$

We will describe the motion of the poles in the case $\omega \geq 0$ (the case $\omega \leq 0$ is completely symmetric). For $0 < \lambda < \sqrt{1-\omega^2}/2$ there are two conjugate poles on the second sheet which, in the w-plane, move towards the point ω on a vertical line. After their collision at $\lambda = \sqrt{1-\omega^2}/2$, they turn into a pair of real poles moving towards $\pm\infty$ (see Fig. 7). The pole moving to the right reaches $w = 1$ at $\lambda = \sqrt{(1-\omega)/2}$ and enters the first sheet of R. We conclude that h_λ has a positive eigenvalue

$$\omega_+(\lambda) = \frac{1}{2}\left(\omega + \sqrt{4\lambda^2 + \omega^2 - 1} + \frac{1}{\omega + \sqrt{4\lambda^2 + \omega^2 - 1}}\right) ,$$

for $\lambda > \sqrt{(1-\omega)/2}$. The pole moving to the left reaches $w = 0$ at $\lambda = 1/2$. This means that this pole reaches $z = \infty$ on the second sheet of R. For $\lambda > 1/2$, the pole continues its route towards $w = -1$, i.e., it comes back from $z = \infty$ towards $z = -1$, still on the second sheet of R. At $\lambda = \sqrt{(1+\omega)/2}$, it reaches $w = -1$ and enters the first sheet. We conclude that h_λ has a negative eigenvalue

$$\omega_-(\lambda) = \frac{1}{2}\left(\omega - \sqrt{4\lambda^2 + \omega^2 - 1} + \frac{1}{\omega - \sqrt{4\lambda^2 + \omega^2 - 1}}\right) ,$$

for $\lambda > \sqrt{(1+\omega)/2}$. The trajectory of these poles in the z cut-plane is shown on Fig. 8. For clarity, only one pole of the conjugate pair is displayed.

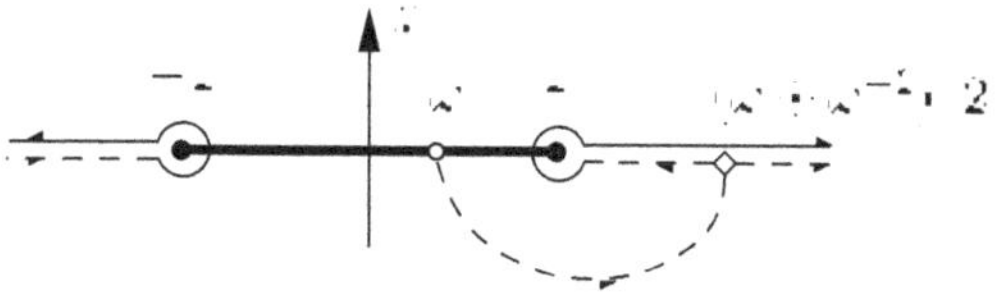

Fig. 8. The trajectories of the poles of F_λ in the z-plane. *Dashed lines* are on the second sheet

Example 4. In Examples 1–3 there were no resonances in the large coupling regime, i.e., the second sheet poles of F_λ kept away from the continuous spectrum as $\lambda \to \infty$. This fact can be understood as follows. If a resonance $\omega(\lambda)$ approaches the real axis as $\lambda \to \infty$, then it follows from (8) that $\mathrm{Im}\, F_{\mathcal{R}}(\omega(\lambda)) = o(\lambda^{-2})$. Since under Assumptions (A1) and (A2) $F_{\mathcal{R}}$ is continuous on $\overline{\mathbb{C}_+}$, we conclude that if $\lim_{\lambda\to\infty} \omega(\lambda) = \overline{\omega} \in \mathbb{R}$, then $\mathrm{Im}\, F_{\mathcal{R}}(\overline{\omega} + \mathrm{i}0) = 0$. Since $\|f(x)\|_{\mathfrak{K}}$ is also continuous on $\overline{X}$, if $\overline{\omega} \in \overline{X}$, then we must have $f(\overline{\omega}) = 0$. Thus the only possible locations of large coupling resonances are the zeros of f in $\overline{X}$. We finish this subsection with an example where such large coupling resonances exist.

Let again $X =]-1, 1[$ and set

$$f(x) \equiv \sqrt{\frac{1}{\pi}}\, x(1 - x^2)^{1/4} .$$

The Borel transform

$$F_{\mathcal{R}}(z) = \frac{1}{\pi} \int_{-1}^{1} \frac{x^2\sqrt{1-x^2}}{x - z}\, \mathrm{d}x ,$$

is easily evaluated by a residue calculation and the change of variable

$$x = (u + u^{-1})/2 .$$

Using the same uniformizing variable w as in Example 3, we get

$$F_{\mathcal{R}}(z) = -\frac{1}{4}\left(1 + \frac{1}{w^2}\right)\frac{1}{w} , \tag{55}$$

and

$$F_\lambda(z) = \frac{-4w^3}{2w^4 - 4\omega w^3 + (2 - \lambda^2)w^2 - \lambda^2} . \tag{56}$$

We shall again restrict ourselves to the case $0 < \omega < 1$. At $\lambda = 0$ the denominator of (56) has a double zero at $w = 0$ and a pair of conjugated zeros at $\omega \pm \mathrm{i}\sqrt{1 - \omega^2}$. As λ increases, the double zero at 0 splits into a pair of real zeros going to $\pm\infty$. The right zero reaches 1 and enters the first sheet at $\lambda = \sqrt{2(1 - \omega)}$. At $\lambda = \sqrt{2(1 + \omega)}$, the left zero reaches -1 and also enters the first sheet. The pair of conjugated zeros move from their original positions towards $\pm\mathrm{i}$ (of course, they remain within the unit disk). For large λ they behave like

$$w = \pm\mathrm{i} + \frac{2\omega}{\lambda^2} - \frac{2\omega(2 \pm 5\mathrm{i}\omega)}{\lambda^4} + O(\lambda^{-6}) .$$

Thus, in the z plane, F_λ has two real poles emerging from $\pm\infty$ on the second sheet and traveling towards ± 1. The right pole reaches 1 at $\lambda = \sqrt{2(1-\omega)}$ and becomes an eigenvalue of h_λ which returns to $+\infty$ as λ further increases. The left pole reaches -1 at $\lambda = \sqrt{2(1+\omega)}$, becomes an eigenvalue of h_λ, and further proceeds towards $-\infty$. On the other hand, the eigenvalue ω of h_0 turns into a pair of conjugated poles on the second sheet which, as $\lambda \to \infty$, tend towards 0 as

$$\omega(\lambda) = \frac{2\omega}{\lambda^2} - \frac{4\omega(1 \pm 2\mathrm{i}\omega)}{\lambda^4} + O(\lambda^{-6}) ,$$

see Fig. 9. We conclude that h_λ has a large coupling resonance approaching 0 as $\lambda \to \infty$.

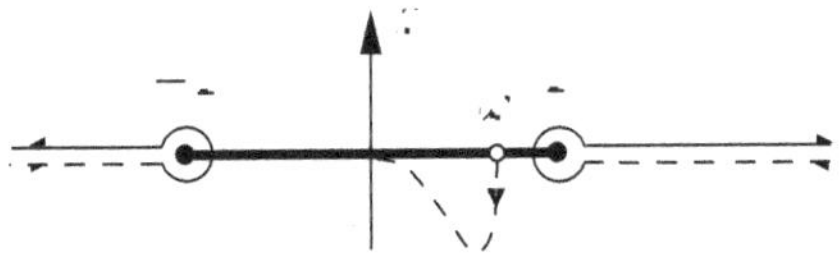

Fig. 9. The trajectories of the poles of F_λ in the z-plane. *Dashed lines* are on the second sheet

4 Fermionic Quantization

4.1 Basic Notions of Fermionic Quantization

This subsection is a telegraphic review of fermionic quantization. For additional information and references the reader may consult Sect. 5 in [2].

Let $\mathfrak{h}$ be a Hilbert space. We denote by $\Gamma(\mathfrak{h})$ the fermionic (antisymmetric) Fock space over $\mathfrak{h}$, and by $\Gamma_n(\mathfrak{h})$ the n-particle sector in $\mathfrak{h}$. $\Phi_\mathfrak{h}$ denotes the vacuum in $\Gamma(\mathfrak{h})$ and $a(f), a^*(f)$ the annihilation and creation operators associated to $f \in \mathfrak{h}$. In the sequel $a^\#(f)$ represents either $a(f)$ or $a^*(f)$. Recall that $\|a^\#(f)\| = \|f\|$. The CAR algebra over $\mathfrak{h}$, CAR($\mathfrak{h}$), is the C^*-algebra of bounded operators on $\Gamma(\mathfrak{h})$ generated by $\{a^\#(f) \,|\, f \in \mathfrak{h}\}$.

Let u be a unitary operator on $\mathfrak{h}$. Its second quantization

$$\Gamma(u)|_{\Gamma_n(\mathfrak{h})} \equiv u \otimes \cdots \otimes u = u^{\otimes n} ,$$

defines a unitary operator on $\Gamma(\mathfrak{h})$ which satisfies

$$\Gamma(u)a^{\#}(f) = a^{\#}(uf)\Gamma(u) .$$

Let h be a self-adjoint operator on $\mathfrak{h}$. The second quantization of e^{ith} is a strongly continuous group of unitary operators on $\Gamma(\mathfrak{h})$. The generator of this group is denoted by $\mathrm{d}\Gamma(h)$,

$$\Gamma(\mathrm{e}^{\mathrm{i}th}) = \mathrm{e}^{\mathrm{i}t\mathrm{d}\Gamma(h)} .$$

$\mathrm{d}\Gamma(h)$ is essentially self-adjoint on $\Gamma(\mathrm{Dom}\, h)$, where $\mathrm{Dom}\, h$ is equipped with the graph norm, and one has

$$\mathrm{d}\Gamma(h)|_{\Gamma_n(\mathrm{Dom}\, h)} = \sum_{k=1}^{n} \underbrace{I \otimes \cdots \otimes I}_{k-1} \otimes h \otimes \underbrace{I \otimes \cdots \otimes I}_{n-k} .$$

The maps

$$\tau^t(a^{\#}(f)) = \mathrm{e}^{\mathrm{i}t\mathrm{d}\Gamma(h)} a^{\#}(f) \mathrm{e}^{-\mathrm{i}t\mathrm{d}\Gamma(h)} = a^{\#}(\mathrm{e}^{\mathrm{i}th} f) ,$$

uniquely extend to a group τ of $*$-automorphisms of $\mathrm{CAR}(\mathfrak{h})$. τ is often called the group of Bogoliubov automorphisms induced by h. The group τ is norm continuous and the pair $(\mathrm{CAR}(\mathfrak{h}), \tau)$ is a C^*-dynamical system. We will call it a CAR dynamical system. We will also call the pair $(\mathrm{CAR}(\mathfrak{h}), \tau)$ the fermionic quantization of $(\mathfrak{h}, h)$.

If two pairs $(\mathfrak{h}_1, h_1)$ and $(\mathfrak{h}_2, h_2)$ are unitarily equivalent, that is, if there exists a unitary $u : \mathfrak{h}_1 \to \mathfrak{h}_2$ such that $uh_1u^{-1} = h_2$, then the fermionic quantizations $(\mathrm{CAR}(\mathfrak{h}_1), \tau_1)$ and $(\mathrm{CAR}(\mathfrak{h}_2), \tau_2)$ are isomorphic – the map $\sigma(a^{\#}(f)) = a^{\#}(uf)$ extends uniquely to a $*$-isomorphism such that $\sigma \circ \tau_1^t = \tau_2^t \circ \sigma$.

4.2 Fermionic Quantization of the WWA

Let h_λ be a WWA on $\mathfrak{h} = \mathbb{C} \oplus \mathfrak{h}_\mathcal{R}$. Its fermionic quantization is the pair $(\mathrm{CAR}(\mathfrak{h}), \tau_\lambda)$, where

$$\tau_\lambda^t(a^{\#}(\phi)) = \mathrm{e}^{\mathrm{i}t\mathrm{d}\Gamma(h_\lambda)} a^{\#}(\phi) \mathrm{e}^{-\mathrm{i}t\mathrm{d}\Gamma(h_\lambda)} = a^{\#}(\mathrm{e}^{\mathrm{i}th_\lambda} \phi) .$$

We will refer to $(\mathrm{CAR}(\mathfrak{h}), \tau_\lambda)$ as the *Simple Electronic Black Box* (SEBB) model. This model has been discussed in the recent lecture notes [2]. The SEBB model is the simplest non-trivial example of the Electronic Black Box model introduced and studied in [3].

The SEBB model is also the simplest non-trivial example of an open quantum system. Set

$$\tau_\mathcal{S}^t(a^{\#}(\alpha)) = a^{\#}(\mathrm{e}^{\mathrm{i}t\omega}\alpha), \qquad \tau_\mathcal{R}^t(a^{\#}(g)) = a^{\#}(\mathrm{e}^{\mathrm{i}th_\mathcal{R}} g) .$$

The CAR dynamical systems $(\mathrm{CAR}(\mathbb{C}), \tau_\mathcal{S})$ and $(\mathrm{CAR}(\mathfrak{h}_\mathcal{R}), \tau_\mathcal{R})$ are naturally identified with subsystems of the non-interacting SEBB $(\mathrm{CAR}(\mathfrak{h}), \tau_0)$.

The system $(\mathrm{CAR}(\mathbb{C}), \tau_{\mathcal{S}})$ is a two-level quantum dot without internal structure. The system $(\mathrm{CAR}(\mathfrak{h}_{\mathcal{R}}), \tau_{\mathcal{R}})$ is a free Fermi gas reservoir. Hence, $(\mathrm{CAR}(\mathfrak{h}_\lambda), \tau_\lambda)$ describes the interaction of a two-level quantum system with a free Fermi gas reservoir.

In the sequel we denote $H_\lambda \equiv \mathrm{d}\Gamma(h_\lambda)$, $H_{\mathcal{S}} \equiv \mathrm{d}\Gamma(\omega)$, $H_{\mathcal{R}} \equiv \mathrm{d}\Gamma(h_{\mathcal{R}})$, and

$$V \equiv \mathrm{d}\Gamma(v) = a^*(f)a(1) + a^*(1)a(f) .$$

Clearly,

$$H_\lambda = H_0 + \lambda V .$$

4.3 Spectral Theory

The vacuum of $\Gamma(\mathfrak{h})$ is always an eigenvector of H_λ with eigenvalue zero. The rest of the spectrum of H_λ is completely determined by the spectrum of h_λ and one may use the results of Sects. 2 and 3.2 to characterize the spectrum of H_λ. We mention several obvious facts. If the spectrum of h_λ is purely absolutely continuous, then the spectrum of H_λ is also purely absolutely continuous except for a simple eigenvalue at zero. H_λ has no singular continuous spectrum iff h_λ has no singular continuous spectrum. Let $\{e_i\}_{i\in I}$ be the eigenvalues of h_λ, repeated according to their multiplicities. The eigenvalues of H_λ are given by

$$\mathrm{sp_p}(H_\lambda) = \left\{ \sum_{i\in I} n_i e_i \,\middle|\, n_i \in \{0,1\}, \sum_{i\in I} n_i < \infty \right\} \cup \{0\} .$$

Until the end of this subsection we will discuss the fermionic quantization of the Radiating Wigner-Weisskopf Atom introduced in Sect. 3.2. The point spectrum of H_0 consists of two simple eigenvalues $\{0, \omega\}$. The corresponding normalized eigenfunctions are

$$\Psi_n = a(1)^n \Phi_{\mathfrak{h}}, \qquad n = 0, 1 .$$

Apart from these simple eigenvalues, the spectrum of H_0 is purely absolutely continuous and $\mathrm{sp_{ac}}(H_0)$ is equal to the closure of the set

$$\left\{ e + \sum_{i=1}^{n} x_i \,\middle|\, x_i \in X,\, e \in \{0, \omega\},\, n \geq 1 \right\} .$$

Let Λ be as in Theorem 9. Then for $0 < |\lambda| < \Lambda$ the spectrum of H_λ is purely absolutely continuous except for a simple eigenvalue 0.

Note that

$$\begin{aligned}(\Psi_1|\mathrm{e}^{-\mathrm{i}tH_\lambda}\Psi_1) &= (a(1)\Phi_{\mathfrak{h}}|\mathrm{e}^{-\mathrm{i}tH_\lambda}a(1)\Phi_{\mathfrak{h}}) \\ &= (a(1)\Phi_{\mathfrak{h}}|a(\mathrm{e}^{-\mathrm{i}th_\lambda})\Phi_{\mathfrak{h}}) = (1|\mathrm{e}^{-\mathrm{i}th_\lambda}1) .\end{aligned}$$

Similarly,

$$(\Psi_1|(H_\lambda - z)^{-1}\Psi_1) = (1|(h_\lambda - z)^{-1}1) .$$

Hence, one may use directly the results (and examples) of Section 3 to describe the asymptotic of $(\Psi_1|\mathrm{e}^{-\mathrm{i}tH_\lambda}\Psi_1)$ and the meromorphic continuation of

$$\mathbb{C}_+ \ni z \mapsto (\Psi_1|(H_\lambda - z)^{-1}\Psi_1) . \tag{57}$$

4.4 Scattering Theory

Let h_λ be a WWA on $\mathfrak{h} = \mathbb{C} \oplus \mathfrak{h}_\mathcal{R}$. The relation

$$\tau_0^{-t} \circ \tau_\lambda^t(a^\#(\phi)) = a^\#(\mathrm{e}^{-\mathrm{i}th_0}\mathrm{e}^{\mathrm{i}th_\lambda}\phi) ,$$

yields that for $\phi \in \mathfrak{h}_{\mathrm{ac}}(h_\lambda)$ the limit

$$\lim_{t\to\infty} \tau_0^{-t} \circ \tau_\lambda^t(a^\#(\phi)) = a^\#(\Omega_\lambda^- \phi) ,$$

exists in the norm topology of $\mathrm{CAR}(\mathfrak{h})$. Denote

$$\tau_{\lambda,\mathrm{ac}} \equiv \tau_\lambda|_{\mathrm{CAR}(\mathfrak{h}_{\mathrm{ac}}(h_\lambda))}, \qquad \tau_{\mathcal{R},\mathrm{ac}} \equiv \tau_\mathcal{R}|_{\mathrm{CAR}(\mathfrak{h}_{\mathrm{ac}}(h_\mathcal{R}))} .$$

By the intertwining property (26) of the wave operator Ω_λ^-, the map

$$\sigma_\lambda^+(a^\#(\phi)) \equiv a^\#(\Omega_\lambda^- \phi) ,$$

satisfies $\sigma_\lambda^+ \circ \tau_{\lambda,\mathrm{ac}}^t = \tau_{\mathcal{R},\mathrm{ac}}^t \circ \sigma_\lambda^+$. Hence, σ_λ^+ is a $*$-isomorphism between the CAR dynamical systems $(\mathrm{CAR}(\mathfrak{h}_{\mathrm{ac}}(h_\lambda)), \tau_{\lambda,\mathrm{ac}})$ and $(\mathrm{CAR}(\mathfrak{h}_{\mathrm{ac}}(h_\mathcal{R})), \tau_{\mathcal{R},\mathrm{ac}})$. This isomorphism is the algebraic analog of the wave operator in Hilbert space scattering theory and is often called the Møller isomorphism.

5 Quantum Statistical Mechanics of the SEBB Model

5.1 Quasi-Free States

This subsection is a direct continuation of Subsect. 4.1. A positive linear functional $\eta : \mathrm{CAR}(\mathfrak{h}) \to \mathbb{C}$ is called a *state* if $\eta(I) = 1$. A physical system $\mathcal{P}$ is described by the CAR dynamical system $(\mathrm{CAR}(\mathfrak{h}), \tau)$ if its physical observables can be identified with elements of $\mathrm{CAR}(\mathfrak{h})$ and if their time evolution is specified by the group τ. The physical states of $\mathcal{P}$ are specified by states on $\mathrm{CAR}(\mathfrak{h})$. If $\mathcal{P}$ is initially in a state described by η and $A \in \mathrm{CAR}(\mathfrak{h})$ is a given observable, then the expectation value of A at time t is $\eta(\tau^t(A))$. This is the usual framework of the algebraic quantum statistical mechanics in the Heisenberg picture. In the Schrödinger picture one evolves the states and keeps the observables fixed, i.e., if η is the initial state, then the state at time t is $\eta \circ \tau^t$. A state η is called τ-invariant (or stationary state, steady state) if $\eta \circ \tau^t = \eta$ for all t.

Let T be a self-adjoint operator on $\mathfrak{h}$ such that $0 \leq T \leq I$. The map

$$\eta_T(a^*(f_n) \cdots a^*(f_1) a(g_1) \cdots a(g_m)) = \delta_{n,m} \det\{(g_i|Tf_j)\} , \qquad (58)$$

uniquely extends to a state η_T on $\mathrm{CAR}(\mathfrak{h})$. This state is usually called the quasi-free gauge-invariant state generated by T. The state η_T is completely determined by its two point function

$$\eta_T(a^*(f)a(g)) = (g|Tf) .$$

Note that if $A \equiv \sum_j f_j(g_j| \cdot)$ is a finite rank operator on $\mathfrak{h}$, then

$$\mathrm{d}\Gamma(A) = \sum_j a^*(f_j)a(g_j) ,$$

and

$$\eta_T(\mathrm{d}\Gamma(A)) = \mathrm{Tr}\,(TA) = \sum_j (g_j|Tf_j) . \qquad (59)$$

Let $(\mathrm{CAR}(\mathfrak{h}), \tau)$ be the fermionic quantization of $(\mathfrak{h}, h)$. The quasi-free state η_T is τ-invariant iff $\mathrm{e}^{ith}T = T\mathrm{e}^{ith}$ for all $t \in \mathbb{R}$. In particular, the quasi-free state generated by $T = \varrho(h)$, where ϱ is a positive bounded Borel function on the spectrum of h, is τ-invariant. The function ϱ is the energy density of this quasi-free state. Let $\beta > 0$ and $\mu \in \mathbb{R}$. Of particular importance in quantum statistical mechanics is the quasi-free state associated with $T = \varrho_{\beta\mu}(h)$, where the energy density $\varrho_{\beta\mu}$ is given by the Fermi-Dirac distribution

$$\varrho_{\beta\mu}(\varepsilon) \equiv \frac{1}{\mathrm{e}^{\beta(\varepsilon - \mu)} + 1} . \qquad (60)$$

We denote this state by $\eta_{\beta\mu}$. The pair $(\mathrm{CAR}(\mathfrak{h}), \tau)$ and the state $\eta_{\beta\mu}$ describe free Fermi gas in thermal equilibrium at inverse temperature β and chemical potential μ.

5.2 Non-Equilibrium Stationary States

In this subsection we assume that h_λ has purely absolutely continuous spectrum. We make this assumption in order to ensure that the system will evolve towards a stationary state. This assumption will be partially relaxed in Subsect. 5.5, where we discuss the effect of eigenvalues of h_λ. We do not make any assumptions on the spectrum of $h_{\mathcal{R}}$.

Let η_T be a quasi-free state on $\mathrm{CAR}(\mathbb{C} \oplus \mathfrak{h}_{\mathcal{R}})$ generated by $T = \alpha \oplus T_{\mathcal{R}}$. We denote by $\eta_{T_{\mathcal{R}}}$ the quasi-free state on $\mathrm{CAR}(\mathfrak{h}_{\mathcal{R}})$ generated by $T_{\mathcal{R}}$. We assume that $\eta_{T_{\mathcal{R}}}$ is $\tau_{\mathcal{R}}$-invariant and denote by $T_{\mathcal{R},\mathrm{ac}}$ the restriction of $T_{\mathcal{R}}$ to the subspace $\mathfrak{h}_{\mathrm{ac}}(h_{\mathcal{R}})$.

Let $\phi_1, \ldots, \phi_n \in \mathfrak{h}$ and

$$A = a^{\#}(\phi_1) \cdots a^{\#}(\phi_n) . \tag{61}$$

Since η_T is τ_0-invariant,

$$\begin{aligned} \eta_T(\tau_\lambda^t(A))) &= \eta_T(\tau_0^{-t} \circ \tau_\lambda^t(A)) \\ &= \eta_T(a^{\#}(\mathrm{e}^{-\mathrm{i}th_0}\mathrm{e}^{\mathrm{i}th_\lambda}\phi_1) \cdots a^{\#}(\mathrm{e}^{-\mathrm{i}th_0}\mathrm{e}^{\mathrm{i}th_\lambda}\phi_n)) , \end{aligned}$$

hence

$$\lim_{t\to\infty} \eta_T(\tau_\lambda^t(A)) = \eta_T(a^{\#}(\Omega_\lambda^- \phi_1) \cdots a^{\#}(\Omega_\lambda^- \phi_n)) .$$

Since the set of observables of the form (61) is dense in $\mathfrak{h}$, we conclude that for all $A \in \mathrm{CAR}(\mathfrak{h})$ the limit

$$\eta_\lambda^+(A) = \lim_{t\to\infty} \eta_T(\tau_\lambda^t(A)) ,$$

exists and defines a state η_λ^+ on $\mathrm{CAR}(\mathfrak{h})$. Note that η_λ^+ is the quasi-free state generated by $T_\lambda^+ \equiv (\Omega_\lambda^-)^* T \Omega_\lambda^-$. Since $\mathrm{Ran}\, \Omega_\lambda^- = \mathfrak{h}_{\mathrm{ac}}(h_0) = \mathfrak{h}_{\mathrm{ac}}(h_{\mathcal{R}})$, we have

$$T_\lambda^+ = (\Omega_\lambda^-)^* T_{\mathcal{R},\mathrm{ac}} \Omega_\lambda^- , \tag{62}$$

and so

$$\eta_\lambda^+ = \eta_{T_{\mathcal{R},\mathrm{ac}}} \circ \sigma_\lambda^+ ,$$

where σ_λ^+ is the Møller isomorphism introduced in Subsect. 4.4. Obviously, η_λ^+ does not depend on the choice of α and on the restriction of $T_{\mathcal{R}}$ to $\mathfrak{h}_{\mathrm{sing}}(h_{\mathcal{R}})$. Since

$$\mathrm{e}^{\mathrm{i}th_\lambda} T_\lambda^+ \mathrm{e}^{-\mathrm{i}th_\lambda} = (\Omega_\lambda^-)^* \mathrm{e}^{\mathrm{i}th_{\mathcal{R}}} T_{\mathcal{R},\mathrm{ac}} \mathrm{e}^{-\mathrm{i}th_{\mathcal{R}}} \Omega_\lambda^- = T_\lambda^+ ,$$

η_λ^+ is τ_λ-invariant.

The state η_λ^+ is called the *non-equilibrium stationary state* (NESS) of the CAR dynamical system $(\mathrm{CAR}(\mathfrak{h}), \tau_\lambda)$ associated to the initial state η_T. Note that if $A \equiv \sum_j \phi_j (\psi_j | \cdot)$, then, according to (59),

$$\eta_\lambda^+(\mathrm{d}\Gamma(A)) = \mathrm{Tr}\,(T \Omega_\lambda^- A \Omega_\lambda^{-*}) = \sum_j (\Omega_\lambda^- \psi_j | T \Omega_\lambda^- \phi_j) . \tag{63}$$

By passing to the GNS representation associated to η_T one can prove the following more general result. Let $\mathcal{N}$ be the set of states on $\mathrm{CAR}(\mathfrak{h})$ which are normal with respect to η_T (the set $\mathcal{N}$ does not depend on the choice of α). Then for any $\eta \in \mathcal{N}$ and $A \in \mathrm{CAR}(\mathfrak{h})$,

$$\lim_{t\to\infty} \eta(\tau_\lambda^t(A)) = \eta_\lambda^+(A) .$$

If $T_{\mathcal{R}} = \varrho(h_{\mathcal{R}})$ for some bounded Borel function ϱ on the spectrum of $h_{\mathcal{R}}$, then the intertwining property of the wave operator implies that $T_\lambda^+ = \varrho(h_\lambda)$ and hence $\eta_\lambda^+ = \eta_{\varrho(h_\lambda)}$. In particular, if the reservoir is initially in thermal equilibrium at inverse temperature $\beta > 0$ and chemical potential $\mu \in \mathbb{R}$, then η_λ^+ is the quasi-free state associated to $(\mathrm{e}^{\beta(h_\lambda - \mu)} + 1)^{-1}$, which is the thermal equilibrium state of $(\mathrm{CAR}(\mathfrak{h}), \tau_\lambda)$ at the inverse temperature β and chemical potential μ. This phenomenon is often called return to equilibrium.

5.3 Subsystem Structure

In the rest of these lecture notes we assume that $h_{\mathcal{R}}$ is multiplication by x on $\mathfrak{h}_{\mathcal{R}} = L^2(X, \mathrm{d}\mu; \mathfrak{K})$, where $X \subset \mathbb{R}$ is an open set and $\mathfrak{K}$ is a separable Hilbert space. The internal structure of $\mathcal{R}$ is specified by an orthogonal decomposition $\mathfrak{K} = \oplus_{k=1}^{M} \mathfrak{K}_k$. We set $\mathfrak{h}_{\mathcal{R}_k} = L^2(X, \mathrm{d}\mu; \mathfrak{K}_k)$ and denote by $h_{\mathcal{R}_k}$ the operator of multiplication by x on $\mathfrak{h}_{\mathcal{R}_k}$. Thus, we can write

$$\mathfrak{h}_{\mathcal{R}} = \bigoplus_{k=1}^{M} \mathfrak{h}_{\mathcal{R}_k}, \qquad h_{\mathcal{R}} = \bigoplus_{k=1}^{M} h_{\mathcal{R}_k} . \tag{64}$$

We interpret (64) as a decomposition of the reservoir $\mathcal{R}$ into M independent subreservoirs $\mathcal{R}_1, \cdots, \mathcal{R}_M$.

According to (64), we write $f = \oplus_{k=1}^{M} f_k$ and we split the interaction v as $v = \sum_{k=1}^{M} v_k$, where

$$v_k = (1|\cdot)f_k + (f_k|\cdot)1 .$$

The wave operators $\Omega_\lambda^\pm$ and the scattering matrix S have the following form.

Proposition 12. *Let $\phi = \alpha \oplus \varphi \in \mathfrak{h}$. Then*

$$(\Omega_\lambda^\pm \phi)(x) = \varphi(x) - \lambda f(x) F_\lambda(x \pm \mathrm{i}0)(\alpha - \lambda(f|(h_{\mathcal{R}} - x \mp \mathrm{i}0)^{-1}\varphi)) . \tag{65}$$

Moreover, for any $\psi \in L^2(X, \mathrm{d}\mu_{\mathrm{ac}}; \mathfrak{K})$ one has $(S\psi)(x) = S(x)\psi(x)$ where $S(x) : \mathfrak{K} \to \mathfrak{K}$ has the form

$$(S\psi)(x) = \psi(x) + 2\pi\mathrm{i}\lambda^2 F_\lambda(x + \mathrm{i}0)\frac{\mathrm{d}\mu_{\mathrm{ac}}}{\mathrm{d}x}(x)(f(x)|\psi(x))_{\mathfrak{K}} f(x) . \tag{66}$$

This result is deduced from Proposition 7 as follows. Let $\mathfrak{h}_{\mathcal{R},f}$ be the cyclic space generated by $h_{\mathcal{R}}$ and f and $\mathrm{d}\mu_{\mathcal{R}}(x) = \|f(x)\|_{\mathfrak{K}}^2 \mathrm{d}\mu(x)$ the spectral measure for $h_{\mathcal{R}}$ and f. Let $U : \mathfrak{h}_{\mathcal{R},f} \to L^2(\mathbb{R}, \mathrm{d}\mu_{\mathcal{R}})$ be defined by $U(Ff) = F$, $F \in L^2(\mathbb{R}, \mathrm{d}\mu_{\mathcal{R}})$. U is unitary, $\tilde{h}_{\mathcal{R}} = U h_{\mathcal{R}} U^{-1}$ is the operator of multiplication by x, and $Uf = \mathbb{1}$. We extend $\tilde{h}_{\mathcal{R}}$ to $\tilde{\mathfrak{h}}_{\mathcal{R}} = L^2(\mathbb{R}, \mathrm{d}\mu_{\mathcal{R}}) \oplus \mathfrak{h}_{\mathcal{R},\psi}^\perp$ by setting $\tilde{h}_{\mathcal{R}} = h_{\mathcal{R}}$ on $\mathfrak{h}_{\mathcal{R},f}^\perp$. Proposition 7 applies to the pair of operators $\tilde{h}_0 = \omega \oplus \tilde{h}_{\mathcal{R}}$ and

$$\tilde{h}_\lambda = \tilde{h}_0 + \lambda((1|\cdot)\mathbb{1} + (\mathbb{1}|\cdot)1) ,$$

acting on $\mathbb{C} \oplus \tilde{\mathfrak{h}}_{\mathcal{R}}$. We denote the corresponding wave operators and S-matrix by $\tilde{\Omega}_\lambda^\pm$ and $\tilde{S}$. We extend U to $\mathfrak{h} = \mathbb{C} \oplus \mathfrak{h}_{\mathcal{R},f} \oplus \mathfrak{h}_{\mathcal{R},f}^\perp$ by setting $U\psi = \psi$ on $\mathbb{C} \oplus \mathfrak{h}_{\mathcal{R},f}^\perp$. Clearly,

$$\Omega_\lambda^\pm = U^{-1}\tilde{\Omega}_\lambda^\pm U, \qquad S = U^{-1}\tilde{S}U ,$$

and an explicit computation yields the statement. We leave the details of this computation as an exercise for the reader.

The formula (65) can be also proven directly following line by line the proof of Proposition 7.

5.4 Non-Equilibrium Thermodynamics

In the sequel we assume that $f \in \mathrm{Dom}\, h_{\mathcal{R}}$. In this subsection we also assume that h_λ has purely absolutely continuous spectrum. The projection onto the subspace $\mathfrak{h}_{\mathcal{R}_k}$ is denoted by $1_{\mathcal{R}_k}$. Set

$$\begin{aligned} \mathfrak{f}_k &\equiv -\frac{\mathrm{d}}{\mathrm{d}t} \mathrm{e}^{\mathrm{i}th_\lambda} h_{\mathcal{R}_k} \mathrm{e}^{-\mathrm{i}th_\lambda}\Big|_{t=0} \\ &= -\mathrm{i}[h_\lambda, h_{\mathcal{R}_k}] = -\mathrm{i}[h_{\mathcal{S}} + \textstyle\sum_j \left(h_{\mathcal{R}_j} + \lambda v_j\right), h_{\mathcal{R}_k}] \\ &= \lambda \mathrm{i}[h_{\mathcal{R}_k}, v_k] \\ &= \lambda \mathrm{i}\left((1|\cdot) h_{\mathcal{R}_k} f_k - (h_{\mathcal{R}_k} f_k|\cdot)1\right) , \end{aligned} \tag{67}$$

and

$$\begin{aligned} \mathfrak{j}_k &\equiv -\frac{\mathrm{d}}{\mathrm{d}t} \mathrm{e}^{\mathrm{i}th_\lambda} 1_{\mathcal{R}_k} \mathrm{e}^{-\mathrm{i}th_\lambda}\Big|_{t=0} \\ &= -\mathrm{i}[h_\lambda, 1_{\mathcal{R}_k}] = -\mathrm{i}[h_{\mathcal{S}} + \textstyle\sum_j \left(h_{\mathcal{R}_j} + \lambda v_j\right), 1_{\mathcal{R}_k}] \\ &= \lambda \mathrm{i}[1_{\mathcal{R}_k}, v_k] \\ &= \lambda \mathrm{i}\left((1|\cdot) f_k - (f_k|\cdot)1\right) . \end{aligned} \tag{68}$$

The observables describing the heat and particle fluxes out of the k-th subreservoir are

$$\mathfrak{F}_k \equiv \mathrm{d}\Gamma(\mathfrak{f}_k) = \lambda \mathrm{i}(a^*(h_{\mathcal{R}_k} f_k) a(1) - a^*(1) a(h_{\mathcal{R}_k} f_k)) ,$$

$$\mathfrak{J}_k \equiv \mathrm{d}\Gamma(\mathfrak{j}_k) = \lambda \mathrm{i}(a^*(f_k) a(1) - a^*(1) a(f_k)) .$$

We assume that the initial state of the coupled system $\mathcal{S}+\mathcal{R}$ is the quasi-free state associated to $T \equiv \alpha \oplus T_{\mathcal{R}}$, where

$$T_{\mathcal{R}} = \bigoplus_{k=1}^{M} T_{\mathcal{R}_k} = \bigoplus_{k=1}^{M} \varrho_k(h_{\mathcal{R}_k}) ,$$

and the ϱ_k are bounded positive Borel functions on X.

Let η_λ^+ be the NESS of $(\mathrm{CAR}(\mathfrak{h}), \tau_\lambda)$ associated to the initial state η_T. According to (62) and (59), the steady state heat current out of the subreservoir $\mathcal{R}_k$ is

$$\begin{aligned} \eta_\lambda^+(\mathfrak{F}_k) &= \mathrm{Tr}\,(T_\lambda^+ \mathfrak{f}_k) = \mathrm{Tr}\,(T_{\mathcal{R}} \Omega_\lambda^- \mathfrak{f}_k (\Omega_\lambda^-)^*) \\ &= \sum_{j=1}^{M} \mathrm{Tr}\,(\varrho_j(h_{\mathcal{R}_j}) 1_{\mathcal{R}_j} \Omega_\lambda^- \mathfrak{f}_k (\Omega_\lambda^-)^* 1_{\mathcal{R}_j}) . \end{aligned}$$

Using (67) we can rewrite this expression as

$$\eta_\lambda^+(\mathfrak{J}_k) = 2\lambda \sum_{j=1}^{M} \mathrm{Im}\,(1_{\mathcal{R}_j}\Omega_\lambda^- h_{\mathcal{R}_k} f_k | \varrho_j(h_{\mathcal{R}_j}) 1_{\mathcal{R}_j}\Omega_\lambda^- 1)\,.$$

Equation (65) yields the relations

$$(\varrho_j(h_{\mathcal{R}_j}) 1_{\mathcal{R}_j}\Omega_\lambda^- 1)(x) = -\lambda \varrho_j(x) F_\lambda(x - \mathrm{i}0) f_j(x)\,,$$

$$(1_{\mathcal{R}_j}\Omega_\lambda^- h_{\mathcal{R}_k} f_k)(x) = \left(\delta_{kj}\, x + \lambda^2 F_\lambda(x-\mathrm{i}0)\, H_k(x-\mathrm{i}0)\right) f_j(x)\,,$$

where we have set

$$H_k(z) \equiv \int_X \frac{x \|f_k(x)\|^2_{\mathfrak{K}_k}}{x - z}\, \mathrm{d}\mu(x)\,.$$

Since $\mathrm{Ran}\,\Omega_\lambda^- = \mathfrak{h}_{\mathrm{ac}}(h_{\mathcal{R}})$, it follows that $(1_{\mathcal{R}_j}\Omega_\lambda^- h_{\mathcal{R}_k} f_k | \varrho_j(h_{\mathcal{R}_j}) 1_{\mathcal{R}_j}\Omega_\lambda^- 1)$ is equal to

$$\lambda \int_X \left(\delta_{kj} x F_\lambda(x+\mathrm{i}0) - \lambda^2 |F_\lambda(x+\mathrm{i}0)|^2 H_k(x+\mathrm{i}0)\right) \|f_j(x)\|^2_{\mathfrak{K}_j}\, \varrho_j(x)\, \mathrm{d}\mu_{\mathrm{ac}}(x)\,.$$

From (18) we deduce that

$$\mathrm{Im}\, H_k(x+\mathrm{i}0) = \pi x \|f_k(x)\|^2_{\mathfrak{K}_k} \frac{\mathrm{d}\mu_{\mathrm{ac}}}{\mathrm{d}x}(x)\,,$$

for Lebesgue a.e. $x \in X$. Equation (19) yields

$$\mathrm{Im}\, F_\lambda(x+\mathrm{i}0) = \pi \lambda^2 |F_\lambda(x+\mathrm{i}0)|^2 \|f(x)\|^2_{\mathfrak{K}} \frac{\mathrm{d}\mu_{\mathrm{ac}}}{\mathrm{d}x}(x)\,.$$

It follows that $\mathrm{Im}\,(1_{\mathcal{R}_j}\Omega_\lambda^- h_{\mathcal{R}_k} f_k | \varrho_j(h_{\mathcal{R}_j}) 1_{\mathcal{R}_j}\Omega_\lambda^- 1)$ is equal to

$$\lambda^3 \pi \int_X \left(\delta_{kj} \|f(x)\|^2_{\mathfrak{K}} - \|f_k(x)\|^2_{\mathfrak{K}_k}\right) \|f_j(x)\|^2_{\mathfrak{K}_j} |F_\lambda(x+\mathrm{i}0)|^2\, x \varrho_j(x) \left(\frac{\mathrm{d}\mu_{\mathrm{ac}}}{\mathrm{d}x}(x)\right)^2 \mathrm{d}x\,.$$

Finally, using the fact that $\|f(x)\|^2_{\mathfrak{K}} = \sum_j \|f_j(x)\|^2_{\mathfrak{K}_j}$, we obtain

$$\eta_\lambda^+(\mathfrak{J}_k) = \sum_{j=1}^{M} \int_X x(\varrho_k(x) - \varrho_j(x)) D_{kj}(x) \frac{\mathrm{d}x}{2\pi}\,, \tag{69}$$

where

$$D_{kj}(x) \equiv 4\pi^2 \lambda^4 \|f_k(x)\|^2_{\mathfrak{h}_k} \|f_j(x)\|^2_{\mathfrak{h}_j} |F_\lambda(x+\mathrm{i}0)|^2 \left(\frac{\mathrm{d}\mu_{\mathrm{ac}}}{\mathrm{d}x}(x)\right)^2\,. \tag{70}$$

Proceeding in a completely similar way we obtain the formula for the steady particle current

$$\eta_\lambda^+(\mathfrak{J}_k) = \sum_{j=1}^{M} \int_X (\varrho_k(x) - \varrho_j(x)) D_{kj}(x) \frac{\mathrm{d}x}{2\pi} \,. \tag{71}$$

The functions D_{kj} can be related to the S-matrix associated to $\Omega_\lambda^\pm$. According to the decomposition (64), the S-matrix (66) can be written as

$$(1_{\mathcal{R}_k} S\psi)(x) = \sum_{j=1}^{M} S_{kj}(x)(1_{\mathcal{R}_j}\psi)(x) \equiv (1_{\mathcal{R}_k}\psi)(x) + \sum_{j=1}^{M} t_{kj}(x)(1_{\mathcal{R}_j}\psi)(x) \,,$$

where

$$t_{kj}(x) = 2\pi \mathrm{i} \lambda^2 \frac{\mathrm{d}\mu_{\mathrm{ac}}}{\mathrm{d}x}(x) F_\lambda(x + \mathrm{i}0) f_k(x) (f_j(x)|\cdot)_{\mathfrak{K}_j} \,,$$

and we derive that

$$D_{kj}(x) = \mathrm{Tr}_{\mathfrak{K}_j}\left(t_{kj}(x)^* t_{kj}(x)\right) \,. \tag{72}$$

Equation (69), (71) together with (72) are the well-known Büttiker-Landauer formulas for the steady currents.

It immediately follows from (69) that

$$\sum_{k=1}^{M} \eta_\lambda^+(\mathfrak{F}_k) = 0 \,,$$

which is the first law of thermodynamics (conservation of energy). Similarly, particle number conservation

$$\sum_{k=1}^{M} \eta_\lambda^+(\mathfrak{J}_k) = 0 \,,$$

follows from (71).

To describe the entropy production of the system, assume that the k-th subreservoir is initially in thermal equilibrium at inverse temperature $\beta_k > 0$ and chemical potential $\mu_k \in \mathbb{R}$. This means that

$$\varrho_k(x) = F(Z_k(x)) \,,$$

where $F(t) \equiv (\mathrm{e}^t + 1)^{-1}$ and $Z_k(x) \equiv \beta_k(x - \mu_k)$. The entropy production observable is then given by

$$\sigma \equiv -\sum_{k=1}^{M} \beta_k(\mathfrak{F}_k - \mu_k \mathfrak{J}_k) \,.$$

The entropy production rate of the NESS η_λ^+ is

$$\mathrm{Ep}(\eta_\lambda^+) = \eta_\lambda^+(\sigma) = \frac{1}{2} \sum_{k,j=1}^{M} \int_X (Z_j - Z_k)(F(Z_k) - F(Z_j)) D_{kj} \frac{\mathrm{d}x}{2\pi} \,. \tag{73}$$

Since the function F is monotone decreasing, $\mathrm{Ep}(\eta_\lambda^+)$ is clearly non-negative. This is the second law of thermodynamics (increase of entropy). Note that in the case of two subreservoirs with $\mu_1 = \mu_2$ the positivity of the entropy production implies that the heat flows from the hot to the cold reservoir. For $k \neq j$ let

$$F_{kj} \equiv \{x \in X \mid \|f_k(x)\|_{\mathfrak{h}_k} \|f_j(x)\|_{\mathfrak{h}_j} > 0\} .$$

The subreservoirs $\mathcal{R}_k$ and $\mathcal{R}_j$ are *effectively coupled* if $\mu_{\mathrm{ac}}(F_{kj}) > 0$. The SEBB model is thermodynamically trivial unless some of the subreservoirs are effectively coupled. If $\mathcal{R}_k$ and $\mathcal{R}_j$ are effectively coupled, then $\mathrm{Ep}(\eta_\lambda^+) > 0$ unless $\beta_k = \beta_j$ and $\mu_k = \mu_j$, that is, unless the reservoirs $\mathcal{R}_k$ and $\mathcal{R}_j$ are in the same thermodynamical state.

5.5 The Effect of Eigenvalues

In our study of NESS and thermodynamics in Subsects. 5.2 and 5.4 we have made the assumption that h_λ has purely absolutely continuous spectrum. If $X \neq \mathbb{R}$, then this assumption does not hold for λ large. For example, if $X =]0, \infty[$, $\omega > 0$, and

$$\lambda^2 > \omega \left(\int_0^\infty \|f(x)\|_{\mathfrak{h}}^2 \, x^{-1} \mathrm{d}\mu(x) \right)^{-1} ,$$

then h_λ will have an eigenvalue in $]-\infty, 0[$. In particular, if

$$\int_0^\infty \|f(x)\|_{\mathfrak{h}}^2 \, x^{-1} \mathrm{d}\mu(x) = \infty ,$$

then h_λ will have a negative eigenvalue for all $\lambda \neq 0$. Hence, the assumption that h_λ has empty point spectrum is very restrictive, and it is important to understand the NESS and thermodynamics of the SEBB model in the case where h_λ has some eigenvalues. Of course, we are concerned only with point spectrum of h_λ restricted to the cyclic subspace generated by the vector 1.

Assume that λ is such that $\mathrm{sp}_{\mathrm{pp}}(h_\lambda) \neq \emptyset$ and $\mathrm{sp}_{\mathrm{sc}}(h_\lambda) = \emptyset$. We make no assumption on the structure of $\mathrm{sp}_{\mathrm{pp}}(h_\lambda)$ (in particular this point spectrum may be dense in some interval). We also make no assumptions on the spectrum of $h_{\mathcal{R}}$.

For notational simplicity, in this subsection we write $\mathfrak{h}_{\mathrm{ac}}$ for $\mathfrak{h}_{\mathrm{ac}}(h_\lambda)$, $\mathbf{1}_{\mathrm{ac}}$ for $\mathbf{1}_{\mathrm{ac}}(h_\lambda)$, etc.

Let T and η_T be as in Subsect. 5.2 and let $\phi, \psi \in \mathfrak{h} = \mathbb{C} \oplus \mathfrak{h}_{\mathcal{R}}$. Then,

$$\eta_T(\tau_\lambda^t(a^*(\phi)a(\psi))) = (\mathrm{e}^{\mathrm{i}th_\lambda}\psi | T\mathrm{e}^{\mathrm{i}th_\lambda}\phi) = \sum_{j=1}^{3} N_j(\mathrm{e}^{\mathrm{i}th_\lambda}\psi, \mathrm{e}^{\mathrm{i}th_\lambda}\phi) ,$$

where we have set

$$N_1(\psi,\phi) \equiv (\mathbf{1}_{\rm ac}\psi|T\mathbf{1}_{\rm ac}\phi)\,,$$

$$N_2(\psi,\phi) \equiv 2{\rm Re}\,(\mathbf{1}_{\rm pp}\psi|T\mathbf{1}_{\rm ac}\phi)\,,$$

$$N_3(\psi,\phi) \equiv (\mathbf{1}_{\rm pp}\psi|T\mathbf{1}_{\rm pp}\phi)\,.$$

Since $\mathrm{e}^{-\mathrm{i}th_0}T = T\mathrm{e}^{-\mathrm{i}th_0}$, we have

$$N_1(\mathrm{e}^{\mathrm{i}th_\lambda}\psi, e^{\mathrm{i}th_\lambda}\phi) = (\mathrm{e}^{-\mathrm{i}th_0}\mathrm{e}^{\mathrm{i}th_\lambda}\mathbf{1}_{\rm ac}\psi|T\mathrm{e}^{-\mathrm{i}th_0}\mathrm{e}^{\mathrm{i}th_\lambda}\mathbf{1}_{\rm ac}\phi)\,,$$

and so

$$\lim_{t\to\infty} N_1(\mathrm{e}^{\mathrm{i}th_\lambda}\psi, e^{\mathrm{i}th_\lambda}\phi) = (\Omega_\lambda^-\psi|T\Omega_\lambda^-\phi)\,.$$

Since $\mathfrak{h}$ is separable, there exists a sequence P_n of finite rank projections commuting with h_λ such that $\mathrm{s}-\lim P_n = \mathbf{1}_{\rm pp}$. The Riemann-Lebesgue lemma yields that for all n

$$\lim_{t\to\infty} \|P_nT\mathrm{e}^{\mathrm{i}th_\lambda}\mathbf{1}_{\rm ac}\phi\| = 0\,.$$

The relation

$$\begin{aligned}N_2(\mathrm{e}^{\mathrm{i}th_\lambda}\psi|\mathrm{e}^{\mathrm{i}th_\lambda}\phi) &= (\mathrm{e}^{\mathrm{i}th_\lambda}\mathbf{1}_{\rm pp}\psi|P_nT\mathrm{e}^{\mathrm{i}th_\lambda}\mathbf{1}_{\rm ac}\phi)\\ &+ (\mathrm{e}^{\mathrm{i}th_\lambda}(I-P_n)\mathbf{1}_{\rm pp}\psi|T\mathrm{e}^{\mathrm{i}th_\lambda}\mathbf{1}_{\rm ac}\phi)\,,\end{aligned}$$

yields that

$$\lim_{t\to\infty} N_2(\mathrm{e}^{\mathrm{i}th_\lambda}\psi, \mathrm{e}^{\mathrm{i}th_\lambda}\phi) = 0\,.$$

Since $N_3(\mathrm{e}^{\mathrm{i}th_\lambda}\psi, \mathrm{e}^{\mathrm{i}th_\lambda}\phi)$ is either a periodic or a quasi-periodic function of t, the limit

$$\lim_{t\to\infty} \eta_T(\tau_\lambda^t(a^*(\phi)a(\psi)))\,,$$

does not exist in general. The resolution of this difficulty is well known – to extract the steady part of a time evolution in the presence of a (quasi-) periodic component one needs to average over time. Indeed, one easily shows that

$$\lim_{t\to\infty}\frac{1}{t}\int_0^t N_3(\mathrm{e}^{\mathrm{i}sh_\lambda}\psi, \mathrm{e}^{\mathrm{i}sh_\lambda}\phi)\mathrm{d}s = \sum_{e\in\mathrm{sp_p}(h_\lambda)} (P_e\psi|TP_e\phi)\,,$$

where P_e denotes the spectral projection of h_λ associated with the eigenvalue e. Hence,

$$\lim_{t\to\infty}\frac{1}{t}\int_0^t \eta_T(\tau_\lambda^s(a^*(\phi)a(\psi)))\mathrm{d}s = \sum_{e\in\mathrm{sp_p}(h_\lambda)} (P_e\psi|TP_e\phi) + (\Omega_\lambda^-\psi|T\Omega_\lambda^-\phi)\,.$$

In a similar way one concludes that for any observable of the form

$$A = a^*(\phi_n)\cdots a^*(\phi_1)a(\psi_1)\cdots a(\psi_m)\,, \tag{74}$$

the limit

$$\lim_{t\to\infty} \frac{1}{t}\int_0^t \eta_T(\tau_\lambda^s(A))\mathrm{d}s = \delta_{n,m} \lim_{t\to\infty} \frac{1}{t}\int_0^t \det\{(\mathrm{e}^{\mathrm{i}sh_\lambda}\psi_i|T\mathrm{e}^{\mathrm{i}sh_\lambda}\phi_j)\}\mathrm{d}s\ ,$$

exists and is equal to the limit

$$\lim_{t\to\infty} \frac{1}{t}\int_0^t \det\left\{(\mathrm{e}^{\mathrm{i}sh_\lambda}\mathbf{1}_{\mathrm{pp}}\psi_i|T\mathrm{e}^{\mathrm{i}sh_\lambda}\mathbf{1}_{\mathrm{pp}}\phi_j) + (\Omega_\lambda^-\mathbf{1}_{\mathrm{ac}}\psi_i|T\Omega_\lambda^-\mathbf{1}_{\mathrm{ac}}\phi_j)\right\}\mathrm{d}s\ , \tag{75}$$

see [43] Sect. VI.5 for basic results about quasi-periodic function on $\mathbb{R}$. Since the linear span of the set of observables of the form (74) is dense in $\mathfrak{h}$, we conclude that for all $A \in \mathrm{CAR}(\mathfrak{h})$ the limit

$$\eta_\lambda^+(A) = \lim_{t\to\infty} \frac{1}{t}\int_0^t \eta_T(\tau_\lambda^s(A))\mathrm{d}s\ ,$$

exists and defines a state η_λ^+ on $\mathrm{CAR}(\mathfrak{h})$. By construction, this state is τ_λ-invariant. η_λ^+ is the NESS of $(\mathrm{CAR}(\mathfrak{h}), \tau_\lambda)$ associated to the initial state η_T. Note that this definition reduces to the previous if the point spectrum of h_λ is empty.

To further elucidate the structure of η_λ^+ we will make use of the decomposition

$$\mathfrak{h} = \mathfrak{h}_{\mathrm{ac}} \oplus \mathfrak{h}_{\mathrm{pp}}\ . \tag{76}$$

The subspaces $\mathfrak{h}_{\mathrm{ac}}$ and $\mathfrak{h}_{\mathrm{pp}}$ are invariant under h_λ and we denote the restrictions of τ_λ to $\mathrm{CAR}(\mathfrak{h}_{\mathrm{ac}})$ and $\mathrm{CAR}(\mathfrak{h}_{\mathrm{pp}})$ by $\tau_{\lambda,\mathrm{ac}}$ and $\tau_{\lambda,\mathrm{pp}}$. We also denote by $\eta_{\lambda,\mathrm{ac}}^+$ and $\eta_{\lambda,\mathrm{pp}}^+$ the restrictions of η_λ^+ to $\mathrm{CAR}(\mathfrak{h}_{\mathrm{ac}})$ and $\mathrm{CAR}(\mathfrak{h}_{\mathrm{pp}})$. $\eta_{\lambda,\mathrm{ac}}^+$ is the quasi-free state generated by $T_\lambda^+ \equiv (\Omega_\lambda^-)^* T\Omega_\lambda^-$. If A is of the form (74) and $\phi_j, \psi_i \in \mathfrak{h}_{\mathrm{pp}}$, then

$$\eta_{\lambda,\mathrm{pp}}^+(A) = \delta_{n,m} \lim_{t\to\infty} \frac{1}{t}\int_0^t \det\{(\mathrm{e}^{\mathrm{i}sh_\lambda}\psi_i|T\mathrm{e}^{\mathrm{i}sh_\lambda}\phi_j)\}\mathrm{d}s\ .$$

Clearly, $\eta_{\lambda,\mathrm{ac}}^+$ is $\tau_{\lambda,\mathrm{ac}}$ invariant and $\eta_{\lambda,\mathrm{pp}}^+$ is $\tau_{\lambda,\mathrm{pp}}$ invariant. Expanding the determinant in (75) one can easily see that $\eta_{\lambda,\mathrm{ac}}^+$ and $\eta_{\lambda,\mathrm{pp}}^+$ uniquely determine η_λ^+.

While the state $\eta_{\lambda,\mathrm{pp}}^+$ obviously depends on the choice of α and on $T_{\mathcal{R}}|_{\mathfrak{h}_{\mathrm{sing}}(h_{\mathcal{R}})}$ in $T = \alpha \oplus T_{\mathcal{R}}$, the state $\eta_{\lambda,\mathrm{ac}}^+$ does not. In fact, if η is any initial state normal w.r.t. η_T, then for $A \in \mathrm{CAR}(\mathfrak{h}_{\mathrm{ac}})$,

$$\lim_{t\to\infty} \eta(\tau_\lambda^t(A)) = \eta_{\lambda,\mathrm{ac}}^+(A)\ .$$

For a finite rank operator $A \equiv \sum_j \phi_j(\psi_j|\cdot)$ one has

$$\eta_\lambda^+(\mathrm{d}\Gamma(A)) = \sum_j \eta_\lambda^+(a^*(\phi_j)a(\psi_j))\ ,$$

and so

$$\eta_\lambda^+(\mathrm{d}\Gamma(A)) = \sum_j \left(\sum_{\mathrm{e}\in \mathrm{sp_p}(h_\lambda)} (P_e\psi_j|TP_e\phi_j) + (\Omega_\lambda^-\psi_j|T\Omega_\lambda^-\phi_j) \right) .$$

The conclusion is that in the presence of eigenvalues one needs to add the term

$$\sum_j \sum_{\mathrm{e}\in \mathrm{sp_p}(h_\lambda)} (P_e\psi_j|TP_e\phi_j) ,$$

to (63), i.e., we obtain the following formula generalizing (63),

$$\eta_\lambda^+(\mathrm{d}\Gamma(A)) = \mathrm{Tr}\left\{ T \left(\sum_{\mathrm{e}\in \mathrm{sp_p}(h_\lambda)} P_eAP_e + \Omega_\lambda^- A \Omega_\lambda^{-*} \right) \right\} . \tag{77}$$

Note that if for some operator q, $A = \mathrm{i}[h_\lambda, q]$ in the sense of quadratic forms on $\mathrm{Dom}\, h_\lambda$, then $P_eAP_e = 0$ and eigenvalues do not contribute to $\eta_\lambda^+(\mathrm{d}\Gamma(A))$. This is the case of the current observables $\mathrm{d}\Gamma(\mathfrak{f}_k)$ and $\mathrm{d}\Gamma(\mathfrak{j}_k)$ of Subsect. 5.4. We conclude that the formulas (69) and (71), which we have previously derived under the assumption $\mathrm{sp_{sing}}(h_\lambda) = \emptyset$, remain valid as long as $\mathrm{sp_{sc}}(h_\lambda) = \emptyset$, i.e., they are not affected by the presence of eigenvalues.

5.6 Thermodynamics in the Non-Perturbative Regime

The results of the previous subsection can be summarized as follows.

If $\mathrm{sp_{sc}}(h_\lambda) = \emptyset$ and $\mathrm{sp_{pp}}(h_\lambda) \neq \emptyset$ then, on the qualitative level, the thermodynamics of the SEBB model is similar to the case $\mathrm{sp_{sing}}(h_\lambda) = 0$. To construct NESS one takes the ergodic averages of the states $\eta_T \circ \tau_\lambda^t$. The NESS is unique. The formulas for steady currents and entropy production are not affected by the point spectra and are given by (69), (71), (73) and (70) or (72) for all $\lambda \neq 0$. In particular, *the NESS and thermodynamics are well defined for all* $\lambda \neq 0$ *and all* ω. One can proceed further along the lines of [2] and study the linear response theory of the SEBB model (Onsager relations, Kubo formulas, etc.) in the non-perturbative regime. Given the results of the previous subsection, the arguments and the formulas are exactly the same as in [2] and we will not reproduce them here.

The study of NESS and thermodynamics is more delicate in the presence of singular continuous spectrum and we will not pursue it here. We wish to point, however, that unlike the point spectrum, the singular continuous spectrum can be excluded in "generic" physical situations. Assume that X is an open set and that the absolutely continuous spectrum of $h_{\mathcal{R}}$ is "well-behaved" in the sense that $\mathrm{Im}\, F_{\mathcal{R}}(x + \mathrm{i}0) > 0$ for Lebesgue a.e. $x \in X$. Then, by the Simon-Wolff theorem 5, h_λ has no singular continuous spectrum for Lebesgue a.e. $\lambda \in \mathbb{R}$. If f is a continuous function and $\mathrm{d}\mu_{\mathcal{R}} = \mathrm{d}x$, then h_λ has no singular continuous spectrum for all λ.

5.7 Properties of the Fluxes

In this subsection we consider a SEBB model without singular continuous spectrum, *i.e*, we assume that $\mathrm{sp_{sc}}(h_\lambda) = \emptyset$ for all λ and ω. We will study the properties of the steady currents as functions of (λ, ω). For this reason, we will again indicate explicitly the dependence on ω.

More precisely, in this subsection we will study the properties of the function

$$(\lambda, \omega) \mapsto \eta^+_{\lambda,\omega}(\mathfrak{F}) , \qquad (78)$$

where $\mathfrak{F}$ is one of the observables $\mathfrak{F}_k$ or $\mathfrak{J}_k$ for a given k. We assume that (A1) holds. For simplicity of exposition we also assume that the functions

$$g_j(t) \equiv \int_X \mathrm{e}^{-\mathrm{i}tx} \|f_j(x)\|^2_{\mathfrak{K}_j} \, \mathrm{d}x ,$$

are in $L^1(\mathbb{R}, \mathrm{d}t)$, that $\|f(x)\|_{\mathfrak{K}}$ is non-vanishing on X, that the energy densities $\varrho_j(x)$ of the subreservoirs are bounded continuous functions on X, and that the functions $(1 + |x|)\varrho_j(x)$ are integrable on X. According to (69), (71) and (70), one has

$$\eta^+_{\lambda,\omega}(\mathfrak{F}) = 2\pi\lambda^4 \sum_{j=1}^{M} \int_X \|f_k(x)\|^2_{\mathfrak{K}_k} \|f_j(x)\|^2_{\mathfrak{K}_j} |F_\lambda(x + \mathrm{i}0)|^2 x^n (\varrho_k(x) - \varrho_j(x)) \, \mathrm{d}x ,$$

where $n = 0$ if $\mathfrak{F} = \mathfrak{J}_k$ and $n = 1$ if $\mathfrak{F} = \mathfrak{F}_k$.

Obviously, the function (78) is real-analytic on $\mathbb{R} \times \mathbb{R} \setminus \overline{X}$ and for a given $\omega \notin \overline{X}$,

$$\eta^+_{\lambda,\omega}(\mathfrak{F}) = O(\lambda^4) , \qquad (79)$$

as $\lambda \to 0$. The function (78) is also real-analytic on $\mathbb{R} \setminus \{0\} \times \mathbb{R}$. For $\omega \in X$, Lemma 3 shows that

$$\lim_{\lambda \to 0} \lambda^{-2} \eta^+_{\omega,\lambda}(\mathfrak{F}) = 2\pi \sum_{j=1}^{M} \frac{\|f_k(\omega)\|^2_{\mathfrak{K}_k} \|f_j(\omega)\|^2_{\mathfrak{K}_j}}{\|f(\omega)\|^2_{\mathfrak{K}}} \omega^n (\varrho_k(\omega) - \varrho_j(\omega)) . \qquad (80)$$

Comparing (79) and (80) we see that in the weak coupling limit we can distinguish two regimes: the "conducting" regime $\omega \in X$ and the "insulating" regime $\omega \notin \overline{X}$. Clearly, the conducting regime coincides with the "resonance" regime for $h_{\lambda,\omega}$ and, colloquially speaking, the currents are carried by the resonance pole. In the insulating regime there is no resonance for small λ and the corresponding heat flux is infinitesimal compared to the heat flux in the "conducting" regime.

For $x \in X$ one has

$$\lambda^4 |F_\lambda(x + \mathrm{i}0)|^2 = \frac{\lambda^4}{(\omega - x - \lambda^2 \mathrm{Re}\, F_{\mathcal{R}}(x + \mathrm{i}0))^2 + \lambda^4 \pi^2 \|f(x)\|^4_{\mathfrak{K}}} .$$

Hence,

$$\sup_{\lambda \in \mathbb{R}} \lambda^4 |F_\lambda(x+\mathrm{i}0)|^2 = \left(\pi \sum_{j=1}^{M} \|f_j(x)\|^2_{\mathfrak{K}_j} \right)^{-2} , \tag{81}$$

and so

$$\lambda^4 \|f_k(x)\|^2_{\mathfrak{K}_k} \|f_j(x)\|^2_{\mathfrak{K}_j} |F_\lambda(x+\mathrm{i}0)|^2 \leq \frac{1}{4\pi^2} .$$

This estimate and the dominated convergence theorem yield that for all $\omega \in \mathbb{R}$,

$$\lim_{|\lambda| \to \infty} \eta^+_{\lambda,\omega}(\mathfrak{F}) = 2\pi \sum_{j=1}^{M} \int_X \frac{\|f_k(x)\|^2_{\mathfrak{K}_k} \|f_j(x)\|^2_{\mathfrak{K}_j}}{|F_{\mathcal{R}}(x+\mathrm{i}0)|^2} x^n (\varrho_k(x) - \varrho_i(x)) \, \mathrm{d}x . \tag{82}$$

Thus, the steady currents are independent of ω in the strong coupling limit. In the same way one shows that

$$\lim_{|\omega| \to \infty} \eta^+_{\lambda,\omega}(\mathfrak{F}) = 0 , \tag{83}$$

for all λ.

The cross-over between the weak coupling regime (80) and the large coupling regime (82) is delicate and its study requires detailed information about the model. We will discuss this topic further in the next subsection.

We finish this subsection with one simple but physically important remark. Assume that the functions

$$C_j(x) \equiv 2\pi \|f_k(x)\|^2_{\mathfrak{K}_k} \|f_j(x)\|^2_{\mathfrak{K}_j} x^n (\varrho_k(x) - \varrho_j(x)) ,$$

are sharply peaked around the points $\overline{x}_j$. This happens, for example, if all the reservoirs are at thermal equilibrium at low temperatures. Then, the flux (78) is well approximated by the formula

$$\eta^+_{\lambda,\omega}(\mathfrak{F}) \simeq \sum_{j=1}^{M} \lambda^4 |F_\lambda(\overline{x}_j)|^2 \int_X C_j(x) \, \mathrm{d}x ,$$

and since the supremum in (81) is achieved at $\omega = x + \lambda^2 \mathrm{Re}\, F_\lambda(x+\mathrm{i}0)$, the flux (78) will be peaked along the parabolic resonance curves

$$\omega = \overline{x}_j + \lambda^2 \mathrm{Re}\, F_\lambda(\overline{x}_j + \mathrm{i}0) .$$

5.8 Examples

We finish these lecture notes with several examples of the SEBB model which we will study using numerical calculations. For simplicity, we will only consider the case of two subreservoirs, i.e., in this subsection $\mathfrak{K} = \mathbb{C}^2 = \mathbb{C} \oplus \mathbb{C}$. We also take

$$f(x) = f_1(x) \oplus f_2(x) \equiv \frac{1}{\sqrt{2}} \begin{pmatrix} f_0(x) \\ 0 \end{pmatrix} \oplus \frac{1}{\sqrt{2}} \begin{pmatrix} 0 \\ f_0(x) \end{pmatrix} ,$$

so that

$$\|f_1(x)\|^2_{\mathfrak{K}_1} = \|f_2(x)\|^2_{\mathfrak{K}_2} = \frac{1}{2}\|f(x)\|^2_{\mathfrak{K}} = \frac{1}{2}|f_0(x)|^2 .$$

Example 1. We consider the fermionic quantization of Example 1 in Subsect. 3.5, i.e., $\mathfrak{h}_{\mathcal{R}} = L^2(]0,\infty[, \mathrm{d}x; \mathbb{C}^2)$ and

$$f_0(x) = \pi^{-1/2}(2x)^{1/4}(1+x^2)^{-1/2} .$$

We put the two subreservoirs at thermal equilibrium

$$\varrho_j(x) = \frac{1}{1+\mathrm{e}^{\beta_j(x-\mu_j)}} ,$$

where we set the inverse temperatures to $\beta_1 = \beta_2 = 50$ (low temperature) and the chemical potentials to $\mu_1 = 0.3$, $\mu_2 = 0.2$. We shall only consider the particle flux ($n = 0$) in this example. The behavior of the heat flux is similar. The function

$$C_2(x) = 2\pi\|f_1(x)\|^2_{\mathfrak{K}_1}\|f_2(x)\|^2_{\mathfrak{K}_2}(\varrho_1(x) - \varrho_2(x)) = \frac{x(\varrho_1(x) - \varrho_2(x))}{\pi(1+x^2)^2} ,$$

plotted in Fig. 10, is peaked at $\overline{x} \simeq 0.25$. In accordance with our discussion in the previous subsection, the particle current, represented in Fig. 11, is sharply peaked around the parabola $\omega = \overline{x} + 2\lambda^2(1-\overline{x})/(1+\overline{x}^2)$ (dark line). The convergence to an ω-independent limit as $\lambda \to \infty$ and convergence to 0 as $\omega \to \infty$ are also clearly illustrated.

Example 2. We consider now the heat flux in the SEBB model corresponding to Example 2 of Subsect. 3.5. Here $\mathfrak{h}_{\mathcal{R}} = L^2(]-1,1[, \mathrm{d}x; \mathbb{C}^2)$,

$$f_0(x) = \sqrt{\frac{2}{\pi}}\,(1-x^2)^{1/4} ,$$

and we choose the high temperature regime by setting $\beta_1 = \beta_2 = 0.1$, $\mu_1 = 0.3$ and $\mu_2 = 0.2$. Convergence of the rescaled heat flux to the weak coupling limit (80) is illustrated in Fig. 12. In this case the function C_2 is given by

$$C_2(x) = \frac{2}{\pi}x(1-x^2)(\varrho_1(x) - \varrho_2(x)) ,$$

and is completely delocalized as shown in Fig. 13.

Even in this simple example the cross-over between the weak and the strong coupling regime is difficult to analyze. This cross-over is non trivial, as can be seen in Fig. 14. Note in particular that the function $\lambda \mapsto \eta^+_{\lambda,\omega}(\mathfrak{F})$ is

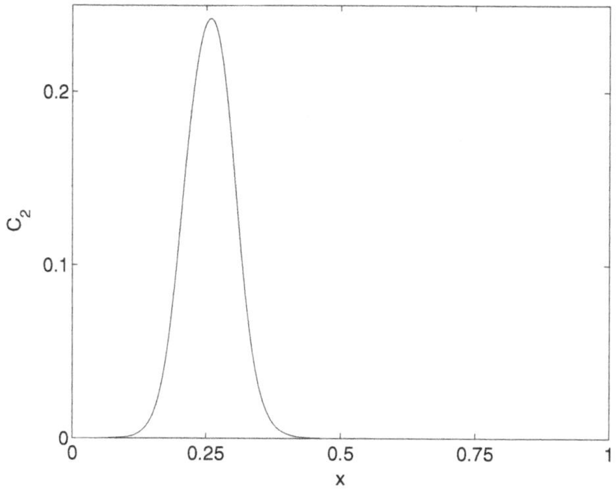

Fig. 10. The function $C_2(x)$ in Example 1

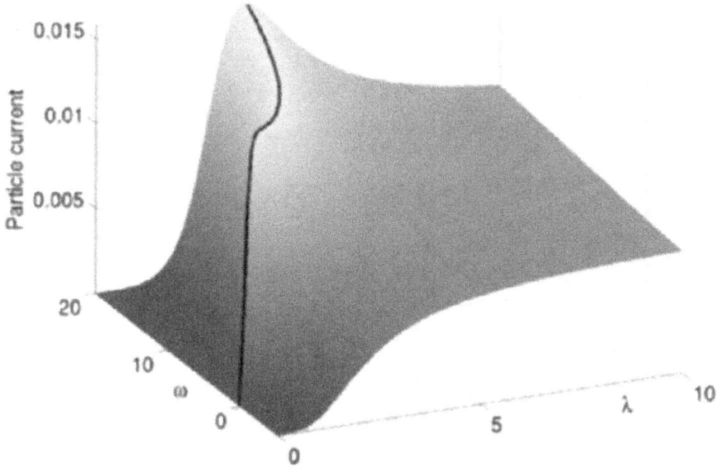

Fig. 11. The particle flux in Example 1

not necessarily monotone and may have several local minima/maxima before reaching its limiting value (82) as shown by the section $\omega = 0.5$ in Fig. 14.

Example 3. In this example we will discuss the large coupling limit. Note that in the case of two subreservoirs (82) can be written as

$$\lim_{|\lambda|\to\infty} \eta^{+}_{\lambda,\omega}(\mathfrak{F}) = \frac{1}{2\pi} \int_X \sin^2\theta(x)\, x^n (\varrho_1(x) - \varrho_2(x))\, \mathrm{d}x\,, \tag{84}$$

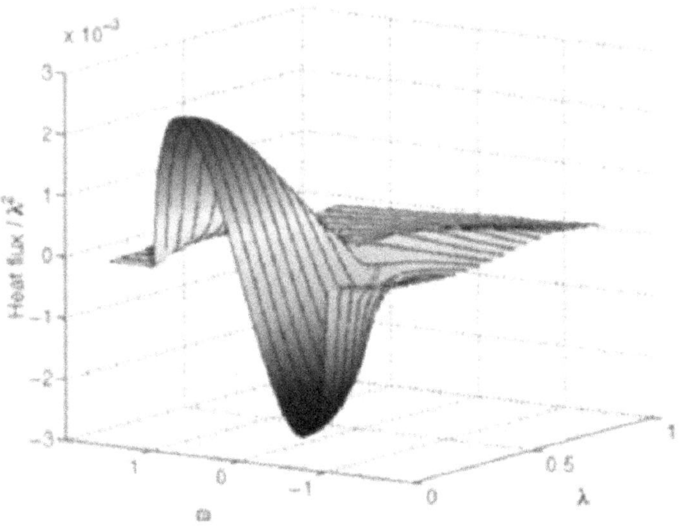

Fig. 12. The rescaled heat flux (weak coupling regime) in Example 2

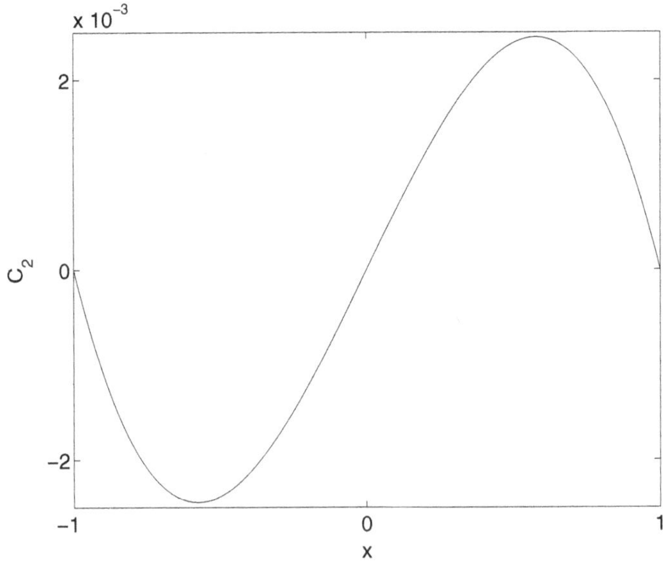

Fig. 13. The function $C_2(x)$ in Example 2

where $\theta(x) \equiv \mathrm{Arg}(F_{\mathcal{R}}(x + \mathrm{i}0))$. Therefore, large currents can be obtained if one of the reservoir, say $\mathcal{R}_1$, has an energy distribution concentrated in a region where $\mathrm{Im}\, F_{\mathcal{R}}(x + \mathrm{i}0) \gg \mathrm{Re}\, F_{\mathcal{R}}(x + \mathrm{i}0)$ while the energy distribution of $\mathcal{R}_2$ is concentrated in a region where $\mathrm{Im}\, F_{\mathcal{R}}(x + \mathrm{i}0) \ll \mathrm{Re}\, F_{\mathcal{R}}(x + \mathrm{i}0)$.

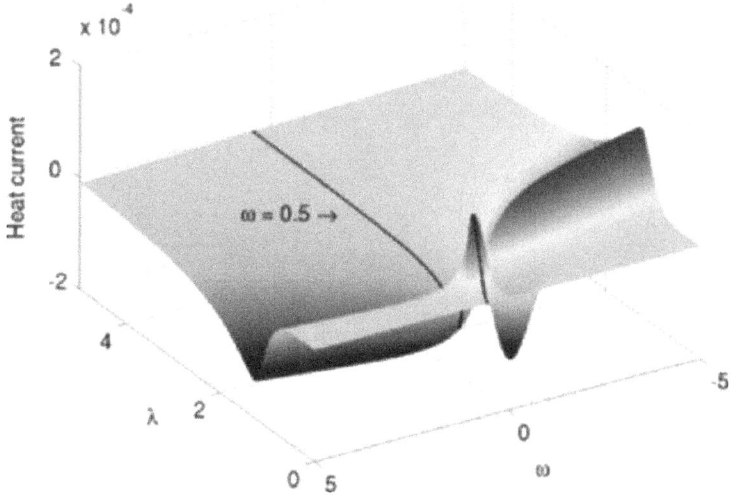

Fig. 14. The heat flux in Example 2

As an illustration, we consider the SEBB model corresponding to Example 3 of Subsect. 3.5, i.e., $\mathfrak{h}_{\mathcal{R}} = L^2(]-1,1[, \mathrm{d}x; \mathbb{C}^2)$ and

$$f_0(x) = \sqrt{\frac{1}{\pi}}\, x(1-x^2)^{1/4} \,.$$

From (55) we obtain that

$$F_{\mathcal{R}}(x+\mathrm{i}0) = -x\left(x^2 - \frac{1}{2}\right) + \mathrm{i}x^2\sqrt{1-x^2} \,.$$

Hence,

$$\sin^2\theta(x) = 4x^2(1-x^2) \,,$$

reaches its maximal value 1 at energy $x = \pm 1/\sqrt{2}$.

We use the following initial states: the first subreservoir has a quasi-monochromatic energy distribution

$$\varrho_1(x) \equiv 3\,\mathrm{e}^{-1000(x-\Omega)^2} \,,$$

at energy $\Omega \in [-1,1]$. The second subreservoir is at thermal equilibrium at low temperature $\beta = 10$ and chemical potential $\mu_2 = -0.9$. Thus, ϱ_2 is well localized near the lower band edge $x = -1$ where $\sin\theta$ vanishes. Figure 15 shows the limiting currents (84) as functions of Ω, with extrema near $\pm 1/\sqrt{2} \simeq \pm 0.7$ as expected.

Another feature of Fig. 15 is worth a comment. As discussed in Example 4 of Subsect. 3.5, this model has a resonance approaching 0 as $\lambda \to \infty$.

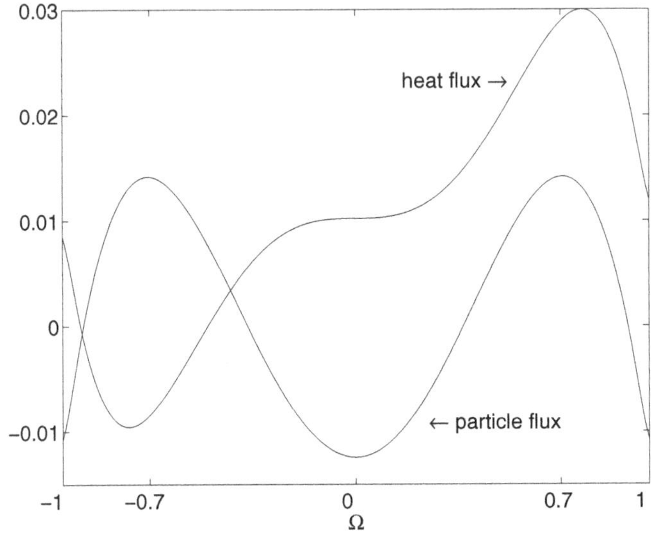

Fig. 15. The limiting particle and heat fluxes in Example 3

However, since $\sin\theta(0) = 0$, the large coupling resonance near zero does not lead to a noticeable flux enhancement. This can be seen in Fig. 15 by noticing that the fluxes at the resonant energy $\Omega = 0$ are the same as at the band edges $\Omega = \pm 1$. It is a simple exercise to show that this phenomenon is related to the fact that the resonance pole of F_λ approaches 0 tangentially to the real axis (see Fig. 9).

In fact, the following argument shows that this behavior is typical. Assume that $F_{\mathcal{R}}(z)$ has a meromorphic continuation from the upper half-plane across X with a zero at $\overline{\omega} \in X$ (we argued in our discussion of Example 3 in Subsect. 3.5 that this is a necessary condition for $\overline{\omega}$ to be a large coupling resonance). Since $\mathrm{Im}\, F_{\mathcal{R}}(x + \mathrm{i}y) \geq 0$ for $y \geq 0$, it is easy to show, using the power series expansion of $F_{\mathcal{R}}$ around $\overline{\omega}$, that $(\partial_z F_{\mathcal{R}})(\overline{\omega}) > 0$. Combining this fact with the Cauchy-Riemann equations we derive

$$\partial_x \mathrm{Re}\, F_{\mathcal{R}}(x + \mathrm{i}0)|_{x=\overline{\omega}} > 0, \qquad \partial_x \mathrm{Im}\, F_{\mathcal{R}}(x + \mathrm{i}0)|_{x=\overline{\omega}} = 0\,,$$

and so

$$\sin\theta(\overline{\omega}) = 0\,.$$

Thus, in contrast with the weak coupling resonances, the strong coupling resonances do not induce large currents.

References

1. Aguilar, J., Combes, J.M.: A class of analytic perturbations for one-body Schrödinger Hamiltonians. Commun. Math. Phys. **22**, 269 (1971).

2. Aschbacher, W., Jakšić, V., Pautrat, Y., Pillet, C.-A.: Topics in quantum statistical mechanics. In *Open Quantum Systems III.* Lecture Notes of the Summer School on Open Quantum Systems held in Grenoble, June 16–July 4, 2003. To be published in *Lecture Notes in Mathematics*, Springer, New York.
3. Aschbacher, W., Jakšić, V., Pautrat, Y., Pillet, C.-A.: Transport properties of ideal Fermi gases. In preparation.
4. Aizenstadt, V.V., Malyshev, V.A.: Spin interaction with an ideal Fermi gas. J. Stat. Phys. **48**, 51 (1987).
5. Arai, A.: On a model of a harmonic oscillator coupled to a quantized, massless, scalar field. J. Math. Phys. **21**, 2539 (1981).
6. Balslev, E., Combes, J.M.: Spectral properties of many-body Schrödinger operators with dilation analytic interactions. Commun. Math. Phys. **22**, 280 (1971).
7. Braun, E.: Irreversible behavior of a quantum harmonic oscillator coupled to a heat bath. Physica A **129**, 262 (1985).
8. Bratteli, O., Robinson, D.W.: *Operator Algebras and Quantum Statistical Mechanics 1.* Springer, Berlin (1987).
9. Bratteli, O., Robinson, D. W.: *Operator Algebras and Quantum Statistical Mechanics 2.* Springer, Berlin (1996).
10. Baez, J.C., Segal, I.E., Zhou, Z.: *Introduction to algebraic and constructive quantum field theory.* Princeton University Press, Princeton NJ (1991).
11. Cycon, H., Froese, R., Kirsch, W., Simon, B.: *Schrödinger Operators.* Springer, Berlin (1987).
12. Carmona, R., Lacroix, J.: *Spectral Theory of Random Schrödinger Operators.* Birkhauser, Boston (1990).
13. Cohen-Tannoudji, C., Dupont-Roc, J., Grynberg, G.: *Atom-Photon Interactions.* John Wiley, New York (1992).
14. Davies, E.B.: The harmonic oscillator in a heat bath. Commun. Math. Phys. **33**, 171 (1973).
15. Davies, E.B.: Markovian master equations. Commun. Math. Phys. **39**, 91 (1974).
16. Davies, E.B.: Markovian master equations II. Math. Ann. **219**, 147 (1976).
17. Davies, E.B.: Dynamics of a multilevel Wigner-Weisskopf atom. J. Math. Phys. **15**, 2036 (1974).
18. De Biévre, S.: *Local states of free Bose fields.* Lect. Notes Phys. **695**, 17–63 (2006).
19. Dereziński, J.: *Introduction to representations of canonical commutation and anticommutation relation.* Lect. Notes Phys. **695**. 65–145 (2006).
20. Dereziński, J.: Fermi golden rule, Feshbach method and embedded point spectrum. Séminaire Equations aux Dérivées Partielles, 1998–1999, Exp. No. XXIII, 13 pp., Sémin. Equation Dériv. Partielles, Ecole Polytech., Palaiseau, (1999).
21. Dereziński, J.: Fermi golden rule and open quantum systems. In *Open Quantum Systems III.* Lecture Notes of the Summer School on Open Quantum Systems held in Grenoble, June 16–July 4, 2003. To be published in *Lecture Notes in Mathematics*, Springer, New York.
22. del Rio, R., Fuentes S., Poltoratskii, A.G.: Coexistence of spectra in rank-one perturbation problems. Bol. Soc. Mat. Mexicana **8**, 49 (2002).
23. Dirac, P.A.M.: The quantum theory of the emission and absorption of radiation. Proc. Roy. Soc. London, Ser. A **114**, 243 (1927).

24. Davidson, R., Kozak, J.J.: On the relaxation to quantum-statistical equilibrium of the Wigner-Weisskopf Atom in a one-dimensional radiation field: a study of spontaneous emission. J. Math. Phys. **11**, 189 (1970).
25. del Rio, R., Makarov, N., Simon, B.: Operators with singular continuous spectrum: II. Rank one operators. Commun. Math. Phys. **165**, 59 (1994).
26. del Rio, R., Simon, B.: Point spectrum and mixed spectral types for rank one perturbations. Proc. Amer. Math. Soc. **125**, 3593 (1997).
27. Fermi, E.: *Nuclear Physics.* Notes compiled by Orear J., Rosenfeld A.H. and Schluter R.A. The University of Chicago Press, Chicago, (1950).
28. Friedrichs, K.O.: *Perturbation of Spectra in Hilbert Space.* AMS, Providence (1965).
29. Frigerio, A., Gorini, V., Pulé, J.V.: Open quasi free systems. J. Stat. Phys. **22**, 409 (1980).
30. Ford, G.W., Kac, M., Mazur, P.: Statistical mechanics of assemblies of coupled oscillators. J. Math. Phys. **6**, 504 (1965).
31. Gordon, A.: Pure point spectrum under 1-parameter perturbations and instability of Anderson localization. Commun. Math. Phys. **164** , 489 (1994).
32. Haag, R.: *Local Quantum Physics.* Springer-Verlag, New York (1993).
33. Haake, F.: *Statistical treatment of open systems by generalized master equation.* Springer Tracts in Modern Physics **66**, Springer, Berlin (1973).
34. Heitler, W.: *The Quantum Theory of Radiation.* Oxford, Oxford University Press (1954).
35. Herbst, I.: Exponential decay in the Stark effect. Commun. Math. Phys. **75**, 197 (1980).
36. Howland, J.S.: Perturbation theory of dense point spectra. J. Func. Anal. **74**, 52 (1987).
37. Jakšić, V.: Topics in spectral theory. In *Open Quantum Systems I.* Lecture Notes of the Summer School on Open Quantum Systems held in Grenoble, June 16–July 4, 2003. To be published in *Lecture Notes in Mathematics*, Springer, New York.
38. Jakšić, V., Last, Y.: A new proof of Poltoratskii's theorem. J. Func. Anal. **215**, 103 (2004).
39. Jakšić, V., Pillet, C.-A.: On a model for quantum friction II: Fermi's golden rule and dynamics at positive temperature. Commun. Math. Phys. **176**, 619 (1996).
40. Jakšić, V., Pillet, C.-A.: On a model for quantum friction III: Ergodic properties of the spin-boson system, Commun. Math. Phys. **178**, 627 (1996).
41. Jakšić, V., Pillet, C.-A.: Statistical mechanics of the FC oscillator. In preparation.
42. Kato, T.: *Perturbation Theory for Linear Operators.* Second edition. Springer, Berlin (1976).
43. Katznelson, A.: *An Introduction to Harmonic Analysis.* Dover, New York, (1976).
44. Koosis, P.: *Introduction to H_p Spaces.* Second edition. Cambridge University Press, New York (1998).
45. Kritchevski, E.: Ph.D. Thesis, McGill University. In preparation.
46. Maassen, H.: Quantum stochastic calculus with integral kernels. II. The Wigner-Weisskopf atom. Proceedings of the 1st World Congress of the Bernoulli Society, Vol. **1**(Tashkent, 1986), 491–494, VNU Sci. Press, Utrecht (1987).

47. Kindenberg, K., West, B.J.: Statistical properties of quantum systems: the linear oscillator. Phys. Rev. A **30**, 568 (1984).
48. Mehta, M.L.: *Random Matrices.* Second edition. Academic Press, New York (1991).
49. Messiah, A.: *Quantum Mechanics. Volume II.* John Wiley & Sons, New York.
50. Okamoto, T., Yajima, K.: Complex scaling technique in non-relativistic qed, Ann. Inst. H. Poincare **42**, 311 (1985).
51. Pillet, C.-A.: Quantum dynamical systems. In *Open Quantum Systems I.* Lecture Notes of the Summer School on Open Quantum Systems held in Grenoble, June 16–July 4, 2003. To be published in *Lecture Notes in Mathematics*, Springer, New York.
52. Poltoratskii, A.G.: The boundary behavior of pseudocontinuable functions. St. Petersburg Math. J. **5**, 389 (1994).
53. Paloviita, A., Suominen, K-A., Stenholm, S.: Weisskopf-Wigner model for wave packet excitation. J. Phys. B **30**, 2623 (1997).
54. Reed, M., Simon, B.: *Methods of Modern Mathematical Physics, I. Functional Analysis.* Second edition. Academic Press, London (1980).
55. Reed, M., Simon, B.: *Methods of Modern Mathematical Physics, II. Fourier Analysis, Self-Adjointness.* Academic Press, London (1975).
56. Reed, M., Simon, B.: *Methods of Modern Mathematical Physics, III. Scattering Theory.* Academic Press, London (1978).
57. Reed, M., Simon, B.: *Methods of Modern Mathematical Physics, IV. Analysis of Operators.* Academic Press, London, (1978).
58. Rudin, W.: *Real and Complex Analysis.* 3rd edition. McGraw-Hill (1987).
59. Simon, B.: Spectral analysis of rank one perturbations and applications. CRM Lecture Notes Vol. **8**, pp. 109–149, AMS, Providence, RI (1995).
60. Simon, B.: Resonances in N-body quantum systems with dilation analytic potential and foundations of time-dependent perturbation theory. Ann. Math. **97**, 247 (1973).
61. Simon, B., Wolff, T.: Singular continuous spectrum under rank one perturbations and localization for random Hamiltonians. Comm. Pure Appl. Math. **39**, 75 (1986).
62. Van Hove, L.: Master equation and approach to equilibrium for quantum systems. In *Fundamental problems in statistical mechanics*, compiled by E.G.D. Cohen, North-Holland, Amsterdam (1962).
63. Weder, R.A.: On the Lee model with dilatation analytic cutoff function. J. Math. Phys. **15**, 20 (1974).
64. Weisskopf, V., Wigner, E.: Berechnung der natürlichen Linienbreite auf Grund der Diracschen Lichttheorie. Zeitschrift für Physik **63**, 54 (1930).
65. De Groot, S.R., Mazur, P.: *Non-Equilibrium Thermodynamics.* North-Holland, Amsterdam (1969).

Non-Relativistic Matter and Quantized Radiation

M. Griesemer*

Department of Mathematics, University of Alabama at Birmingham, Birmingham, AL 35294, USA
marcel@math.uab.edu

Abstract. This is a didactic review of spectral and dynamical properties of atoms and molecules at energies below the ionization threshold, the focus being on recent work in which the author was involved. As far as possible, the results are described using a simple model with *one* electron only, and with *scalar* bosons. The main ideas are explained but no complete proofs are given. The full-fledged standard model of non-relativistic QED and various of its aspects are described in the appendix.

*Supported in part by U.S. National Science Foundation grant DMS 01-00160.

M. Griesemer: *Non-Relativistic Matter and Quantized Radiation*, Lect. Notes Phys. **695**, 217–248 (2006)
www.springerlink.com

1 Introduction

An atom or molecule in an excited state with energy below the ionization threshold will eventually relax to its ground state by dissipating excess energy in the form of radiation. This process of *relaxation to the ground state* is one of the basic phenomena responsible for the production of all visible light. It involves a range of energies within a few electron volts; a scale where the electron-positron pair creation and the production of ultraviolet radiation is highly suppressed. In a first mathematical study of relaxation to the ground state it is therefore reasonable and legitimate to work with a model where the electron-positron pair creation is entirely neglected and an ultraviolet cutoff is imposed on the electron-photon interaction. These simplifying assumptions lead to a mathematically well defined model of matter, often called standard-model of non-relativistic quantum electrodynamics, or Pauli-Fierz model. Since the numerical predictions of this model are in good agreement with measurement data it is a viable physical model. Yet, only little mathematically rigorous work on this model had been done before the middle of the 1990s, when several groups of researchers started to investigate various of its aspects. Most influential, perhaps, were the papers of Hübner and Spohn [33–35] on spectral and scattering theory, of Bach et al. on spectral analysis [7–10], of Dereziński and Gérard on scattering theory [14, 15], and of Jakšić and Pillet on thermal relaxation [37–40]. The present article reviews recent work on the phenomenon of relaxation to the ground state for states with total energy below the threshold energy for ionization. It is guided by papers of Lieb, Loss and Griesemer, and of Fröhlich, Schlein and Griesemer [17,18,24,25]. The focus is on the existence of an ionization threshold and the localization of the electrons below this energy, the existence of a ground state, and the existence and completeness of many-photon scattering states (asymptotic completeness of Rayleigh scattering).

The results to be discussed on existence of a ground state and on the localization of photons with energy below the ionization threshold, unlike previous results, hold for all values of the physical parameters such as the fine structure constant and the ultraviolet cutoff. This is crucial for moving on to physically more realistic models without ultraviolet regularization. The analysis of electron-photon scattering is based on methods and ideas from the scattering theory of N-body quantum systems [22, 23, 50, 55]. On the one hand the electron-photon dynamics is easier to analyze than the full N-body problem since there is no photon-photon interaction. On the other hand the number of photons is not constant! In fact, it might even diverge as time $t \to \infty$. This divergence is avoided by imposing a cutoff on the interaction between electrons and low-energy photons (infrared-cutoff). It is one of the main open challenges in the mathematical analysis of matter interacting with quantized radiation to prove asymptotic completeness for Rayleigh scattering without this infrared-cutoff.

This article is organized as follows. In Sect. 2.1 we begin with the description of a simple, but non-trivial model of matter and radiation. There is only one electron, besides the static nuclei, and the radiation is described by scalar bosons.

Sections 2.4, 2.5, 2.7, and 2.8 describe the main results of the papers [17,24,25] and [18], respectively. Sections 2.2 and 2.3 summarize mathematical and physical background, and Sects. 2.9 and 2.10 are devoted to side issues in the aforementioned papers.

Section 3 outlines the modification of results and proofs that are necessary to accommodate $N > 1$ electrons, and Sect. 4 ends this review with concluding remarks and a discussion of selected open problems.

There is a self-contained appendix on the standard model of non-relativistic QED.

Acknowledgments

Most of the content of this review article I learned from my collaborators Jürg Fröhlich, Elliott H. Lieb, Michael Loss, and Benjamin Schlein. I am indebted to all of them. I thank David Hasler for his careful proofreading.

2 Matter and Radiation

All electrons of an atom or molecule are well localized near the nuclei if the total energy is below the ionization threshold. Therefore the number of electrons is inessential for the phenomena to be described mathematically in this section. To simplify notation and presentation we restrict ourselves to one-electron systems; the generalization to $N > 1$ electrons is described in Sect. 3.

2.1 A Simple Mathematical Model

The main features of quantum electrodynamics that are responsible for the phenomena to be studied, are the peculiar form of interaction between light and matter, through creation and annihilation of photons, and the fact that photons are massless relativistic particles. The spin of the electron and the helicity of the photons do not play an essential role in most of our analysis. For the purpose of this introduction we therefore neglect these subtleties and present a caricature of QED which only retains the aforementioned main features. The full-fledged standard model is described in the appendix.

We first introduce our models for matter and radiation separately before describing the composed system and the interaction.

A (pure) state of a quantum particle, henceforth called *electron*, is described by a normalized vector $\psi \in L^2(\mathbb{R}^3, \mathbb{C})$, $\int_A |\psi(x)|^2 dx$ being the probability to find the particle in the region $A \subset \mathbb{R}^3$. Its time evolution is generated by a Schrödinger operator

$$H_{\mathrm{at}} = -\Delta + V \tag{1}$$

where $-\Delta$ is the positive Laplacian and V is the operator of multiplication with a real-valued function $V(x)$, $x \in \mathbb{R}^3$. We assume that $V \in L^2_{\mathrm{loc}}(\mathbb{R}^3)$ and that there exist constants $\alpha < 1$ and β such that

$$\langle \varphi, V_- \varphi \rangle \leq \alpha \langle \varphi, (-\Delta)\varphi \rangle + \beta \langle \varphi, \varphi \rangle \tag{2}$$

for all $\varphi \in C_0^\infty(\mathbb{R}^3)$, where $V_- = \max(-V, 0)$. Hence the operator $-\Delta + V$ is symmetric and bounded from below, which allows us to define a self-adjoint Hamiltonian H_{at} by the Friedrichs' extension of $-\Delta + V$.

We shall be most interested in the case where

$$V(x) = V_Z(x) := -\sum_{j=1}^{K} \frac{Z_j}{|x - R_j|}, \tag{3}$$

Z_j, $j = 1 \ldots K$, are positive integers, and $R_j \in \mathbb{R}^3$. The function (3) is the potential energy (or the scalar potential in Coulomb gauge) of one electron at $x \in \mathbb{R}^3$ in the field of K nuclei with positions $R_1, \ldots, R_K$ and atomic numbers $Z_1, \ldots, Z_K$.

The Hamiltonian (1) with V given by (3) describes a molecule with one electron and static nuclei in units where the unit of length is $\hbar^2/(2me^2) = r_B/2$ and the unit of energy is $2e^2/r_B = 4$ Ry (see Appendix A). Here $r_B = \hbar^2/(me^2)$ is the Bohr radius, $-e$ is the charge of the electron and m is its mass. This Hamiltonian is self-adjoint with domain $D(H_{\mathrm{at}}) = H^2(\mathbb{R}^3)$, the Sobolev space of twice weakly differentiable L^2-functions [41, 47].

A pure state of the radiation field is described by a normalized vector in the *bosonic Fock space over* $L^2(\mathbb{R}^3)$. This is the space

$$\mathcal{F} = \bigoplus_{n \geq 0} \mathcal{S}_n L^2(\mathbb{R}^{3n}; \mathbb{C})$$

where $\mathcal{S}_0 L^2(\mathbb{R}^0) := \mathbb{C}$, and $\mathcal{S}_n$ denotes the orthogonal projection onto the subspace of square integrable functions $f(k_1, \ldots, k_n)$ that are symmetric with respect to permutations of the n arguments $k_1, \ldots, k_n \in \mathbb{R}^3$. Such a function describes a state of n bosons, henceforth called *photons*, with wave vectors $k_1, \ldots, k_n$. The vector $|\mathrm{vac}\rangle = (1, 0, 0, \ldots) \in \mathcal{F}$ is called the *vacuum* vector. With $\mathcal{F}_{\mathrm{fin}}$ we denote the subspace of sequences $\varphi = (\varphi)_{n \geq 0} \in \mathcal{F}$ with $\varphi_n = 0$ for all but finitely many $n \in \mathbb{N}$.

The energy of a state $\varphi = (\varphi_n)_{n=0}^\infty \in \mathcal{F}$ is measured by the Hamiltonian H_f defined by

$$
\begin{aligned}
(H_f\varphi)_0 &= 0 \\
(H_f\varphi)_n(k_1,\ldots,k_n) &= \sum_{j=1}^{n} \omega(k_j)\varphi_n(k_1,\ldots,k_n)\,, \qquad n \geq 1\,,
\end{aligned}
\tag{4}
$$

where $\omega(k) = |k|$. The domain of H_f is the largest set of vectors for which (4) defines a vector in $\mathcal{F}$.

The interaction between photons and electrons comes about in a process of creation and annihilation of photons. To describe it mathematically, creation and annihilation operators are needed. Given $h \in L^2(\mathbb{R}^3)$ and $\varphi \in \mathcal{F}_{\text{fin}}$ we define $a^*(h)\varphi$ by

$$[a^*(h)\varphi]_n = \sqrt{n}S_n(h \otimes \varphi_{n-1})\,.$$

The operator $a^*(h)$ is called a *creation operator*. It adds a photon with wave function h to the state φ. The *annihilation operator* $a(h)$ is the adjoint of the closure of $a^*(h)$. These operators satisfy the canonical commutation relations

$$[a(g), a^*(h)] = (g, h)\,, \qquad [a^\sharp(g), a^\sharp(h)] = 0\,, \tag{5}$$

where $a^\sharp(\cdot)$ stands for either $a(\cdot)$ or $a^*(\cdot)$.

A further important operator on $\mathcal{F}$ is the number operator N_f, defined by

$$(N_f\varphi)_n = n\varphi_n$$

and $D(N_f) = \{\varphi \in \mathcal{F} : \sum n^2\|\varphi_n\|^2\} < \infty$.

A state of the composed system of electron and photons is described by a vector $\Psi \in \mathcal{H}_{\text{at}} \otimes \mathcal{F}$, that is, by a sequence $(\psi_n)_{n=0}^{\infty}$ where ψ_n is a square integrable function

$$\psi_n(x, k_1, \ldots, k_n)\,,$$

describing a state of one electron and n photons. It is often helpful to use that $\mathcal{H}_{\text{at}} \otimes \mathcal{F} \simeq L^2(\mathbb{R}^3; \mathcal{F})$ and to consider Ψ as a square integrable function $x \mapsto \Psi(x)$ with values in $\mathcal{F}$. Then $\|\Psi(x)\|_{\mathcal{F}}^2$ is the probability density for finding the electron at position $x \in \mathbb{R}^3$.

For the generator of the time-evolution $t \mapsto \Psi_t$ we choose the Hamiltonian

$$H \equiv H_g = H_{\text{at}} \otimes 1 + 1 \otimes H_f + gH_{\text{int}}\,, \tag{6}$$

the interaction H_{int} being given by

$$
\begin{aligned}
(H_{\text{int}}\Psi)(x) &= [a(G_x) + a^*(G_x)]\Psi(x) \\
G_x(k) &= \mathrm{e}^{-ik\cdot x}\kappa(k)\,,
\end{aligned}
$$

where $g \in \mathbb{R}$ and $\kappa \in C_0^\infty(\mathbb{R}^3)$. It is easy to prove that H_{int} is operator-bounded with respect to $H_{g=0}$ with bound zero, and hence, for every $g \in \mathbb{R}$, H_g is bounded below and self-adjoint on the domain of $H_{g=0}$ by the Kato-Rellich theorem [47].

Both g and κ measure the strength of interaction between electron and photons. Given κ, some of the results in the following sections hold for $|g|$ small enough only, others for small $|g| \neq 0$. The value $\kappa(k)$ of the *form-factor* κ measures the strength of interaction between electron and radiation with wave vector k. There is no interaction for k outside the support of κ, which is the case, e.g., for $|k|$ larger than the *ultraviolet cutoff* $\Lambda := \sup\{|k| : \kappa(k) \neq 0\}$.

We reiterate that this model is a *caricature* of the standard model of quantum electrodynamics for atoms and molecules interacting with quantized radiation. With QED it has in common that it describes a non-relativistic particle, the electron, interacting with massless relativistic bosons in a momentum conserving process of creation and annihilation of such bosons.

It would make our toy model physically more realistic if we assumed $\kappa(k) \sim |k|^{-1/2}$ for small $|k|$. But then H has no ground state [44], a problem that does not occur in the standard model of non-relativistic QED. Therefore we assume that κ is non-singular near $k = 0$.

2.2 Spectrum and Eigenfunctions of H_{at}

As a preparation for the following sections we recall a few facts concerning the spectrum of Schrödinger operators $H_{\text{at}} = -\Delta + V$ and the decay of their eigenfunctions.

Suppose that V satisfies assumption (2) and let H_{at} be defined in terms of the Friedrichs' extension of the symmetric operator $-\Delta + V$ on $C_0^\infty(\mathbb{R}^3)$. Let $D_R = C_0^\infty(|x| > R)$, the space of smooth, compactly supported functions with support outside the ball $B_R(0)$, and let

$$\Sigma_{\text{at}} := \lim_{R\to\infty} \left(\inf_{\varphi \in D_R,\, \|\varphi\|=1} \langle \varphi, H_{\text{at}}\varphi \rangle \right) . \tag{7}$$

By a theorem due to Arne Persson [2, 46]

$$\Sigma_{\text{at}} = \inf \sigma_{\text{ess}}(H_{\text{at}}) \tag{8}$$

where $\sigma_{\text{ess}}(H_{\text{at}})$ denotes the essential spectrum of H_{at}, i.e. the complement, within the spectrum, of the isolated eigenvalues of finite multiplicity. For the Coulomb potential (3), $V_Z(x) \to 0$ as $|x| \to \infty$ and hence $\Sigma_{\text{at}} = 0$. Furthermore $\sigma_{\text{ess}}(H_{\text{at}}) = [0, \infty)$, and by a simple variational argument, H_{at} has infinitely many eigenvalues below 0 [48].

Eigenfunctions of H_{at} with energy below Σ_{at} decay exponentially with increasing $|x|$: for every eigenvalue $E < \Sigma_{\text{at}}$ and every $\beta > 0$ with $E + \beta^2 < \Sigma_{\text{at}}$ there exists a constant C_β such that

$$|\psi(x)| \leq C_\beta e^{-\beta|x|} , \qquad \text{a.e. on } \mathbb{R}^3 \tag{9}$$

for all normalized eigenfunctions that belong to E. Of course, the actual decay of ψ will not be isotropic unless V is spherically symmetric. There is the better, but non-explicit, bound $|\psi(x)| \leq C_\varepsilon \mathrm{e}^{-(1-\varepsilon)\rho(x)}$ where $\rho(x)$ is the geodesic distance from x to the origin with respect to a certain metric $\mathrm{d}s^2 = c_E(x/|x|)\mathrm{d}x^2$ in $\mathbb{R}^3$ [2]. Since $c_E(x/|x|) \geq \Sigma_{\mathrm{at}} - E$ (if $\Sigma_{\mathrm{at}} < \infty$), the isotropic bound (9) follows from this stronger result.

For proving (9) it suffices to show that

$$\mathrm{e}^{\beta|\cdot|}\psi \in L^2(\mathbb{R}^3) \tag{10}$$

whenever $E + \beta^2 < \inf \sigma_{\mathrm{ess}}(H_{\mathrm{at}})$. The point-wise bound (9) then follows from a general result on point-wise bounds for (weak) solutions of second order elliptic equations [2, 21]. The L^2-bound (10), in turn, is easily derived from the characterization (7), (8) for $\inf \sigma_{\mathrm{ess}}(H_{\mathrm{at}})$ [36].

We now turn again to H_g, the Hamiltonian (6) describing matter and radiation. Most properties of H_g to be discussed in the following sections hold for all $g \in \mathbb{R}$, including $g = 0$. Hence they generalize properties of

$$H_{g=0} = H_{\mathrm{at}} \otimes 1 + 1 \otimes H_f \ .$$

By general spectral theory $\sigma(H_0) = \overline{\sigma(H_{\mathrm{at}}) + \sigma(H_f)}$. Furthermore, it is easy to see from the definition of H_f that $\sigma(H_f) = [0, \infty)$ and that 0 is the only eigenvalue of H_f, the vacuum $|\mathrm{vac}\rangle$ being its eigenvector. It follows that $\sigma(H_0) = [\inf \sigma(H_{\mathrm{at}}), \infty)$ and that H_0 and H_{at} have the same eigenvalues. The corresponding eigenvectors are the products $\psi \otimes |\mathrm{vac}\rangle$, where ψ is an eigenvector of H_{at}.

2.3 Physical Phenomena and Mathematical Description

The experimental evidence on isolated atoms in contact with radiation is easiest described in an idealized setup or "Gedankenexperiment". Consider an atom in a universe that is otherwise free of matter. For simplicity, we assume that the atom has only one electron. There may be radiation near the atom initially but no "external fields" or sources of radiation shall be present. Independent of its initial state, this system of atom *and* radiation will eventually approach one of only two qualitatively distinct final states: the "bound state" or the "ionized state".

In the bound state the atom is in its ground state, the state of least energy, where the electron is confined to within a small neighborhood of the nucleus. All excess energy has been radiated off. This radiation is very far away from the atom and escaping at the speed of light.

In the ionized state the electron and the nucleus are spatially separated with increasing distance. In addition, there may be radiation going off to infinity.

Of course, the ionized state can only be attained if the total energy of matter and radiation initially is hight enough to overcome the attraction between nucleus and electron. High energy however, does not guarantee ionization, as the excess energy may just as well turn into radiation. On the other hand, if the total energy initially is not sufficient for ionization, then the atom will certainly relax to its ground state. Other conceivable scenarios, like relaxation to a stationary state with non-minimal energy, or the permanent radiation without total loss of the excess energy have never been observed. Bohr's *stationary states*, with the exception of the ground state, are unstable, and radiation is only emitted in the very short period of transition to the ground state.

The goal is to give a proof of the phenomenon of relaxation to the ground state in the model introduced in the previous section. In view of the experimental evidence described above, we expect that this model has the following mathematical properties.

Existence of the ionization threshold. There exists a threshold energy $\Sigma \geq \inf \sigma(H)$, such that the electrons described by states in the spectral subspace $E_{(-\infty,\Sigma)}(H)\mathcal{H}$ are well localized near the nuclei, while states with energy above Σ may be ionized.

Existence of a ground state. There exists a state of least energy (ground state), or, equivalently, $\inf \sigma(H)$ is an eigenvalue of H.

Absence of excited stationary states. The operator H has no eigenvalues above $\inf \sigma(H)$.

Asymptotic completeness of Rayleigh scattering (ACR). In the limit $t \to \infty$ the time evolution $e^{-iHt}\Psi$ of every state $\Psi \in E_{(-\infty,\Sigma)}(H)\mathcal{H}$ is well approximated by a superposition of states of the form

$$a^*(h_{1,t}) \cdot \dots \cdot a^*(h_{n,t}) e^{-iE_0 t}\Psi_0 \, , \tag{11}$$

where $h_{i,t}(k) = e^{-i|k|t}h_i(k)$, Ψ_0 is a normalized ground state of H and E_0 is its energy. The vector (11) describes a state composed of the atom in its ground state and n freely propagating photons with wave functions $h_{1,t}, \dots, h_{n,t}$. In the limit $t \to \infty$ they will be far away from the atom.

The papers [17,18,24,25] are devoted to proving the above four properties for the standard model of QED was well as for the model introduced in Sect. 2.1. One exception is asymptotic completeness for Rayleigh scattering where the results concern the model of Sect. 2.1 only, with the additional important simplification that $\kappa(k) = 0$ for $|k| \leq \sigma$ where $\sigma > 0$ is arbitrarily small but positive. This assumptions that photons with energy near zero don't interact with the electron is usually referred to as an *infrared cutoff*, IR-cutoff, for short. Sometimes the constant σ is called infrared cutoff as well. The significance of the IR-cutoff is that it allows us to control the number of bosons that are being produced in the course of time: the number of bosons with energy below σ stays constant and the number of bosons with energy above σ can be bounded from above in terms of the total energy. The assumption of an IR-cutoff is *not* expected to be necessary for the validity of asymptotic completeness of Rayleigh scattering as formulated above; but as of now, no convincing mathematical argument is known that would substantiate this belief.

The above list of physical phenomena is limited to the coarsest properties of atoms interacting with radiation. Even below the ionization threshold, there are many other phenomena that are worth rediscovering in the standard model, and in part this has already been done. Most important, perhaps, are the occurrence of sharp lines in the spectrum of the emitted radiation (Bohr frequencies), the resonances in lieu of Bohr's stationary state and their extended life time, and the correspondence principle at energies near the ionization threshold. We shall come back to some of these phenomena in later sections, when we review known results or comment on open problems. One must keep in mind, however, the limitations of our model. Quantitative predictions will be of limited accuracy when relativity or high-energy photons play a significant role.

2.4 Exponential Decay and Ionization Threshold

The *ionization threshold* Σ of an atom or molecule with only one electron is the least energy that this system can achieve in a state where the electron has been moved "infinitely far away" from the nuclei. The electron is outside the ball $|x| < R$ with probability one, if its wave function $\Psi(x)$ vanishes in this ball. Therefore we define

$$\Sigma = \lim_{R\to\infty} \left(\inf_{\Psi\in D_R, \|\Psi\|=1} \langle \Psi, H\Psi \rangle \right) \tag{12}$$

where

$$D_R = \{\Psi \in D(H) | \Psi(x) = 0 \text{ if } |x| < R\} .$$

Note the analogy with Persson's characterization (7) of $\inf \sigma_{\text{ess}}(H_{\text{at}})$. Here, however, $\Sigma \neq \inf \sigma_{\text{ess}}(H)$ unless $\Sigma = \inf \sigma(H)$. In general $\Sigma \geq \inf \sigma(H)$ and, if $V(x) \to 0$ as $|x| \to \infty$, then $\Sigma = \inf \sigma(H - V)$, which is greater than $\inf \sigma(H)$ for $V = V_Z$ [25] (see also Sect. 2.5).

According to (12), the electron described by a state Ψ with energy below Σ cannot be arbitrarily far away from the origin. In fact, by [24], for all $\lambda, \beta \in \mathbb{R}$ with $\lambda + \beta^2 < \Sigma$

$$\left\| (e^{\beta|x|} \otimes 1) E_\lambda(H) \right\| < \infty . \tag{13}$$

This shows that the probability(-density) $\|\Psi(x)\|^2_{\mathcal{F}}$ to find the electron at the point x decays exponentially fast, as $|x| \to \infty$, at least in the averaged sense

$$\int e^{2\beta|x|} \|\Psi(x)\|^2_{\mathcal{F}} \, dx < \infty . \tag{14}$$

There is an obvious similarity between (14) and (10) that is not accidental. In [24] the bound (13) is derived from an abstract result for semi-bounded self-adjoint operators H in Hilbert spaces of the form $L^2(\mathbb{R}^n) \otimes \mathcal{F}$, where $\mathcal{F}$ is an arbitrary complex Hilbert space. The only assumptions are that $fD(|H|^{1/2}) \subset D(|H|^{1/2})$ and that

$$f^2H + Hf^2 - 2fHf = -2|\nabla f|^2 \tag{15}$$

for smooth, bounded functions $f(x)$ with bounded first derivatives. Equation (15) holds for $H = -\Delta \otimes 1$ and since the left-hand side of (15) formally equals $[f, [f, H]]$, it follows that (15) holds for are large class of self-adjoint operators H whose principal symbol is given by the Laplacian. The result (10) thus emerges as a special case of (13).

Even though the above assumptions on H are largely independent of H_f and gH_{int}, the result depends on these operators! The binding energy $\Sigma - \inf\sigma(H)$ depends of g and hence so does the decay rate one obtains for the ground state.

It is well known, since the work of Agmon [2], that eigenfunctions of second order elliptic equations decay exponentially in energetically forbidden regions, that is, in regions where the differential operator, as a quadratic form, is strictly larger than the eigenvalue [2]. The result (13) shows that this idea can be brought to bear in a much more general framework, including many models of non-relativistic QED. It is clear that our proof of (13) can be generalized, along the lines of [2], to yield non-isotropic bounds, as well as exponential bounds for QED-models where H_{at} is a more general, uniformly elliptic second order differential operator.

We owe the strategy for proving (13) to Bach et al. [7], where this bound is established for $|g|$ sufficiently small and $\lambda + \beta^2 < \Sigma_{\text{at}} - \text{const}|g|$.

2.5 Existence of a Ground State

By the main result of [25], $\inf\sigma(H)$ is an eigenvalue of H whenever

$$\inf\sigma(H) < \Sigma\,. \tag{16}$$

This reduces a difficult spectral problem to a variational problem: the problem of finding a state $\Psi \in D(H)$ with $\langle\Psi, H\Psi\rangle < \Sigma\langle\Psi,\Psi\rangle$. There are two important classes of potentials V for which this variational problem can be solved without much effort: if $V(x) \to \infty$ as $|x| \to \infty$, then obviously $\Sigma = \infty$ and hence $\inf\sigma(H) < \Sigma$. On the other hand, if $V(x) \to 0$ as $|x| \to \infty$ then $\Sigma = \inf\sigma(H - V)$ where the nuclei are removed in $H - V$. Using the translation invariance of $H - V$ one shows that

$$\inf\sigma(H) \leq \inf\sigma(H - V) + \inf\sigma(-\Delta + V)\,.$$

It follows that $\inf\sigma(H) < \Sigma$ whenever $V(x) \to 0$, $(|x| \to \infty)$ and $\inf\sigma(-\Delta + V) < 0$. Since this is the case for the Coulomb potential $V = V_Z$, all one-electron atoms and molecules have a ground state.

We next sketch the proof that (16) guarantees existence of a ground state. To begin with we recall that a hypothetical eigenvector Ψ of H, with eigenvalue $\inf \sigma(H)$, minimizes the quadratic form $\Psi \mapsto \langle \Psi, H\Psi \rangle$ subject to the constraint $\|\Psi\| = 1$. It is thus natural to establish existence of Ψ by proving relative compactness for a suitable minimizing sequence. The problem with this approach is that a generic minimizing sequence will tend weakly to zero. The fact that $\inf \sigma(H)$ belongs to the essential spectrum alone implies that there are infinitely many energy minimizing sequences with this defect. Our task is thus, first, to choose a suitable minimizing sequence, and second, to prove its relative compactness.

We choose the elements of our minimizing sequence to be the ground states Ψ_m of modified Hamiltonians H_m ($m \to 0$) in which the photon energy $\omega(k)$ is altered to be $\omega_m(k) = \sqrt{k^2 + m^2}$. That is, we give the photons a positive mass m. For m small enough the binding assumption (16) is inherited by H_m, which we use to show that $\inf \sigma(H_m)$ is indeed an eigenvalue. As a matter of fact, $\inf \sigma(H_m)$ is separated from the essential spectrum of H_m by a gap of size m [18]. The sequence of ground states $(\Psi_m)_{m>0}$ is a minimizing sequence for H that can be assumed to be weakly convergent. It remains to show that the weak limit is not the zero vector in $\mathcal{H}$.

We first argue that it suffices to prove relative compactness of the sequence of L^2-functions $\psi_{m,n}(x, k_1, \ldots, k_n)$ restricted to large balls $B \subset \mathbb{R}^{(3+3n)}$. This follows from the exponential decay w.r.to x, from $\psi_{m,n}(x, k_1, \ldots, k_n) = 0$ if $|k_i| > \Lambda$, and from the bound $\sup_m \langle \Psi_m, N_f \Psi_m \rangle < \infty$. Then we use the compactness of the embedding

$$W^{1,p}(B) \hookrightarrow L^2(B) , \qquad \text{for } 2 > p > \frac{2 \cdot (3+3n)}{2 + (3+3n)}$$

due to Rellich-Kondrachov [1]. We thus need to show that $\sup_m \|\nabla \psi_{m,n}\|_p < \infty$, which we derive from a refinement of the argument that we used to prove the bound on $\langle \Psi_m, N_f \Psi_m \rangle$, and from the H-boundedness of $-\Delta_x$.

There was a large number of previous papers on the existence of a ground state for models similar to the one discussed here [4, 5, 7, 9, 19, 30, 31, 52]. Among these, the best result is due to Bach et al [9]. It established existence of a ground state when $\inf \sigma(H_{\mathrm{at},N})$ is an isolated eigenvalue and the fine-structure constant α is small enough. Most importantly, in this paper for the first time existence of a ground state is proven in the standard model without an IR-regularization.

2.6 Relaxation to the Ground State is a Scattering Phenomenon

As discussed in Sect. 2.3 every state $\Psi \in \mathrm{Ran} E_{(-\infty,\Sigma)}(H)$ is expected to "relax to the ground state by emission of photons". In mathematical terms, this means that $e^{-iHt}\Psi$, in the distant future, $t \to \infty$, is well approximated in norm by a linear combination of vectors of the form

$$a^*(h_{1,t}) \dots a^*(h_{n,t}) e^{-iE_0 t} \Psi_0 \tag{17}$$

where $h_{i,t}(k) = e^{-i\omega t} h_i(k)$, Ψ_0 is the ground state, and E_0 is its energy. This has a chance to be correct only if H has no other eigenvalues below Σ. If it has, then *relaxation to a bound state* may occur, which means that Ψ_0 in (17) may be *any* eigenvector of H with eigenvalue below Σ. In this weaker form, the above assertion is called *Asymptotic Completeness for Rayleigh scattering.* The problem of proving absence of excited eigenvalues is independent of the scattering problem and its discussion is deferred to a later section.

Before proving completeness of the scattering states one needs to address the problem of their existence. An example of a *scattering state* is a vector $\Psi_+ \in \mathcal{H}$ for which there exist $n \geq 1$ photons $h_1, \dots, h_n$ and an eigenvector Ψ, $H\Psi = E\Psi$ such that

$$e^{-iHt}\Psi_+ \simeq a^*(h_{1,t}) \dots a^*(h_{n,t}) e^{-iEt} \Psi , \qquad t \to \infty$$

in the sense that norm of the difference vanishes in the limit $t \to \infty$. The scattering state Ψ_+ is said to *exists* if the limit

$$\Psi_+ = \lim_{t \to \infty} e^{iHt} a^*(h_{1,t}) \dots a^*(h_{n,t}) e^{-iEt} \Psi \qquad t \to \infty \tag{18}$$

exists. Let $\mathcal{H}_+$ denote the closure of the space spanned by vectors Ψ_+ of the form (18). All elements of $\mathcal{H}_+$ are called scattering states and asymptotic completeness of Rayleigh scattering is the property that

$$\mathcal{H}_+ \supset \mathrm{Ran} E_{(-\infty,\Sigma)}(H) . \tag{19}$$

Existence of scattering states is established in [17], and (19) is proven in [18] for the model introduced in Sect. 2.1, assuming either an infrared cutoff on the interaction or that the photon dispersions relation $\omega(k)$ is bounded from below by a positive constant, which excludes $\omega(k) = |k|$. The latter assumption serves the same purpose as the infrared cutoff, and it is satisfied, e.g., for massive bosons where $\omega(k) = \sqrt{k^2 + m^2}$ with $m > 0$. As of today, there is no proof of (19) without a form of infrared cutoff or another drastically simplifying assumption [3, 51].

2.7 Existence of Scattering States

Generalizing (18), we ask whether the limits

$$\Psi_+ = \lim_{t \to \infty} e^{iHt} a^\sharp(h_{1,t}) \dots a^\sharp(h_{n,t}) e^{-iHt} \Psi \tag{20}$$

exist for given $\Psi \in \mathcal{H}$, and $h_i \in L^2(\mathbb{R}^3)$, where $a^\sharp(h_{i,t})$ is a creation or an annihilation operator. The scattering states Ψ_- obtained in the limit $t \to -\infty$ are physically interesting as well, but the problem of their existence is mathematically equivalent to the existence of (20). Beginning with the easiest case, $n = 1$, let us ask whether the limits

$$a^\sharp_+(h)\Psi = \lim_{t\to\infty} \mathrm{e}^{iHt} a^\sharp(h_t)\mathrm{e}^{-iHt}\Psi \tag{21}$$

exist. If the photons are *massless*, as they are in nature, the answer depends on the electron dispersion relation and on the energy distribution of Ψ. For *massive* photons, however, $a^\sharp_+(h)\Psi$ exists for all $\Psi \in D(H)$ and all $h \in C_0^\infty(\mathbb{R}^3)$, and the proof is short and easy [32]: by the Cauchy criterion the limit (21) exists if the time derivative of the right-hand side is (absolutely) integrable. A straightforward computation using $a^*(h_t) = \mathrm{e}^{-iH_f t} a^*(h)\mathrm{e}^{iH_f t}$ and (5) shows that

$$\int_1^\infty \left\| \frac{d}{dt}\mathrm{e}^{iHt} a^*(h_t)\mathrm{e}^{-iHt}\Psi \right\| \mathrm{d}t = \int_1^\infty \left\|(G_x, h_t)\mathrm{e}^{-iHt}\Psi\right\| \mathrm{d}t \tag{22}$$

where

$$(G_x, h_t) = \int \mathrm{e}^{ik\cdot x - i\omega t}\overline{\kappa(k)}h(k)\,\mathrm{d}k \tag{23}$$

and $\omega(k) = \sqrt{k^2 + m^2}$. Since the Hessian of ω is strictly positive,

$$\sup_x |(G_x, h_t)| \leq \text{const}\; t^{-3/2}\,, \qquad t \geq 1$$

by a standard result on oscillatory integrals [49]. Hence $\|(G_x, h_t)\mathrm{e}^{-iHt}\Psi\| \leq$ const $t^{-3/2}$ and (22) is finite which proves the existence of $a^*_+(h)\Psi$. For *massless* photons, however,

$$\sup_x |(G_x, h_t)| \sim \text{const}\; t^{-1}\,, \qquad t \geq 1$$

which is not integrable and we need to estimate the integrand of (22) more carefully. To begin with, we note that the phase in (23) is non-stationary away from the "wave front" $|x| = t$. Hence

$$\sup_{x:||x|/t-1|\geq\varepsilon} |(G_x, h_t)| \leq \frac{C_n}{t^n}\,, \qquad t \geq 1$$

for every integer n [49], and it remains to estimate

$$\int_0^\infty \frac{\mathrm{d}t}{t} \left\|\chi_{[1-\varepsilon,1+\varepsilon]}(|x|/t)\mathrm{e}^{-iHt}\Psi\right\| . \tag{24}$$

Finiteness of (24) requires, in particular, that the electrons do not propagate at the speed of light, which is true in nature, but not precluded for the dynamics generated by the non-relativistic Schrödinger operator H_{at}. The easiest case occurs when the electrons are in a bound state $\Psi \in \mathrm{Ran}E_\lambda(H)$, $\lambda < \Sigma$, where, by (13),

$$\sup_t \|e^{\beta|x|}e^{-iHt}\Psi\| \leq \left\|e^{\beta|x|}E_\lambda(H)\right\| < \infty \tag{25}$$

for some $\beta > 0$. Then obviously the integrand in (24) decays exponentially in time, and hence the limit (21) exists. If Ψ is not in a bound state but its energy is insufficient for an electron to reach the speed $1-\varepsilon$, then (24) is still finite, at least if H is the Hamiltonian (37) of the standard model [17]. More precisely, (24) is finite for all Ψ in a dense subspace of $\mathrm{Ran}E_\lambda(H)$ with $\lambda < \Sigma + m/2$, and for ε in (24) small enough. Here $m/2 = mc^2/2$ is the non-relativistic kinetic energy of a particle at the speed of light. The assumption $\lambda < \Sigma + m/2$ thus ensures that no electron can reach the speed of light.

The asymptotic field operators have the important property that

$$\begin{aligned} a_+^*(g)\mathrm{Ran}E_\lambda(H) &\subset \mathrm{Ran}E_{\lambda+M}(H) \\ a_+(h)\mathrm{Ran}E_\lambda(H) &\subset \mathrm{Ran}E_{\lambda-m}(H) \end{aligned} \tag{26}$$

if $\mathrm{supp}(g) \subset \{k : |k| \leq M\}$ and $\mathrm{supp}(h) \subset \{k : |k| \geq m\}$. Using (26) and the existence of the limit (21) we prove existence of the limit (20) and that

$$\lim_{t\to\infty} e^{iHt}a^\sharp(h_{1,t})\dots a^\sharp(h_{n,t})e^{-iHt}\Psi = a_+^\sharp(h_1)\dots a_+^\sharp(h_n)\Psi \tag{27}$$

if $\psi \in \mathrm{Ran}E_\lambda(H)$, $\lambda + \sum_j M_j < \Sigma$, where $M_j = \sup\{|k| : h_j(k) \neq 0\}$ and the sum $\sum_j M_j$ extends over all creation operators in (27). The main technical difficulty in this last step is the unboundedness of the asymptotic field operators (21).

From the above discussion it is clear that it is physically more sensible to describe the time evolution of the electron by a relativistic Hamiltonian such as

$$H_{\mathrm{at}}^{\mathrm{rel}} = \sqrt{-\Delta + 1} + V(X)\,, \tag{28}$$

in place of (1). Then (24) is finite for all Ψ in any spectral subspace $E_\lambda(H)\mathcal{H}$ with $\lambda < \infty$. Hence the asymptotic operators $a_+^\sharp(h)$ exist on a dense subspace of $\mathcal{H}$. The main results in [16–18] apply to both electron-Hamiltonian, (1) and (28).

To conclude this discussion of scattering states we remark that for Rayleigh scattering it suffices to prove existence of the asymptotic field-operators $a_+^\sharp(h)$ on $\mathrm{Ran}E_{(-\infty,\Sigma)}(H)$, which follows from (25). The improved results discussed thereafter are important in the study of photon scattering at a free electron (Compton scattering) [16].

2.8 Asymptotic Completeness

A characterization of ACR that is mathematically more convenient than (19) is achieved by mapping the freely propagating photons $h_{i,t}$ in (18) into an auxiliary Fock space $\mathcal{F}$. We attach $\mathcal{F}$ to the Hilbert space $\mathcal{H}$ by defining an extended Hilbert space $\tilde{\mathcal{H}} = \mathcal{H} \otimes \mathcal{F}$. The appropriate time evolution on $\tilde{\mathcal{H}}$ is generated by the extended Hamiltonian $\tilde{H} = H \otimes 1 + 1 \otimes H_f$. Furthermore we define an identification operator $I : D \subset \tilde{\mathcal{H}} \to \mathcal{H}$ on a dense subspace D of $\tilde{\mathcal{H}}$ by

$$I\Psi \otimes |\text{vac}\rangle = \Psi$$
$$I\Psi \otimes a^*(h_1)\cdots a^*(h_n)|\text{vac}\rangle = a^*(h_1)\cdots a^*(h_n)\Psi$$

and linear extension. Since $\mathrm{e}^{-iH_f t}|\text{vac}\rangle = |\text{vac}\rangle$ and $a^*(h_t) = \mathrm{e}^{-iH_f t}a^*(h)\mathrm{e}^{iH_f t}$ we can use I to write

$$a^*(h_{1,t})\cdots a^*(h_{n,t})\mathrm{e}^{-iHt}\Psi = I\mathrm{e}^{-i\tilde{H}t}\big[\Psi \otimes a^*(h_1)\cdots a^*(h_n)|\text{vac}\rangle\big] \, .$$

If Ψ is an eigenvector of H and P_B denotes the orthogonal projection onto the closure of the span of all eigenvectors, it follows that

$$\begin{aligned} a_+^*(h_1)\cdots a_+^*(h_n)\Psi &= \lim_{t\to\infty} \mathrm{e}^{iHt}a^*(h_{1,t})\cdots a^*(h_{n,t})\mathrm{e}^{-iHt}\Psi \\ &= \Omega_+\big[\Psi \otimes a^*(h_1)\cdots a^*(h_n)|\text{vac}\rangle\big] \end{aligned}$$

where

$$\Omega_+ = s - \lim_{t\to\infty} \mathrm{e}^{iHt} I \mathrm{e}^{-i\tilde{H}t} P_B \otimes 1$$

is the *wave operator*. Thus existence of scattering states becomes equivalent to existence of the wave operator Ω_+, and, since $\mathrm{Ran}\Omega_+ = \mathcal{H}_+$, asymptotic completeness as defined in (19) becomes

$$E_{(-\infty,\Sigma)}(H) \subset \mathrm{Ran}\Omega_+ \, . \tag{29}$$

(It turns out that Ω_+ is a partial isometry and hence $\mathrm{Ran}\Omega_+$ is closed.)

The reader familiar with quantum mechanical scattering theory is cautioned not to think of ACR as a form of asymptotic completeness for potential scattering. The comparison dynamics generated by $\tilde{H}$ is *not* the free dynamics for all bosons. The bosons in the first factor of $\tilde{\mathcal{H}}$ still fully interact with the electrons. Rayleigh scattering is more similar to N-body quantum scattering with the additional complication that the number of particles is fluctuating.

Asymptotic completeness of Rayleigh scattering, as described in Sect. 2.6, makes two assertions. First, any initial state $\Psi \in \mathrm{Ran}E_{(-\infty,\Sigma)}(H)$, that is not an eigenvector of H, in the course of time will relax to a bound state by emission of photons. Second, the asymptotic dynamics of the emitted radiation is well approximated by the free photon dynamics. Our proof of ACR contains two main technical ingredients that address these issues. In both of them we need to assume that either the photons are massive, i.e., $\omega(k) = \sqrt{k^2 + m^2}$, or that an infrared cutoff $\sigma > 0$ is imposed on the interaction. In the second case $m = \sigma/2$ in the following.

Our first main ingredient is an estimate on the ballistic spacial expansion of the system for states with energy distribution away from $S = \sigma_{pp}(H) + \mathbb{N}m$.

We show that S is closed and countable and that for each $\lambda \in \mathbb{R}\backslash S$ there is an open interval $\Delta \ni \lambda$ and a positive constant C_λ such that

$$\langle \Psi_t, \mathrm{d}\Gamma(y^2)\Psi_t \rangle \geq C_\lambda t^2 \ , \qquad t \to \infty \tag{30}$$

for all $\Psi \subset \mathrm{Ran} E_\Delta(H)$. The proof is based on the positivity of the commutator obtained by differentiating the left hand side twice with respect to time. This positive commutator estimate, often called Mourre estimate, is proven by induction in energy steps of size m along a strategy very similar to the proof of the Mourre estimate for N-body Schrödinger operators [36]. We generalize the Mourre estimate in [14] to accommodate our model.

The second main ingredient is a *propagation estimate* for the asymptotic dynamics of escaping photons. Explicitly we show that

$$\int_1^\infty \frac{\mathrm{d}t}{t} \langle \Psi_t, fF\mathrm{d}\Gamma(P_t)Ff\Psi_t \rangle \leq C\|\Psi\|^2 \tag{31}$$

for all $\Psi \in \mathcal{H}$, where

$$P_t = (\nabla\omega - y/t) \cdot \chi(|y| \geq t^\delta)(\nabla\omega - y/t)$$

and $0 < \delta < 1$. Here f is an energy cutoff, $F = F(\mathrm{d}\Gamma(y^2/t^2\lambda^2))$ a space cutoff, and $\lambda > 0$ a parameter that is chosen sufficiently large eventually. The left-hand side of (31) compares the average photon velocity, y/t, with the group velocity, $\nabla\omega$, for photons in the region $\{|y| \geq t^\delta\}$. This includes all photons that escape the electron ballistically. The finiteness of C thus confirms that the dynamics of outgoing radiation is approaching the free photon dynamics in the limit $t \to \infty$.

Asymptotic completeness had previously been established for a model with $\Sigma = \infty$ (confined electrons), and massive photons $\omega(k) = \sqrt{k^2 + m^2}$, $m > 0$, by Dereziński and Gérard [14]. The methods in [14] could probably be extended to prove ACR for our system. Instead of doing so, we chose to give an entirely new prove of AC based on the relatively elementary propagation estimate (31), and using (30) as the only dynamical consequence of the Mourre estimate. Our work is inspired by the Graf-Schenker proof of asymptotic completeness for N-body quantum systems [22].

2.9 Absence of Excited States

At present the knowledge on absence of eigenvalues above $\inf \sigma(H_g)$ is far less complete then, e.g., our knowledge regarding existence of a ground state. Known results on absence of eigenvalues are derived under the assumption that $g > 0$ is small enough [7, 8, 10], and to ensure that no new eigenvalues emerge near $\inf \sigma(H)$ an infrared cutoff is imposed [18]. There is a further assumption, the Fermi golden rule condition, which ensures that eigenvalues of $H_{g=0}$ dissolve for $g \neq 0$. This assumption can be checked in any explicitly given model.

For the model introduced in Sect. 2.1, with the assumption of an infrared cutoff, the following results hold true. For any given $\varepsilon > 0$ and for $|g| > 0$ small enough, depending on ε,

$$\sigma_{pp}(H_g) \cap (\inf \sigma(H_g), \Sigma_{\text{at}} - \varepsilon) = \emptyset ,$$

where $\Sigma_{\text{at}} = \inf \sigma(H_{\text{at}})$ [10, 18]. This result, combined with ACR from the previous section implies that $\text{Ran}\, \Omega_+ \supset \text{Ran}\, E_{(\inf \sigma(H), \Sigma_{\text{at}} - \varepsilon)}(H)$ where the projector P_B in the definition of Ω_+ is the projector onto the ground state. That is, every $\psi \in \text{Ran}\, E_{(\inf \sigma(H), \Sigma_{\text{at}} - \varepsilon)}(H)$ relaxes to the ground state in the sense of Sect. 2.6, (17).

It has been asserted in [8] that the methods of [7, 10] can be used to show absolute continuity of the spectrum of H above and away from Σ_{at} for small $|g|$. This is presumably correct but a proof is missing.

The strategy for proving absence of eigenvalues in a given spectral interval $\Delta \subset \mathbb{R}$ is clear and simple: One tries to find a symmetric operator A on $\mathcal{H}$, such that

$$E_\Delta(H)[iH, A]E_\Delta(H) \geq C E_\Delta(H)$$

with a positive constant C. Since, formally, $\langle \Psi, [iH, A]\Psi \rangle = 0$ for every eigenvector Ψ of H, it immediately follows that $\sigma_{pp}(H) \cap \Delta = \emptyset$. The main problems, of course, are to find a suitable conjugate operator A, and to make these formal arguments rigorous.

2.10 Relaxation to the Ground State

An important consequence of AC for Rayleigh scattering and the absence of eigenvectors besides a unique ground state Ψ_0, is *relaxation to the ground state.* To explain this let $\mathcal{A}$ denote the C^*-algebra generated by all operators of the form

$$B \otimes \mathrm{e}^{i\phi(h)} , \qquad B \in \mathbf{B}(\mathcal{H}_{\text{at}}),\ h \in C_0^\infty(\mathbb{R}^3) ,$$

where $\phi(h) = a(h) + a^*(h)$. We say that $\Psi_t = \mathrm{e}^{-iHt}\Psi$ *relaxes to the ground state* Ψ_0, if

$$\lim_{t\to\infty} \langle \Psi_t, A\Psi_t \rangle = \langle \Psi_0, A\Psi_0 \rangle \langle \Psi, \Psi \rangle \tag{32}$$

for all $A \in \mathcal{A}$. Suppose H has a unique ground state Ψ_0 and let $\mathcal{H}_+$ denote the space of scattering states over Ψ_0. That is, $\mathcal{H}_+$ is the closure of the span of all vectors of the form

$$a_+^*(h_1) \cdots a_+^*(h_n)\Psi_0 .$$

Then, by a short computation, all states in $\mathcal{H}_+$ relax to the ground state Ψ_0 [17]. Since the assumptions of Sect. 2.9 imply AC in the form

$$\mathcal{H}_+ \supset \text{Ran} E_{(\inf \sigma(H), \Sigma_{\text{at}} - \varepsilon)}(H_g)$$

for given $\varepsilon > 0$ and small enough coupling $|g|$, it follows that all states in $\text{Ran} E_{(\inf \sigma(H), \Sigma_{\text{at}} - \varepsilon)}(H_g)$ relax to the ground state in the sense of Equation 32.

3 N-Electron Atoms and Molecules

We now briefly describe how the results of the previous sections are generalized to the case of $N > 1$ electrons. For simplicity we neglect spin and Pauli principle. A (pure) state of N electrons is described by a vector $\psi \in L^2(\mathbb{R}^{3N})$, and the Schrödinger operator for N electrons in the field of K static nuclei is given by

$$H_{\mathrm{at},N} = \sum_{j=1}^{N} (-\Delta_{x_j} + V_Z(x_j)) + \sum_{i<j} \frac{1}{|x_i - x_j|}$$

where $x_j \in \mathbb{R}^3$ is the position of the jth electron. The coupling of the electrons to the radiation field is done by a straightforward generalization of (6). The Hamiltonian of the entire system is given by

$$\begin{aligned} H_{g,N} &= H_{\mathrm{at},N} \otimes 1 + 1 \otimes H_f + gH_{\mathrm{int}} \\ H_{\mathrm{int}} &= \sum_{j=1}^{N} \left[a(G_{x_j}) + a^*(G_{x_j})\right], \qquad G_{x_j}(k) = \mathrm{e}^{-ik\cdot x_j}\kappa(k) \end{aligned}$$

and acts on $L^2(\mathbb{R}^{3N}) \otimes \mathcal{F}$. Again, H_g is self-adjoint on $D(H_{g=0})$.

Spectrum and Eigenfunctions of $H_{\mathrm{at},N}$

Like H_{at}, $H_{\mathrm{at},N}$ is a Schrödinger operator of the general form $-\Delta + V$, and hence $\inf \sigma_{\mathrm{ess}}(H_{\mathrm{at},N})$ is given by Persson's theorem:

$$\inf \sigma_{\mathrm{ess}}(H_{\mathrm{at},N}) = \lim_{R\to\infty} \left(\inf_{\varphi \in D_R,\, \|\varphi\|=1} \langle \varphi, H_{\mathrm{at},N}\varphi \rangle \right) =: \Sigma_{\mathrm{at},N}$$

where $D_R = C_0^\infty(|X| > R)$. Using the decay of the two-body potentials and the electron-electron repulsion one shows that $\Sigma_{\mathrm{at},N} = \inf \sigma(H_{\mathrm{at},N-1})$, which leads to

$$\inf \sigma_{\mathrm{ess}}(H_{\mathrm{at},N}) = \inf \sigma(H_{\mathrm{at},N-1}) , \tag{33}$$

a special case of the more general HVZ-Theorem [36, 48]. For $Z > N - 1$ the system described by $H_{\mathrm{at},N-1}$ has a net positive charge and can bind at least one more electron. It follows, by a simple variational argument, that $\inf \sigma(H_{\mathrm{at},N}) < \inf \sigma(H_{\mathrm{at},N-1})$, which, by (33) implies that $\inf \sigma(H_{\mathrm{at},N})$ is an eigenvalue of $H_{\mathrm{at},N}$. In fact, $H_{\mathrm{at},N}$ has infinitely many (discrete) eigenvalues below $\inf \sigma_{\mathrm{ess}}(H_{\mathrm{at},N})$ [48]. The continuous part of the spectrum of $H_{\mathrm{at},N}$ is the interval $[\inf \sigma_{\mathrm{ess}}(H_{\mathrm{at},N}), \infty)$, and this interval may contain further eigenvalues below 0 [54]. For a discussion of the structure of the continuous spectrum, the reader is referred to [36].

The results (9), (10) on the decay of eigenfunctions hold for Schrödinger operators in arbitrary dimensions, hence in particular for $H_{\mathrm{at},N}$. Better, non-isotropic exponential bounds are known too, but they are expressed in terms of a geodesic distance $\rho(X)$ w.r.to a metric in $\mathbb{R}^{3N}$ that depends on the spectra of the Hamiltonians $H_{\mathrm{at},k}$, for $k < N$ [2,36]. Explicit expressions for ρ are known for $N \leq 3$, and for atoms under the (unproven) assumption that the ionization energy increases monotonically as the electrons, one by one, are removed from the atom [12].

This concludes our discussion of $H_{\mathrm{at},N}$ and we return to the composed system of N electrons and radiation.

Exponential Decay and Ionization Thresholds

The ionization threshold Σ is the least energy that an atom or molecule can achieve in a state where one or more electrons have been moved "infinitely far away" from the nuclei. In an N-particle configuration $X \in \mathbb{R}^{3N}$, one or more electrons are far away from the (static) nuclei if and only if $|X|$ is large. In this respect there is no difference between $N = 1$ and $N > 1$ besides the dimension of the configuration space. Since this dimension is irrelevant for the proof of (13), our result on exponential decay for $N > 1$ and its proof are straightforward generalizations of result and proof for $N = 1$. Let $D_R := \{\Psi \in D(H) | \, \Psi(X) = 0 \text{ if } |X| < R\}$ and let

$$\Sigma_N = \lim_{R\to\infty} \left(\inf_{\Psi\in D_R, \|\psi\|=1} \langle \Psi, H_N \Psi \rangle \right) . \tag{34}$$

Then for all real numbers λ and β with $\lambda + \beta^2 < \Sigma$,

$$\left\| (\mathrm{e}^{\beta|X|} \otimes 1) E_\lambda(H_N) \right\| < \infty . \tag{35}$$

In the case of only one electron subject to an external potential V that vanishes at infinity, such as V_Z, we saw that $\Sigma_{N=1} = \inf \sigma(H_1 - V)$. The proper generalization to $N > 1$ is analog to the HVZ theorem for N-particle Schrödinger operators. We show that

$$\Sigma_N = \min_{N' \geq 1} \{ E^V_{N-N'} + E^0_{N'} \} \tag{36}$$

where $E^0_{N'}$ is the least energy of N' electrons with no nuclei present, $Z = 0$ [24]. Like the HVZ theorem, (33), for the bottom of the essential spectrum of $H_{\mathrm{at},N}$, (36) requires the decay of the interaction between material particles with increasing spacial separation. While this decay is obvious for the instantaneous Coulomb interaction, it is more tedious to quantify for the interaction mediated trough the quantized radiation field. The main problem in proving (36), however, is to control the error which arises when the field energy is split up into two parts, one associated with the N' electrons far out and one with the other $N - N'$ electrons. This error is proportional to the number of photons, as measured by the number operator N_f, which in turn is *not bounded* with respect to the total energy and thus not under control.

To deal with this problem we first prove (36) with an IR-cutoff $\sigma > 0$ in the interaction and then we show that (36) is obtained in the limit $\sigma \to 0$ [24,25].

The characterization (36) of the ionization threshold is important for proving that $\Sigma_N > \inf \sigma(H_N)$.

Existence of a Ground State

The dimension of the electron configuration space is inessential for proving that (16) guarantees the existence of a ground state. Therefore $\inf \sigma(H_N)$ is an eigenvalue of H_N whenever

$$\inf \sigma(H_N) < \Sigma_N .$$

However, it is much harder to verify this condition for $N > 1$. The only easy case occurs for spatially confining external potentials where $\Sigma_N = \infty$. In a tour de force Lieb and Loss recently showed that

$$E_N^Z < \min_{N' \geq 1} \{E_{N-N'}^Z + E_{N'}^0\}$$

for all atoms and molecules with $Z > N - 1$ [43]. Combined with (36) this proves that $E_N^Z < \Sigma_N$ and hence that E_N^Z is an eigenvalue of H_N^Z indeed.

4 Concluding Remarks and Open Problems

The results we have described on localization of the electron, existence of a ground state and existence of scattering states are established in [17,24,25] within the standard model of QED for non-relativistic electrons (see Appendix A). Asymptotic completeness is proved for the dipole approximation of that model, Hamiltonian (59), an approximation that is physically reasonable for confined electrons [18]. We don't expect serious obstacles in proving ACR for the standard model (with IR cutoff), but to do so appears prohibitive in view of the additional work due to the interaction terms quadratic in creation and annihilation operators. The most important and most interesting open problem in connection with Rayleigh scattering is to prove completeness without IR cutoff. This has been done so far only for the explicitly soluble model of a harmonically bound electron coupled to radiation in dipole approximation [3], and for perturbations thereof [51]. Steps toward ACR for more general electron Hamiltonians have been undertaken by Gérard [20].

The problem of the emitted low energy radiation can be understood as one aspect of the more general question of the intensity of the radiation in Rayleigh scattering as a function of the frequency. Experimentally, sharp spectral lines with frequencies ω given by Bohr's condition are observed. This condition says that $\hbar\omega$ is the difference between the energies of two stationary states, that is, between two eigenvalues of H_{at}. From quantum theory this phenomenon is expected to be a consequence of the smallness of α, which allows one to compute transition amplitudes in leading order perturbation theory. Rigorous work in this direction is currently being done by Bach, Fröhlich and Pizzo [6]. One also expects that eigenvalues of H_{at} show up as resonances in the spectrum of H_g and that the eigenvectors of $H_{g=0}$ are meta-stable states for the dynamics generated by H_g, if $g > 0$, with a life-time inversely proportional to the resonance width. Both these expectations have been confirmed by work of Bach et al, and by Mück [9,45]. While the existence of resonances and meta-stable states is consistent with the experimentally observed spectral lines, it does not fully account for them. It remains to be shown that, for small α, first order transitions between meta-stable states dominate the process of relaxation to the ground state and hence that the intensity is largest for radiation obeying Bohr's frequency condition. A related question is the one about a confirmation of the correspondence principle within QED. By the correspondence principle, the frequency of radiation emitted by a highly excited atom agrees with the angular frequency of a classical point charge on the corresponding Bohr orbit. This principle together with Bohr's frequency condition determines the distribution of eigenvalues of highly excited states. A rigorous derivation of the correspondence principle would therefore confirm – but not prove – the domination of Bohr frequencies at least in the low energy spectrum.

Many further questions arise once we allow for total energies *above* the ionization threshold Σ. Then the atom can become ionized and the dynamics of the removed electrons is close to the free one. The first task is thus to study the scattering of photons at a freely moving electron, the so-called Compton scattering. This has been done in [16], where we established asymptotic completeness for Compton scattering for energies below a threshold energy that limits the speed of the electron from above to one-third of the speed of light. To do so, we had to impose an infrared cutoff, for otherwise no dresses one-electron states exist.

The natural next step is to combine Rayleigh with Compton scattering to obtain a complete classification of the long time asymptotics of matter coupled to radiation. This would include the photo effect as well as the occurrence of Bremsstrahlung.

There are also very interesting and difficult open questions related to the binding energy $\Sigma - E_N$ even for $N = 1$. From Sect. 2.5 we know that

$$E_{N=1} \leq \Sigma + \inf \sigma(-\Delta + V) \, .$$

That is, if $V \to 0$, the binding energy $\Sigma - E_{N=1}$ with coupling to the radiation field is at least as large as the binding energy $-\inf \sigma(-\Delta + V)$ without radiation. Physical intuition tells us that this binding energy should actually increase due to the coupling to radiation: the radiation field accompanying the electron, by the energy-mass equivalence, adds to the inertia of the electron, that is, makes it heavier and thus easier to bind. This *mass renormalization* can explicitly be computed in the dipole approximation and this has been used to prove enhanced binding by Hiroshima and Spohn [29]. Without dipole approximation the mass renormalization is not known explicitly and enhanced binding has been established so far only for small α [26,27]. It is an interesting and challenging problem to establish enhanced binding without dipole approximation and for arbitrary α and Λ.

Once there are two or more electrons, one would like to know, first of all, whether two electrons attract or repel each another in our model of matter. Of course, equal charges repel each other but this argument neglects the effect of the quantized radiation field, which is attractive. Two charges close to each other will share part of their radiation field. Since this reduces the energy to produce it, binding is encouraged. The questions is thus whether this binding effect may overcome the Coulomb repulsion.

A Non-Relativistic QED of Atoms and Molecules

The purpose of this appendix is to describe atoms and molecules within UV-regularized, non-relativistic quantum electrodynamics in Coulomb gauge. We shall also comment on suitable choices of units, on representations of the theory that avoid the use of polarization vectors, and on the dipole approximation. For further information the reader is referred to [7,11,13,53].

A.1 Formal Description of the Model

To write down the model quickly and in a form familiar from physics books we shall be somewhat formal at first, using operator-valued distributions and avoiding domain questions.

The Hilbert space of pure states of N electrons and an arbitrary number of transversal photons is the tensor product $\mathcal{H} = \mathcal{H}_{\rm at} \otimes \mathcal{F}$ where

$$\mathcal{H}_{\rm at} := \wedge_{i=1}^{N} L^2(\mathbb{R}^3;\mathbb{C}^2)\ , \qquad \mathcal{F} := \oplus_{n=0}^{\infty} \otimes_s^n L^2(\mathbb{R}^3;\mathbb{C}^2)\ ,$$

$\otimes_s^{n=0} L^2 := \mathbb{C}$, and where $\otimes_s^n L^2(\mathbb{R}^3;\mathbb{C}^2)$, $n \geq 1$, stands for the symmetrized tensor product of n copies of $L^2(\mathbb{R}^3;\mathbb{C}^2)$. The vector $\Omega := (1,0,0,\dots) \in \mathcal{F}$ is called *vacuum*. The one-particle wave functions in $\mathcal{H}_{\rm at}$ and $\mathcal{F}$ are $\mathbb{C}^2$-valued to account for the two spin and the two polarization states of the electrons and transversal photons, respectively.

The Hamiltonian of an atom or molecule with static nuclei is a self-adjoint operator in $\mathcal{H}$ of the form

$$H = \sum_{j=1}^{N} \frac{1}{2m}\left[\sigma_j \cdot (-i\nabla_{x_j} + \sqrt{\alpha} A_\Lambda(x_j))\right]^2 + \alpha V_R \otimes 1 + 1 \otimes H_f\ , \tag{37}$$

where $\sigma_j = (\sigma_{j,x}, \sigma_{j,y}, \sigma_{j,z})$ denotes the triple of Pauli matrices acting on the spin degrees of freedom of the jth electron and $x_j \in \mathbb{R}^3$ is the position of the jth electron. The constant $m > 0$ is the (bare) mass of an electron and $\alpha = e^2/(\hbar c) = e^2$ is the fine structure constant. In our units $\hbar = 1 = c$. Another common form of H is obtained by using that

$$\left[\sigma \cdot (-i\nabla_x + \sqrt{\alpha}A_\Lambda(x))\right]^2 = (-i\nabla_x + \sqrt{\alpha}A_\Lambda(x))^2 + \sqrt{\alpha}\sigma \cdot B(x) \tag{38}$$

where $B(x) = \mathrm{curl} A(x)$.

The operator V_R acts by multiplication with the electrostatic potential

$$V_R(x) = -\sum_{j=1}^{K}\sum_{i=1}^{N} \frac{Z_j}{|x_i - R_j|} + \sum_{i<j} \frac{1}{|x_i - x_j|} \tag{39}$$

of the electrons in the field of K static nuclei with positions $R_1, \dots, R_K \in \mathbb{R}^3$ and atomic numbers $Z_1, \dots, Z_K \in \mathbb{Z}_+$. We use the short-hands $x = (x_1, \dots, x_N)$ and $R = (R_1, \dots, R_N)$.

The operators H_f and $A_\Lambda(x)$, for fixed $x \in \mathbb{R}^3$, are operators on Fock space. H_f has been defined in Sect. 2.1 and $A_\Lambda(x)$ can be expressed in the form

$$A_\Lambda(x) = (2\pi)^{-3/2} \sum_{\lambda=1,2} \int_{|k|\le\Lambda} \frac{\mathrm{d}^3k}{\sqrt{2|k|}} \left\{\varepsilon_\lambda(k)^* \mathrm{e}^{ik\cdot x} a_\lambda(k) + \varepsilon_\lambda(k)\mathrm{e}^{-ik\cdot x} a_\lambda^*(k)\right\} . \tag{40}$$

The *polarization vectors* $\varepsilon_\lambda(k) \in \mathbb{C}^3$, $\lambda \in \{1, 2\}$, are orthogonal to the wave vector k and normalized

$$\varepsilon_\lambda^*(k) \cdot \varepsilon_\mu(k) = \delta_{\lambda\mu} , \qquad \varepsilon_\lambda(k) \cdot k = 0 . \tag{41}$$

In addition we assume that $\varepsilon_\lambda(tk) = \varepsilon_\lambda(k)$ for all $t > 0$. The operators $a_\lambda^*(k)$ and $a_\lambda(k)$ are *creation- and annihilation operators* in $\mathcal{F}$. These are operator-valued distributions, formally defined by $a_\lambda(k)\Omega = 0$ for all $k \in \mathbb{R}^3$, $\lambda \in \{1, 2\}$, and by the canonical commutation relation

$$[a_\lambda(k), a_\mu^*(q)] = \delta_{\lambda\mu}\delta(k - q) , \qquad [a_\lambda^\sharp(k), a_\mu^\sharp(q)] = 0 . \tag{42}$$

A rigorous definition of $A_\Lambda(x)$ will be given in the next section. The constant $\Lambda > 0$ in (40) is the *ultraviolet cutoff*. Photons with $|k| > \Lambda$ do not interact with the electrons under the dynamics generated by H. This is nonphysical but necessary to define $A_{\Lambda,i}(x)$ on a dense subspace of $\mathcal{F}$. For $\Lambda = \infty$ not even the vacuum would be in the domain of $A(x)$. In fact, by a formal computation using the properties of $a_\lambda(k)$ and $a_\lambda^*(k)$, $\|A_i(x)\Omega\|^2 = \mathrm{const} \int_{|k|\le\Lambda} |k|^{-1}\mathrm{d}^3k \to \infty$ as $\Lambda \to \infty$.

In the QED of Feynman, Schwinger and Tomanaga, removing the UV cutoff requires a renormalization of mass, charge and field strength, a procedure that is mathematically not sufficiently well understood yet.

A.2 Atomic Units and Perturbation Theory

To work in the small-α regime it is convenient to choose the UV-cutoff Λ and the nuclear positions $R_i \in \mathbb{R}^3$ fixed on scales of energy and length where the units are proportional to the Rydberg energy $m\alpha^2/2 = mc^2\alpha^2/2$ and the Bohr radius $1/(m\alpha) = \hbar^2/me^2$. We shall therefore rewrite the Hamiltonian in these units. It is instructive to begin by first scaling electron position and photon momentum independently. Let $U : \mathcal{H} \to \mathcal{H}$ be defined by $(U\varphi)_n(x, k_1, \ldots, k_n) = \eta^{3/2}\mu^{3n/2}\varphi_n(\eta x, \mu k_1, \ldots, \mu k_n)$. Then

$$\begin{aligned}\mu^{-1}UHU^* =& \sum_{j=1}^{N} \frac{1}{2m\eta^2\mu} \left[\sigma_j \cdot (-i\nabla_{x_j} + \sqrt{\alpha}\eta\mu A_{\Lambda/\mu}(\eta\mu x_j))\right]^2 \\ &+ \frac{\alpha}{\eta\mu} V_{R/\eta} \otimes 1 + 1 \otimes H_f,\end{aligned} \tag{43}$$

which is most easily verified using the definition of $A_\Lambda(x)$ given in the next section. In order that $2m\eta^2\mu = 1$ and $\eta\mu = \alpha$ we choose $\eta = (2m\alpha)^{-1}$ and $\mu = 2m\alpha^2$. Next we express the UV cutoff and the nuclear positions in these units, that is we replace

$$\Lambda/\mu \to \Lambda\,, \qquad R/\eta \to R\,, \tag{44}$$

a non-unitary change of the Hamiltonian! Thus in the new units the Hamiltonian reads

$$\sum_{i=1}^{N} \left[\sigma_i \cdot (-i\nabla_i + \alpha^{3/2} A_\Lambda(\alpha x_i))\right]^2 + V_R \otimes 1 + 1 \otimes H_f \tag{45}$$

where the dependence on α is concentrated in electron-photon interaction $\alpha^{3/2}A_\Lambda(\alpha x)$. The papers by Bach et al. concern the Hamiltonian (45), many others concern (37). When comparing results that are valid for small α only, one must keep in mind that these Hamiltonians are not equivalent, not even for atoms: the substitution $\Lambda \to \Lambda\alpha^2$, which occurs in (44), corresponds to the change $m \mapsto m/\alpha^2$ of the electron mass, as follows from (43) with $\eta = \mu^{-1} = \alpha^2$.

A.3 Fock-Spaces, Creation- and Annihilation Operators

We next give a rigorous definition of the quantized vector potential $A_\Lambda(x)$ and we shall comment on the self-adjointness of H. In order to prepare the ground for the next section we define Fock space, creation- and annihilation operators in larger generality then needed here. A good reference for this section is [11].

Given a complex Hilbert space $\mathfrak{h}$ the bosonic Fock space over $\mathfrak{h}$,

$$\mathcal{F} = \mathcal{F}(\mathfrak{h}) = \oplus_{n\geq 0} \mathcal{S}_n(\otimes^n \mathfrak{h}) \tag{46}$$

is the space of sequences $\varphi = (\varphi_n)_{n\geq 0}$, with $\varphi_0 \in \mathbb{C}$, $\varphi_n \in \mathcal{S}_n(\otimes^n \mathfrak{h})$, and $\sum_{n\geq 0} \|\varphi_n\|^2 < \infty$. Here $\mathcal{S}_n$ denotes the orthogonal projection onto the subspace of symmetrized tensor products of n vectors in $\mathfrak{h}$. The inner product in $\mathcal{F}$ is defined by

$$\langle \varphi, \psi \rangle = \sum_{n\geq 0} (\varphi_n, \psi_n) ,$$

where (φ_n, ψ_n) denotes the inner product in $\otimes^n \mathfrak{h}$. We use $\mathcal{F}_{\mathrm{fin}}$ to denote the dense subspace of vectors $\varphi \in \mathcal{F}$ with $\varphi_n = 0$ for all but finitely many $n \in \mathbb{N}$.

Given $h \in \mathfrak{h}$ the creation operator $a^*(h) : \mathcal{F}_{\mathrm{fin}} \subset \mathcal{F} \to \mathcal{F}$ is defined by

$$[a^*(h)\varphi]_n = \sqrt{n}\mathcal{S}_n(h \otimes \varphi_{n-1}) \tag{47}$$

and the annihilation operator $a(h) : \mathcal{F}_{\mathrm{fin}} \subset \mathcal{F} \to \mathcal{F}$ is the restricted to $\mathcal{F}_{\mathrm{fin}}$ of the adjoint of $a^*(h)$. The operators $a(h)$ and $a^*(h)$ satisfy the canonical commutation relations (CCR)

$$[a(g), a^*(h)] = (g, h) , \qquad [a^\sharp(g), a^\sharp(h)] = 0 .$$

In particular, $[a(h), a^*(h)] = \|h\|^2$ which implies that $\|a(h)\varphi\| + \|\varphi\|$ and $\|a^*(h)\varphi\| + \|\varphi\|$ are equivalent norms. It follows that the closures of $a(h)$ and $a^*(h)$ have the same domain. On this domain $a^*(h)$ is the adjoint of $a(h)$ [11, Theorem 5.2.12]. The operator

$$\phi(h) = \frac{1}{\sqrt{2}}(a(h) + a^*(h)) \tag{48}$$

is essentially self-adjoint on $\mathcal{F}_{\mathrm{fin}}$ [11]. It is useful to note that

$$[\phi(g), \phi(h)] = i\,\mathrm{Im}(g, h) .$$

In the case of QED, $\mathfrak{h} = L^2(\mathbb{R}^3; \mathbb{C}^2)$ with inner product $(g, h) = \sum_{\lambda=1,2} \int \overline{g_\lambda(k)} h_\lambda(k)\, \mathrm{d}^3k$ and $A_i(x) = \phi(G_{x,i})$, $G_{x,i} \in \mathfrak{h}$, $i = 1, 2, 3$, being the components of

$$G_x(k, \lambda) = \frac{\kappa(k)}{|k|^{1/2}} \varepsilon_\lambda(k) \mathrm{e}^{-ik\cdot x} , \tag{49}$$

where $\kappa(k) = (2\pi)^{-3/2} \chi_{|k|\leq \Lambda}(k)$. More generally we may allow κ to be any real-valued, spherically symmetric function with $\kappa/\sqrt{\omega} \in L^2(\mathbb{R}^3)$. In particular $|\kappa(-k)| = |\kappa(k)|$ which implies that $[A_i(x), A_j(y)] = 0$ for all $x, y \in \mathbb{R}^3$ and $i, j \in \{1, 2, 3\}$.

The Hamiltonian (37) is defined and symmetric on the dense subspace

$$\mathcal{D} = \left[\, \wedge_{i=1}^N C_0^\infty(\mathbb{R}^3; \mathbb{C}^2)\right] \otimes \mathcal{F}_{\mathrm{fin}}(C_0^\infty(\mathbb{R}^3; \mathbb{C}^2)) \tag{50}$$

where $\mathcal{F}_{\mathrm{fin}}(C_0^\infty(\mathbb{R}^3; \mathbb{C}^2))$ is the space of vectors $\varphi = (\varphi_n)_{n\geq 0} \in \mathcal{F}_{\mathrm{fin}}$ with $\varphi_n \in \otimes^n C_0^\infty(\mathbb{R}^3; \mathbb{C}^2)$. Let $H_0 = -\Delta \otimes 1 + 1 \otimes H_f$. Then H_0 is essentially self-adjoint on $\mathcal{D}$, self-adjoint on $D(H_0) = D(-\Delta \otimes 1) \cap D(1 \otimes H_f)$, and

$H - H_0$ is bounded relative to H_0. It follows that the closure of $H\lceil\mathcal{D}$ is defined on $D(H_0)$ and symmetric on this domain. Since H is self-adjoint on $D(H_0)$, according to Hiroshima [28], we conclude that H is essentially self-adjoint on $\mathcal{D}$. Alternatively, the Hamiltonian (37) may be self-adjointly realized in terms of the Friedrichs' extension of $H\lceil\mathcal{D}$, since this operator is bounded from below, or, by using the theorem of Kato-Rellich for Λ/α small enough [9]. ($\alpha\Lambda$ small enough for the Hamiltonian (45).)

A.4 Avoiding Polarization Vectors

The fact that the polarization vectors are necessarily discontinuous as functions of $\hat{k} = k/|k| \in S^2$, by a well-known result of H. Hopf, may lead to annoying technical problems [25]. To show how these problems can be avoided we construct a representation of H that does not depend on a choice of polarization vectors [18, 42]. This representation is based on a description of single-photon states by vectors in

$$\mathfrak{h}_T := \{h \in L^2(\mathbb{R}^3;\mathbb{C}^3) | h(k)\cdot k = 0 \quad \text{for all} \quad k\}\,, \tag{51}$$

the space of transversal photons. Let $u : L^2(\mathbb{R}^3;\mathbb{C}^2) \to \mathfrak{h}_T$ be the unitary map

$$u : (h_1,h_2) \mapsto \sum_{\lambda=1,2} h_\lambda \varepsilon_\lambda^* \tag{52}$$

where $\{\varepsilon_\lambda\}_{\lambda=1,2}$ are the polarization vectors employed in the definition of H, an let $U : \mathcal{F} \to \mathcal{F}(\mathfrak{h}_T)$ be defined by

$$Ua^*(h)U^* = a^*(uh)\,, \qquad U\Omega = \Omega\,.$$

It follows that $UA_i(x)U^* = \phi(uG_{x,i})$ where

$$\begin{aligned}(uG_{x,j})(k) &= \sum_{\lambda=1,2} \frac{\kappa(k)}{|k|^{1/2}} \mathrm{e}^{-ik\cdot x} \varepsilon_\lambda(k)_j \varepsilon_\lambda(k)^* \\ &= \frac{\kappa(k)}{|k|^{1/2}} \mathrm{e}^{-ik\cdot x}(e_j - \hat{k}(\hat{k}\cdot e_j))\,,\end{aligned} \tag{53}$$

$\{e_1,e_2,e_3\}$ being the canonical basis of $\mathbb{R}^3$. Note that $\phi(uG_{x,j})$ is *one* operator and not a triple of operators even though $k \mapsto uG_{x,j}(k)$ is a vector-valued function.

The Hamilton operator $H_T = UHU^*$ is the desired new representation of H. It has the form of H in (37) with the only difference that the form-factor of $A(x)$ is now given by (53), a function in $C^\infty(\mathbb{R}^3\backslash\{0\})$.

By choosing other unitary mappings u from $L^2(\mathbb{R}^3;\mathbb{C}^2)$ onto $\mathfrak{h}_T$ one may define more equivalent representations of QED. For example the map

$$u_2 : (h_1,h_2) \mapsto \sum_{\lambda=1,2} h_\lambda \varepsilon_\lambda^* \wedge \hat{k} \tag{54}$$

leads to the representation of H where the quantized vector potential is defined in terms of the form factor

$$(u_2 G_{x,i})(k) = \frac{\kappa(k)}{|k|^{1/2}} \mathrm{e}^{-ik\cdot x}(e_i \wedge \hat{k}) \,, \tag{55}$$

the choice preferred in [42].

For the mathematical analysis of systems of electrons interacting with photons it is often necessary to localize the photons in their position space. That is, the photon wave function $h \in \mathfrak{h}_T$ is mapped to $J(i\nabla_k)h$ where $J \in C_0^\infty(\mathbb{R}^3)$. Now $J(i\nabla_k)h \notin \mathfrak{h}_T$ unless $J = 0$ or $h = 0$, and projecting $J(i\nabla_k)h$ back to $\mathfrak{h}_T$ would destroy the localization accomplished by the operator $J(i\nabla_k)$. The solution to this problem is to work on the enlarged one-boson Hilbert space $\mathfrak{h}_{\text{ext}} = L^2(\mathbb{R}^3; \mathbb{C}^3) = \mathfrak{h}_T \oplus \mathfrak{h}_L$ which also includes the space of longitudinal photons $\mathfrak{h}_L = \{h | k \wedge h(k) = 0\}$. The Hilbert space for the entire system becomes $\mathcal{H}_{\text{ext}} = \mathcal{H}_{\text{at}} \otimes \mathcal{F}(\mathfrak{h}) \simeq \mathcal{H} \otimes \mathcal{F}(\mathfrak{h}_L)$ and we define a Hamiltonian on $\mathcal{H}_{\text{ext}}$ by

$$\begin{aligned} H_{\text{ext}} &= H_T \otimes 1 + 1 \otimes H_{f,L} \\ &= \sum_{j=1}^{N} \frac{1}{2m} [\sigma_j \cdot (-i\nabla_{x_j} + \alpha^{1/2} A(x))]^2 + \alpha V_R + H_f \end{aligned}$$

where $A_j(x) = \phi(uG_{x,j})$ as above, but now $uG_{x,j}$ is considered as an element of $\mathfrak{h}_{\text{ext}} = L^2(\mathbb{R}^3; \mathbb{C}^3)$. The fake longitudinal bosons from $\mathfrak{h}_L$ do not interact with the electrons and hence do not affect the dynamical properties of the system. However, by definition of H_{ext} they contribute additively to the total energy and need to be projected out at the end of any analysis of the energy spectrum.

To conclude this section we return to a more formal representation of $A(x)$ by expanding photon wave functions in terms of δ-distributions. Let $\delta_k(q) = \delta(q - k)$. We define, formally,

$$a_j^\sharp(k) = a^\sharp(e_j \delta_k) \,, \qquad a^\sharp(k) = (a_1^\sharp(k), a_2^\sharp(k), a_3^\sharp(k)) \,.$$

From the expansion $h(k) = \sum_{j=1}^3 \int h_j(q) e_j \delta_k(q) \, \mathrm{d}^3 q$ and the (semi)-linearity of $a^\sharp(h)$ we obtain

$$\begin{aligned} a(h) &= \sum_{j=1}^{3} \int \overline{h_j(k)} a_j(k) \, \mathrm{d}^3 k = \int \overline{h(k)} \cdot a(k) \, \mathrm{d}^3 k \\ a^*(h) &= \sum_{j=1}^{3} \int h_j(k) a_j^*(k) \, \mathrm{d}^3 k = \int h(k) \cdot a^*(k) \, \mathrm{d}^3 k \,. \end{aligned}$$

In particular, in the representation defined by (53),

$$A(x) = \int \frac{\kappa(k)}{|k|^{1/2}} P(k)\{\mathrm{e}^{ik\cdot x}a(k) + \mathrm{e}^{-ik\cdot x}a^*(k)\}\,\mathrm{d}^3k$$

where $P(k)$ denotes the orthogonal projection onto the plane perpendicular to k. If (55) is used then

$$A(x) = \int \frac{\kappa(k)}{|k|^{1/2}} \hat{k} \wedge \{\mathrm{e}^{ik\cdot x}a(k) + \mathrm{e}^{-ik\cdot x}a^*(k)\}\,\mathrm{d}^3k\,.$$

A.5 The Dipole Approximation

In the dipole approximation of QED the quantized vector potential $A(x)$ in the Hamilton (37) is replaced by $A(0)$. By (38) the Hamiltonian (37) then reduces to

$$H_{\mathrm{dip}} = \sum_{j=1}^{N} \frac{1}{2m}(-i\nabla_{x_j} + \alpha^{1/2}A(0))^2 + \alpha V_R + H_f \tag{56}$$

where the interaction with the electron spin has dropped out. Without loss of generality we may now describe the electrons by vectors in the smaller space $\mathcal{H}_{\mathrm{at}} = \wedge_{i=1}^{N} L^2(\mathbb{R}^3)$ of spin-less N-fermion systems.

The "constant" vector potential in (56) may be gauged away with the help of the operator-valued gauge transformation

$$U = \exp\left(\alpha^{1/2}\sum_{i=1}^{N} x_i \cdot A(0)\right), \tag{57}$$

also known as Pauli-Fierz transformation. Since $U(-i\nabla_{x_j})U^* = -i\nabla_{x_j} - \alpha^{1/2}A(0)$, $UA(0)U^* = A(0)$, and

$$UH_fU^* = H_f + \sqrt{\alpha}\sum_{j=1}^{N} x_j \cdot E(0) + \alpha\|\kappa\|^2\left(\sum_{j=1}^{N} x_j\right)^2$$

where $E(0) = -i[H_f, A(0)]$ is the quantized electric field, we arrive at

$$\begin{aligned} UH_{\mathrm{dip}}U^* =& \sum_{j=1}^{N}\left(-\frac{1}{2m}\Delta_{x_j} + \sqrt{\alpha}x_j \cdot E(0)\right) \\ &+ \alpha V_R + H_f + \alpha\|\kappa\|^2\left(\sum_{j=1}^{N} x_j\right)^2 . \end{aligned} \tag{58}$$

The dipole approximation seems justified when all electrons are localized in a small neighborhood of the origin $x = 0$, that is, when the total energy is below the ionization threshold. It then seems equally justified to drop the last term in (58) and to multiply $x \cdot E(0)$ with a space cutoff $g \in C_0^\infty(\mathbb{R}^3)$; the later serves to ensure that the Hamiltonian H remains semi-bounded (after dropping the last term). This leads us to

$$\tilde{H}_{\rm dip} = \sum_{j=1}^{N} \left(-\frac{1}{2m} \Delta_{x_j} + \sqrt{\alpha} g(x_j) x_j \cdot E(0) \right) + \alpha V_R + H_f \; , \qquad (59)$$

which is also called dipole approximation of (37). It has the advantage, over (56), to be linear in creation and annihilation operators, which may simplify the analysis.

The Pauli-Fierz transformation (57) is very useful in the analysis of the original Hamiltonian (37) as well. Its effect is to replace $A(x)$ by $A(x) - A(0) = \phi(G_x - G_0)$ where

$$|G_x(k) - G_0(k)| = \left| \frac{\kappa(k)}{\sqrt{|k|}} (\mathrm{e}^{ik \cdot x} - 1) \right| \leq |k|^{1/2} |\kappa(k)| |x| \; .$$

Thus the IR-singularity of the form-factor in UHU^* is reduced by one power of $|k|$ at the expense of the unbounded factor $|x|$. This factor, however, is compensated by the exponential decay whenever the total energy is below the ionization threshold (see Sect. 2.4).

References

1. Robert A. Adams. *Sobolev spaces.* Academic Press [A subsidiary of Harcourt Brace Jovanovich, Publishers], New York-London, 1975. Pure and Applied Mathematics, Vol. 65.
2. Shmuel Agmon. *Lectures on exponential decay of solutions of second-order elliptic equations: bounds on eigenfunctions of N-body Schrödinger operators,* volume 29 of *Mathematical Notes.* Princeton University Press, Princeton, NJ, 1982.
3. Asao Arai. Rigorous theory of spectra and radiation for a model in quantum electrodynamics. *J. Math. Phys.*, 24(7):1896–1910, 1983.
4. Asao Arai and Masao Hirokawa. On the existence and uniqueness of ground states of a generalized spin-boson model. *J. Funct. Anal.*, 151(2):455–503, 1997.
5. Asao Arai and Masao Hirokawa. Ground states of a general class of quantum field Hamiltonians. *Rev. Math. Phys.*, 12(8):1085–1135, 2000.
6. V. Bach, J. Fröhlich, and A. Pizzo. private communication by Jürg Fröhlich, July 2004.
7. Volker Bach, Jürg Fröhlich, and Israel Michael Sigal. Quantum electrodynamics of confined nonrelativistic particles. *Adv. Math.*, 137(2):299–395, 1998.
8. Volker Bach, Jürg Fröhlich, and Israel Michael Sigal. Renormalization group analysis of spectral problems in quantum field theory. *Adv. Math.*, 137(2):205–298, 1998.

9. Volker Bach, Jürg Fröhlich, and Israel Michael Sigal. Spectral analysis for systems of atoms and molecules coupled to the quantized radiation field. *Comm. Math. Phys.*, 207(2):249–290, 1999.
10. Volker Bach, Jürg Fröhlich, Israel Michael Sigal, and Avy Soffer. Positive commutators and the spectrum of Pauli-Fierz Hamiltonian of atoms and molecules. *Comm. Math. Phys.*, 207(3):557–587, 1999.
11. Ola Bratteli and Derek W. Robinson. *Operator algebras and quantum statistical mechanics. 2.* Texts and Monographs in Physics. Springer-Verlag, Berlin, second edition, 1997. Equilibrium states. Models in quantum statistical mechanics.
12. R. Carmona and B. Simon. Pointwise bounds on eigenfunctions and wave packets in N-body quantum systems. V. Lower bounds and path integrals. *Comm. Math. Phys.*, 80(1):59–98, 1981.
13. Claude Cohen-Tannoudji, Jacques Dupont-Roc, and Gilbert Grynberg. *Photons and Atoms - Introduction to Quantum Electrodynamics.* Wiley-Interscience, February 1997.
14. J. Dereziński and C. Gérard. Asymptotic completeness in quantum field theory. Massive Pauli-Fierz Hamiltonians. *Rev. Math. Phys.*, 11(4):383–450, 1999.
15. J. Dereziński and C. Gérard. Spectral scattering theory of spatially cut-off $P(\phi)_2$ Hamiltonians. *Comm. Math. Phys.*, 213(1):39–125, 2000.
16. J. Fröhlich, M. Griesemer, and B. Schlein. Asymptotic completeness for Compton scattering. to appear in Communications in Mathematical Physics.
17. J. Fröhlich, M. Griesemer, and B. Schlein. Asymptotic electromagnetic fields in models of quantum-mechanical matter interacting with the quantized radiation field. *Adv. Math.*, 164(2):349–398, 2001.
18. J. Fröhlich, M. Griesemer, and B. Schlein. Asymptotic completeness for Rayleigh scattering. *Ann. Henri Poincaré*, 3(1):107–170, 2002.
19. C. Gérard. On the existence of ground states for massless Pauli-Fierz Hamiltonians. *Ann. Henri Poincaré*, 1(3):443–459, 2000.
20. Christian Gerard. On the scattering theory of massless nelson models. mp-arc 01-103, March 2001.
21. David Gilbarg and Neil S. Trudinger. *Elliptic partial differential equations of second order.* Classics in Mathematics. Springer-Verlag, Berlin, 2001. Reprint of the 1998 edition.
22. Gian Michele Graf and Daniel Schenker. Classical action and quantum N-body asymptotic completeness. In *Multiparticle quantum scattering with applications to nuclear, atomic and molecular physics (Minneapolis, MN, 1995)*, pp. 103–119. Springer, New York, 1997.
23. G.M. Graf. Asymptotic completeness for N-body short-range quantum systems: A new proof. *Comm. Math. Phys.*, 132:73–102, 1990.
24. M. Griesemer. Exponential decay and ionization thresholds in non-relativistic quantum electrodynamics. *J. Funct. Anal.*, 210(2):321–340, 2004.
25. Marcel Griesemer, Elliott H. Lieb, and Michael Loss. Ground states in non-relativistic quantum electrodynamics. *Invent. Math.*, 145(3):557–595, 2001.
26. Christian Hainzl. Enhanced binding through coupling to a photon field. In *Mathematical results in quantum mechanics (Taxco, 2001)*, volume 307 of *Contemp. Math.*, pp. 149–154. Amer. Math. Soc., Providence, RI, 2002.
27. Christian Hainzl, Vitali Vougalter, and Semjon A. Vugalter. Enhanced binding in non-relativistic QED. *Comm. Math. Phys.*, 233(1):13–26, 2003.

28. F. Hiroshima. Self-adjointness of the Pauli-Fierz Hamiltonian for arbitrary values of coupling constants. *Ann. Henri Poincaré*, 3(1):171–201, 2002.
29. F. Hiroshima and H. Spohn. Enhanced binding through coupling to a quantum field. *Ann. Henri Poincaré*, 2(6):1159–1187, 2001.
30. Fumio Hiroshima. Ground states of a model in nonrelativistic quantum electrodynamics. I. *J. Math. Phys.*, 40(12):6209–6222, 1999.
31. Fumio Hiroshima. Ground states of a model in nonrelativistic quantum electrodynamics. II. *J. Math. Phys.*, 41(2):661–674, 2000.
32. R. Hoegh-Krohn. Asymptotic fields in some models of quantum field theory. ii, iii. *J. Mathematical Phys.*, 11:185–188, 1969.
33. Matthias Hübner and Herbert Spohn. The spectrum of the spin-boson model. In *Mathematical results in quantum mechanics (Blossin, 1993)*, volume 70 of *Oper. Theory Adv. Appl.*, pp. 233–238. Birkhäuser, Basel, 1994.
34. Matthias Hübner and Herbert Spohn. Radiative decay: nonperturbative approaches. *Rev. Math. Phys.*, 7(3):363–387, 1995.
35. Matthias Hübner and Herbert Spohn. Spectral properties of the spin-boson Hamiltonian. *Ann. Inst. H. Poincaré Phys. Théor.*, 62(3):289–323, 1995.
36. W. Hunziker and I. M. Sigal. The quantum N-body problem. *J. Math. Phys.*, 41(6):3448–3510, 2000.
37. V. Jakšić and C.-A. Pillet. On a model for quantum friction. I. Fermi's golden rule and dynamics at zero temperature. *Ann. Inst. H. Poincaré Phys. Théor.*, 62(1):47–68, 1995.
38. V. Jakšić and C.-A. Pillet. On a model for quantum friction. II. Fermi's golden rule and dynamics at positive temperature. *Comm. Math. Phys.*, 176(3):619–644, 1996.
39. Vojkan Jakšić and Claude-Alain Pillet. On a model for quantum friction. III. Ergodic properties of the spin-boson system. *Comm. Math. Phys.*, 178(3):627–651, 1996.
40. Vojkan Jakšić and Claude-Alain Pillet. Spectral theory of thermal relaxation. *J. Math. Phys.*, 38(4):1757–1780, 1997. Quantum problems in condensed matter physics.
41. Tosio Kato. Fundamental properties of Hamiltonian operators of Schrödinger type. *Trans. Amer. Math. Soc.*, 70:195–211, 1951.
42. E. Lieb and M. Loss. A note on polarization vectors in quantum electrodynamics. math-ph/0401016, Jan. 2004.
43. Elliott H. Lieb and Michael Loss. Existence of atoms and molecules in non-relativistic quantum electrodynamics. *Adv. Theor. Math. Phys.*, 7(4):667–710, 2003.
44. J. Lorinczi, R. A. Minlos, and H. Spohn. The infrared behaviour in Nelson's model of a quantum particle coupled to a massless scalar field. *Ann. Henri Poincaré*, 3(2):269–295, 2002.
45. Matthias Mück. Construction of metastable states in quantum electrodynamics. *Rev. Math. Phys.*, 16(1):1–28, 2004.
46. Arne Persson. Bounds for the discrete part of the spectrum of a semi-bounded Schrödinger operator. *Math. Scand.*, 8:143–153, 1960.
47. Michael Reed and Barry Simon. *Methods of modern mathematical physics. II. Fourier analysis, self-adjointness.* Academic Press [Harcourt Brace Jovanovich Publishers], New York, 1975.

48. Michael Reed and Barry Simon. *Methods of modern mathematical physics. IV. Analysis of operators.* Academic Press [Harcourt Brace Jovanovich Publishers], New York, 1978.
49. Michael Reed and Barry Simon. *Methods of modern mathematical physics. III.* Academic Press [Harcourt Brace Jovanovich Publishers], New York, 1979. Scattering theory.
50. I.M. Sigal and A. Soffer. The N-particle scattering problem: asymptotic completeness for short-range systems. *Annals of Math.*, 126:35–108, 1987.
51. Herbert Spohn. Asymptotic completeness for Rayleigh scattering. *J. Math. Phys.*, 38(5):2281–2296, 1997.
52. Herbert Spohn. Ground state of a quantum particle coupled to a scalar Bose field. *Lett. Math. Phys.*, 44(1):9–16, 1998.
53. Herbert Spohn. *Dynamics of Charged Particles and their Radiation Field.* Cambridge University Press, August 2004.
54. Walter Thirring. *A course in mathematical physics. Vol. 3.* Springer-Verlag, New York, 1981. Quantum mechanics of atoms and molecules, Translated from the German by Evans M. Harrell, Lecture Notes in Physics, 141.
55. D. Yafaev. Radiation condition and scattering theory for N-particle Hamiltonians. *Comm. Math. Phys.*, pp. 523–554, 1993.

Dilute, Trapped Bose Gases and Bose-Einstein Condensation

R. Seiringer

Department of Physics, Jadwin Hall, Princeton University, P.O. Box 708, Princeton, NJ 08544, USA
rseiring@math.princeton.edu

1 Introduction

The recent experimental success in creating Bose-Einstein condensates of alkali atoms, honored by the Nobel prize awards in 2001 [1,5], led to renewed interest in the mathematical description of interacting Bose gases. These

R. Seiringer: *Dilute, Trapped Bose Gases and Bose-Einstein Condensation*, Lect. Notes Phys. **695**, 249–274 (2006)
www.springerlink.com

lectures will give a detailed account of part of the author's joint work with E.H. Lieb and J. Yngvason in this field [10, 12]. In particular, a complete, self-contained proof of Bose-Einstein condensation (BEC) for dilute gases in traps will be given.

1.1 The Model

The Hilbert space under consideration is the subspace of totally symmetric functions in $\bigotimes_{i=1}^N L^2(\mathbb{R}^3, \mathrm{d}\boldsymbol{x}_i)$, which we denote by $\mathcal{H}_N$. The Hamiltonian, acting on this space, is given by

$$H_{N,a} = \sum_{i=1}^{N} \big(- \Delta_i + V(\boldsymbol{x}_i)\big) + \sum_{1\leq i<j\leq N} v(\boldsymbol{x}_i - \boldsymbol{x}_j) \,. \tag{1}$$

Units are chosen such that $\hbar = 2m = 1$, where $\hbar$ denotes Planck's constant divided by 2π, and m is the particle mass. The external potential is a real-valued, locally bounded function $V \in L^\infty_{\rm loc}(\mathbb{R}^3)$, satisfying $\lim_{|\boldsymbol{x}|\to\infty} V(\boldsymbol{x}) = \infty$. It is then no restriction to assume $V \geq 0$. The interaction potential v is assumed to be positive (i.e., repulsive), radial and of compact support. We do not demand it to be integrable, it is allowed to have a hard core, which reduces the domain of definition of $H_{N,a}$ to wave functions in $\mathcal{H}_N$ that vanish whenever two particles are closer together then the size of the hard core. The parameter a in the Hamiltonian is the scattering length of v, which we define next.

1.2 Scattering Length; Length Scales

The scattering length of v is defined via the zero-energy scattering equation

$$- \Delta f(\boldsymbol{x}) + \tfrac{1}{2} v(\boldsymbol{x}) f(\boldsymbol{x}) = 0 \,. \tag{2}$$

The factor $\frac{1}{2}$ in front of v is due to the reduced mass of the two-body problem. The solution to (2), if normalized by $f(\boldsymbol{x}) \to 1$ as $|\boldsymbol{x}| \to \infty$, equals $1 - a/|\boldsymbol{x}|$ for $|\boldsymbol{x}| \geq R_0$, the range of v. The positive number a is called the *scattering length*. It is a measure of the effective range of the interaction v.

Example. If v is the potential for hard spheres, i.e., $v(\boldsymbol{x}) = \infty$ for $|\boldsymbol{x}| \leq a$ and $v(\boldsymbol{x}) = 0$ for $|\boldsymbol{x}| > a$, then the solution to (2) is given by $f(\boldsymbol{x}) = \max\{1 - a/|\boldsymbol{x}|, 0\}$, and a is the scattering length of v.

The scattering length a can also be obtained from a variational principle (see the Appendix of [18]). From this it follows that $a < (8\pi)^{-1} \int v(\boldsymbol{x}) \mathrm{d}\boldsymbol{x}$, where the right side need not be finite, of course.

In the following, we want to change a with N in order to have a *dilute* system, meaning that the mean particle distance is much bigger that the range of the interaction, measured by a. To this end we write v as

$$v(\boldsymbol{x}) = \left(\frac{a_1}{a}\right)^2 v_1(a_1\boldsymbol{x}/a) \tag{3}$$

for a given v_1 with scattering length a_1. Then v has scattering length a, which is now a parameter that can be varied, keeping v_1 and a_1 fixed. This explains also the notation $H_{N,a}$ of our Hamiltonian above.

The fact that (3) is the physically relevant scaling can also be seen from the following argument. Consider the particles confined in a box of side length L, and interacting via v. The Hamiltonian is unitarily equivalent, via scaling $\boldsymbol{x} \to L\boldsymbol{x}$, to L^{-2} times the Hamiltonian in a box of side length 1, but with interaction potential $L^2 v(L\boldsymbol{x})$. Making the system dilute means making L big, which is the same as making a small in (3).

Despite the fact that we are considering dilute systems, or actually *because* of that, perturbation theory is *not* applicable! This can immediately be seen from the scaling (3). Since a is going to be small, v is a very hard potential of short range. This is the opposite of a mean-field limit, where perturbation theory would apply. In the case considered here, the interaction energy is mostly *kinetic* energy!

Now what can we expect for the ground state energy of a gas of N particles in a box of side length L? One might guess that it is roughly the lowest energy of a pair of particles, which can easily be computed to be $8\pi a/L^3$ for $a/L \ll 1$, times the number of pairs, $\frac{1}{2}N(N-1)$. This gives

$$(N-1)4\pi a\rho \, , \tag{4}$$

where $\rho = N/L^3$ is the density. Equation (4) was first obtained by Lenz [6] who considered the case of a particle interacting with fixed scatterers of diameter a. Note that perturbation theory with a constant trial function would give $\frac{1}{2}(N-1)\rho \int v$, which, even if finite, is always larger than the expected energy $(N-1)4\pi a\rho$ since $\int v > 8\pi a$ as mentioned above.

1.3 Quantities of Interest

The main quantities we are interested in are the following.

- Ground state energy, defined by

$$E^{\rm QM}(N,a) = \inf \operatorname{spec} H_{N,a} \, . \tag{5}$$

Besides N and a it depends on V and v_1, of course, but these potentials are assumed to be fixed once and for all. The ground state wave function, Ψ_0, is determined by Schrödinger's equation

$$H_{N,a}\Psi_0 = E^{\rm QM}(N,a)\Psi_0 \, . \tag{6}$$

It is unique, up to constant phase factor, which can be chosen such that $\Psi_0 \geq 0$.

- Ground state density matrix, defined by the kernel

$$\gamma_N^{(1)}(\boldsymbol{x},\boldsymbol{x}') = N\int \Psi_0(\boldsymbol{x},\boldsymbol{x}_2,\ldots,\boldsymbol{x}_N)\Psi_0(\boldsymbol{x}',\boldsymbol{x}_2,\ldots,\boldsymbol{x}_N)^*\mathrm{d}\boldsymbol{x}_2\cdots\mathrm{d}\boldsymbol{x}_N\ . \qquad (7)$$

The * denotes complex conjugation. The kernel (7) defines a positive trace class operator on $L^2(\mathbb{R}^3,\mathrm{d}\boldsymbol{x})$, also denoted by $\gamma_N^{(1)}$, with $\mathrm{Tr}[\gamma_N^{(1)}]=N$.

Definition of *Bose-Einstein condensation* [19]: BEC (in the ground state) means that the largest eigenvalue, $\lambda_{\max}$, of $\gamma_N^{(1)}$ is of order N as $N\to\infty$. The eigenvalues of $\gamma_N^{(1)}$ are interpreted as occupation numbers of the corresponding eigenstates, and an eigenvalue of order N means that the state is macroscopically occupied. If there is one eigenvalue of order N, and all the others are of lower order, than the eigenfunction corresponding to $\lambda_{\max}$ is called the "condensate wave function". If $\lambda_{\max}/N\to 1$ as $N\to\infty$, we say that there is *complete* BEC.

The definition of BEC above is readily generalized to positive temperature states, but here we consider only the zero temperature case, i.e., the ground state. (See the remark after Corollary 2, however.)

- Ground state density, given by the diagonal

$$\rho^{\mathrm{QM}}(\boldsymbol{x}) = \gamma_N^{(1)}(\boldsymbol{x},\boldsymbol{x})\ , \qquad (8)$$

and momentum density, given by the Fourier transform

$$\widetilde{\rho}^{\mathrm{QM}}(\boldsymbol{k}) = (2\pi)^{-3}\int \gamma_N^{(1)}(\boldsymbol{x},\boldsymbol{x}')\mathrm{e}^{\mathrm{i}\boldsymbol{k}\cdot(\boldsymbol{x}-\boldsymbol{x}')}\mathrm{d}\boldsymbol{x}\mathrm{d}\boldsymbol{x}'\ . \qquad (9)$$

2 The Gross-Pitaevskii Functional

In order to be able to state our results on properties of the quantities described above, we first have to introduce the Gross-Pitaevskii (GP) model [4,20]. This model is commonly used as an approximation to the complicated many-body problem described above [2]. It is the purpose of this presentation to rigorously justify the use of this approximation and clarify under what circumstances it can be expected to yield reliable results.

We start by introducing the GP functional, and state the main properties of its minimizer.

2.1 Definition; Existence and Uniqueness of a Minimizer

For $\phi \in H^1(\mathbb{R}^3, \mathrm{d}\boldsymbol{x})$ and $g \in \mathbb{R}$, define the Gross-Pitaevskii functional by

$$\mathcal{E}_g^{\mathrm{GP}}[\phi] = \int_{\mathbb{R}^3} \left[|\nabla\phi(\boldsymbol{x})|^2 + V(\boldsymbol{x})|\phi(\boldsymbol{x})|^2 + g|\phi(\boldsymbol{x})|^4 \right] \mathrm{d}\boldsymbol{x} \,. \tag{10}$$

The corresponding ground state energy is given by

$$E^{\mathrm{GP}}(g) = \inf \left\{ \mathcal{E}_g^{\mathrm{GP}}[\phi] \,:\, \|\phi\|_2 = 1 \right\} . \tag{11}$$

It is finite for $g \geq 0$. We have the following existence and uniqueness result for the minimization problem (11).

Proposition 1 (Minimizers of $\mathcal{E}_g^{\mathrm{GP}}$) *For each $g \geq 0$ there exists a unique (up to a constant phase) minimizing function for (10) under the condition $\|\phi\|_2 = 1$, denoted by ϕ^{GP}. It fulfills the non-linear Schrödinger equation*

$$-\Delta\phi(\boldsymbol{x}) + V(\boldsymbol{x})\phi(\boldsymbol{x}) + 2g|\phi(\boldsymbol{x})|^2\phi(\boldsymbol{x}) = \mu_g^{\mathrm{GP}}\phi(\boldsymbol{x}) \,, \tag{12}$$

where μ_g^{GP} is the chemical potential, given by

$$\mu_g^{\mathrm{GP}} = E^{\mathrm{GP}}(g) + g \int |\phi^{\mathrm{GP}}(\boldsymbol{x})|^4 \mathrm{d}\boldsymbol{x} \,. \tag{13}$$

Proof. The proof of Prop. 1 is fairly standard. Any minimizing sequence ϕ_n with $\|\phi_n\|_2 = 1$ fulfills

$$\int \left[|\nabla\phi_n(\boldsymbol{x})|^2 + V(\boldsymbol{x})|\phi_n(\boldsymbol{x})|^2 \right] \mathrm{d}\boldsymbol{x} < C \tag{14}$$

for some $C > 0$ independent of n. Since $-\Delta + V$ has a compact resolvent [21, Thm. XIII.65], (14) implies that the sequence ϕ_n lies in a compact subset of $L^2(\mathbb{R}^3)$. It therefore converges strongly to some $\phi \in L^2(\mathbb{R}^3)$, which minimizes $\mathcal{E}_g^{\mathrm{GP}}$ because of weak lower-semicontinuity of this functional.

The variational (12) follows from stationarity of

$$\mathcal{E}_g^{\mathrm{GP}}[\phi + \varepsilon f] + \mu\|\phi + \varepsilon f\|_2^2 \tag{15}$$

at $\varepsilon = 0$ for an appropriate Lagrange multiplier μ, which satisfies (13), as can be seen from multiplying (12) by ϕ and integrating.

We are left with proving uniqueness. Actually, uniqueness of $|\phi|$ follows from strict convexity of $|\phi|^2 \mapsto \mathcal{E}_g^{\mathrm{GP}}[|\phi|]$. To see that also ϕ is unique, note that $\mathcal{E}_g^{\mathrm{GP}}[|\phi|] \leq \mathcal{E}_g^{\mathrm{GP}}[\phi]$, hence if ϕ is a minimizer, so is $|\phi|$. Hence $|\phi|$ satisfies (12), and since it is positive, it must be the ground state of $-\Delta + V + 2g|\phi|^2$, which is unique [21, Thm. XIII.47]. Since we already know that $|\phi|$ is unique, it follows that $\phi = |\phi|$ up to a constant phase factor. □

We suppress the dependence on g in ϕ^{GP} for simplicity of notation.

2.2 Properties of the Minimizer

Proposition 2 (Properties of ϕ^{GP}) *The arbitrary phase factor in the minimizer of $\mathcal{E}_g^{\mathrm{GP}}$ can be chosen such that $\phi^{\mathrm{GP}} > 0$. We have $\phi^{\mathrm{GP}} \in C^1$ and $\mathrm{e}^{t|\cdot|}\phi^{\mathrm{GP}} \in L^\infty$ for all $t \in \mathbb{R}$. Moreover,*

$$\|\phi^{\mathrm{GP}}\|_\infty^2 \leq \frac{\mu_g^{\mathrm{GP}}}{2g} \,. \tag{16}$$

Proof. The strict positivity follows from the fact that ϕ^{GP} is the ground state of $-\Delta + V + 2g|\phi^{\mathrm{GP}}|^2$ [21, Thm. XIII.47]. From elliptic regularity [9, Thm. 10.2] we infer that $\phi^{\mathrm{GP}} \in C^1$. The exponential decay is standard (see, e.g., [12, Lemma A.5]).

The bound (16) follows from the maximum principle: let $\mathcal{B} \subset \mathbb{R}^3$ denote the set where $|\phi^{\mathrm{GP}}(\boldsymbol{x})|^2 > \mu_g^{\mathrm{GP}}/(2g)$. Since $V > 0$, the GP equation implies that $\Delta\phi^{\mathrm{GP}} > 0$ on $\mathcal{B}$, i.e., ϕ^{GP} is subharmonic there. Hence it achieves its maximum on the boundary of $\mathcal{B}$, where $\phi^{\mathrm{GP}} = \mu_g^{\mathrm{GP}}/(2g)$, so $\mathcal{B}$ is empty. □

In the following, we will always assume that the arbitrary phase factor in ϕ^{GP} is chosen such that $\phi^{\mathrm{GP}} > 0$.

The following property concerning the *linearized problem* has been used in the proofs above. It will be important later, so we emphasize it in the following Corollary.

Proposition 3 (Linearized problem) *The minimizer ϕ^{GP} is the unique ground state of the operator*

$$-\Delta + V(\boldsymbol{x}) + 2g|\phi^{\mathrm{GP}}(\boldsymbol{x})|^2 \,, \tag{17}$$

with ground state energy μ_g^{GP}.

2.3 Thomas-Fermi Limit

The GP functional has a well-defined limit as $g \to \infty$, at least if the external potential V is reasonably well behaved at infinity. This is usually referred to as the Thomas-Fermi (TF) limit, because of a formal analogy with the Thomas-Fermi functional for atoms and molecules, which also has no gradient term. Consider, for simplicity, the case when V is homogeneous of some order $s > 0$, i.e., $V(\lambda\boldsymbol{x}) = \lambda^s V(\boldsymbol{x})$ for all $\lambda > 0$. A simple scaling shows that, for

$$\widetilde{\phi}(\boldsymbol{x}) = g^{3/(s+3)}\phi(g^{1/(s+3)}\boldsymbol{x}) \,, \tag{18}$$

we have

$$\begin{aligned}&\mathcal{E}_g^{\mathrm{GP}}[\phi]\\ &= g^{s/(s+3)} \int_{\mathbb{R}^3} \left[g^{-(s+2)/(s+3)}|\boldsymbol{\nabla}\widetilde{\phi}(\boldsymbol{x})|^2 + V(\boldsymbol{x})|\widetilde{\phi}(\boldsymbol{x})|^2 + |\widetilde{\phi}(\boldsymbol{x})|^4\right] \mathrm{d}\boldsymbol{x} \,. \end{aligned}\tag{19}$$

This shows that for large g the gradient term in the GP functional is negligible and the functional simplifies in this case to

$$\phi \mapsto \int_{\mathbb{R}^3} \left[V(\boldsymbol{x})|\phi(\boldsymbol{x})|^2 + g|\phi(\boldsymbol{x})|^4\right] \mathrm{d}\boldsymbol{x} \ , \tag{20}$$

called the TF functional. Its ground state energy, $E^{\mathrm{TF}}(g)$, has the simple scaling property $E^{\mathrm{TF}}(g) = g^{s/(s+3)} E^{\mathrm{TF}}(1)$.

3 Main Results

3.1 Convergence of Energy and Density Matrix

With the preliminaries of the previous section in hand, we can now state our main results. We are interested in the ground state energy $E^{\mathrm{QM}}(N,a)$ of the N-particle Hamiltonian (1), for large N and small a. In fact, as it will turn out below, $a \sim N^{-1}$ is the case of interest. That is, for some fixed $g > 0$, we will set $a = (4\pi)^{-1} g N^{-1}$, or, more generally, we will assume that $4\pi N a \to g$ as $N \to \infty$. (The factor 4π is chosen for convenience.) Moreover, we will also derive results for the corresponding ground states or, more generally, for *approximate ground states*. For fixed $g \geq 0$ we call a sequence $\Psi_N \in \mathcal{H}_N$ an approximate ground state if $\|\Psi_N\|_2 = 1$ and

$$\lim_{N\to\infty,\, 4\pi N a \to g} \langle \Psi_N | H_{N,a} \Psi_N \rangle \, E^{\mathrm{QM}}(N,a)^{-1} = 1 \ . \tag{21}$$

Given such an approximate ground state, we define its reduced n-particle density matrix in analogy with (7) by the kernel

$$\gamma_N^{(n)}(\boldsymbol{x}_1,\ldots,\boldsymbol{x}_n,\boldsymbol{y}_1,\ldots,\boldsymbol{y}_n) = \binom{N}{n} \int_{\mathbb{R}^{3(N-n)}} \Psi_N(\boldsymbol{x}_1,\ldots,\boldsymbol{x}_N)\Psi_N^*(\boldsymbol{y}_1,\ldots,\boldsymbol{y}_n,\boldsymbol{x}_{n+1},\ldots,\boldsymbol{x}_N) \prod_{j=n+1}^{N} \mathrm{d}\boldsymbol{x}_j \ . \tag{22}$$

Note that since Ψ_N is totally symmetric in the particle coordinates, it doesn't matter over which variables we integrate. The $\gamma_N^{(n)}$ defined in (22) are trace class operators on $\mathcal{H}_n$, and the normalization is chosen such that $\mathrm{Tr}[\gamma_N^{(n)}] = \binom{N}{n}$. Our main results are the following:

Theorem 1 (Convergence of the energy). *If $4\pi N a \to g$ as $N \to \infty$, then*

$$\lim_{N\to\infty} \frac{1}{N} E^{\mathrm{QM}}(N,a) = E^{\mathrm{GP}}(g) \ . \tag{23}$$

We stated Theorem 1 for bounded Na, but it can actually be extended to the TF limit $Na \to \infty$, as long as $a^3\overline{\rho} \to 0$, where $\overline{\rho}$ denotes the mean density. See [13] for details.

Theorem 2 (Convergence of one-particle density matrix). *For given $g \geq 0$ let $\phi^{\rm GP}$ be the unique minimizer of $\mathcal{E}_g^{\rm GP}$, and let $\gamma_N^{(1)}$ denote the one-particle reduced density matrix of an approximate ground state of $H_{N,a}$. If $4\pi Na \to g$ as $N \to \infty$, then*

$$\lim_{N\to\infty} \frac{1}{N}\gamma_N^{(1)} = |\phi^{\rm GP}\rangle\langle\phi^{\rm GP}| \tag{24}$$

in trace class norm.

Note that Theorem 2 proves complete BEC for the ground state in the dilute limit considered. The density matrix factorizes into a product, like for non-interacting systems, but this does not mean that there are no significant interactions taking place. In fact, $\phi^{\rm GP}$ depends on $g \sim 4\pi Na$. This is a truly remarkable feature. As a consequence of the diluteness of the system, the results are independent of the v_1 in (3), the scattering length is the only relevant quantity.

The proof of these two theorems will be given in Sect. 4. Theorem 1 was proved in [12], and Theorem 2 in [10]. For both proofs the methods and results of [17] are of substantial importance.

3.2 Corollaries

The following is easily deduced from Thm. 2 [10].

Corollary 1 (Convergence of densities) *If $4\pi Na \to g$ as $N \to \infty$, then*

$$\lim_{N\to\infty} \frac{1}{N}\rho^{\rm QM}(\boldsymbol{x}) = |\phi^{\rm GP}(\boldsymbol{x})|^2 \tag{25}$$

and

$$\lim_{N\to\infty} \frac{1}{N}\widehat{\rho}^{\rm QM}(\boldsymbol{k}) = |\widehat{\phi}^{\rm GP}(\boldsymbol{k})|^2 \tag{26}$$

strongly in $L^1(\mathbb{R}^3)$. Here $\widehat{\phi}^{\rm GP}$ denotes the Fourier transform of $\phi^{\rm GP}$.

Moreover, Thm. 2 can be readily extended to n-particle density matrices for $n > 1$ [10].

Corollary 2 (n-particle density matrices) *If $4\pi Na \to g$ as $N \to \infty$, then*

$$\lim_{N\to\infty} \binom{N}{n}^{-1} \gamma_N^{(n)} = \underbrace{|\phi^{\rm GP}\rangle\langle\phi^{\rm GP}| \otimes \cdots \otimes |\phi^{\rm GP}\rangle\langle\phi^{\rm GP}|}_{n \text{ times}} \tag{27}$$

in trace class norm.

Remark. Cor. 2 shows that there is complete BEC of all n-particle density matrices. This is true not only for the ground state, but for any approximate ground state. Moreover, instead of vector states one may consider more general states, given by N-particle density matrices, as approximate ground states. In particular, the assertions above are true for all Gibbs states

$$\Gamma^{\beta}_{N,a} := \frac{\exp(-\beta H_{N,a})}{\mathrm{Tr}[\exp(-\beta H_{N,a})]}\,, \tag{28}$$

where $\beta = 1/T > 0$ is the inverse temperature. Here we assumed that the trace in the denominator is finite, which is guaranteed for external potentials V that increase at least logarithmically in $|\boldsymbol{x}|$ at infinity. One can show [23] that, if $4\pi Na \to g$ for some $g \geq 0$, then

$$\lim_{N\to\infty} \frac{1}{N}\left(\mathrm{Tr}[H_{N,a}\Gamma^{\beta}_{N,a}] - E^{\mathrm{QM}}(N,a)\right) = 0 \tag{29}$$

for all *fixed* $\beta > 0$, i.e., (28) is an approximate ground state in the sense defined above. Cor. 2 thus implies complete BEC of the reduced density matrices of (28), i.e., they converge to the right hand side of (27). It is important to note, however, that if we take the $N \to \infty$ limit in a *fixed* trap potential V (as we do), then the (mean) density and hence also the relevant temperature scale goes to infinity. Measured on this scale, any *fixed* T becomes infinitesimally small as $N \to \infty$. Fixing β thus really amounts in our case to taking a *zero-temperature limit.* To obtain a true effect of the temperature one has to scale it appropriately with N. E.g., for a harmonic trap potential the relevant temperature scale would be $T \sim N^{1/3}$ [2].

Note that for (29) to hold true it is essential to restrict ourselves to the bosonic subspace in (28), as we do here. Without this restriction (29) will not be true, as can be seen from the non-interacting case $g = 0$.

4 The Proof

In the following, we will give a complete and self-contained proof of Theorems 1 and 2 above. We follow the ideas in the original proofs in [10,12,17], but present various simplifications.

4.1 Heuristics and Ideas

The main point that has to be understood is a separation of scales. The range of the interaction, which is of order $a \sim N^{-1}$, is much smaller than the mean particle distance, which is $\sim N^{-1/3}$ in a fixed volume. To explain the main idea, let us for the moment double the kinetic energy, and add and subtract a term $\sum_j \phi^{\mathrm{GP}}(\boldsymbol{x}_j)^2$ to the Hamiltonian. We then split it into two parts,

$$H^{(1)} = \sum_{j=1}^{N} \big(-\Delta_j + V(\boldsymbol{x}_j) + 2g\phi^{\rm GP}(\boldsymbol{x}_j)^2 \big) \,, \tag{30}$$

and

$$H^{(2)} = -\sum_{j=1}^{N} \Delta_j + \sum_{1\le i<j\le N} v(\boldsymbol{x}_i - \boldsymbol{x}_j) - 2g\sum_{j=1}^{N} \phi^{\rm GP}(\boldsymbol{x}_j)^2 \,. \tag{31}$$

Now let us try to minimize these two operators separately. The ground state energy of $H^{(1)}$ is, according to Prop. 3, $N\mu_g^{\rm GP}$, and in order to have an approximate ground state, the one-particle density matrix must be given by $N|\phi^{\rm GP}\rangle\langle\phi^{\rm GP}|$ to leading order. This is exactly what we claim in Theorem 2.

What can one expect for the ground state energy of $H^{(2)}$? Assuming that the gas is locally homogeneous and following the heuristics in Subsect. 1.2, the ground state energy of the first two terms in (31) should approximately equal $N4\pi a\rho$, with ρ being the density. This density will not be constant, however, because of the last term in (31). We can thus expect the ground state energy of $H^{(2)}$ to be equal to

$$4\pi a \int \rho(\boldsymbol{x})^2 {\rm d}\boldsymbol{x} - 2g \int \rho(\boldsymbol{x})\phi^{\rm GP}(\boldsymbol{x})^2 {\rm d}\boldsymbol{x} \,. \tag{32}$$

If $g = 4\pi Na$, (32) gives exactly $-Ng\int|\phi^{\rm GP}|^4$ when minimized over ρ. Adding this to the ground state energy of $H^{(1)}$ we obtain the desired result.

The main problem in the preceding argumentation is that we used the kinetic energy twice. However, because of the different relevant scales mentioned above, it turns out that in order to get the right energy for $H^{(2)}$ it suffices to consider only the part of a particle's kinetic energy when it is very close to at least one other particle, in fact much closer than the mean particle distance. The remaining part, when the particle is sufficiently far away from all the other particles, is used in $H^{(1)}$.

4.2 Splitting of the Energy

We are now going to make the above ideas precise. Fix some $R > 0$ and $0 < \varepsilon < 1$. They will be chosen later to depend on N in a definite way. We introduce the short hand notation

$$\boldsymbol{X}_i = (\boldsymbol{x}_1, \ldots, \boldsymbol{x}_{i-1}, \boldsymbol{x}_{i+1}, \ldots, \boldsymbol{x}_N) \tag{33}$$

and

$${\rm d}\boldsymbol{X}_i = \prod_{j=1,\, j\neq i}^{N} {\rm d}\boldsymbol{x}_j \,. \tag{34}$$

For fixed $\boldsymbol{X}_j$, let $\mathcal{O}_j \subset \mathbb{R}^3$ denote the set

$$\mathcal{O}_j = \left\{ \boldsymbol{x}_j \in \mathbb{R}^3 \ : \ \min_{k,\, k \neq j} |\boldsymbol{x}_j - \boldsymbol{x}_k| \leq R \right\} , \tag{35}$$

and let $\mathcal{O}_j^c$ denote its complement. Given any $\Psi \in \mathcal{H}_N$ with $\|\Psi\|_2 = 1$, we can write the expectation value of our Hamiltonian (1) as

$$\langle \Psi | H_{N,a} | \Psi \rangle = E_\Psi^{(1)} + E_\Psi^{(2)} , \tag{36}$$

where

$$E_\Psi^{(1)} = \sum_{j=1}^{N} \int_{\mathbb{R}^{3(N-1)}} \mathrm{d}\boldsymbol{X}_j \left[(1-\varepsilon) \int_{\mathcal{O}_j^c} \mathrm{d}\boldsymbol{x}_j \, |\boldsymbol{\nabla}_j \Psi|^2 + \frac{\varepsilon}{2} \int_{\mathbb{R}^3} \mathrm{d}\boldsymbol{x}_j \, |\boldsymbol{\nabla}_j \Psi|^2 \right.$$
$$\left. + \int_{\mathbb{R}^3} \mathrm{d}\boldsymbol{x}_j \left(V(\boldsymbol{x}_j) + 2g\phi^{\mathrm{GP}}(\boldsymbol{x}_j)^2 \right) |\Psi|^2 \right] \tag{37}$$

and

$$E_\Psi^{(2)} = \sum_{j=1}^{N} \int_{\mathbb{R}^{3(N-1)}} \mathrm{d}\boldsymbol{X}_j \left[(1-\varepsilon) \int_{\mathcal{O}_j} \mathrm{d}\boldsymbol{x}_j \, |\boldsymbol{\nabla}_j \Psi|^2 + \frac{\varepsilon}{2} \int_{\mathbb{R}^3} \mathrm{d}\boldsymbol{x}_j \, |\boldsymbol{\nabla}_j \Psi|^2 \right.$$
$$\left. + \int_{\mathbb{R}^3} \mathrm{d}\boldsymbol{x}_j \Big(\tfrac{1}{2} \sum_{i=1,\, i \neq j}^{N} v(\boldsymbol{x}_i - \boldsymbol{x}_j) - 2g\phi^{\mathrm{GP}}(\boldsymbol{x}_j)^2 \Big) |\Psi|^2 \right] . \tag{38}$$

In the following two subsections, we will investigate the two terms (37) and (38) separately. It is always understood that, for a given fixed g, $4\pi Na - g = o(1)$ as $N \to \infty$.

4.3 Part I. A generalized Poincaré Inequality

We are going to need the following Lemma. It is related to the generalized Poincaré inequalities studied in [15], and was used in [23] in the case of magnetic fields. For a measurable set $\mathcal{O} \subset \mathbb{R}^3$ we denote by $\mathcal{O}^c$ its complement, and by $|\mathcal{O}|$ its Lebesgue measure.

Lemma 1. *Let $V \in L^\infty_{\mathrm{loc}}(\mathbb{R}^3; \mathbb{R})$, and assume that $\lim_{|\boldsymbol{x}| \to \infty} V(\boldsymbol{x}) = \infty$. Let*

$$E = \inf \operatorname{spec} \Big[-\Delta + V(\boldsymbol{x}) \Big] , \tag{39}$$

and let P denote the projector in $\mathcal{H} = L^2(\mathbb{R}^3, \mathrm{d}\boldsymbol{x})$ onto the corresponding ground state. Let

$$\Delta E = \inf \operatorname{spec} \big[-\Delta + V(\boldsymbol{x}) \big] \restriction_{(1-P)\mathcal{H}} - E \tag{40}$$

denote the gap in the spectrum above the ground state energy, which is positive because of the discrete spectrum of the operator under consideration.

For all $\varepsilon > 0$ *there exists a* $\delta > 0$ *such that for all* $\mathcal{O} \subset \mathbb{R}^3$ *with* $|\mathcal{O}^c| < \delta$ *and for all* $f \in H^1(\mathbb{R}^3)$

$$\varepsilon \int_{\mathbb{R}^3} |\nabla f|^2 + \int_{\mathcal{O}} |\nabla f|^2 + \int_{\mathbb{R}^3} V|f|^2 \geq E \, \|f\|^2_{L^2(\mathbb{R}^3)} + \Delta E \, \|f - Pf\|^2_{L^2(\mathbb{R}^3)} \,. \quad (41)$$

Proof. It is no restriction to assume that $V \geq 0$. As in [15] we will use a compactness argument. Suppose that the Lemma is wrong. Then there exists an $\varepsilon_0 > 0$ and a sequence of pairs $(f_n, \mathcal{O}_n)$, such that $\lim_{n\to\infty} |\mathcal{O}_n^c| = 0$, $\|f_n\|_{L^2(\mathbb{R}^3)} = 1$, and

$$\lim_{n\to\infty} \left[\varepsilon_0 \int_{\mathbb{R}^3} |\nabla f_n|^2 + \int_{\mathcal{O}_n} |\nabla f_n|^2 + \int_{\mathbb{R}^3} V|f_n|^2 - \Delta E \, \|f_n - Pf_n\|^2_{L^2(\mathbb{R}^3)} \right] \leq E \,. \quad (42)$$

Now both ∇f_n and f_n are bounded sequences in $L^2(\mathbb{R}^3)$, so we can pass to a subsequence that converges weakly in $L^2(\mathbb{R}^3)$ to ∇f and f, respectively. We may also assume that $\sum_n |\mathcal{O}_n^c|$ is finite. Defining Σ_N by $\Sigma_N = \mathbb{R}^3 \setminus \bigcup_{n\geq N} \mathcal{O}_n^c$ we have $\Sigma_N \subset \mathcal{O}_n$ for $n \geq N$. Using weak lower semicontinuity of the norms in question, we therefore get

$$\begin{aligned} \liminf_{n\to\infty} & \left[\varepsilon_0 \int_{\mathbb{R}^3} |\nabla f_n|^2 + \int_{\mathcal{O}_n} |\nabla f_n|^2 + \int_{\mathbb{R}^3} V|f_n|^2 \right] \\ & \geq \sup_N \left[\varepsilon_0 \int_{\mathbb{R}^3} |\nabla f|^2 + \int_{\Sigma_N} |\nabla f|^2 + \int_{\mathbb{R}^3} V|f|^2 \right] \\ & = \left[(1+\varepsilon_0) \int_{\mathbb{R}^3} |\nabla f|^2 + \int_{\mathbb{R}^3} V|f|^2 \right] > E \int_{\mathbb{R}^3} |f|^2 \,. \end{aligned} \quad (43)$$

Now since V goes to infinity at infinity, $-\Delta + V$ has a compact resolvent (cf., e.g., [21, Thm. XIII.65]). Since

$$\int_{\mathbb{R}^3} \left(|\nabla f_n|^2 + V|f_n|^2 \right) < C \quad (44)$$

for some $C < \infty$ independent of n, we can conclude that f_n is contained in a compact subset of $L^2(\mathbb{R}^3)$, and thus $f_n \to f$ strongly in $L^2(\mathbb{R}^3)$. This implies that $\|f\|_2 = 1$, and also

$$\lim_{n\to\infty} \|f_n - Pf_n\|_{L^2(\mathbb{R}^3)} = \|f - Pf\|_{L^2(\mathbb{R}^3)} \,. \quad (45)$$

Together with (43) this contradicts (42). □

Using more sophisticated methods, as in [15], it is possible to investigate the relation between ε and δ. This is needed to get precise error estimates, but we shall not do this here.

We now derive a lower bound on $E_\Psi^{(1)}$. For fixed $\boldsymbol{X}_j$ define f_j by

$$f_j(\boldsymbol{x}_j) = \Psi(\boldsymbol{x}_1, \ldots, \boldsymbol{x}_j, \ldots, \boldsymbol{x}_N) , \tag{46}$$

and let

$$W(\boldsymbol{x}) = V(\boldsymbol{x}) + 2g\phi^{\mathrm{GP}}(\boldsymbol{x})^2 . \tag{47}$$

Note that $W \geq 0$. We have

$$E_\Psi^{(1)} \geq \sum_{j=1}^{N} (1-\varepsilon) \int_{\mathbb{R}^{3(N-1)}} \mathrm{d}\boldsymbol{X}_j F_j , \tag{48}$$

with

$$F_j = \int_{\mathcal{O}_j} |\boldsymbol{\nabla} f_j|^2 + \frac{\varepsilon}{2} \int_{\mathbb{R}^3} |\boldsymbol{\nabla} f_j|^2 + \int_{\mathbb{R}^3} W |f_j|^2 . \tag{49}$$

We now apply Lemma 1 to F_j. Note that $|\mathcal{O}_j^c| \leq N \frac{4\pi}{3} R^3$, and that

$$\sum_{j=1}^{N} \int_{\mathbb{R}^{3(N-1)}} \mathrm{d}\boldsymbol{X}_j \|f_j - Pf_j\|_2^2 = \mathrm{Tr}\left[\gamma_\Psi^{(1)} \left(1 - |\phi^{\mathrm{GP}}\rangle\langle\phi^{\mathrm{GP}}|\right)\right] , \tag{50}$$

where $\gamma_\Psi^{(1)}$ denotes the one-particle density matrix of Ψ. Thus, if $R \ll N^{-1/3}$, we can choose $\varepsilon = o(1)$ as $N \to \infty$ such that

$$E_\Psi^{(1)} \geq N \left(\mu_g^{\mathrm{GP}} + C \mathrm{Tr}\left[\tfrac{1}{N} \gamma_\Psi^{(1)} \left(1 - |\phi^{\mathrm{GP}}\rangle\langle\phi^{\mathrm{GP}}|\right)\right]\right) (1 - o(1)) . \tag{51}$$

The constant $C > 0$ is then given by the spectral gap of the operator (17) above its ground state energy.

4.4 Part II. Dyson Lemma, Box Method, Temple's Inequality

To obtain a lower bound on $E_\Psi^{(2)}$, we start with a lemma that was originally proved by Dyson for the case of hard spheres [3], and later generalized by Lieb and Yngvason [11, 17] to the case of arbitrary repulsive potentials with finite range. It allows us to replace v by a "soft" potential, at the cost of sacrificing kinetic energy and increasing the range.

Lemma 2. *Let $v(\boldsymbol{x}) \geq 0$ be radial, with finite range R_0. Let $U(\boldsymbol{x}) \geq 0$ be any radial function satisfying $\int U(\boldsymbol{x}) \mathrm{d}\boldsymbol{x} \leq 4\pi$ and $U(\boldsymbol{x}) = 0$ for $|\boldsymbol{x}| < R_0$. Let $\mathcal{B} \subset \mathbb{R}^3$ be a set that is star-shaped with respect to $\mathbf{0}$ (e.g., convex with $\mathbf{0} \in \mathcal{B}$). Then for all functions $\psi \in H^1(\mathcal{B})$*

$$\int_{\mathcal{B}} \left[|\boldsymbol{\nabla}\psi(\boldsymbol{x})|^2 + \tfrac{1}{2} v(\boldsymbol{x}) |\psi(\boldsymbol{x})|^2\right] \mathrm{d}\boldsymbol{x} \geq a \int_{\mathcal{B}} U(\boldsymbol{x}) |\psi(\boldsymbol{x})|^2 \mathrm{d}\boldsymbol{x} . \tag{52}$$

Proof. Actually, we will show that (52) holds with $|\nabla\psi(\boldsymbol{x})|^2$ replaced by the (smaller) radial kinetic energy, $|\partial\psi(\boldsymbol{x})/\partial r|^2$, with $r = |\boldsymbol{x}|$. Hence it suffices to prove the analog of (52) for the integral along each radial line with fixed angular variables. Along such a line we write $\psi(\boldsymbol{x}) = u(r)/r$ with $u(0) = 0$. We consider first the special case when U is a delta-function at some radius $R \geq R_0$, i.e.,

$$U(\boldsymbol{x}) = \frac{1}{R^2}\delta(|\boldsymbol{x}| - R) . \tag{53}$$

For such U the analog of (52) along the radial line is

$$\int_0^{R_1} \left[(u'(r) - u(r)/r)^2 + \tfrac{1}{2}v(r)|u(r)|^2\right]dr \geq \begin{cases} 0 & \text{if } R_1 < R \\ a|u(R)|^2/R^2 & \text{if } R_1 \geq R \ , \end{cases} \tag{54}$$

where R_1 is the length of the radial line segment in $\mathcal{B}$. The case $R_1 < R$ is trivial, because $v \geq 0$ by assumption. If $R \leq R_1$ we consider the integral on the the left side of (54) from 0 to R instead of R_1 and minimize it under the boundary condition that $u(0) = 0$ and $u(R) =$ fixed constant. Since everything is homogeneous in u we may normalize this value to $u(R) = R - a$. This minimization problem leads to the zero energy scattering (2). Because $v(r) = 0$ for $r > R_0$ the solution, u_0, satisfies $u_0(r) = r - a$ for $r > R_0$. By partial integration,

$$\begin{aligned} &\int_0^{R} \left[(u_0'(r) - u_0(r)/r)^2 + \tfrac{1}{2}v(r)|u_0(r)|^2\right]dr \\ &= a|R - a|/R \geq a|R - a|^2/R^2 \ . \end{aligned} \tag{55}$$

But $|R - a|^2/R^2$ is precisely the right side of (54) if u satisfies the normalization condition. This proves (54).

The derivation of (52) for the special case (53) implies the general case, because every U can be written as a superposition of δ-functions, and $\int U(\boldsymbol{x})\mathrm{d}\boldsymbol{x} \leq 4\pi$ by assumption. □

Now consider, for fixed $\boldsymbol{X}_j$ in the square bracket in (38), the Voronoi cells Ω_i around $\boldsymbol{x}_i$, $i \neq j$. I.e.,

$$\Omega_i = \{\boldsymbol{x} \in \mathbb{R}^3 \, : \, |\boldsymbol{x} - \boldsymbol{x}_i| \leq |\boldsymbol{x} - \boldsymbol{x}_k| \text{ for all } k \neq j\} \ . \tag{56}$$

In each cell Ω_i, intersected with a ball of radius R centered at $\boldsymbol{x}_i$, we will use Lemma 2 for the $\boldsymbol{x}_j$ integration. Assuming that U has a range $\leq R$, this gives the lower bound

$$\begin{aligned} E_\Psi^{(2)} \geq \sum_{j=1}^{N} \int_{\mathbb{R}^{3(N-1)}} \mathrm{d}\boldsymbol{X}_j \Bigg[\frac{\varepsilon}{2} \int_{\mathbb{R}^3} \mathrm{d}\boldsymbol{x}_j \, |\nabla_j \Psi|^2 \\ + \int_{\mathbb{R}^3} \mathrm{d}\boldsymbol{x}_j \left[a(1-\varepsilon)U(\boldsymbol{x}_j - \boldsymbol{x}_{k(j)}) - 2g\phi^{\mathrm{GP}}(\boldsymbol{x}_j)^2\right] |\Psi|^2 \Bigg] \ . \end{aligned} \tag{57}$$

Here $\boldsymbol{x}_{k(j)}$ denotes the nearest neighbor of $\boldsymbol{x}_j$ among the points $\boldsymbol{x}_k$, $k \neq j$, and we neglected the interaction between non-nearest neighbors, which is alright for a lower bound.

Hence we converted the "hard" potential v into a "soft" potential U, at the expense of the local kinetic energy inside $\mathcal{O}_j$ and the neglect of other than nearest neighbor interaction. Note that the right side of (57), with Ψ of the form $\prod_j \phi^{\mathrm{GP}}(\boldsymbol{x}_j)$, would now give the desired answer for the energy. Hence we can hope that some kind of perturbation theory might be applicable now. One way to make perturbation theory precise is Temple's inequality [25] (see also [21, Thm. XIII.5]). However, this is only useful if the particle number is not too large. To ensure this condition, we have to break up space into small boxes. This has also the advantage that ϕ^{GP} in the last term in (57) can be approximated by a constant in each box, so we have effectively a homogeneous system.

We proceed by applying the box method, as in [12, 17]. More precisely, we divide $\mathbb{R}^3$ into boxes of side length ℓ, labeled by α, and distribute our N particles over these boxes. Taking Neumann boundary conditions in each box and minimizing the energy with respect to all distributions of the particles, this can only lower the energy.

Let

$$\rho_\alpha = \sup_{\boldsymbol{x}\in\alpha} \phi^{\mathrm{GP}}(\boldsymbol{x})^2 \,, \tag{58}$$

and let $E_\varepsilon(n, \ell)$ denote the ground state energy of

$$\sum_{i=1}^{n} \left(-\frac{\varepsilon}{2}\Delta_i + a(1-\varepsilon)U(\boldsymbol{x}_i - \boldsymbol{x}_{k(i)}) \right) \tag{59}$$

with Neumann boundary conditions on $L^2([0,\ell]^{3n})$, i.e., for all particles confined to a box of side length ℓ. Neglecting the (positive) interaction among particles in different boxes, we obtain

$$E_\Psi^{(2)} \geq \inf_{\{n_\alpha\}} \sum_\alpha \left(E_\varepsilon(n_\alpha, \ell) - 2g\rho_\alpha n_\alpha \right) \,, \tag{60}$$

where the infimum is taken over all distribution of the N particles into the boxes α. Each box contains n_α particles, and $\sum_\alpha n_\alpha = N$. We now need a lower bound on $E_\varepsilon(n, \ell)$. We follow closely [17], and choose U to be

$$U(\boldsymbol{x}) = \begin{cases} 3(R^3 - R_0^3)^{-1} & \text{for } R_0 < |\boldsymbol{x}| < R \,, \\ 0 & \text{otherwise} \,. \end{cases} \tag{61}$$

The result is:

Lemma 3. *If* $\ell > 2R$ *and* $\varepsilon\pi^2 > 8\pi a n(n-1)/\ell$, *then*

$$E_\varepsilon(n,\ell) \geq (1-\varepsilon)4\pi a \frac{n(n-1)}{\ell^3}\left(1-\frac{2R}{\ell}\right)^3\left(1-(n-2)\frac{4\pi}{3}\frac{(2R)^3}{\ell^3}\right) \cdot \left(1-\frac{6na\ell^2}{(R^3-R_0)^3\left(\varepsilon\pi^2-8\pi a n(n-1)/\ell\right)}\right) . \quad (62)$$

Proof. We start by recalling *Temple's inequality* [21,25] for the expectation value of an operator $H = H_0 + W$ in the ground state $\langle\cdot\rangle_0$ of H_0. It is a simple consequence of the operator inequality

$$(H-E_0)(H-E_1) \geq 0 \quad (63)$$

for the two lowest eigenvalues, $E_0 < E_1$, of H and reads

$$E_0 \geq \langle H\rangle_0 - \frac{\langle W^2\rangle_0 - \langle W\rangle_0^2}{E_1 - \langle H\rangle_0} \quad (64)$$

provided $E_1 - \langle H\rangle_0 > 0$. Furthermore, if $W \geq 0$ we may replace E_1 in (64) by the second lowest eigenvalue of H_0. We can also neglect the second term in the numerator, $\langle W\rangle_0^2$, for a lower bound.

We now apply this to the operator in (59), with $H_0 = -\frac{1}{2}\varepsilon\sum_i \Delta_i$. The ground state energy of H_0 is zero because of the Neumann boundary conditions, and the second lowest eigenvalue is $\frac{1}{2}\varepsilon(\pi/\ell)^2$. We will use the bound

$$\langle W^2\rangle_0 \leq \frac{3na}{R^3-R_0^3}\langle W\rangle_0 \quad (65)$$

which follows from $U^2 = 3(R^3-R_0^3)^{-1}U$ together with the Cauchy-Schwarz inequality. This gives

$$E_\varepsilon(n,\ell) \geq \langle W\rangle_0\left(1-\frac{3na}{R^3-R_0^3}\,\frac{2}{\varepsilon\pi^2/\ell^2-2\langle W\rangle_0}\right) . \quad (66)$$

To proceed we need upper and lower bounds on

$$\langle W\rangle_0 = \frac{(1-\varepsilon)a}{\ell^{3n}}\sum_{j=1}^n \int_{[0,\ell]^{3n}} U(\boldsymbol{x}_j - \boldsymbol{x}_{k(j)})\mathrm{d}\boldsymbol{x}_1\cdots\mathrm{d}\boldsymbol{x}_n . \quad (67)$$

Since $\int U(\boldsymbol{x})\mathrm{d}\boldsymbol{x} = 4\pi$, an upper bound is obviously

$$\langle W\rangle_0 \leq \frac{4\pi a}{\ell^3}n(n-1) . \quad (68)$$

For the lower bound, we have to take into account the boundary of the box and the fact that the interaction is only among nearest neighbors. We have

$$
\begin{aligned}
&\langle W\rangle_0 \left[(1-\varepsilon)\frac{a}{\ell^{3n}}\right]^{-1} \\
&\geq \sum_{i\neq j}\int_{[0,\ell]^{3n}} U(\boldsymbol{x}_j-\boldsymbol{x}_i) \prod_{k,\,k\neq(i,j)} \theta(|\boldsymbol{x}_k-\boldsymbol{x}_j|-2R)\mathrm{d}\boldsymbol{x}_1\cdots\mathrm{d}\boldsymbol{x}_n \\
&\geq \sum_{i\neq j}\int_{[0,\ell]^{3n}} U(\boldsymbol{x}_j-\boldsymbol{x}_i)\left(1-\sum_{k,\,k\neq(i,j)}\theta(2R-|\boldsymbol{x}_k-\boldsymbol{x}_j|)\right)\mathrm{d}\boldsymbol{x}_1\cdots\mathrm{d}\boldsymbol{x}_n \,.
\end{aligned}
\tag{69}
$$

Considering only the integration over $\boldsymbol{x}_j \in [R, \ell - R]^3$, this gives as a lower bound

$$
\langle W\rangle_0 \geq (1-\varepsilon)\frac{4\pi a}{\ell^3}n(n-1)\left(1-\frac{2R}{\ell}\right)^3\left(1-(n-2)\frac{4\pi}{3}\frac{(2R)^3}{\ell^3}\right) \,. \tag{70}
$$

Inserting (68) and (70) into (66) proves (62). □

Now fix some $\eta > 0$, and suppose that $n \leq \eta N\ell^3$. We are going to choose $R = N^{-\delta}$ and $\ell = N^{-\gamma}$, with $1/5 < 3\delta/5 < \gamma < 1/3$. Since $a \sim N^{-1}$ this ensures that a/R and R/ℓ are $o(1)$ as $N \to \infty$, and also that all the other error terms in (62) are small, provided n satisfies the bound mentioned above. Namely,

- $n(R/\ell)^3 \leq \eta NR^3 = \eta N^{1-3\delta}$
- $na\ell^2/R^3 \leq \eta Na\ell^5/R^3 = \eta NaN^{5\delta-3\gamma}$
- $n^2a/\ell \leq \eta^2 NaN\ell^5 = \eta^2 NaN^{1-5\gamma}$

Note that $\delta > 1/3$ means that $NR^3 = o(1)$, which is exactly the condition we needed in the previous subsection.

Let $\overline{n}_\alpha$ be a minimizing configuration of the n_α's in (60). Let Λ_η denote the collection of those boxes α where $\overline{n}_\alpha \leq \eta N\ell^3$. We choose $\varepsilon = o(1)$, but not too small, namely $\varepsilon > (8/\pi)\eta^2 NaN^{1-5\gamma}$ to ensure positivity in the denominator in (62). Then we can use (62) to estimate, for $\alpha \in \Lambda_\eta$,

$$
\begin{aligned}
E_\varepsilon(\overline{n}_\alpha,\ell) - 2g\rho_\alpha\overline{n}_\alpha &\geq 4\pi a\frac{\overline{n}_\alpha(\overline{n}_\alpha-1)}{\ell^3}(1-o(1)) - 2g\rho_\alpha\overline{n}_\alpha \\
&\geq -Ng\rho_\alpha^2\ell^3\left(1+\frac{1}{N\ell^3\rho_\alpha}\right)^2(1+o(1)) \,,
\end{aligned}
\tag{71}
$$

where we also used that $g - 4\pi Na = o(1)$ by assumption. Note that $(N\ell^3)^{-1} = o(1)$ by our choice of ℓ.

For $\alpha \notin \Lambda_\eta$, we use *superadditivity* of $E_\varepsilon(n,\ell)$ in n, i.e., that $E_\varepsilon(n+m,\ell) \geq E_\varepsilon(n,\ell) + E_\varepsilon(m,\ell)$, which follows from the positivity of the interaction potential U. It implies that

$$
E_\varepsilon(n,\ell) \geq \left\lfloor\frac{n}{\eta N\ell^3}\right\rfloor E_\varepsilon(\eta N\ell^3,\ell) \geq \frac{n}{2\eta N\ell^3}E_\varepsilon(\eta N\ell^3,\ell) \,, \tag{72}
$$

where $[\cdot]$ denotes the integer part. We can now use again (62). Since the effective particle number is now $\eta N\ell^3$, we can use the same arguments as above, together with $(N\ell^3)^{-1} = o(1)$, to obtain

$$(72) \geq \tfrac{1}{2} n g \eta (1 - o(1)) . \tag{73}$$

This implies that if we choose η such that

$$\tfrac{1}{2} g\eta > 2g \sup_{\alpha} \rho_\alpha , \tag{74}$$

or equivalently

$$\eta > 4\|\phi^{\mathrm{GP}}\|_\infty^2 , \tag{75}$$

then the contribution to (60) from boxes with $n \geq \eta N\ell^3$ is positive (for N large enough) and can be neglected for a lower bound. We might even estimate it from below by (71), which is negative. We thus obtain

$$E_\Psi^{(2)} \geq \sum_\alpha -Ng\rho_\alpha^2 \ell^3 \left(1 + \frac{1}{N\ell^3 \rho_\alpha}\right)^2 (1 + o(1)) , \tag{76}$$

where the sum is now over *all* boxes. Now $N\ell^3 \to \infty$ as $N \to \infty$, and

$$\sum_\alpha \rho_\alpha^2 \ell^3 = \int |\phi^{\mathrm{GP}}(\boldsymbol{x})|^4 \mathrm{d}\boldsymbol{x} (1 + o(1)) \tag{77}$$

because the sum is a Riemann sum for the corresponding integral.

The small problem with small ρ_α in the denominator in (76) can be avoided by applying the above method only to boxes with $\rho_\alpha > \hat{\varepsilon}$ for some $\hat{\varepsilon} > 0$, and estimating the contribution from the other boxes to (60) simply by $-2Ng\hat{\varepsilon}$. With $\hat{\varepsilon} = o(1)$ appropriately chosen, this shows that

$$E_\Psi^{(2)} \geq -Ng \int |\phi^{\mathrm{GP}}(\boldsymbol{x})|^4 \mathrm{d}\boldsymbol{x}\, (1 + o(1)) . \tag{78}$$

4.5 Upper Bound to the Energy

It remains to derive an upper bound to the ground state energy of (1). We use the variational principle. As N-particle trial function we choose

$$\Psi(\boldsymbol{x}_1, \ldots, \boldsymbol{x}_N) = F(\boldsymbol{x}_1, \ldots, \boldsymbol{x}_N) \prod_{i=1}^N \phi^{\mathrm{GP}}(\boldsymbol{x}_i) , \tag{79}$$

where ϕ^{GP} is the minimizer of (10), for some fixed $g \geq 0$ such that $4\pi Na \to g$ as $N \to \infty$. Here F is the Dyson wave function defined in [12]. It is given by

$$F(\boldsymbol{x}_1, \ldots, \boldsymbol{x}_N) = \prod_{i=1}^N F_i(\boldsymbol{x}_1, \ldots, \boldsymbol{x}_i) , \tag{80}$$

with

$$F_i(\boldsymbol{x}_1, \ldots, \boldsymbol{x}_i) = f(t_i(\boldsymbol{x}_1, \ldots, \boldsymbol{x}_i)) . \tag{81}$$

Here $t_i = \min\{|\boldsymbol{x}_i - \boldsymbol{x}_j|, 1 \leq j \leq i-1\}$ is the distance of $\boldsymbol{x}_i$ to its nearest neighbor among the points $\boldsymbol{x}_1, \ldots, \boldsymbol{x}_{i-1}$, and f is a function of $t \geq 0$. It is chosen to be

$$f(t) = \begin{cases} f_0(t)/f_0(b) & \text{for} \quad t < b \\ 1 & \text{for} \quad t \geq b , \end{cases} \tag{82}$$

where f_0 is the solution of the zero energy scattering equation (2) for the interaction potential v, and b is some cut-off parameter of the order of the mean particle distance, $b \sim N^{-1/3}$. Note that f is a monotone increasing function [18]. The function F is a suitable generalization of the function Dyson used in [3] to obtain an upper bound on the ground state energy of a homogeneous Bose gas of hard spheres. The calculation of the expectation value of the Hamiltonian with the trial function (79), or rather of a good upper bound on it, was given in [12] and is presented here.

Note that our trial function (79) is not symmetric in the particle coordinates. The expectation value $\langle \Psi | H_{N,a} | \Psi \rangle / \langle \Psi | \Psi \rangle$ is still an upper bound to the bosonic ground state energy, however, since the Hamiltonian is symmetric and has a unique, positive, ground state wave function. This implies that the bosonic ground state energy is equal to the *absolute* ground state energy [3,7], i.e., the ground state energy without symmetry restrictions.

We start by computing the kinetic energy of our trial state. Using partial integration, one easily sees that

$$\int_{\mathbb{R}^{3N}} \Psi \Delta_k \Psi = \int_{\mathbb{R}^{3N}} |\Psi|^2 \phi^{\mathrm{GP}}(\boldsymbol{x}_k)^{-1} \Delta_k \phi^{\mathrm{GP}}(\boldsymbol{x}_k) - \int_{\mathbb{R}^{3N}} |\Psi|^2 F^{-2} |\boldsymbol{\nabla}_k F|^2 . \tag{83}$$

Now define $\varepsilon_{ik}(\boldsymbol{x}_1, \ldots, \boldsymbol{x}_N)$ by

$$\varepsilon_{ik} = \begin{cases} 1 \ \text{for } i = k , \\ -1 \ \text{for } t_i = |\boldsymbol{x}_i - \boldsymbol{x}_k| , \\ 0 \ \text{otherwise} . \end{cases} \tag{84}$$

Let $\boldsymbol{n}_i$ be the unit vector in the direction of $\boldsymbol{x}_i - \boldsymbol{x}_{j(i)}$, where $\boldsymbol{x}_{j(i)}$ denotes the nearest neighbor of $\boldsymbol{x}_i$ among the points $(\boldsymbol{x}_1, \ldots, \boldsymbol{x}_{i-1})$. Then

$$F^{-1} \boldsymbol{\nabla}_k F = \sum_i F_i^{-1} \varepsilon_{ik} \boldsymbol{n}_i f'(t_i) , \tag{85}$$

and after summation over k we obtain

$$\begin{aligned} F^{-2} \sum_k |\boldsymbol{\nabla}_k F|^2 &= \sum_{i,j,k} \varepsilon_{ik} \varepsilon_{jk} (\boldsymbol{n}_i \cdot \boldsymbol{n}_j) F_i^{-1} F_j^{-1} f'(t_i) f'(t_j) \\ &\leq 2 \sum_i F_i^{-2} f'(t_i)^2 + 2 \sum_{k \leq i < j} |\varepsilon_{ik} \varepsilon_{jk}| F_i^{-1} F_j^{-1} f'(t_i) f'(t_j) . \end{aligned} \tag{86}$$

The expectation value of the Hamiltonian can thus be bounded as follows:

$$\frac{\langle\Psi|H_{N,a}|\Psi\rangle}{\langle\Psi|\Psi\rangle} \leq 2\sum_{i=1}^{N} \frac{\int |\Psi|^2 F_i^{-2} f'(t_i)^2}{\int |\Psi|^2} + \sum_{j<i} \frac{\int |\Psi|^2 v(\boldsymbol{x}_i - \boldsymbol{x}_j)}{\int |\Psi|^2}$$
$$+2\sum_{k\leq i<j} \frac{\int |\Psi|^2 |\varepsilon_{ik}\varepsilon_{jk}| F_i^{-1} F_j^{-1} f'(t_i) f'(t_j)}{\int |\Psi|^2}$$
$$+\sum_{i=1}^{N} \frac{\int |\Psi|^2 \left(-\phi^{\rm GP}(\boldsymbol{x}_i)^{-1}\Delta_i \phi^{\rm GP}(\boldsymbol{x}_i) + V(\boldsymbol{x}_i)\right)}{\int |\Psi|^2} . \tag{87}$$

For $i < p$, let $F_{p,i}$ be the value that F_p would take if the point $\boldsymbol{x}_i$ were omitted from consideration as a possible nearest neighbor. Note that $F_{p,i}$ is independent of $\boldsymbol{x}_i$. Analogously we define $F_{p,ij}$ by omitting $\boldsymbol{x}_i$ and $\boldsymbol{x}_j$. The functions F_i occur both in the numerator and the denominator so we need estimates from below and above. Since f is monotone increasing,

$$F_p = \min\{F_{p,ij}, f(|\boldsymbol{x}_p - \boldsymbol{x}_j|), f(|\boldsymbol{x}_p - \boldsymbol{x}_i|)\} , \tag{88}$$

and we have, using $0 \leq f \leq 1$,

$$F_{p,ij}^2 f(|\boldsymbol{x}_p - \boldsymbol{x}_i|)^2 f(|\boldsymbol{x}_p - \boldsymbol{x}_j|)^2 \leq F_p^2 \leq F_{p,ij}^2 . \tag{89}$$

Hence, for $j < i$, we have the upper bound

$$F_{j+1}^2 \cdots F_{i-1}^2 F_{i+1}^2 \cdots F_N^2 \leq F_{j+1,j}^2 \cdots F_{i-1,j}^2 F_{i+1,ij}^2 \cdots F_{N,ij}^2 , \tag{90}$$

and the lower bound

$$F_j^2 \cdots F_N^2 \geq F_{j+1,j}^2 \cdots F_{i-1,j}^2 F_{i+1,ij}^2 \cdots F_{N,ij}^2 \tag{91}$$
$$\left(1 - \sum_{k=1,\, k\neq i,j}^{N} (1 - f(|\boldsymbol{x}_j - \boldsymbol{x}_k|)^2)\right)\left(1 - \sum_{k=1,\, k\neq i}^{N} (1 - f(|\boldsymbol{x}_i - \boldsymbol{x}_k|)^2)\right) .$$

We now consider the first two terms on the right side of (87). In the numerator of the first term we use, for each fixed i, the estimate

$$f'(t_i)^2 \leq \sum_{j=1}^{i-1} f'(\boldsymbol{x}_i - \boldsymbol{x}_j)^2 , \tag{92}$$

and in the second term we use $F_i \leq f(|\boldsymbol{x}_i - \boldsymbol{x}_j|)$. For fixed i and j we can eliminate $\boldsymbol{x}_i$ and $\boldsymbol{x}_j$ from the rest of the integrand by using the bound (90) in the numerator and (91) in the denominator to do the $\boldsymbol{x}_i$ and $\boldsymbol{x}_j$ integrations. Note that, using partial integration, (2) implies that

$$\int \left(f'(|\boldsymbol{x}|)^2 + \tfrac{1}{2}v(|\boldsymbol{x}|)f(|\boldsymbol{x}|)^2\right) \mathrm{d}\boldsymbol{x} = 4\pi a(1 - a/b) f_0(b)^{-2} \tag{93}$$

for $b \geq R_0$. Now, for N large, $b \gg a$ and in particular $b \geq R_0$, and hence $f_0(b) = 1 - a/b = 1 - O(N^{-2/3})$. Using the Cauchy-Schwarz inequality, we obtain

$$\int \left(2f'(\boldsymbol{x}_i - \boldsymbol{x}_j)^2 + v(\boldsymbol{x}_i - \boldsymbol{x}_j) f(\boldsymbol{x}_i - \boldsymbol{x}_j)^2\right) \phi^{\mathrm{GP}}(\boldsymbol{x}_i)^2 \phi^{\mathrm{GP}}(\boldsymbol{x}_j)^2 \mathrm{d}\boldsymbol{x}_i \mathrm{d}\boldsymbol{x}_j$$
$$\leq 8\pi a \int \phi^{\mathrm{GP}}(\boldsymbol{x})^4 \mathrm{d}\boldsymbol{x} \left(1 + O(N^{-2/3})\right) . \tag{94}$$

In the denominator, we estimate

$$\int \left(1 - \sum_{p=1,\, p\neq i}^{N} (1 - f(|\boldsymbol{x}_p - \boldsymbol{x}_i|)^2) \right) \phi^{\mathrm{GP}}(\boldsymbol{x}_i)^2 \mathrm{d}\boldsymbol{x}_i \geq 1 - N \|\phi^{\mathrm{GP}}\|_\infty^2 I \,, \tag{95}$$

where we set $I = \int (1 - f(|\boldsymbol{x}|)^2)\mathrm{d}\boldsymbol{x}$. Using that $f(|\boldsymbol{x}|) \geq [1 - a/|\boldsymbol{x}|]_+$ (see [18]), we get that $I \leq (4\pi/3)ab^2$. Moreover, $\|\phi^{\mathrm{GP}}\|_\infty$ is bounded uniformly in g, by Proposition 2. The same factor comes from the $\boldsymbol{x}_j$-integration. The remaining factors are identical in numerator and denominator and hence we conclude that the first and second term in (87) are above bounded by

$$\sum_{i=1}^{N} (i-1) 8\pi a \int \phi^{\mathrm{GP}}(\boldsymbol{x})^4 \mathrm{d}\boldsymbol{x} \left(1 + O(N^{-2/3})\right)$$
$$\leq 4\pi a N^2 \int \phi^{\mathrm{GP}}(\boldsymbol{x})^4 \mathrm{d}\boldsymbol{x} \left(1 + O(N^{-2/3})\right) . \tag{96}$$

A similar argument is now applied to the third term of (87). We omit the details. The result is an upper bound

$$\frac{2}{3} N^3 \frac{K^2 \|\phi^{\mathrm{GP}}\|_\infty^4}{(1 - N\|\phi^{\mathrm{GP}}\|_\infty^2 I)^2} \,, \tag{97}$$

with K given by $K = \int f(|\boldsymbol{x}|) f'(|\boldsymbol{x}|)\mathrm{d}\boldsymbol{x}$. Using again that $[1 - a/|\boldsymbol{x}|]_+ \leq f(|\boldsymbol{x}|) \leq 1$ as well as partial integration, we can estimate $K \leq 4\pi a b(1 + O(a/b))$, hence (97) is, for bounded Na, bounded by $Nb^2 \sim N^{1/3}$.

Finally, consider the last term on the right side of (87). The GP equation (12) implies that it is equal to

$$\sum_{i=1}^{N} \frac{\int |\Psi|^2 \left(\mu_g^{\mathrm{GP}} - 2g|\phi^{\mathrm{GP}}(\boldsymbol{x}_i)|^2\right)}{\int |\Psi|^2} \,. \tag{98}$$

Moreover, by Prop. 2, $\mu_g^{\mathrm{GP}} - 2g|\phi^{\mathrm{GP}}(\boldsymbol{x}_i)|^2$ is positive, and hence we can proceed as above. After eliminating $\boldsymbol{x}_i$ from the integrands in the numerator and the denominator and we get the upper bound

$$(98) \leq N \left(\mu_g^{\mathrm{GP}} - 2g \int \phi^{\mathrm{GP}}(\boldsymbol{x})^4 \mathrm{d}\boldsymbol{x}\right) \left(1 + O(N^{-2/3})\right) . \tag{99}$$

Collecting all the estimates, we thus obtain, using (13),

$$E^{\mathrm{QM}}(N,(4\pi N)^{-1}g) \leq N E^{\mathrm{GP}}(g)\left(1+O(N^{-2/3})\right) , \tag{100}$$

uniformly in g on compact intervals.

4.6 Putting Everything Together

We now show how the results of the previous subsections lead to a proof of Theorems 1 and 2. Inserting the lower bounds (51) and (78) to $E_{\Psi}^{(1)}$ and $E_{\Psi}^{(2)}$ in (36) we obtain, for some constant $C>0$ (depending only on g),

$$\frac{1}{N}\left\langle \Psi \left| H_{N,a} \right| \Psi \right\rangle \geq \left(E^{\mathrm{GP}}(g) + C \,\mathrm{Tr}\left[\tfrac{1}{N}\gamma_N^{(1)}\left(1-|\phi^{\mathrm{GP}}\rangle\langle\phi^{\mathrm{GP}}|\right)\right]\right)(1-o(1)) \tag{101}$$

as $N\to\infty$, if $4\pi Na\to g$. Here we have also used (13). Together with the upper bound (100) this implies that

$$\lim_{N\to\infty}\frac{1}{N}E^{\mathrm{QM}}(N,(4\pi N)^{-1}g) = E^{\mathrm{GP}}(g) , \tag{102}$$

uniformly in g on compact intervals, and also that

$$\lim_{N\to\infty}\frac{1}{N}\langle\phi^{\mathrm{GP}}|\gamma_N^{(1)}|\phi^{\mathrm{GP}}\rangle = 1 \tag{103}$$

for $\gamma_N^{(1)}$ the reduced one-particle density matrix of an approximate ground state in the sense described in the beginning of Subsect. 3.1.

Equation (102) proves Theorem 1. Moreover, since $\frac{1}{N}\gamma_N^{(1)}$ is a positive trace class operator with trace 1, general arguments show that (103) implies (24) (see, e.g., [24, Thm. 2.20]). We show this directly, using the following simple lemma.

Lemma 4. *Let $0\leq\gamma\leq 1$, with $\mathrm{Tr}[\gamma]=1$, and let $P=|w\rangle\langle w|$ be a rank-one projection. Then*

$$\|\gamma-P\|_1 \leq \left(1+\sqrt{2}\right)\left(1-\langle w|\gamma|w\rangle\right)^{1/2} . \tag{104}$$

Proof. Let $Q=1-P$. Using the triangle inequality and the Cauchy-Schwarz inequality,

$$\begin{aligned}
\|\gamma-P\|_1 &\leq \|(\gamma-P)P\|_1 + \|(\gamma-P)Q\|_1 = \|(\gamma-P)P\|_1 + \|\gamma Q\|_1 \\
&\leq \|\gamma-P\|_2\|P\|_2 + \|\gamma^{1/2}\|_2\|\gamma^{1/2}Q\|_2 = \|\gamma-P\|_2 + \|Q\gamma Q\|_1^{1/2} \\
&= \left(1+\mathrm{Tr}[\gamma^2]-2\mathrm{Tr}[\gamma P]\right)^{1/2} + \left(1-\mathrm{Tr}[\gamma P]\right)^{1/2} \\
&\leq \left(\sqrt{2}+1\right)\left(1-\mathrm{Tr}[\gamma P]\right)^{1/2} .
\end{aligned} \tag{105}$$

□

Since, by (103), $1 - \langle\phi^{\rm GP}|\frac{1}{N}\gamma_N^{(1)}|\phi^{\rm GP}\rangle \to 0$ as $N \to \infty$, this implies that $\frac{1}{N}\gamma_N^{(1)} \to |\phi^{\rm GP}\rangle\langle\phi^{\rm GP}|$ in trace class norm, as claimed in Theorem 2.

Remark. The proof of complete BEC can be extended to the TF case, e.g., the case when $g = 4\pi Na \to \infty$ as $N \to \infty$, as long as g grows slowly enough with N. More precisely, the proof above shows that $\mathrm{Tr}\left[\gamma_N^{(1)}\left(1 - |\phi^{\rm GP}\rangle\langle\phi^{\rm GP}|\right)\right]$ can be bounded from above by the difference in the upper and lower bound to the ground state energy of $H_{N,a}$ (which was shown to be $o(N)$ for fixed g) divided by the gap above the ground state energy of the operator in (17), which also depends on g. Hence, if g grows slowly enough to ensure that this quantity is still $o(N)$, this proves complete BEC.

5 Extensions

Using (and extending) the methods presented in this paper, a number of additional and more complicated problems have been addressed successfully. We describe them here briefly.

5.1 The Bose Gas in Two Dimensions

The same questions as above can be asked for bosons confined to a two-dimensional plane instead of moving in three-dimensional space. The problem is a little bit more complicated than in three dimensions, mainly because the formula for the ground state energy per particle of a dilute gas, $4\pi a\rho$ in three dimensions (see (4)), is more complicated in two dimensions. It is given by

$$\frac{4\pi\rho}{|\ln(a^2\rho)|}, \tag{106}$$

where ρ and a are now the two-dimensional density and scattering length, respectively. In particular, the ground state energy is much bigger that just the energy of a pair of particles times the number of pairs, so the same reasoning as on page 251 does not lead to the right formula. Nevertheless, a proof of (106) can be found in [18].

The extension to inhomogeneous systems was done in [13]. The corresponding Gross-Pitaevskii theory in two dimensions is essentially the same as in three-dimensions, the only difference is that the coupling parameter g has to be chosen differently and, in particular, depending non-trivially on the density.

5.2 The Rotating Gas and Superfluidity

Particles in a rotating trap can be described by a Hamiltonian similar to (1). If the angular velocity vector is given by some $\boldsymbol{\Omega} \in \mathbb{R}^3$, the one-particle part of the Hamiltonian, $-\Delta + V(\boldsymbol{x})$, has to be replaced by

$$-\Delta - \boldsymbol{\Omega} \cdot \boldsymbol{L} + V(\boldsymbol{x}) , \tag{107}$$

where $\boldsymbol{L} = -\mathrm{i}\boldsymbol{x} \wedge \boldsymbol{\nabla}$ denotes the angular momentum operator. Here we assume that V is axially symmetric, i.e., it commutes with $\boldsymbol{\Omega} \cdot \boldsymbol{L}$. In the same way, one can define a modified GP functional for rotating systems. One big difference to the case of non-rotating systems can already by observed in the GP functional: for large enough g there will *not* be a unique minimizer anymore [22]. This is true for any $\boldsymbol{\Omega} \neq \boldsymbol{0}$. The reason for this non-uniqueness is the appearance of quantized vortices which, when more than one, necessarily break the axial symmetry present in the system.

It is conjectured that the GP functional also correctly describes the ground state properties of dilute, rotating Bose gases. This, however, has so far been proved only in the case of small enough $|\boldsymbol{\Omega}|$ (for fixed g), where there is no symmetry breaking and no vortices are present. This result has a natural interpretation as superfluidity of the system [14]. For larger rotation speeds, the problem is still open.

One new feature that arises in the study of rotating systems is the fact that it becomes essential to restrict to the *symmetric subspace* of the total Hilbert space, e.g., to consider bosons. This did not really matter for non-rotating systems, because in this case the absolute ground state (i.e., without symmetry restrictions) agrees with the bosonic one. This need not be the case for rotating systems and, in fact, it is not for a certain range of the parameters N, $\boldsymbol{\Omega}$ and a where the above mentioned symmetry breaking occurs. In [23] it is shown that the absolute ground state can differ significantly from the bosonic one. It is also shown that the absolute ground state of dilute, rotating systems *cannot* in general be described by the GP functional, but rather by a modified GP *density matrix* functional, which depends on one-particle density matrices rather than single wave functions.

5.3 One-Dimensional Behavior of Bose Gases in Elongated Traps

Recently it has become experimentally feasible to create trap potentials that are highly elongated in one direction. In this way, it has become possible to observe properties of gases that are typical for one-dimensional systems. In order to describe this situation, we can think of our trap potential as consisting of two parts,

$$V(\boldsymbol{x}) = \frac{1}{r^2} V^{\perp}(\boldsymbol{x}^{\perp}/r) + \frac{1}{L^2} V^{\parallel}(z/L) \tag{108}$$

where $\boldsymbol{x} = (\boldsymbol{x}^{\perp}, z)$, and r and L denote the length scales of the confinement in perpendicular and longitudinal direction. It is then possible to study the limit $r \ll L$. In [16] it is shown that under certain conditions on the parameters, this system can be well described by a one-dimensional Bose gas with repulsive δ-function interaction. This one-dimensional system is an exactly solvable model which was first studied in [8]. E.g., the motion in the perpendicular directions is frozen because of the energy gap in the strong confinement, and the system behaves essentially one-dimensional. We refer to [16] for details.

References

1. E.A. Cornell and C.E. Wieman, *Bose-Einstein condensation in a dilute gas, the first 70 years and some recent experiments*, in: Les Prix Nobel 2001 (The Nobel Foundation, Stockholm, 2002), pp. 87–108. Reprinted in: Rev. Mod. Phys. **74**, 875–893 (2002); Chem. Phys. Chem. **3**, 476–493 (2002)
2. F. Dalfovo, S. Giorgini, L.P. Pitaevskii, and S. Stringari, *Theory of Bose-Einstein condensation in trapped gases*, Rev. Mod. Phys. **71**, 463–512 (1999)
3. F.J. Dyson, *Ground-State Energy of a Hard-Sphere Gas*, Phys. Rev. **106**, 20–26 (1957)
4. E.P. Gross, *Structure of a Quantized Vortex in Boson Systems*, Nuovo Cimento **20**, 454–466 (1961). *Hydrodynamics of a superfluid condensate*, J. Math. Phys. **4**, 195–207 (1963)
5. W. Ketterle, *When atoms behave as waves: Bose-Einstein condensation and the atom laser*, in: Les Prix Nobel 2001 (The Nobel Foundation, Stockholm, 2002), pp. 118–154. Reprinted in: Rev. Mod. Phys. **74**, 1131–1151 (2002); Chem. Phys. Chem. **3**, 736–753 (2002)
6. W. Lenz, *Die Wellenfunktion und Geschwindigkeitsverteilung des entarteten Gases*, Z. Phys. **56**, 778–789 (1929)
7. E.H. Lieb, *Simplified Approach to the Ground State Energy of an Imperfect Bose Gas*, Phys. Rev. **130**, 2518–2528 (1963)
8. E.H. Lieb and W. Liniger, *Exact Analysis of an Interacting Bose Gas. I. The General Solution and the Ground State*, Phys. Rev. **130**, 1605–1616 (1963). E.H. Lieb, *Exact Analysis of an Interacting Bose Gas. II. The Excitation Spectrum*, Phys. Rev. **130**, 1616–1624 (1963)
9. E.H. Lieb and M. Loss, *Analysis*, 2nd ed., Amer. Math. Soc., Providence (2001)
10. E.H. Lieb and R. Seiringer, *Proof of Bose-Einstein Condensation for Dilute Trapped Gases*, Phys. Rev. Lett. **88**, 170409-1–4 (2002)
11. E.H. Lieb, R. Seiringer, J.P. Solovej, and J. Yngvason, *The Ground State of the Bose Gas*, in: Current Developments in Mathematics, 2001, 131–178, International Press, Cambridge (2002). See also arXiv:math-ph/0405004
12. E.H. Lieb, R. Seiringer, and J. Yngvason, *Bosons in a Trap: A Rigorous Derivation of the Gross-Pitaevskii Energy Functional*, Phys. Rev. A **61**, 043602-1–13 (2000)
13. E.H. Lieb, R. Seiringer, and J. Yngvason, *A Rigorous Derivation of the Gross-Pitaevskii Energy Functional for a Two-Dimensional Bose Gas*, Commun. Math. Phys. **224**, 17–31 (2001)
14. E.H. Lieb, R. Seiringer, and J. Yngvason, *Superfluidity in Dilute Trapped Bose Gases*, Phys. Rev. B **66**, 134529-1–6 (2002)
15. E.H. Lieb, R. Seiringer, and J. Yngvason, *Poincaré Inequalities in Punctured Domains*, Ann. Math. **158**, 1067–1080 (2003)
16. E.H. Lieb, R. Seiringer, and J. Yngvason, *One-Dimensional Bosons in Three-Dimensional Traps*, Phys. Rev. Lett. **91**, 150401-1–4 (2003). *One-Dimensional*

Behavior of Dilute, Trapped Bose Gases, Commun. Math. Phys. **244**, 347–393 (2004)
17. E.H. Lieb and J. Yngvason, *Ground State Energy of the Low Density Bose Gas*, Phys. Rev. Lett. **80**, 2504–2507 (1998)
18. E.H. Lieb and J. Yngvason, *The Ground State Energy of a Dilute Two-Dimensional Bose Gas*, J. Stat. Phys. **103**, 509–526 (2001)
19. O. Penrose and L. Onsager, *Bose-Einstein Condensation and Liquid Helium*, Phys. Rev. **104**, 576–584 (1956)
20. L.P. Pitaevskii, *Vortex lines in an imperfect Bose gas*, Sov. Phys. JETP **13**, 451–454 (1961)
21. M. Reed and B. Simon, *Methods of Modern Mathematical Physics IV. Analysis of Operators*, Academic Press (1978)
22. R. Seiringer, *Gross-Pitaevskii Theory of the Rotating Bose Gas*, Commun. Math. Phys. **229**, 491–509 (2002)
23. R. Seiringer, *Ground state asymptotics of a dilute, rotating gas*, J. Phys. A: Math. Gen. **36**, 9755–9778 (2003)
24. B. Simon, *Trace ideals and their application*, Cambridge University Press (1979)
25. G. Temple, *The theory of Rayleigh's principle as applied to continuous systems*, Proc. Roy. Soc. London A **119**, 276–293 (1928)

Perturbation Theory for QED Calculations of High-Z Few-electron Atoms

V.M. Shabaev

Department of Physics, St. Petersburg State University, Oulianovskaya 1
Petrodvorets, 198504, St. Petersburg, Russia
shabaev@pcqnt1.phys.spbu.ru

1 Introduction

Experimental investigations of heavy few-electron ions (see, e.g., [1]) have triggered a great interest to accurate quantum electrodynamic (QED) calculations of these systems. At present, such calculations are feasible only by perturbation theory in two small parameters α and $1/Z$, where $\alpha \approx 1/137$ is the fine structure constant and Z is the nuclear charge number. For heavy few-electron ions, the parameters α and $1/Z$ characterize the QED and interelectronic-interaction corrections, respectively. To derive formal expressions for these corrections in a systematic way one needs to employ special methods. One of such methods was first developed by Gell-Mann and Low [2] and Sucher [3]. This method is based on introducing an adiabatically damped factor, $\exp(-\lambda|t|)$, in the interaction Hamiltonian and expressing the energy shift in terms of so-called adiabatic S_λ matrix elements. The Gell-Mann–Low–Sucher formula for the energy shift of a single level gained wide spreading in the literature related to high-Z few-electron atoms [4–9]. This is mainly due to a simple formulation of the method. However, practical calculations showed that the presence of the adiabatic factor strongly complicates derivations of formal expressions for the energy shift in second and higher orders of the perturbation theory. In addition, this method requires special investigation of the renormalization procedure since the adibatic S_λ-matrix suffers from ultraviolet divergences. This is due to the fact that the adiabatically damped factor, $\exp(-\lambda|t|)$, is non-covariant and, therefore, the ultraviolet divergences can not be removed from S_λ if $\lambda \neq 0$. For the case of a single level, this problem can be disregarded since, from the physical point of view, one may expect the divergenes to cancel each other in the expression for the energy shift. However, this is not the case if one considers degenerate levels. We can not expect that the standard renormalization procedure makes the secular operator finite in the ultraviolet limit [5, 6]. Similar difficulties occur in the evolution operator method developed in [10–14]. Some modifications of these methods gaining on their extention to quasidegenerate states were recently considered in [15, 16]. We note also that at present there is no formalism based on the Gell-Mann–Low–Sucher or the evolution operator method which would be suitable for calculation of the transition or scattering amplitudes.

In this paper we consider another method to construct the perturbation theory for high-Z few-electron systems. This method, which was developed in [17–21] and described in details in [22], provides a solution of all the problems appearing in the other formalisms indicated above. In particular, it is equally suitable for calculations energy levels of single, degenerate, and quasidegenerate states as well as for calculations of the transition and scattering amplitudes. It was successfully employed in many practical calculations (see, e.g., [22–30] and references therein). Since one of the key elements of the method consists in using two-time Green functions, in what follows we will call it the two-time Green function (TTGF) method.

It should be noted that there are also some other methods employing Green functions. In particular, in [6, 31, 32] two-time Green functions were used to construct quasipotential equations for high-Z few-electron systems. This corresponds to the perturbation theory in the Brillouin-Wigner form. In contrast to that, the method considered here yields the perturbation theory in the Rayleigh-Schrödinger form, which is much more convenient for calculations of high-Z few-electron atoms.

Below we formulate the basic principles of the TTGF method. The relativistic unit system ($\hbar = c = 1$) and the Heaviside charge unit ($\alpha = \frac{e^2}{4\pi}, e < 0$) are used in the paper.

2 Energy Levels

In high-Z few-electron atoms the number of electrons, denoted by N, is much smaller than the nuclear charge number Z. For this reason, the interaction of the electrons with each other and with the quantized electromagnetic field is much smaller (by factors $1/Z$ and α, respectively) than the interaction of the electrons with the Coulomb field of the nucleus. Therefore, we can assume that in zeroth approximation the electrons interact only with the Coulomb field of the nucleus and obey the Dirac equation

$$(-\mathrm{i}\boldsymbol{\alpha}\cdot\boldsymbol{\nabla}+\beta m+V_{\mathrm{C}}(\mathbf{x}))\psi_n(\mathbf{x})=\varepsilon_n\psi_n(\mathbf{x})\,. \tag{1}$$

The interaction of the electrons with each other and with the quantized electromagnetic field is accounted for by perturbation theory. In this way we obtain quantum electrodynamics in the Furry picture. We note that we could start also with the Dirac equation with an effective potential $V_{\mathrm{eff}}(\mathbf{x})$ (e.g., a local version of the Hartree-Fock potential) which approximately accounts for the interelectronic interaction. Then the interaction with the potential $\delta V(\mathbf{x}) = V_{\mathrm{C}}(\mathbf{x}) - V_{\mathrm{eff}}(\mathbf{x})$ must be accounted for by perturbation theory. However, for simplicity, in what follows we will consider that in zeroth approximation the electrons interact only with the Coulomb field of the nucleus.

We will consider the perturbation theory with the standard QED vacuum. The transition to the formalism in which closed shells are regarded as belonging to the vacuum can be performed in a usual manner (see, e.g., [22]).

Before to introduce the two-time Green function and formulate the perturbation theory, we consider standard equations for the $2N$-time Green function in quantum electrodynamics.

2.1 $2N$-Time Green Function

In principle, the complete information about the energy levels of an N-electron atom can be derived from the Green function defined as

$$G(x_1',\dots x_N';x_1,\dots x_N)=\langle 0|T\psi(x_1')\cdots\psi(x_N')\overline{\psi}(x_N)\cdots\overline{\psi}(x_1)|0\rangle\,, \tag{2}$$

where $\psi(x)$ is the electron-positron field operator in the Heisenberg representation, $\overline{\psi}(x)=\psi^\dagger\gamma^0$, and T is the time-ordered product operator. It can be shown (see, e.g., [33, 34]) that in the interaction representation the Green function is given by

$$\begin{aligned}
&G(x_1',\dots x_N';x_1,\dots x_N)\\
&=\frac{\langle 0|T\psi_{\mathrm{in}}(x_1')\cdots\psi_{\mathrm{in}}(x_N')\overline{\psi}_{\mathrm{in}}(x_N)\cdots\overline{\psi}_{\mathrm{in}}(x_1)\exp\{-\mathrm{i}\int \mathrm{d}^4z\,\mathcal{H}_I(z)\}|0\rangle}{\langle 0|T\exp\{-\mathrm{i}\int \mathrm{d}^4z\,\mathcal{H}_I(z)\}|0\rangle}\\
&=\Big\{\sum_{m=0}^{\infty}\frac{(-\mathrm{i})^m}{m!}\int \mathrm{d}^4y_1\cdots\mathrm{d}^4y_m\,\langle 0|T\psi_{\mathrm{in}}(x_1')\cdots\psi_{\mathrm{in}}(x_N')\\
&\times\overline{\psi}_{\mathrm{in}}(x_N)\cdots\overline{\psi}_{\mathrm{in}}(x_1)\mathcal{H}_I(y_1)\cdots\mathcal{H}_I(y_m)|0\rangle\Big\}\\
&\times\Big\{\sum_{l=0}^{\infty}\frac{(-\mathrm{i})^l}{l!}\int \mathrm{d}^4z_1\cdots\mathrm{d}^4z_l\,\langle 0|T\mathcal{H}_I(z_1)\cdots\mathcal{H}_I(z_l)|0\rangle\Big\}^{-1}
\end{aligned} \tag{3}$$

where

$$\mathcal{H}_I(x) = \frac{e}{2}\,[\overline{\psi}_{\rm in}(x)\gamma_\mu, \psi_{\rm in}(x)]A^\mu_{in}(x) - \frac{\delta m}{2}\,[\overline{\psi}_{\rm in}(x), \psi_{\rm in}(x)] \qquad (4)$$

is the interaction Hamiltonian. The commutators in (4) refer to operators only. The first term in (4) describes the interaction of the electron-positron field with the quantized electromagnetic field and the second one is the mass renormalization counterterm. It is assumed that the interaction of the electrons with the Coulomb field of the nucleus is included in the unperturbed Hamiltonian, i.e. the Furry picture is used.

The Green function G is constructed by perturbation theory according to (3) with the aid of the Wick theorem (see, e.g., [33]). The individual terms of the perturbation series are conveniently represented by so-called Feynman diagrams. Some of these diagrams contain ultraviolet divergences and, therefore, must be regularized. It can be shown that in calculation of any physical quantity the divergent parts either cancel each other or incorporate into the renormalized (physical) values of the electron charge and mass. Alternatively, from the very beginning one can formulate the theory in terms of the renormalized field operators, the renormalized electron charge, and the renormalized Green functions. It results in appearing additional counterterms in the Feynman rules. Both schemes can easily be adopted within the approach considered here.

The spectral representation for G shows that it contains the complete information about the energy levels of the atomic system. However, it is a hard task to extract this information directly from G because it depends on $2(N-1)$ relative times. It is much more convenient to employ the two-time Green function

$$\widetilde{G}(t', t) \equiv G(t'_1 = t'_2 = \cdots t'_N \equiv t'; t_1 = t_2 = \cdots t_N \equiv t)\,, \qquad (5)$$

which also contains the complete information about the energy levels.

2.2 Two-Time Green Function and Its Analytical Properties

Let us introduce the Fourier transform of the two-time Green function by

$$\begin{aligned} &\mathcal{G}(E; \mathbf{x}'_1, \ldots \mathbf{x}'_N; \mathbf{x}_1, \ldots \mathbf{x}_N)\delta(E - E') \\ &= \frac{1}{2\pi\mathrm{i}}\frac{1}{N!}\int_{-\infty}^{\infty} \mathrm{d}x^0 \mathrm{d}x'^0\, \exp\left(\mathrm{i}E'x'^0 - \mathrm{i}Ex^0\right) \\ &\quad\times \langle 0|T\psi(x'^0, \mathbf{x}'_1)\cdots\psi(x'^0, \mathbf{x}'_N)\overline{\psi}(x^0, \mathbf{x}_N)\cdots\overline{\psi}(x^0, \mathbf{x}_1)|0\rangle\,, \end{aligned} \qquad (6)$$

where, as in (2), the Heisenberg representation for the electron-positron field operators is used. Defined by (6) for real E, the Green function $\mathcal{G}$ can be continued analytically to the complex E plane. Analytical properties of this type of Green functions in the complex E plane were studied in various fields of physics (see, e.g., [35–38]). To consider these properties we derive the spectral representation for $\mathcal{G}$. Using the time-shift transformation rule for the Heisenberg operators

$$\psi(x^0, \mathbf{x}) = \exp(\mathrm{i}Hx^0)\psi(0, \mathbf{x}) \exp(-\mathrm{i}Hx^0) \tag{7}$$

and the equations

$$H|n\rangle = E_n|n\rangle , \qquad \sum_n |n\rangle\langle n| = I , \tag{8}$$

where H is the Hamiltonian of the system in the Heisenberg representation, we find

$$\begin{aligned}
&\mathcal{G}(E; \mathbf{x}'_1, \ldots, \mathbf{x}'_N; \mathbf{x}_1, \ldots, \mathbf{x}_N)\delta(E - E') \\
&= \frac{1}{2\pi\mathrm{i}}\frac{1}{N!}\int_{-\infty}^{\infty} \mathrm{d}x^0 \mathrm{d}x'^0 \ \exp(\mathrm{i}E'x'^0 - \mathrm{i}Ex^0) \\
&\times \Big\{ \theta(x'^0 - x^0) \sum_n \exp[\mathrm{i}(E_0 - E_n)(x'^0 - x^0)] \langle 0|\psi(0, \mathbf{x}'_1) \cdots \psi(0, \mathbf{x}'_N)|n\rangle \\
&\times \langle n|\overline{\psi}(0, \mathbf{x}_N) \cdots \overline{\psi}(0, \mathbf{x}_1)|0\rangle + (-1)^{N^2}\theta(x^0 - x'^0) \\
&\times \sum_n \exp[\mathrm{i}(E_0 - E_n)(x^0 - x'^0)] \langle 0|\overline{\psi}(0, \mathbf{x}_N) \cdots \overline{\psi}(0, \mathbf{x}_1)|n\rangle \\
&\times \langle n|\psi(0, \mathbf{x}'_1) \cdots \psi(0, \mathbf{x}'_N)|0\rangle \Big\} .
\end{aligned} \tag{9}$$

Assuming, for simplicity, $E_0 = 0$ (it corresponds to choosing the vacuum energy as the origin of reference) and taking into account that

$$\begin{aligned}
&\int_{-\infty}^{\infty} \mathrm{d}x^0 \mathrm{d}x'^0 \, \theta(x'^0 - x^0) \exp[-\mathrm{i}E_n(x'^0 - x^0)] \exp[\mathrm{i}(E'x'^0 - Ex^0)] \\
&\qquad\qquad = 2\pi\delta(E' - E)\frac{\mathrm{i}}{E - E_n + \mathrm{i}0} ,
\end{aligned} \tag{10}$$

$$\begin{aligned}
&\int_{-\infty}^{\infty} \mathrm{d}x^0 \mathrm{d}x'^0 \, \theta(x^0 - x'^0) \exp[-\mathrm{i}E_n(x^0 - x'^0)] \exp[\mathrm{i}(E'x'^0 - Ex^0)] \\
&\qquad\qquad = -2\pi\delta(E' - E)\frac{\mathrm{i}}{E + E_n - \mathrm{i}0} ,
\end{aligned} \tag{11}$$

we obtain

$$\mathcal{G}(E) = \sum_n \frac{\Phi_n \overline{\Phi}_n}{E - E_n + i0} - (-1)^N \sum_n \frac{\Xi_n \overline{\Xi}_n}{E + E_n - i0} , \tag{12}$$

where the variables $\mathbf{x}'_1, \ldots, \mathbf{x}'_N, \mathbf{x}_1, \ldots, \mathbf{x}_N$ are implicit and

$$\Phi_n(\mathbf{x}_1, \ldots \mathbf{x}_N) = \frac{1}{\sqrt{N!}}\langle 0|\psi(0, \mathbf{x}_1) \cdots \psi(0, \mathbf{x}_N)|n\rangle , \tag{13}$$

$$\Xi_n(\mathbf{x}_1, \ldots \mathbf{x}_N) = \frac{1}{\sqrt{N!}}\langle n|\psi(0, \mathbf{x}_1) \cdots \psi(0, \mathbf{x}_N)|0\rangle . \tag{14}$$

In (12) the summation runs over all bound and continuum states of the system of the interacting fields. Let us introduce the functions

$$A(E;\mathbf{x}'_1,\dots,\mathbf{x}'_N;\mathbf{x}_1,\dots,\mathbf{x}_N)=\sum_n \delta(E-E_n)\times\Phi_n(\mathbf{x}'_1,\dots,\mathbf{x}'_N)\overline{\Phi}_n(\mathbf{x}_1,\dots,\mathbf{x}_N)\,, \quad (15)$$

$$B(E;\mathbf{x}'_1,\dots,\mathbf{x}'_N;\mathbf{x}_1,\dots,\mathbf{x}_N)=\sum_n \delta(E-E_n)\times\Xi_n(\mathbf{x}'_1,\dots,\mathbf{x}'_N)\overline{\Xi}_n(\mathbf{x}_1,\dots,\mathbf{x}_N)\,. \quad (16)$$

These functions satisfy the conditions

$$\int_{-\infty}^{\infty} \mathrm{d}E\; A(E;\mathbf{x}'_1,\dots,\mathbf{x}'_N;\mathbf{x}_1,\dots,\mathbf{x}_N)=\frac{1}{N!}\langle 0|\psi(0,\mathbf{x}'_1)\cdots\psi(0,\mathbf{x}'_N)\times\overline{\psi}(0,\mathbf{x}_N)\cdots\overline{\psi}(0,\mathbf{x}_1)|0\rangle\,, \quad (17)$$

$$\int_{-\infty}^{\infty} \mathrm{d}E\; B(E;\mathbf{x}'_1,\dots,\mathbf{x}'_N;\mathbf{x}_1,\dots,\mathbf{x}_N)=\frac{1}{N!}\langle 0|\overline{\psi}(0,\mathbf{x}_N)\cdots\overline{\psi}(0,\mathbf{x}_1)\times\psi(0,\mathbf{x}'_1)\cdots\psi(0,\mathbf{x}'_N)|0\rangle\,. \quad (18)$$

In terms of these functions, (12) has the form

$$\mathcal{G}(E)=\int_0^{\infty}\mathrm{d}E'\,\frac{A(E')}{E-E'+\mathrm{i}0}-(-1)^N\int_0^{\infty}\mathrm{d}E'\,\frac{B(E')}{E+E'-\mathrm{i}0}\,, \quad (19)$$

where we have omitted the variables $\mathbf{x}_1,\dots,\mathbf{x}_N,\mathbf{x}'_1,\dots,\mathbf{x}'_N$ and have taken into account that $A(E')=B(E')=0$ for $E'<0$ since $E_n\geq 0$. In fact, due to charge conservation, only states with an electric charge of eN contribute to A in the sum over n in the right-hand side of (15) and only states with an electric charge of $-eN$ contribute to B in the sum over n in the right-hand side of (16). This can easily be shown by using the following commutation relations

$$[Q,\psi(x)]=-e\psi(x)\,,\qquad [Q,\overline{\psi}(x)]=e\overline{\psi}(x)\,, \quad (20)$$

where Q is the charge operator in the Heisenberg representation. Therefore, (19) can be written as

$$\mathcal{G}(E)=\int_{E^{(+)}_{\mathrm{min}}}^{\infty}\mathrm{d}E'\,\frac{A(E')}{E-E'+\mathrm{i}0}-(-1)^N\int_{E^{(-)}_{\mathrm{min}}}^{\infty}\mathrm{d}E'\,\frac{B(E')}{E+E'-\mathrm{i}0}\,, \quad (21)$$

where $E^{(+)}_{\mathrm{min}}$ is the minimal energy of states with electric charge eN and $E^{(-)}_{\mathrm{min}}$ is the minimal energy of states with electric charge $-eN$. So far we considered $\mathcal{G}(E)$ for real E. Equation (21) shows that the Green function $\mathcal{G}(E)$ is the sum of Cauchy-type integrals. Using the fact that the integrals $\int_{E^{(+)}_{\mathrm{min}}}^{\infty}\mathrm{d}E\; A(E)$

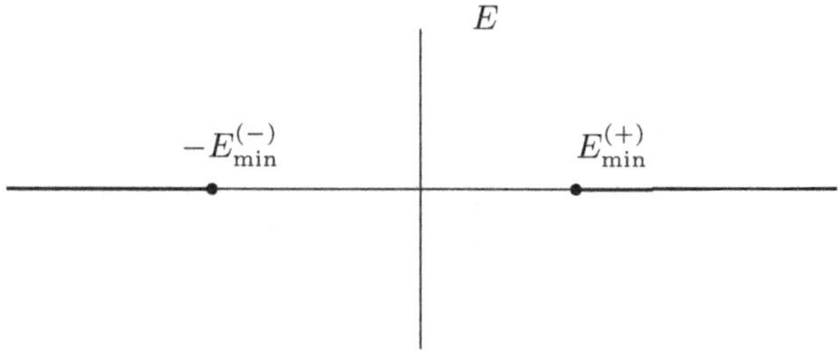

Fig. 1. Singularities of the two-time Green function in the complex E plane

and $\int_{E_{\rm min}^{(-)}}^{\infty} \mathrm{d}E\, B(E)$ converge (see (17), (18)), one can show with the help of standard mathematical methods that the equation

$$\mathcal{G}(E) = \int_{E_{\rm min}^{(+)}}^{\infty} \mathrm{d}E' \, \frac{A(E')}{E - E'} - (-1)^N \int_{E_{\rm min}^{(-)}}^{\infty} \mathrm{d}E' \, \frac{B(E')}{E + E'} \tag{22}$$

defines an analytical function of E in the complex E plane with the cuts $(-\infty, -E_{\rm min}^{(-)}]$ and $[E_{\rm min}^{(+)}, \infty)$ (see Fig. 1). This equation provides the analytical continuation of the Green function to the complex E plane. According to (21), to get the Green function for real E we have to approach the right-hand cut from the upper half-plane and the left-hand cut from the lower half-plane.

In what follows we will be interested in bound states of the system. According to (12)–(22), the bound states correspond to the poles of the function $\mathcal{G}(E)$ on the right-hand real semiaxis. If the interaction between the electron-positron field and the electromagnetic field is switched off, the poles corresponding to bound states are isolated. Switching on the interaction between the fields transforms the isolated poles into branch points. This is caused by the fact that due to zero photon mass the bound states are no longer isolated points of the spectrum. Disregarding the instability of excited states, the singularities of the Green function $\mathcal{G}(E)$ are shown in Fig. 2. The poles corresponding to the bound states lie on the upper boundary of the cut starting

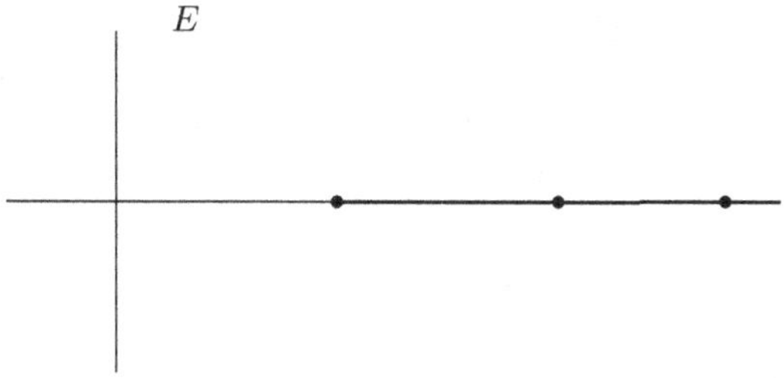

Fig. 2. Singularities of the two-time Green function in the bound-state region, disregarding the instability of excited states

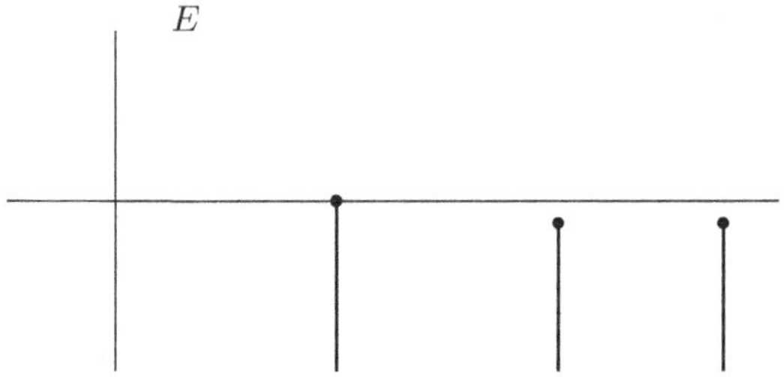

Fig. 3. Singularities of the two-time Green function in the bound-state region if the cuts are turned down, to the second sheet of the Riemann surface. The instability of excited states is taken into account

from the pole corresponding to the ground state. It is natural to assume that $\mathcal{G}(E)$ can be continued analytically under the cut, to the second sheet of the Riemann surface. As a result of this continuation the singularities of $\mathcal{G}(E)$ can be turned down. In fact due to instability of excited states the energies of these states have small imaginary components and, therefore, the related poles lie slightly below the right-hand real semiaxis (Fig. 3). However, in calculations of the energy levels and the transition and scattering amplitudes of non-resonance processes we can neglect the instability of excited states and, therefore, assume that the poles lie on the real axis. The imaginary parts of the energies must be taken into account if one considers resonance scattering processes.

To formulate the perturbation theory for calculations of the energy levels and the transition and scattering amplitudes we will need to isolate the poles corresponding to the bound states from the related cuts. It can be done by introducing a non-zero photon mass μ which is generally assumed to be larger than the energy shift (or the energy splitting) of the level (levels) under consideration and much smaller than the distance to other levels. The singularities of $\mathcal{G}(E)$ with non-zero photon mass, including one- and two-photon spectra, are shown in Fig. 4. As one can see from this figure, introducing the photon mass makes the poles corresponding to the bound states to be isolated.

In every finite order of perturbation theory the singularities of the Green function $\mathcal{G}(E)$ in the complex E-plane are defined by the unperturbed Hamiltonian. It can be shown (see [22] and references therein) that to n-th order of perturbation theory the Green function has poles of all orders till $n+1$ at the unperturbed positions of the bound state energies.

2.3 Energy Shift of a Single Level

In this section we derive the energy shift $\Delta E_a = E_a - E_a^{(0)}$ of a single isolated level a of an N-electron atom due to the perturbative interaction. The unperturbed energy $E_a^{(0)}$ is equal to the sum of the one-electron Dirac-Coulomb

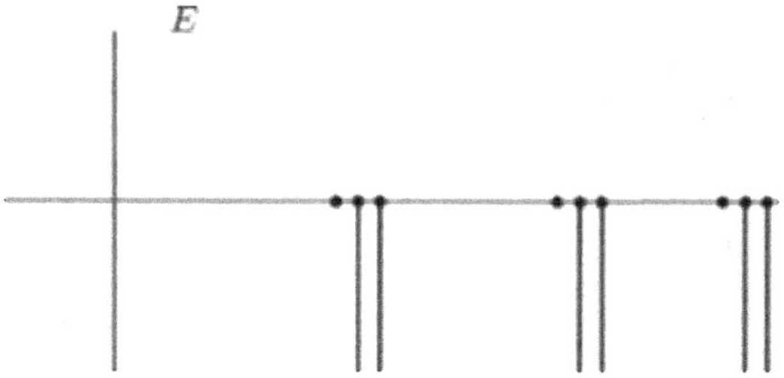

Fig. 4. Singularities of the two-time Green function in the bound state region for a non-zero photon mass, including one- and two-photon spectra, if the cuts are turned down, to the second sheet of the Riemann surface. The instability of excited states is disregarded

energies

$$E_a^{(0)} = \varepsilon_{a_1} + \cdots + \varepsilon_{a_N} \; , \tag{23}$$

which are defined by the Dirac equation (1). In the simplest case the unperturbed wave function $u_a(\mathbf{x}_1, \ldots, \mathbf{x}_N)$ is a one-determinant function

$$u_a(\mathbf{x}_1, \ldots, \mathbf{x}_N) = \frac{1}{\sqrt{N!}} \sum_P (-1)^P \psi_{Pa_1}(\mathbf{x}_1) \cdots \psi_{Pa_N}(\mathbf{x}_N) \; , \tag{24}$$

where ψ_n are the one-electron Dirac wave functions defined by (1) and P is the permutation operator. In the general case the unperturbed wave function is a linear combination of the one-determinant functions

$$u_a(\mathbf{x}_1, \ldots, \mathbf{x}_N) = \sum_b C_a^b \frac{1}{\sqrt{N!}} \sum_P (-1)^P \psi_{Pb_1}(\mathbf{x}_1) \cdots \psi_{Pb_N}(\mathbf{x}_N) \; . \tag{25}$$

We introduce the Green function $g_{aa}(E)$ by

$$\begin{aligned} g_{aa}(E) &= \langle u_a | \mathcal{G}(E) \gamma_1^0 \cdots \gamma_N^0 | u_a \rangle \\ &\equiv \int d\mathbf{x}_1 \cdots d\mathbf{x}_N d\mathbf{x}_1' \cdots d\mathbf{x}_N' \; u_a^\dagger(\mathbf{x}_1', \ldots, \mathbf{x}_N') \\ &\quad \times \mathcal{G}(E, \mathbf{x}_1', \ldots, \mathbf{x}_N'; \mathbf{x}_1, \ldots, \mathbf{x}_N) \gamma_1^0 \cdots \gamma_N^0 u_a(\mathbf{x}_1, \ldots, \mathbf{x}_N) \; . \end{aligned} \tag{26}$$

From the spectral representation for $\mathcal{G}(E)$ (see (12)–(22)) we have

$$g_{aa}(E) = \frac{A_a}{E - E_a} + \text{terms that are regular at } E \sim E_a \; , \tag{27}$$

where

$$\begin{aligned} A_a = \frac{1}{N!} \int & d\mathbf{x}_1 \cdots d\mathbf{x}_N d\mathbf{x}_1' \cdots d\mathbf{x}_N' \; u_a^\dagger(\mathbf{x}_1', \ldots, \mathbf{x}_N') \langle 0 | \psi(0, \mathbf{x}_1') \cdots \psi(0, \mathbf{x}_N') | a \rangle \\ & \times \langle a | \psi^\dagger(0, \mathbf{x}_N) \cdots \psi^\dagger(0, \mathbf{x}_1) | 0 \rangle u_a(\mathbf{x}_1, \ldots, \mathbf{x}_N) \; . \end{aligned} \tag{28}$$

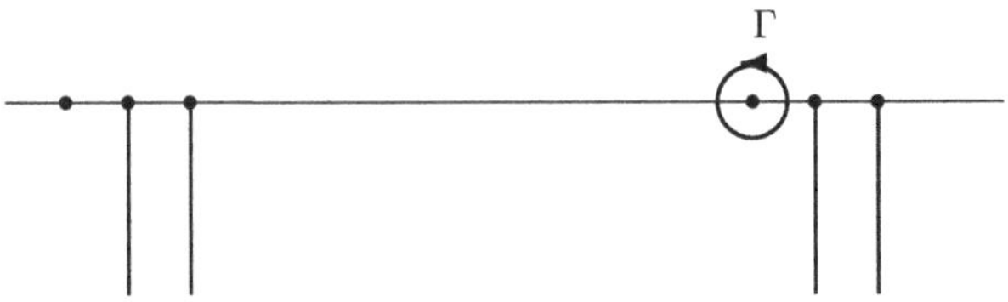

Fig. 5. The contour Γ surrounds the pole corresponding to the level under consideration and keeps outside all other singularities. For simplicity, only one- and two-photon spectra are displayed

We assume here that a non-zero photon mass μ is introduced to isolate the pole corresponding to the bound state a from the related cut. We consider that the photon mass is larger than the energy shift under consideration and much smaller than the distance to other levels. To generate the perturbation series for E_a it is convenient to use a contour integral formalism developed first in operator theory by Szökefalvi-Nagy and Kato [39–42]. Choosing a contour Γ in the complex E plane in a way that it surrounds the pole corresponding to the level a and keeps outside all other singularities (see Fig. 5), we have

$$\frac{1}{2\pi \mathrm{i}} \oint_{\Gamma} \mathrm{d}E \; E g_{aa}(E) = E_a A_a \; , \tag{29}$$

$$\frac{1}{2\pi \mathrm{i}} \oint_{\Gamma} \mathrm{d}E \; g_{aa}(E) = A_a \; . \tag{30}$$

Here we have assumed that the contour Γ is oriented anticlockwise. Dividing (29) by (30), we obtain

$$E_a = \frac{\dfrac{1}{2\pi \mathrm{i}} \displaystyle\oint_{\Gamma} \mathrm{d}E \; E g_{aa}(E)}{\dfrac{1}{2\pi \mathrm{i}} \displaystyle\oint_{\Gamma} \mathrm{d}E \; g_{aa}(E)} \tag{31}$$

It is convenient to transform (31) to a form that directly yields the energy shift $\Delta E_a = E_a - E_a^{(0)}$. To zeroth order, substituting the operators

$$\psi_{\mathrm{in}}(0,\mathbf{x}) = \sum_{\varepsilon_n > 0} b_n \psi_n(\mathbf{x}) + \sum_{\varepsilon_n < 0} d_n^{\dagger} \psi_n(\mathbf{x}) \; , \tag{32}$$

$$\overline{\psi}_{\mathrm{in}}(0,\mathbf{x}) = \sum_{\varepsilon_n > 0} b_n^{\dagger} \overline{\psi}_n(\mathbf{x}) + \sum_{\varepsilon_n < 0} d_n \overline{\psi}_n(\mathbf{x}) \tag{33}$$

into (13) and (14) instead of $\psi(0,\mathbf{x})$ and $\overline{\psi}(0,\mathbf{x})$, respectively, and considering the states $|n\rangle$ in (13) and (14) as unperturbed states in the Fock space, from (12)–(14) and (26) we find

$$g_{aa}^{(0)} = \frac{1}{E - E_a^{(0)}} \,. \tag{34}$$

Denoting $\Delta g_{aa} = g_{aa} - g_{aa}^{(0)}$, from (31) we obtain the desired formula [17]

$$\Delta E_a = \frac{\frac{1}{2\pi \mathrm{i}} \oint_\Gamma \mathrm{d}E \, (E - E_a^{(0)}) \Delta g_{aa}(E)}{1 + \frac{1}{2\pi \mathrm{i}} \oint_\Gamma \mathrm{d}E \, \Delta g_{aa}(E)} \,. \tag{35}$$

The Green function $\Delta g_{aa}(E)$ is constructed by perturbation theory

$$\Delta g_{aa}(E) = \Delta g_{aa}^{(1)}(E) + \Delta g_{aa}^{(2)}(E) + \cdots \,, \tag{36}$$

where the superscript denotes the order in a small parameter (for instance, α). If we represent the energy shift as a series

$$\Delta E_a = \Delta E_a^{(1)} + \Delta E_a^{(2)} + \cdots \,, \tag{37}$$

formula (35) yields

$$\Delta E_a^{(1)} = \frac{1}{2\pi \mathrm{i}} \oint_\Gamma \mathrm{d}E \, \Delta E \, \Delta g_{aa}^{(1)}(E) \,, \tag{38}$$

$$\begin{aligned} \Delta E_a^{(2)} &= \frac{1}{2\pi \mathrm{i}} \oint_\Gamma \mathrm{d}E \, \Delta E \, \Delta g_{aa}^{(2)}(E) \\ &\quad - \left(\frac{1}{2\pi \mathrm{i}} \oint_\Gamma \mathrm{d}E \, \Delta E \, \Delta g_{aa}^{(1)}(E) \right) \left(\frac{1}{2\pi \mathrm{i}} \oint_\Gamma \mathrm{d}E \, \Delta g_{aa}^{(1)}(E) \right) , \end{aligned} \tag{39}$$

where $\Delta E \equiv E - E_a^{(0)}$.

Deriving (31) and (35) we have assumed that a non-zero photon mass μ is introduced. This allows taking all the cuts outside the contour Γ as well as regularizing the infrared singularities of individual contributions. As was noted in the previous subsection, the singularities of the two-time Green function in the complex E plane are defined by the unperturbed Hamiltonian if it is constructed by perturbation theory. In particular, it means that in n-th order of perturbation theory $g_{aa}(E)$ has poles of all orders till $n+1$ at the position of the unperturbed energy level under consideration. Therefore, in calculations by perturbation theory it is sufficient to consider the photon mass as a very small parameter which provides a separation of the pole from the related cut. At the end of the calculations after taking into account a whole gauge invariant set of Feynman diagrams we can put $\mu \to 0$. The possibility of taking the limit $\mu \to 0$ follows, in particular, from the fact that the contour Γ can be shrunk continuosly to the point $E = E_a^{(0)}$ (see Fig. 5).

Generally speaking, the energy shift of an excited level derived by formula (35) contains an imaginary component which is caused by its instability. This component determines the width of the spectral line in the Lorentz approximation.

2.4 Perturbation Theory for Degenerate and Quasidegenerate Levels

We are interested in the atomic levels with energies $E_1, \dots, E_s$ arising from unperturbed degenerate or quasidegenerate levels with energies $E_1^{(0)}, \dots, E_s^{(0)}$. As usual, we assume that the energy shifts of the levels under consideration or their splitting caused by the interaction are much smaller than the distance to other levels. The unperturbed eigenstates form an s-dimensional subspace Ω. We denote the projector on Ω by

$$P^{(0)} = \sum_{k=1}^{s} P_k^{(0)} = \sum_{k=1}^{s} u_k u_k^{\dagger} \,, \tag{40}$$

where $\{u_k\}_{k=1}^{s}$ are the unperturbed wave functions which, in a general case, are linear combinations of one-determinant functions (see (25)). We project the Green function $\mathcal{G}(E)$ on the subspace Ω

$$g(E) = P^{(0)} \mathcal{G}(E) \gamma_1^0 \dots \gamma_N^0 P^{(0)} \,, \tag{41}$$

where, as in (26), the integration over the electron coordinates is implicit. As in the case of a single level, to isolate the poles of $g(E)$ corresponding to the bound states under consideration, we introduce a non-zero photon mass μ. We assume that the photon mass μ is larger than the energy distance between the levels under consideration and much smaller than the distance to other levels. In this case we can choose a contour Γ in the complex E plane in a way that it surrounds all the poles corresponding to the considered states $(E_1, \dots E_s)$ and keeps outside all other singularities, including the cuts starting from the lower-lying bound states (see Fig. 6). In addition, if we neglect the instability of the states under consideration, the spectral representation (see (12)–(22)) gives

$$g(E) = \sum_{k=1}^{s} \frac{\varphi_k \varphi_k^{\dagger}}{E - E_k} + \text{ terms that are regular inside of } \Gamma \,, \tag{42}$$

where

$$\varphi_k = P^{(0)} \Phi_k \,, \qquad \varphi_k^{\dagger} = \Phi_k^{\dagger} P^{(0)} \,. \tag{43}$$

As in the case of a single level, in zeroth approximation one easily finds

$$g^{(0)}(E) = \sum_{k=1}^{s} \frac{P_k^{(0)}}{E - E_k^{(0)}} \,. \tag{44}$$

We introduce the operators K and P by

$$K \equiv \frac{1}{2\pi \mathrm{i}} \oint_{\Gamma} \mathrm{d}E \; E g(E) \,, \tag{45}$$

$$P \equiv \frac{1}{2\pi \mathrm{i}} \oint_{\Gamma} \mathrm{d}E \; g(E) \,. \tag{46}$$

Fig. 6. The contour Γ surrounds the poles corresponding to the quasidegenerate levels under consideration and keeps outside all other singularities. For simplicity, only one-photon spectra are displayed

Using (42), we obtain

$$K = \sum_{i=1}^{s} E_i \varphi_i \varphi_i^{\dagger} , \tag{47}$$

$$P = \sum_{i=1}^{s} \varphi_i \varphi_i^{\dagger} . \tag{48}$$

We note here that, generally speaking, the operator P is not a projector (in particular, $P^2 \neq P$). If the perturbation goes to zero, the vectors $\{\varphi_i\}_{i=1}^{s}$ approach the correct linearly independent combinations of the vectors $\{u_i\}_{i=1}^{s}$. Therefore, it is natural to assume that the vectors $\{\varphi_i\}_{i=1}^{s}$ are also linearly independent. It follows that one can find such vectors $\{v_i\}_{i=1}^{s}$ that

$$\varphi_i^{\dagger} v_k = \delta_{ik} . \tag{49}$$

Indeed, let

$$\varphi_i = \sum_{j=1}^{s} a_{ij} u_j , \qquad v_k = \sum_{l=1}^{s} x_{kl} u_l . \tag{50}$$

The biorthogonality condition (49) gives

$$\sum_{j=1}^{s} a_{ij} x_{kj} = \delta_{ik} . \tag{51}$$

Since the determinant of the matrix $\{a_{ij}\}$ is nonvanishing due to the linear independence of $\{\varphi_i\}_{i=1}^{s}$, the system (51) has a unique solution for any fixed $k = 1, \dots, s$. From (47)–(49) we have

$$P v_k = \sum_{i=1}^{s} \varphi_i \varphi_i^{\dagger} v_k = \varphi_k , \tag{52}$$

$$K v_k = \sum_{i=1}^{s} E_i \varphi_i \varphi_i^{\dagger} v_k = E_k \varphi_k . \tag{53}$$

Hence we obtain the equation for v_k, E_k [17]

$$Kv_k = E_k P v_k \,. \tag{54}$$

According to (49) the vectors v_k are normalized by the condition

$$v_{k'}^{\dagger} P v_k = \delta_{k'k} \,. \tag{55}$$

The solvability of (54) yields an equation for the atomic energy levels

$$\det(K - EP) = 0 \,. \tag{56}$$

The generalized eigenvalue problem (54) with the normalization condition (55) can be transformed by the substitution $\psi_k = P^{\frac{1}{2}} v_k$ to the ordinary eigenvalue problem ("Schrödinger-like equation") [21]

$$H\psi_k = E_k \psi_k \tag{57}$$

with the ordinary normalization condition

$$\psi_k^{\dagger} \psi_{k'} = \delta_{kk'} \,, \tag{58}$$

where $H \equiv P^{-\frac{1}{2}}(K)P^{-\frac{1}{2}}$.

The energy levels are determined from the equation

$$\det(H - E) = 0 \,. \tag{59}$$

Generally speaking, the energies determined by this equation contain imaginary components which are due to the instability of excited states. In the case when the imaginary components are much smaller than the energy distance between the levels (or the levels have different quantum numbers), they define the widths of the spectral lines in the Lorentz approximation. In the opposite case, when the imaginary components are comparable with the energy distance between the levels which have the same quantum numbers, the spectral line shape depends on the process of the formation of the states under consideration even in the resonance approximation (see [22,24] for details). If the instability of excited states can be disregarded, we assume $H \equiv (H + H^{\dagger})/2$ in (57), (59).

The operators K and P are constructed by formulas (45) and (46) using perturbation theory

$$K = K^{(0)} + K^{(1)} + K^{(2)} + \cdots \,, \tag{60}$$

$$P = P^{(0)} + P^{(1)} + P^{(2)} + \cdots \,, \tag{61}$$

where the superscript denotes the order in a small parameter. The operator H is

$$H = H^{(0)} + H^{(1)} + H^{(2)} + \cdots \,, \tag{62}$$

where

$$H^{(0)} = K^{(0)} , \tag{63}$$

$$H^{(1)} = K^{(1)} - \frac{1}{2}P^{(1)}K^{(0)} - \frac{1}{2}K^{(0)}P^{(1)} , \tag{64}$$

$$\begin{aligned} H^{(2)} = K^{(2)} &- \frac{1}{2}P^{(2)}K^{(0)} - \frac{1}{2}K^{(0)}P^{(2)} \\ &- \frac{1}{2}P^{(1)}K^{(1)} - \frac{1}{2}K^{(1)}P^{(1)} \\ &+ \frac{3}{8}P^{(1)}P^{(1)}K^{(0)} + \frac{3}{8}K^{(0)}P^{(1)}P^{(1)} \\ &+ \frac{1}{4}P^{(1)}K^{(0)}P^{(1)} . \end{aligned} \tag{65}$$

It is evident that in zeroth order

$$K_{ik}^{(0)} = E_i^{(0)}\delta_{ik} , \tag{66}$$

$$P_{ik}^{(0)} = \delta_{ik} , \tag{67}$$

$$H_{ik}^{(0)} = E_i^{(0)}\delta_{ik} . \tag{68}$$

To derive (54)–(57) we have introduced a non-zero photon mass μ which was assumed to be larger than the energy distance between the levels under consideration and much smaller than the distance to other levels. At the end of the calculations after taking into account a whole gauge invariant set of Feynman diagrams, we can put $\mu \to 0$. The possibility of taking this limit in the case of quasidegenerate states follows from the fact that the cuts can be drawn to the related poles by a deformation of the contour Γ as shown in Fig. 7.

3 Transition Probabilities

Let us consider the transition of an atom from an initial state a to a final state b with the emission of a photon with momentum $\mathbf{k}_f$ and polarization ϵ_f. The transition probability is given as

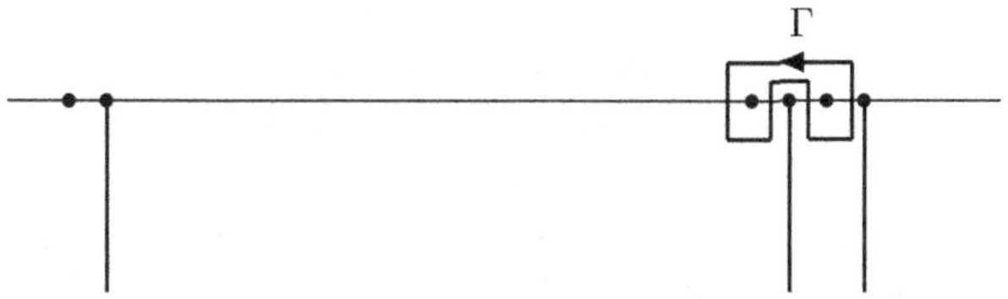

Fig. 7. A deformation of the contour Γ that allows drawing the cuts to the related poles in the case of quasidegenerate states when $\mu \to 0$. For simplicity, only one-photon spectra are displayed

$$dW = 2\pi|\tau_{\gamma_f,b;a}|^2\delta(E_b + k_f^0 - E_a)\mathrm{d}\mathbf{k}_f \,, \tag{69}$$

where $\tau_{\gamma_f,b;a}$ is the transition amplitude which is connected with the S-matrix element by

$$S_{\gamma_f,b;a} = 2\pi\mathrm{i}\tau_{\gamma_f,b;a}\delta(\varepsilon_b + k_f^0 - \varepsilon_a) \tag{70}$$

and $k_f^0 \equiv |\mathbf{k}_f|$. According to the standard reduction technique (see, e.g., [33, 34]), the transition amplitude is

$$S_{\gamma_f,b;a} = -\mathrm{i}Z_3^{-\frac{1}{2}} \int \mathrm{d}^4y \, \frac{\epsilon_f^{\nu *} \exp{(\mathrm{i}k_f \cdot y)}}{\sqrt{2k_f^0(2\pi)^3}} \langle b|j_\nu(y)|a\rangle \,. \tag{71}$$

Here $j_\nu(y)$ is the electron-positron current operator in the Heisenberg representation, $|a\rangle$ and $|b\rangle$ are the vectors of the initial and final states in the Heisenberg representation, Z_3 is a renormalization constant, $a \cdot b \equiv a_\nu b^\nu$, $\epsilon_f = (0, \boldsymbol{\epsilon}_f)$, and $k_f = (k_f^0, \mathbf{k}_f)$. Employing the equation

$$j^\nu(y) = \exp{(\mathrm{i}Hy^0)} j^\nu(0, \mathbf{y}) \exp{(-\mathrm{i}Hy^0)} \,, \tag{72}$$

we obtain

$$\begin{aligned} S_{\gamma_f,b;a} &= -\mathrm{i}Z_3^{-\frac{1}{2}} \int \mathrm{d}^4y \, \exp{[\mathrm{i}(E_b + k_f^0 - E_a)y^0]} A_f^{\nu *}(\mathbf{y}) \langle b|j_\nu(0,\mathbf{y})|a\rangle \\ &= -2\pi\mathrm{i}Z_3^{-\frac{1}{2}} \delta(E_b + k_f^0 - E_a) \int \mathrm{d}\mathbf{y} \, A_f^{\nu *}(\mathbf{y}) \langle b|j_\nu(0,\mathbf{y})|a\rangle \,, \end{aligned} \tag{73}$$

where

$$A_f^\nu(\mathbf{x}) = \frac{\epsilon_f^\nu \exp{(\mathrm{i}\mathbf{k}_f \cdot \mathbf{x})}}{\sqrt{2k_f^0(2\pi)^3}} \tag{74}$$

is the wave function of the emitted photon. Since $|a\rangle$ and $|b\rangle$ are bound states, (73) as well as the standard reduction technique [33, 34] cannot be used for a direct evaluation of the amplitude. The desired calculation formula can be derived within the TTGF formalism [18–20].

To formulate the method for a general case, we assume that in zeroth approximation the state a belongs to an s_a-dimensional subspace of unperturbed degenerate states Ω_a and the state b belongs to an s_b-dimensional subspace of unperturbed degenerate states Ω_b. We denote the projectors onto these subspaces by $P_a^{(0)}$ and $P_b^{(0)}$, respectively. We denote the exact states originating from Ω_a by $|n_a\rangle$ and the exact states originating from Ω_b by $|n_b\rangle$. We also assume that on an intermediate stage of the calculations a non-zero photon mass μ is introduced. It is considered to be larger than the energy splitting of the initial and final states under consideration and much smaller than the distance to other levels.

We introduce

$$
\begin{aligned}
&\mathcal{G}_{\gamma_f}(E',E;\mathbf{x}'_1,\ldots\mathbf{x}'_N;\mathbf{x}_1,\ldots\mathbf{x}_N)\delta(E'+k^0-E)\\
&=\frac{1}{2\pi\mathrm{i}}\frac{1}{2\pi}\frac{1}{N!}\int_{-\infty}^{\infty}\mathrm{d}x^0\mathrm{d}x'^0\int\mathrm{d}^4y\ \exp{(\mathrm{i}E'x'^0-\mathrm{i}Ex^0)}\exp{(\mathrm{i}k^0y^0)}\\
&\times A_f^{\nu*}(\mathbf{y})\langle 0|T\psi(x'^0,\mathbf{x}'_1)\cdots\psi(x'^0,\mathbf{x}'_N)\\
&\times j_\nu(y)\overline{\psi}(x^0,\mathbf{x}_N)\cdots\overline{\psi}(x^0,\mathbf{x}_1)|0\rangle\ ,
\end{aligned}
\tag{75}
$$

where, as in the previous section, $\psi(x)$ is the electron-positron field operator in the Heisenberg representation. Let us investigate the singularities of $\mathcal{G}_{\gamma_f}$ in the region $E'\sim E_b^{(0)}$ and $E\sim E_a^{(0)}$. Using the transformation rules

$$
\begin{aligned}
\psi(x^0,\mathbf{x})&=\exp{(\mathrm{i}Hy^0)}\psi(x^0-y^0,\mathbf{x})\exp{(-\mathrm{i}Hy^0)}\ ,\\
j(y^0,\mathbf{y})&=\exp{(\mathrm{i}Hy^0)}j(0,\mathbf{y})\exp{(-\mathrm{i}Hy^0)}\ ,
\end{aligned}
\tag{76}
$$

we obtain

$$
\begin{aligned}
&\mathcal{G}_{\gamma_f}(E',E;\mathbf{x}'_1,\ldots\mathbf{x}'_N;\mathbf{x}_1,\ldots\mathbf{x}_N)\delta(E'+k^0-E)\\
&=\frac{1}{2\pi\mathrm{i}}\frac{1}{2\pi}\frac{1}{N!}\int_{-\infty}^{\infty}\mathrm{d}t\mathrm{d}t'\int\mathrm{d}^4y\ \exp{(\mathrm{i}E't'-\mathrm{i}Et)}\exp{[\mathrm{i}(E'+k^0-E)y^0]}\\
&\times A_f^{\nu*}(\mathbf{y})\langle 0|T\psi(t',\mathbf{x}'_1)\cdots\psi(t',\mathbf{x}'_N)\\
&\times j_\nu(0,\mathbf{y})\overline{\psi}(t,\mathbf{x}_N)\cdots\overline{\psi}(t,\mathbf{x}_1)|0\rangle\\
&=\frac{1}{2\pi\mathrm{i}}\delta(E'+k^0-E)\frac{1}{N!}\int_{-\infty}^{\infty}\mathrm{d}t\mathrm{d}t'\int\mathrm{d}\mathbf{y}\ \exp{(\mathrm{i}E't'-\mathrm{i}Et)}\\
&\times A_f^{\nu*}(\mathbf{y})\langle 0|T\psi(t',\mathbf{x}'_1)\cdots\psi(t',\mathbf{x}'_N)\\
&\times j_\nu(0,\mathbf{y})\overline{\psi}(t,\mathbf{x}_N)\cdots\overline{\psi}(t,\mathbf{x}_1)|0\rangle\ .
\end{aligned}
\tag{77}
$$

Using again the time-shift transformation rules, we obtain

$$
\begin{aligned}
&\mathcal{G}_{\gamma_f}(E',E;\mathbf{x}'_1,\ldots\mathbf{x}'_N;\mathbf{x}_1,\ldots\mathbf{x}_N)\\
&=\frac{1}{2\pi\mathrm{i}}\frac{1}{N!}\int_{-\infty}^{\infty}\mathrm{d}t\mathrm{d}t'\int\mathrm{d}\mathbf{y}\ \exp{(\mathrm{i}E't'-\mathrm{i}Et)}\sum_{n_1,n_2}A_f^{\nu*}(\mathbf{y})\\
&\times\exp{(-\mathrm{i}E_{n_1}t')}\exp{(\mathrm{i}E_{n_2}t)}\theta(t')\theta(-t)\langle 0|T\psi(0,\mathbf{x}'_1)\cdots\psi(0,\mathbf{x}'_N)|n_1\rangle\\
&\times\langle n_1|j_\nu(0,\mathbf{y})|n_2\rangle\langle n_2|\overline{\psi}(0,\mathbf{x}_N)\cdots\overline{\psi}(0,\mathbf{x}_1)|0\rangle+\cdots\ .
\end{aligned}
\tag{78}
$$

Here we have assumed $E_0=0$, as in the previous section. Taking into account the identities

$$
\begin{aligned}
\int_0^{\infty}\mathrm{d}t\ \exp{[\mathrm{i}(E'-E_{n_1})t]}&=\frac{\mathrm{i}}{E'-E_{n_1}+\mathrm{i}0}\ ,\\
\int_{-\infty}^{0}\mathrm{d}t\ \exp{[\mathrm{i}(-E+E_{n_2})t]}&=\frac{\mathrm{i}}{E-E_{n_2}+\mathrm{i}0}\ ,
\end{aligned}
\tag{79}
$$

we find

$$
\begin{aligned}
&\mathcal{G}_{\gamma_f}(E', E; \mathbf{x}'_1, \ldots \mathbf{x}'_N; \mathbf{x}_1, \ldots \mathbf{x}_N) \\
&= \frac{\mathrm{i}}{2\pi} \frac{1}{N!} \sum_{n_1, n_2} \int \mathrm{d}\mathbf{y}\, A_f^{\nu *}(\mathbf{y}) \frac{1}{E' - E_{n_1} + \mathrm{i}0} \frac{1}{E - E_{n_2} + \mathrm{i}0} \\
&\times \langle 0 | T\psi(0, \mathbf{x}'_1) \cdots \psi(0, \mathbf{x}'_N) | n_1 \rangle \langle n_1 | j_\nu(0, \mathbf{y}) | n_2 \rangle \\
&\times \langle n_2 | \overline{\psi}(0, \mathbf{x}_N) \cdots \overline{\psi}(0, \mathbf{x}_1) | 0 \rangle + \cdots . \qquad (80)
\end{aligned}
$$

We are interested in the analytical properties of $\mathcal{G}_{\gamma_f}$ as a function of the two complex variables E' and E in the region $E' \sim E_b^{(0)}$, $E \sim E_a^{(0)}$. These properties can be studied using the double spectral representation of this type of Green function (see [22, 38]). As it follows from the spectral representation, the terms which are omitted in (80) are regular functions of E' or E if $E' \sim E_b^{(0)}$ and $E \sim E_a^{(0)}$, and, for a non-zero photon mass μ, the Green function $\mathcal{G}_{\gamma_f}(E', E)$ has isolated poles in the variables E' and E at the points $E' = E_{n_b}$ and $E = E_{n_a}$, respectively. Let us now introduce a Green function $g_{\gamma_f, b; a}(E', E)$ by

$$
g_{\gamma_f, b; a}(E', E) = P_b^{(0)} \mathcal{G}_{\gamma_f}(E', E) \gamma_1^0 \cdots \gamma_N^0 P_a^{(0)} , \qquad (81)
$$

where, as in (26), the integration over the electron coordinates is implicit. It can be written as

$$
\begin{aligned}
g_{\gamma_f, b; a}(E', E) = \frac{\mathrm{i}}{2\pi} \sum_{n_a=1}^{s_a} \sum_{n_b=1}^{s_b} & \frac{1}{E' - E_{n_b}} \frac{1}{E - E_{n_a}} \\
& \times \varphi_{n_b} \int \mathrm{d}\mathbf{y}\, A_f^{\nu *}(\mathbf{y}) \langle n_b | j_\nu(0, \mathbf{y}) | n_a \rangle \varphi_{n_a}^\dagger \\
& + \text{terms that are regular functions of } E' \text{ or } E \text{ if } E' \sim E_b^{(0)} \\
& \quad \text{and } E \sim E_a^{(0)} , \qquad (82)
\end{aligned}
$$

where the vectors φ_k are defined by (43). Let the contours Γ_a and Γ_b surround the poles corresponding to the initial and final levels, respectively, and keep outside other singularities of $g_{\gamma_f, b; a}(E', E)$ including the cuts starting from the lower-lying bound states. Comparing (82) with (73) and taking into account the biorthogonality condition (49), we obtain the desired formula [18]

$$
S_{\gamma_f, b; a} = Z_3^{-1/2} \delta(E_b + k_f^0 - E_a) \oint_{\Gamma_b} \mathrm{d}E' \oint_{\Gamma_a} \mathrm{d}E\, v_b^\dagger g_{\gamma_f, b; a}(E', E) v_a , \qquad (83)
$$

where by a we imply one of the initial states and by b one of the final states under consideration. The vectors v_k are determined from (54)–(55).

In the case of a single initial state (a) and a single final state (b), the vectors v_a and v_b simply become normalization factors. So, for the initial state,

$$v_a^* P_a v_a = v_a^* \frac{1}{2\pi \mathrm{i}} \oint_{\Gamma_a} \mathrm{d}E \, g_{aa}(E) v_a = 1 \tag{84}$$

and, therefore,

$$|v_a|^2 = \left[\frac{1}{2\pi \mathrm{i}} \oint_{\Gamma_a} \mathrm{d}E \, g_{aa}(E) \right]^{-1} . \tag{85}$$

Choosing

$$v_a = \left[\frac{1}{2\pi \mathrm{i}} \oint_{\Gamma_a} \mathrm{d}E \, g_{aa}(E) \right]^{-1/2} , \; v_b = \left[\frac{1}{2\pi \mathrm{i}} \oint_{\Gamma_b} \mathrm{d}E \, g_{bb}(E) \right]^{-1/2} , \tag{86}$$

we obtain

$$\begin{aligned} S_{\gamma_f, b; a} &= Z_3^{-1/2} \delta(E_b + k_f^0 - E_a) \oint_{\Gamma_b} \mathrm{d}E' \oint_{\Gamma_a} \mathrm{d}E \, g_{\gamma_f, b; a}(E', E) \\ &\times \left[\frac{1}{2\pi \mathrm{i}} \oint_{\Gamma_b} \mathrm{d}E \, g_{bb}(E) \right]^{-1/2} \left[\frac{1}{2\pi \mathrm{i}} \oint_{\Gamma_a} \mathrm{d}E \, g_{aa}(E) \right]^{-1/2} . \end{aligned} \tag{87}$$

The Green function $g_{\gamma_f, b; a}$ is constructed by perturbation theory after the transition in (75) to the interaction representation and using the Wick theorem [22].

4 Conclusion

We have formulated the perturbation theory for calculations of the energy levels and the transitions probabilities in high-Z few-electron atoms. The TTGF method can also be used for calculations of scattering processes. The corresponding formulas for the photon scattering by an atom and for the radiative recombination of an electron with an atom are presented in [22]. Application of the method to resonance scattering processes yields a systematic theory for the spectral line shape. For a detailed consideration of these processes within the TTGF method we refer to [22].

References

1. H.F. Beyer, H.-J. Kluge, V.P. Shevelko: *X-ray Radiation of Highly Charged Ions* (Springer, Berlin 1997)
2. M. Gell-Mann, F. Low: Phys. Rev. **84**, 350 (1951)
3. J. Sucher: Phys. Rev. **107**, 1448 (1957)
4. L.N. Labzowsky: Zh. Eksp. Teor. Fiz. **59**, 168 (1970) [Sov. Phys. JETP **32**, 94 (1970)]
5. M.A. Braun, A.D. Gurchumelia: Teor. Mat. Fiz. **45**, 199 (1980) [Theor. Math. Phys. **45**, N2 (1980)]

6. M.A. Braun, A.D. Gurchumelia, U.I. Safronova: *Relativistic Atom Theory* (Nauka, Moscow 1984)
7. L. Labzowsky, G. Klimchitskaya, Yu. Dmitriev: *Relativistic Effects in Spectra of Atomic Systems* (IOP Publishing, Bristol 1993)
8. P.J. Mohr, G. Plunien, G. Soff: Phys. Rep. **293**, 227 (1998)
9. J. Sapirstein: Rev. Mod. Phys. **70**, 55 (1998)
10. J. Hubbard: Proc. Roy. Soc. A **240**, 539 (1957)
11. S.S. Schweber: *An Introduction to Relativistic Quantum Field Theory* (Harper & Row, New York 1961)
12. A.N. Vasil'ev, A.L. Kitanin: Teor. Mat. Fiz. **24**, 219 (1975) [Theor. Math. Phys. **24**, N2 (1975)]
13. S.A. Zapryagaev, N.L. Manakov, V.G. Pal'chikov: *Theory of One- and Two-Electron Multicharged Ions* (Energoatomizdat, Moscow 1985)
14. J. Sapirstein: Phys. Scr. **36**, 801 (1987)
15. Yu.Yu. Dmitriev, T.A. Fedorova: Phys. Lett. A **245**, 555 (1998)
16. I. Lindgren, S. Salomonson, B. Åsén: Phys. Rep. **389**, 161 (2004)
17. V.M. Shabaev: The quantum-electrodynamical perturbation theory in the Raylegh-Schrödinger form for the calculation of the energy levels of atomic systems. In: *Many-particles Effects in Atoms*, ed by U.I. Safronova (AN SSSR, Nauchnyi Sovet po Spektroskopii, Moscow 1988) pp 15–23
18. V.M. Shabaev: The quantum-electrodynamical perturbation theory in the Raylegh-Schrödinger form for the calculation of the transition probabilities and the cross-sections of various processes. In: *Many-particles Effects in Atoms*, ed by U.I. Safronova (AN SSSR, Nauchnyi Sovet po Spektroskopii, Moscow 1988) pp 24–33
19. V.M. Shabaev: Izv. Vuz. Fiz. **33**, 43 (1990) [Sov. Phys. Journ. **33**, 660 (1990)]
20. V.M. Shabaev: Teor. Mat. Fiz. **82**, 83 (1990) [Theor. Math. Phys. **82**, 57 (1990)]
21. V.M. Shabaev: J. Phys. A **24**, 5665 (1991)
22. V.M. Shabaev: Phys. Rep. **356**, 119 (2002)
23. V.M. Shabaev, I.G. Fokeeva: Phys. Rev. A **49**, 4489 (1994)
24. V.M. Shabaev: Phys. Rev. A **50**, 4521 (1994)
25. V.A. Yerokhin, A.N. Artemyev, T. Beier et al: Phys. Rev. A **60**, 3522 (1999)
26. A.N. Artemyev, T. Beier, G. Plunien et al: Phys. Rev. A **62**, 022116 (2000)
27. V.M. Shabaev, V.A. Yerokhin, T. Beier et al: Phys. Rev. A **61**, 052112 (2000)
28. E.-O. Le Bigot, P. Indelicato, V.M. Shabaev: Phys. Rev. A **63**, 040501(R) (2001)
29. V.M. Shabaev: Phys. Rev. A **64**, 052104 (2001)
30. P. Indelicato, V.M. Shabaev, A.V. Volotka: Phys. Rev. A **69**, 062506 (2004)
31. M.A. Braun: Teor. Mat. Fiz. **72**, 394 (1987) [Theor. Math. Phys. **72**, 958 (1987)]
32. V.M. Shabaev: Teor. Mat. Fiz. **63**, 394 (1985) [Theor. Math. Phys. **63**, 588 (1985)]
33. C. Itzykson, J.-B. Zuber: *Quantum Field Theory* (McGraw-Hill, New York 1980)
34. J.D. Bjorken, D. Drell: *Relativistic Quantum Fields* (McGraw-Hill, New York 1965)
35. V.L. Bonch-Bruevich, S.V. Tyablikov: *The Green Function Method in Statistical Mechanics* (North-Holland Publishing Company, Amsterdam 1962)
36. A.B. Migdal: *Theory of Finite Fermi Systems and Properties of Atomic Nuclei* (Nauka, Moscow 1983)

37. A.A. Logunov, A.N. Tavkhelidze: Nuovo Cim. **29**, 380 (1963)
38. R.N. Faustov: Teor. Mat. Fiz. **3**, 240 (1970) [Theor. Math. Phys. **3**, N2 (1970)]
39. B. Sz-Nagy: Comm. Math. Helv. **19**, 347 (1946/47)
40. T. Kato: Progr. Theor. Phys. **4**, 514 (1949)
41. T. Kato: *Perturbation Theory for Linear Operators* (Springer, Berlin 1966)
42. M. Reed, B. Simon: *Methods of Modern Mathematical Physics*, vol. 4 (Academic Press, New York 1978)

The Relativistic Electron-Positron Field: Hartree-Fock Approximation and Fixed Electron Number*

H. Siedentop

Mathematisches Institut, Ludwig-Maximilians-Universität München,
Theresienstraße 39, 80333 München, Germany
h.s@lmu.de

*The material has been partly presented in short form at the Fourth Conference on Operator Algebras and Mathematical Physics, June 26 – July 4, 2003, in Sinai, Romania, and in a longer form at the summer school on "Large Coulomb Systems and Quantum Electrodynamics" at the Sophus Lie Center, Nordfjordeid, Norway, August 11–18, 2003. Thanks goes to the European Union for partial support through the IHP network "Analysis and Quantum", contract HPRN-CT-2002-00277. Special thanks go to Matthias Huber for critical reading of the manuscript.

H. Siedentop: *The Relativistic Electron-Positron Field: Hartree-Fock Approximation and Fixed Electron Number*, Lect. Notes Phys. **695**, 297–324 (2006)
www.springerlink.com

Abstract. We give an overview over recent investigations concerning the relativistic electron-positron field. Starting from basic definition, we review the derivation of the Lamb shift and the Hartree-Fock approximation.

1 Introduction

Atoms, molecules, and solid states containing nuclei with high positive charge eZ show certain effects that can only be explained in the context of relativistic quantum mechanics. Of course, this is well known to spectroscopists: the spectral lines of hydrogen have a fine structure that is not seen in non-relativistic quantum mechanics. However, sometimes such differences would be obvious in daily life: neither would gold have the shiny golden appearance that it has, nor would we be able to start our cars with the usual lead accumulator, if they behaved as non-relativistic quantum mechanics predicts. A relativistic treatment of these systems is warranted.

On the other hand, the treatment of relativistic multi-particle quantum systems is notoriously difficult. Even the naive Hamiltonian $D_0 \otimes 1 + 1 \otimes D_0$ of two non-interacting free Dirac particles has the whole line as spectrum, a situation that is unchanged when adding the interaction with nuclei and among the particles. The concept of bound states is lost. One speaks of continuum dissolution or the Brown-Ravenhall disease (Brown and Ravenhall [4]). We will attempt to present some progress that has been made over the years to overcome this problem and to formulate a mathematically consistent relativistic quantum mechanics.

2 The Basic Notation

2.1 The Basic Hilbert Space and the Dirac Operator

The basic Hilbert space is

$$\mathfrak{H} := L^2(G, \mathrm{d}x) , \tag{1}$$

i.e., the square integrable functions of the space-spin variable

$$x := (\mathfrak{x}, \sigma) \in G := \mathbb{R}^3 \times \{1, \dots, 4\} .$$

In the following we will consider Dirac operators

$$D_{\varphi,\mathfrak{A}} := c\boldsymbol{\alpha} \cdot \left(\frac{\hbar}{\mathrm{i}}\nabla + \frac{e}{c}\mathfrak{A}\right) + mc^2\beta - e\varphi$$

for a particle of charge $-e$ in an electric field $-e\nabla\varphi$ and a magnetic field $\nabla \times \mathfrak{A}$. The 4×4 matrices $\boldsymbol{\alpha}$ and β are the four Dirac matrices

$$\beta = \begin{pmatrix} 1 & 0 \\ 0 & -1 \end{pmatrix}, \quad \boldsymbol{\alpha} = \begin{pmatrix} 0 & \boldsymbol{\sigma} \\ \boldsymbol{\sigma} & 0 \end{pmatrix}$$

in standard representation where $\boldsymbol{\sigma} = (\sigma_1, \sigma_2, \sigma_3)$ are the three Pauli matrices

$$\sigma_1 = \begin{pmatrix} 0 & 1 \\ 1 & 0 \end{pmatrix}, \quad \sigma_2 = \begin{pmatrix} 0 & -\mathrm{i} \\ \mathrm{i} & 0 \end{pmatrix}, \quad \sigma_3 = \begin{pmatrix} 1 & 0 \\ 0 & -1 \end{pmatrix}.$$

We will make the following basic assumptions about the electric and magnetic potential φ and $\mathfrak{A}$:

1. The quadratic form domain $\mathfrak{Q}(D_{\varphi,\mathfrak{A}})$ of $D_{\varphi,\mathfrak{A}}$ is equal to $\mathfrak{Q}(D_0) = H^{1/2}(G)$, the one of the free Dirac operator D_0.
2. The resolvent difference $(\mathrm{i} + D_{\varphi,\mathfrak{A}})^{-1} - (\mathrm{i} + D_0)^{-1}$ is compact.

The latter assumption implies according to Weyl that the essential spectra of the free Dirac operator and the one with the electro-magnetic field agree, i.e., we have

$$\sigma_{\mathrm{ess}}(D_{\varphi,\mathfrak{A}}) = (-\infty, -mc^2] \cup [mc^2, \infty). \tag{2}$$

An important class of examples that exhibits these properties is given by Dirac operators $D_Z := D_{eZ/|\cdot|,\mathbf{0}}$ of one-electron atoms and ions with $e^2 Z < 2/\pi$. This follows from a simple application of Kato's inequality (Kato [16])

$$|\nabla| \geq (2/\pi)|\cdot|^{-1}.$$

(Nenciu [21] proved that these Dirac operators have a distinguished self-adjoint realization in $\mathfrak{H}$, if $Ze^2 < 1$.) Physically Z is the atomic number, $-e$ is the electron charge. The constant $\alpha := e^2/(\hbar c)$ is called the Sommerfeld fine structure constant. It has the approximate value $1/137$.

2.2 The Electron and Positron Subspace

As opposed to non-relativistic quantum mechanics the state space of an electron is not a concept that can be defined independently of the environment of the electron. It rather depends on the electromagnetic potential $\varphi_q, \mathfrak{A}_q$ in which it moves. We will, however, suppress – for notational simplicity – the dependence on the magnetic potential whenever we consider the purely electric case. This does not mean that the magnetic field is irrelevant. In fact many properties crucially depend on it (see, e.g., Lieb, Siedentop, and Solovej [17]). We emphasize that φ_q and $\mathfrak{A}_q$ are not necessarily the external potentials only; in fact these potentials might also take into account other electrons and positrons. Given φ_q and $\mathfrak{A}_q$, we define the one electron space as

$$\mathfrak{H}_e := \mathfrak{H}_+ := \Lambda_+(\mathfrak{H}) = [\chi_{[0,\infty)}(D_{\varphi_q,\mathfrak{A}_q})](\mathfrak{H}). \tag{3}$$

The one positron space is

$$\mathfrak{H}_p := C\mathfrak{H}_- := C\Lambda_-(\mathfrak{H}) = C(1 - \Lambda_+)(\mathfrak{H}) \tag{4}$$

where C is the charge conjugation operator $(C\psi)(\mathfrak{x}) = \mathrm{i}\beta\alpha_2\overline{\psi(\mathfrak{x})}$.

The positron and the electron can be distinguished by the charge. Correspondingly the Fock space is

$$\mathfrak{F} := \bigoplus_{n,m\in\mathbb{N}_0} \mathfrak{F}^{n,m} := \bigoplus_{n,m\in\mathbb{N}_0} \underbrace{\left(\bigwedge_{\nu=1}^{n} \mathfrak{H}_e\right) \otimes \left(\bigwedge_{\mu=1}^{m} \mathfrak{H}_p\right)}_{=:\mathfrak{F}^{n,m}} \tag{5}$$

where we set $\bigwedge_{\nu=1}^{0} \mathfrak{H}_e := \bigwedge_{\nu=1}^{0} := \mathbb{C}$. We emphasize again that this construction depends on the potential. Therefore, also the electron creation and annihilation operators a^* and a and the corresponding positron operators b^* and b will depend on the electromagnetic potential $\varphi_q, \mathfrak{A}_q$ as we see in Subsect. 2.3.

The question which electric potential φ_q and which magnetic vector potential $\mathfrak{A}_q$ (and thus $\mathfrak{H}_+$) should be used is not settled in the physics literature. If the free Dirac operator is used one speaks of the free picture; if the external field is included one speaks of the Furry picture. We refer to Sucher [26–28] for this terminology and details of these choices. Mittleman [19] stated that the potential that gives the highest ground state energy should be used.

2.3 Creation and Annihilation Operators

We will define these operators explicitly for given $f \in L^2(\mathbb{R}^3)\otimes\mathbb{C}^4$. It suffices to do this component wise.

Electron Annihilation Operators $a(f)$

On the subspace $\mathfrak{F}^{(n+1,m)}$ it acts as

$$\begin{aligned}&(a(f)\psi)^{(n,m)}(x_1,\dots,x_n;y_1,\dots,y_m)\\ &\quad= \sqrt{n+1}\int_G \mathrm{d}x\,\overline{(\Lambda_+ f)(x)}\,\psi^{(n+1,m)}(x,x_1,\dots,x_n;y_1,\dots,y_m)\,.\end{aligned} \tag{6}$$

Electron Creation Operators $a^*(f)$

On the subspace $\mathfrak{F}^{(n-1,m)}$ into $\mathfrak{F}^{(n,m)}$ it acts as

$$\begin{aligned}&(a^*(f)\psi)^{(n,m)}(x_1,\dots,x_n;y_1,\dots,y_m)\\ &\quad= \frac{1}{\sqrt{n}}\sum_{j=1}^{n}(-1)^{j+1}\Lambda_+ f(x_j)\psi^{(n-1,m)}(x_1,\dots,\hat{x}_j,\dots,x_n;y_1,\dots,y_m)\,.\end{aligned} \tag{7}$$

As usual the hat indicates that the corresponding argument is omitted. The operator $a^*(f)$ is the adjoint of $a(f)$ and the map $f \mapsto a^*(f)$ is linear.

Positron Annihilation Operators $b(f)$

On the subspace $\mathfrak{F}^{(n,m+1)}$ it acts as

$$\begin{aligned}(b(g)\psi)^{(n,m)}(x_1,\dots,x_n;y_1,\dots,y_m)\\ = (-1)^n\sqrt{m+1}\int_G \mathrm{d}y\overline{[C\Lambda_- g](y)}\psi^{(n,m+1)}(x_1,\dots,x_n;y,y_1,\dots,y_m)\,. \quad (8)\end{aligned}$$

Positron Creation Operators $b^*(f)$

On the subspace $\mathfrak{F}^{(n,m-1)}$ it acts as

$$\begin{aligned}(b^*(g)\psi)^{(n,m)}(x_1,\dots,x_n;y_1,\dots,y_m)\\ = \frac{(-1)^n}{\sqrt{m}}\sum_{k=1}^{m}(-1)^{k+1}[C\Lambda_- g](y_k)\psi^{(n,m-1)}(x_1,\dots,x_n;y_1,\dots,\hat{y}_k,\dots,y_m)\,. \quad (9)\end{aligned}$$

We observe that, due to the anti-linearity of C, the map $g \mapsto b(g)$ is linear and the map $g \mapsto b^*(g)$ is anti-linear.

The Canonical Anti-Commutation Relations

These operators fulfill the canonical anti-commutation relations

$$\begin{aligned}\{a(f),a(g)\} &= \{a^*(f),a^*(g)\} = 0\,, &(10)\\ \{a(f),a^*(g)\} &= (f,\Lambda_+ g)\,, &(11)\\ \{b(f),b(g)\} &= \{b^*(f),b^*(g)\} = 0\,, &(12)\\ \{b^*(f),b(g)\} &= (f,\Lambda_- g)\,, &(13)\\ \{b^\#(f),a^\#(g)\} &= 0 &(14)\end{aligned}$$

where a # indicates a starred or unstarred operator.

The Field Operators

The field operators of the electron-positron field are defined as

$$\Psi(f) := a(f) + b^*(f)\,. \quad (15)$$

They fulfill

$$\begin{aligned}\{\Psi(f_1),\Psi(f_2)\} &= \{\Psi^*(f_1),\Psi^*(f_2)\} = 0\,, &(16)\\ \{\Psi(f_1),\Psi^*(f_2)\} &= (f_1,f_2)\mathbf{1}\,, &(17)\end{aligned}$$

for $f_1, f_2 \in \mathfrak{H}$.

We have also

$$\|\Psi(f)\psi\|^2 + \|\Psi^*(f)\psi\|^2 = (\psi, \{\Psi(f), \Psi^*(f)\}\psi)_{\mathfrak{F}} = \|f\|^2\|\psi\|^2 , \tag{18}$$

and hence, after some computation,

$$\|\Psi(f)\| = \|\Psi^*(f)\| = \|f\| . \tag{19}$$

We remark that the vacuum vector $|0\rangle := 1 \in \mathfrak{F}^{(0,0)} = \mathbb{C}$ satisfies the equations

$$\Psi(f)|0\rangle = 0, \ \forall f \in \mathfrak{H}_+ , \tag{20}$$

and

$$\Psi^*(f)|0\rangle = 0, \ \forall f \in \mathfrak{H}_- . \tag{21}$$

This also characterizes the vacuum state up to a constant of modulus one.

3 The Hamiltonian

3.1 The Unrenormalized Hamiltonian

It is convenient to pick an orthonormal basis $\ldots, e_{-2}, e_{-1}, e_1, e_2, \ldots \in H^1(G)$ where we assume that positive indices refer to elements in $\mathfrak{H}_+$ and negative indices refer to $\mathfrak{H}_-$, although the following construction will not depend on the choice. The formal or unrenormalized Hamiltonian of the Dirac field of electrons interacting via Coulomb forces with the nucleus and with each other is

$$\mathbb{H}_{\mathrm{ur}} := \sum_{m,n} D_{m.n}\Psi(e_m)^*\Psi(e_n) + \frac{\alpha}{2}\sum_{k,l,m,n} W_{k,l;m,n}\Psi^*(e_k)\Psi(e_l)\Psi^*(e_m)\Psi(e_n) \tag{22}$$

where

$$D_{m,n} := (e_m, D_\varphi e_n)$$

and

$$W_{k,l;m,n} := \int_G \mathrm{d}x \int_G \mathrm{d}y \overline{e_k(x)} e_l(y) \overline{e_m(x)} e_n(y) |\mathfrak{x} - \mathfrak{y}|^{-1} .$$

Although the Hamiltonian (22) looks superficially as the one of the non-relativistic electron field, it is very much different since $D_{m,n}$ is the matrix of an operator which is unbounded from below; in particular, a ground state does not exist. The purpose of these notes is to replace this formal expression with an expression that has a definite meaning in the *no-particle* setting, in the *one-particle* setting, and for so called *quasi-free states* in the general setting.

It is common to write $a(x) := \sum_n a(e_n)e_n(x)$, $a^*(x) := \sum_n a^*(e_n)\overline{e_n(x)}$, $b(x) := \sum_n b(e_n)\overline{e_n(x)}$, and $b^*(x) := \sum_n b^*(e_n)e_n(x)$. Analogously one defines

$$\Psi(x) := a(x) + b^*(x) = \sum_n (a(e_n) + b^*(e_n))e_n(x) .$$

With the help of this definition, one can write a second quantized integral operator A on $\mathfrak{H}$ with kernel $A(x,y)$ as follows

$$\begin{aligned} \mathbb{A}_{\mathrm{ur}} &= \sum_{m,n} (e_m, Ae_n)\Psi^*(e_m)\Psi(e_n) \\ &= \int_G \mathrm{d}x \sum_{m,n} \int_G \mathrm{d}y \overline{e_m(x)} A(x,y) e_n(y) \Psi^*(e_m)\Psi(e_n) \\ &= \int_G \mathrm{d}x \int_G \mathrm{d}y A(x,y)\Psi^*(x)\Psi(y) . \end{aligned} \tag{23}$$

Local operators such as multiplication and differential operators are brought in this form using the delta function and yields for the unrenormalized Hamiltonian

$$\mathbb{H}_{\mathrm{ur}} := \int_{\mathbb{R}^3} \mathrm{d}\mathfrak{x} \boldsymbol{\Psi}^*(\mathfrak{x}) D_{\varphi,\mathfrak{x}} \boldsymbol{\Psi}(\mathfrak{x}) + \alpha \underbrace{\frac{1}{2} \int_{\mathbb{R}^3} \mathrm{d}\mathfrak{x} \int_{\mathbb{R}^3} \mathrm{d}\mathfrak{y} \frac{\boldsymbol{\Psi}^*(\mathfrak{x})\boldsymbol{\Psi}(\mathfrak{x})\boldsymbol{\Psi}^*(\mathfrak{y})\boldsymbol{\Psi}(\mathfrak{y})}{|\mathfrak{x}-\mathfrak{y}|}}_{=:\mathbb{W}_{\mathrm{ur}}} \tag{24}$$

where $\boldsymbol{\Psi}(\mathfrak{x}) = (\Psi(\mathfrak{x},1), \dots, \Psi(\mathfrak{x},4))^t$.

3.2 The Normal Ordered Hamiltonian

The normal or Wick ordered Hamiltonian $\mathbb{H}$ is obtained from the unordered one by writing all field operators in terms of electron-positron annihilation-creation operators and anti-commuting all starred terms to the left of unstarred ones.

One-Particle Operators

The above rule gives

$$\begin{aligned} &: \Psi^*(f)\Psi(g) : = : (a^*(f) + b(f))(a(g) + b^*(g)) : \\ &= a^*(f)a(g) + a^*(f)b^*(g) + b(f)a(g) - b^*(g)b(f) . \end{aligned} \tag{25}$$

Thus,

$$\int_{\mathbb{R}^3} \boldsymbol{\Psi}^*(\mathfrak{x}) D_{\varphi,\mathfrak{x}} \boldsymbol{\Psi}(\mathfrak{x}) \mathrm{d}\mathfrak{x} = \mathbb{D}_\varphi + \sum_{\sigma,\tau} [D_{\varphi,\mathfrak{x},\sigma,\tau} \Lambda_-(\mathfrak{x}',\sigma;\mathfrak{x},\tau)]_{\mathfrak{x}'=\mathfrak{x}} \mathrm{d}\mathfrak{x} , \tag{26}$$

i.e., the normal ordered Hamiltonian

$$\mathbb{D}_\varphi := \int_{\mathbb{R}^3} : \boldsymbol{\Psi}^*(\mathfrak{x}) D_{\varphi,\mathfrak{x}} \boldsymbol{\Psi}(\mathfrak{x}) : \tag{27}$$

differs from the unordered one by a physically uninteresting constant.

Two-Particle Operators

The above general rule gives

$$\mathbb{W}_{\mathrm{ur}} = \mathbb{W} + \frac{1}{2}\int_G \mathrm{d}x \int_G \mathrm{d}y : \Psi^*(x)\Psi(y) : \underbrace{\frac{\Lambda_+(x,y) - \Lambda_-(x,y)}{|\mathfrak{x} - \mathfrak{y}|}}_{=:-2X^{\varphi_q}(x,y)}$$
$$+ \int_G \mathrm{d}x : \Psi^*(x)\Psi(x) : \int_{\mathbb{R}^3} \mathrm{d}\mathfrak{y} |\mathfrak{x} - \mathfrak{y}|^{-1} \underbrace{\sum_\tau \Lambda_-(\mathfrak{y},\tau;\mathfrak{y},\tau)}_{=:\rho^{\varphi_q}(\mathfrak{y})}$$
$$+ \frac{1}{2}\int_G \mathrm{d}x \int_G \mathrm{d}y \frac{\Lambda_+(x,y)\Lambda_-(x,y)}{|\mathfrak{x} - \mathfrak{y}|} + \frac{1}{2}\int_G \mathrm{d}x \int_G \mathrm{d}y \frac{\Lambda_-(x,x)\Lambda_-(y,y)}{|\mathfrak{x} - \mathfrak{y}|}, \tag{28}$$

where $\mathbb{W} := \int_{\mathbb{R}^3} \mathrm{d}\mathfrak{x} \int_{\mathbb{R}^3} \mathrm{d}\mathfrak{y} : \psi^*(\mathfrak{x})\psi(\mathfrak{x})\psi^*(\mathfrak{y})\psi(\mathfrak{y}) : |\mathfrak{x} - \mathfrak{y}|^{-1}$ is the totally normal ordered interaction. The last two terms are unimportant constants. The second term on the right is the exchange energy of the electron-positron field with the Dirac sea and its complement; the third term is the electrostatic interaction energy with the Dirac sea. (Λ_- is the density matrix of the Dirac sea electrons and ρ^{φ_q} is its density.)

Thus, modulo physically unimportant constants, we have

$$\mathbb{H}_{\mathrm{ur}} = \mathbb{D}_\varphi + \alpha\mathbb{W} + \alpha \int_{\mathbb{R}^3} \mathrm{d}\mathfrak{x} |\cdot|^{-1} * \rho^{\varphi_q}(\mathfrak{x}) : \boldsymbol{\Psi}^*(\mathfrak{x})\boldsymbol{\Psi}(\mathfrak{x}) :$$
$$- \alpha \int_G \mathrm{d}x \int_G \mathrm{d}y X^{\varphi_q}(x,y) : \Psi^*(x)\Psi(y) : . \tag{29}$$

The last two terms are one particle operators. They will influence the fine structure of spectral lines of single particles. Their effect will be investigated in Chap. 4. The many body character of the theory is contained solely in the last term of the first line. Its behavior is far from being understood. The last two terms of the right hand will be responsible for the vacuum polarization. We will discuss this effect in Sect. 4 in a one-body context. Since it affects the fine details of spectral lines only, one may – as we will do in Sect. 5 – consider a model without these terms and concentrate on the expectations of the first line in quasi-free states.

3.3 Physical Principles for Manipulating Hamiltonians

We will use three guiding principles to transform expressions for the energy into other physically equivalent ones as formulated and justified by Weisskopf [30], p. 6: *"The following three properties of the vacuum electrons are assumed to be irrelevant:*

W1 *The energy of the vacuum electrons in field free space.*

W2 *The charge and current density of the vacuum electrons in field free space.*

W3 *A field independent electric and magnetic polarizability that is constant in space and time."*

These principles can be viewed as a condensate of previous results that started with the work of Dirac [8, 9] and Heisenberg [14].

4 The Vacuum Polarization of a Nucleus

4.1 One Electron in the Field of a Nucleus

In this chapter we will consider expectations of $\mathbb{D}+\alpha\mathbb{W}$ in one-electron states. We wish to show that this quadratic form defines a self-adjoint Hamiltonian that is bounded from below.

Throughout this chapter we will assume that the nuclear electric potential $\varphi = |\cdot|^{-1} * n$ is given by a nonnegative spherically symmetric charge density $n \in L^1(\mathbb{R}^3) \cap L^{3/2+\delta}(\mathbb{R}^3)$ for some positive δ. For technical convenience we will also assume that $\hat{n} \in C_0^\infty(\mathbb{R}^3)$. As usual we call $\int n = eZ$ the total charge of the nucleus and Z its atomic number.

This assumption implies immediately corresponding integrability properties for the electric potential.

Lemma 1. *The nuclear potential fulfills $\varphi \in L^{3+\epsilon}(\mathbb{R}^3) \cap L^\infty(\mathbb{R}^3)$ for all positive ϵ and is spherically symmetric.*

Proof. To show this, we decompose the Coulomb kernel and set

$$I := \chi_{B_R(0)}|\cdot|^{-1}, \quad A := |\cdot|^{-1} - I \, .$$

Boundedness: using the above decomposition we have

$$||\cdot|^{-1} * n(\mathfrak{x})| = I * n(\mathfrak{x}) + A * n(\mathfrak{x}) \leq c_{\delta,R}\|n\|_{3/2+\delta} + eZ/R. \tag{30}$$

We remark for later purposes that $c_{\delta,R} \to 0$ as $R \to 0$.

$L^{3+\epsilon}$-property: we estimate

$$\||\cdot|^{-1}*n\|_{3+\epsilon} \leq \|I*n\|_{3+\epsilon}+\|A*n\|_{3+\epsilon} \leq c(\|I\|_{\frac{\epsilon+3}{\epsilon+1}}\|n\|_{1+\frac{\epsilon}{3}}+\|A\|_{3+\epsilon}\|n\|_1) \tag{31}$$

using again the above decomposition and Young's inequality. □

Equipped with the three principles W1 through W3 of Weisskopf we begin to modify the one particle interaction of the Hamiltonian. In a first step we will subtract the potential $|\cdot|^{-1} * \rho^0$ and the exchange energy X^0 of the free Dirac operator. This yields the Hamiltonian

$$\mathbb{D}_\varphi + \alpha\mathbb{W} + \alpha \int_{\mathbb{R}^3} \mathrm{d}\mathfrak{x} |\cdot|^{-1} * (\rho^{\varphi_q} - \rho^0)(\mathfrak{x}) : \boldsymbol{\Psi}^*(\mathfrak{x})\boldsymbol{\Psi}(\mathfrak{x}) : \\ - \alpha \int_G \mathrm{d}x \int_G \mathrm{d}y \underbrace{(X^{\varphi_q}(x,y) - X^0(x,y))}_{X(x,y):=} : \Psi^*(x)\Psi(y) : . \quad (32)$$

The subtraction from ρ^{φ_q} can be viewed as a (pre-)charge renormalization; the subtraction from X^{φ_q} is called a mass renormalization.

We now choose $\varphi_q := \varphi$. In this case the mass renormalized exchange energy operator X turns out to be well defined. However, the density difference $\rho^\varphi - \rho^0$ is still pointwise infinite. To proceed we will renormalize the charge once more in a way that does not change the physics by subtracting a field independent counter term γ_c of the difference $Q^\varphi := \Lambda_-^\varphi - \Lambda_-^0$. To this end, we write γ_c as a function of $\boldsymbol{\xi} := \mathfrak{x} - \mathfrak{y}$ as the first variable and $\boldsymbol{\eta} := (\mathfrak{x}+\mathfrak{y})/2$ as the second variable and Fourier transform in the second variable with Fourier variable $\mathfrak{k}$. We call the resulting integral kernel $\tilde{\gamma}_c(\boldsymbol{\xi}, \mathfrak{k})$. It is

$$\tilde{\gamma}_c(\boldsymbol{\xi}, \mathfrak{k}) := \hat{\varphi}(\mathfrak{k})\mathfrak{k}^2 F_0(\xi, \mathfrak{k}/|\mathfrak{k}|) ; \quad (33)$$

formally,

$$F_0(\boldsymbol{\xi}, \mathfrak{k}/|\mathfrak{k}|) = -\frac{1}{16\pi^3} \int_{\mathbb{R}^3} \mathrm{d}\mathfrak{p} \frac{1 + \mathfrak{p}^2 \sin^2\theta}{(1 + \mathfrak{p}^2)^{5/2}} \mathrm{e}^{\mathrm{i}\mathfrak{p}\cdot\boldsymbol{\xi}} , \quad (34)$$

where θ is the angle between $\mathfrak{k}$ and $\mathfrak{p}$. This density matrix is – for $\boldsymbol{\xi} = 0$ – obviously translationally invariant and not time dependent and can therefore be subtracted according to W3. It is also obvious that it diverges on the diagonal.

The fully charge renormalized density matrix of the polarized Dirac sea is $\Lambda_-^\varphi - \Lambda_-^0 - \gamma_c$ and its density is

$$\rho(\mathfrak{x}) := \mathrm{tr}_{\mathbb{C}^4}(\Lambda_-^\varphi - \Lambda_-^0 - \gamma_c)(\mathfrak{x}, \mathfrak{x}) . \quad (35)$$

Thus, the completely one-particle renormalized Hamiltonian is

$$\boxed{\begin{aligned} \mathbb{H} :=& \mathbb{D}_\varphi + \alpha \int_{\mathbb{R}^3} \mathrm{d}\mathfrak{x} |\cdot|^{-1} * \rho(\mathfrak{x}) : \boldsymbol{\Psi}^*(\mathfrak{x})\boldsymbol{\Psi}(\mathfrak{x}) : \\ & - \alpha \int_G \mathrm{d}x \int_G \mathrm{d}y X(x,y) : \Psi^*(x)\Psi(y) : + \alpha\mathbb{W} . \end{aligned}} \quad (36)$$

Our interest is to show that the quadratic form

$$\begin{aligned} \mathcal{E}_1 := \mathfrak{H}_+(G) \cap H^1(G) &\to \mathbb{R} \\ \psi &\mapsto (\psi, \mathbb{H}\psi) = \big(\psi, (D_\varphi + \alpha|\cdot|^{-1} * \rho - \alpha X)\psi\big) \end{aligned} \quad (37)$$

is well defined and bounded from below, since this will define a distinguished self-adjoint Hamiltonian in the one-electron sector. This will follow from

Theorem 1 (Hainzl and Siedentop [13]). *The operators* $|\cdot|^{-1} * \rho$ *and* X *are relatively bounded with respect to the form* $(\psi, D_\varphi \psi)$ *on* $\mathfrak{H}_+(G) \cap H^1(G)$ *with form bound zero.*

We will prove the theorem in a sequence of Lemmata showing the claim for X and pieces of $|\cdot|^{-1} * \rho$ separately. These pieces come naturally when analyzing the singular structure of $Q^\varphi := \Lambda_-^\varphi - \Lambda_-^0$: using the Cauchy formula (Kato [16]) and the resolvent equation we obtain

$$Q^\varphi = \frac{1}{2\pi} \int_{-\infty}^{\infty} \mathrm{d}\eta \left(\frac{1}{D_\varphi + \mathrm{i}\eta} - \frac{1}{D_0 + \mathrm{i}\eta} \right) = \sum_{n=1}^{4} \alpha^n Q_n \,, \tag{38}$$

where the index n indicates the number of φ's in the expression, i.e.,

$$Q_n := \begin{cases} \frac{1}{2\pi} \int_{-\infty}^{\infty} \mathrm{d}\eta \frac{1}{D_0+\mathrm{i}\eta} \left(\varphi \frac{1}{D_0+\mathrm{i}\eta} \right)^n & n \leq 3 \\ \frac{1}{2\pi} \int_{-\infty}^{\infty} \mathrm{d}\eta \frac{1}{D_0+\mathrm{i}\eta} \varphi \frac{1}{D_0+\mathrm{i}\eta} \varphi \frac{1}{D_\varphi+\mathrm{i}\eta} \varphi \frac{1}{D_0+\mathrm{i}\eta} \varphi \frac{1}{D_0+\mathrm{i}\eta} & n = 4 \end{cases} \tag{39}$$

We remark that the density of Q_2 vanishes: the terms linear in the Dirac matrices vanish after summation over the spin variable σ, since the Dirac matrices are traceless; the remaining terms are odd in η and vanish after integration over η. Thus, Q_2 does not contribute to the vacuum polarization density. The two terms

$$\rho_3(\mathfrak{x}) := \mathrm{tr}_{\mathbb{C}^4} Q_3(\mathfrak{x}, \mathfrak{x}), \;\; \rho_4(\mathfrak{x}) := \mathrm{tr}_{\mathbb{C}^4} Q_4(\mathfrak{x}, \mathfrak{x}) \tag{40}$$

will turn out to be well defined and yields an electric potential which is form bounded with respect to $(f, D_\varphi f)$ with form bound zero.

Lemma 2. *The quadratic form* $P_4[\psi] := \mathrm{tr}(\chi Q_4)$ *is bounded relative to* $(\psi, D_\varphi \psi)$ *on* $\mathfrak{H}_+ \cap H^1(G)$ *with form bound zero, where* $\chi := |\cdot|^{-1} * \psi^2$, *i.e., for every* $\epsilon > 0$ *there exists a constant* M *such that for all* $\psi \in \mathfrak{H}_+ \cap H^1(G)$ *we have*

$$|P_4[\psi]| \leq \epsilon(\psi, D_\varphi \psi) + M(\psi, \psi) \,.$$

Proof. We have

$$\begin{aligned} |\,\mathrm{tr}(\chi Q_4)| &\leq \frac{1}{2\pi} \int_{\mathbb{R}} \mathrm{d}\eta \left\| \chi \frac{1}{D_0 + \mathrm{i}\eta} \right\|_5 \left\| \varphi \frac{1}{D_0 + \mathrm{i}\eta} \right\|_5^3 \left\| \varphi \frac{1}{D_\varphi + \mathrm{i}\eta} \right\|_5 \\ &\leq \frac{1}{2\pi} \int_{\mathbb{R}} \mathrm{d}\eta \left\| \chi \frac{1}{D_0 + \mathrm{i}\eta} \right\|_5 \left\| \varphi \frac{1}{D_0 + \mathrm{i}\eta} \right\|_5^4 \left\| (D_0 + \mathrm{i}\eta) \frac{1}{D_\varphi + \mathrm{i}\eta} \right\|_\infty . \end{aligned} \tag{41}$$

The last factor is bounded by $1 + \|e\varphi(D_\varphi + \mathrm{i}\eta)^{-1}\|$ which is finite, since φ is bounded and $0 \notin \sigma(D_\varphi)$. Thus, according to Hölder, it suffices to bound $\int_{\mathbb{R}} \mathrm{d}\eta \|\chi(D_0 + \mathrm{i}\eta)^{-1}\|_5^5$ and $\int_{\mathbb{R}} \mathrm{d}\eta \|\varphi(D_0 + \mathrm{i}\eta)^{-1}\|_5^5$. By an inequality of Seiler and Simon [25] we have

$$\int_{\mathbb{R}} \mathrm{d}\eta \left\| \varphi \frac{1}{D_0 + \mathrm{i}\eta} \right\|_5^5 \leq \int_{\mathbb{R}} \mathrm{d}\eta \left\| \varphi \frac{1}{\sqrt{-\Delta + m^2 + \eta^2}} \right\|_5^5 \leq c\|\varphi\|_5^5 \int_{\mathbb{R}^4} \frac{\mathrm{d}\mathfrak{k}}{\sqrt{\mathfrak{k}^2 + m^2}^5} \tag{42}$$

which is finite since $\varphi \in L^5(\mathbb{R}^3)$ (see (30) and (31)).

To estimate $\int_{\mathbb{R}} \mathrm{d}\eta \|\chi(D_0 + \mathrm{i}\eta)^{-1}\|_5^5$ we proceed as in (42) and then use (30) with $\epsilon = 6$ followed by the Sobolev inequality

$$\||\cdot|^{-1} * \psi^2\| \leq c\|I\|_{9/7}\|\psi\|_3^2 + c\|A\|_9\|\psi\|_2^2 \leq c_R(\psi, |\mathfrak{p}|\psi) + c\|A\|_9\|\psi\|_2^2 \,, \tag{43}$$

where $c_R \to 0$ as $R \to 0$. This proves that P_4 is relatively form bounded with respect to $|D_\varphi|$ with form bound 0, since for bounded potential $|\mathfrak{p}| \leq c(|D_\varphi| + 1)$. □

Lemma 3. *The quadratic form* $P_3[\psi] := \mathrm{tr}(\chi Q_3)$ *is bounded.*

Proof. We have

$$(\chi, \rho_3) = (\hat{\chi}, \widehat{\rho_3}) = (2\pi)^{-3/2} \int_{\mathbb{R}^3} \mathrm{d}\mathfrak{p}_1 \int_{\mathbb{R}^3} \mathrm{d}\mathfrak{p}_2 \sum_{\sigma=1}^{4} \hat{\chi}(\mathfrak{p}_1 - \mathfrak{p}_2) \hat{Q}_3(\mathfrak{p}_1, \sigma; \mathfrak{p}_2, \sigma) \,, \tag{44}$$

where ρ_3 is defined by (39) and (40). The "eigenfunctions" of the free Dirac operator in momentum space are

$$u_\tau(\mathfrak{p}) := \begin{cases} \frac{1}{N_+(\mathfrak{p})} \begin{pmatrix} \sigma \cdot \mathfrak{p} \mathbf{e}_\tau \\ -(1 - E(\mathfrak{p}))\mathbf{e}_\tau \end{pmatrix} & \tau = 1, 2 \,, \\ \frac{1}{N_-(\mathfrak{p})} \begin{pmatrix} \sigma \cdot \mathfrak{p} \, \mathbf{e}_\tau \\ -(1 + E(\mathfrak{p}))\mathbf{e}_\tau \end{pmatrix} & \tau = 3, 4 \,. \end{cases}$$

with $\mathbf{e}_\tau := (1, 0)^t$ for $\tau = 1, 3$ and $\mathbf{e}_\tau := (0, 1)^t$ for $\tau = 2, 4$ and

$$N_+(\mathfrak{p}) = \sqrt{2E(\mathfrak{p})(E(\mathfrak{p}) - 1)}, \;\; N_-(\mathfrak{p}) = \sqrt{2E(\mathfrak{p})(E(\mathfrak{p}) + 1)} \,. \tag{45}$$

The indices 1 and 2 refer to positive "eigenvalue" $E(\mathfrak{p})$ and the indices 3 and 4 to negative $-E(\mathfrak{p})$. (See, e.g., Evans et al. [10].) Using Plancherel's theorem we get

$$\begin{aligned} (\psi, P_3\psi) = \frac{1}{(2\pi)^7} \int_{\mathbb{R}^3} \mathrm{d}\mathfrak{p}_1 \int_{\mathbb{R}^3} \mathrm{d}\mathfrak{p}_2 \int_{\mathbb{R}^3} \mathrm{d}\mathfrak{p}_3 \int_{\mathbb{R}^3} \mathrm{d}\mathfrak{p}_4 \sum_{\tau_1,\tau_2,\tau_3,\tau_4=1}^{4} \\ \hat{\chi}(\mathfrak{p}_1 - \mathfrak{p}_2)\hat{\varphi}(\mathfrak{p}_2 - \mathfrak{p}_3)\hat{\varphi}(\mathfrak{p}_3 - \mathfrak{p}_4)\hat{\varphi}(\mathfrak{p}_4 - \mathfrak{p}_1) \times \\ \times \langle u_{\tau_1}(\mathfrak{p}_1)|u_{\tau_2}(\mathfrak{p}_2)\rangle \langle u_{\tau_2}(\mathfrak{p}_2)|u_{\tau_3}(\mathfrak{p}_3)\rangle \langle u_{\tau_3}(\mathfrak{p}_3)|u_{\tau_4}(\mathfrak{p}_4)\rangle \langle u_{\tau_4}(\mathfrak{p}_4)|u_{\tau_1}(\mathfrak{p}_1)\rangle \\ \times \int_{-\infty}^{\infty} \mathrm{d}\eta \frac{1}{(\mathrm{i}a_{\tau_1}E(\mathfrak{p}_1) - \eta)(\mathrm{i}a_{\tau_2}E(\mathfrak{p}_2) - \eta)(\mathrm{i}a_{\tau_3}E(\mathfrak{p}_3) - \eta)(\mathrm{i}a_{\tau_4}E(\mathfrak{p}_4) - \eta)} \end{aligned} \tag{46}$$

with $a_\tau = 1$ for $\tau = 1,2$ and $a_\tau = -1$ for $\tau = 3,4$. The integral over η is seen to vanish by Cauchy's theorem, if all four a_{τ_j} have the same sign. In fact we have to distinguish only two cases, namely three of the a_{τ_j} are equal and two of the a_{τ_j} are equal.

Therefore, we will only treat two different cases in the following. The remaining ones work analogously.

We begin with $a_{\tau_1} = -1$ and $a_{\tau_2} = a_{\tau_3} = a_{\tau_4} = 1$. In that case the first factor in (46) reads

$$\begin{aligned}&\sum_{\tau_1=3,4} \langle u_{\tau_1}(\mathfrak{p}_1)|u_{\tau_2}(\mathfrak{p}_2)\rangle \sum_{\tau_2=1,2} \langle u_{\tau_2}(\mathfrak{p}_2)|u_{\tau_3}(\mathfrak{p}_3)\rangle \\ &\qquad \times \sum_{\tau_3=1,2} \langle u_{\tau_3}(\mathfrak{p}_3)|u_{\tau_4}(\mathfrak{p}_4)\rangle \sum_{\tau_4=1,2} \langle u_{\tau_4}(\mathfrak{p}_4)|u_{\tau_1}(\mathfrak{p}_1)\rangle = \\ &\mathrm{tr}_{\mathbb{C}^2}\Bigg[\frac{\sigma\cdot\mathfrak{p}_1\sigma\cdot\mathfrak{p}_2 + (1+E(\mathfrak{p}_1))(1-E(\mathfrak{p}_2))}{N_-(\mathfrak{p}_1)^2 N_+(\mathfrak{p}_2)^2 N_+(\mathfrak{p}_3)^2 N_+(\mathfrak{p}_4)^2}\left[\sigma\cdot\mathfrak{p}_2\sigma\cdot\mathfrak{p}_3 + (1-E(\mathfrak{p}_2))(1-E(\mathfrak{p}_3))\right] \\ &\times \left[\sigma\cdot\mathfrak{p}_3\sigma\cdot\mathfrak{p}_4 + (1-E(\mathfrak{p}_3))(1-E(\mathfrak{p}_4))\right]\left[\sigma\cdot\mathfrak{p}_4\sigma\cdot\mathfrak{p}_1 + (1-E(\mathfrak{p}_4))(1+E(\mathfrak{p}_1))\right]\Bigg].\end{aligned} \tag{47}$$

We estimate the modulus of (47) and obtain

$$\begin{aligned}|(47)| &\le \mathrm{c}\frac{\mathrm{tr}_{\mathbb{C}^2}\,|\sigma\cdot\mathfrak{p}_4\sigma\cdot\mathfrak{p}_1 + (1-E(\mathfrak{p}_4))(1+E(\mathfrak{p}_1))|}{N_-(\mathfrak{p}_1)N_+(\mathfrak{p}_4)} \\ &\le \mathrm{c}\frac{|\mathfrak{p}_4\cdot\mathfrak{p}_1 - (E(\mathfrak{p}_4)-1)(1+E(\mathfrak{p}_1))| + |\mathfrak{p}_4\wedge\mathfrak{p}_1|}{N_-(\mathfrak{p}_1)N_+(\mathfrak{p}_4)}.\end{aligned}$$

(Here and in the following c is a generic positive constant.) Since

$$\begin{aligned}&-\frac{1}{2\pi}\int_{-\infty}^{\infty} \mathrm{d}\eta \frac{1}{(-\mathrm{i}E(\mathfrak{p}_1)-\eta)(\mathrm{i}E(\mathfrak{p}_2)-\eta)(\mathrm{i}E(\mathfrak{p}_3)-\eta)(\mathrm{i}E(\mathfrak{p}_4)-\eta)} \\ &\qquad = \frac{1}{(E(\mathfrak{p}_2)+E(\mathfrak{p}_1))(E(\mathfrak{p}_3)+E(\mathfrak{p}_1))(E(\mathfrak{p}_4)+E(\mathfrak{p}_1))}\end{aligned}$$

our term of interest (46) is bounded by a constant times

$$\begin{aligned}&\int_{\mathbb{R}^3}\mathrm{d}\mathfrak{p}_1\int_{\mathbb{R}^3}\mathrm{d}\mathfrak{p}_2\int_{\mathbb{R}^3}\mathrm{d}\mathfrak{p}_3\int_{\mathbb{R}^3}\mathrm{d}\mathfrak{p}_4 |\hat{\chi}(\mathfrak{p}_1-\mathfrak{p}_2)\hat{\varphi}(\mathfrak{p}_2-\mathfrak{p}_3)\hat{\varphi}(\mathfrak{p}_3-\mathfrak{p}_4)\hat{\varphi}(\mathfrak{p}_4-\mathfrak{p}_1)| \\ &\times \frac{|\mathfrak{p}_4\cdot\mathfrak{p}_1 - (E(\mathfrak{p}_4)-1)(E(\mathfrak{p}_1)+1)| + |\mathfrak{p}_4\wedge\mathfrak{p}_1|}{N_-(\mathfrak{p}_1)N_+(\mathfrak{p}_4)(E(\mathfrak{p}_2)+E(\mathfrak{p}_1))(E(\mathfrak{p}_3)+E(\mathfrak{p}_1))(E(\mathfrak{p}_4)+E(\mathfrak{p}_1))}.\end{aligned} \tag{48}$$

Substituting $\mathfrak{p}_2 \to \mathfrak{p}_1 + \mathfrak{p}_2$ turns (48) into

$$\int_{\mathbb{R}^3} d\mathfrak{p}_1 \int_{\mathbb{R}^3} d\mathfrak{p}_2 \int_{\mathbb{R}^3} d\mathfrak{p}_3 \int_{\mathbb{R}^3} d\mathfrak{p}_4$$
$$|\hat{\chi}(-\mathfrak{p}_2)| \frac{|\hat{\varphi}(\mathfrak{p}_2 + \mathfrak{p}_1 - \mathfrak{p}_3)\hat{\varphi}(\mathfrak{p}_3 - \mathfrak{p}_4)\hat{\varphi}(\mathfrak{p}_4 - \mathfrak{p}_1)|}{N_-(\mathfrak{p}_1)N_+(\mathfrak{p}_4)}$$
$$\times \frac{|\mathfrak{p}_4 \cdot \mathfrak{p}_1 - (E(\mathfrak{p}_4) - 1)(E(\mathfrak{p}_1) + 1)| + |\mathfrak{p}_4 \wedge \mathfrak{p}_1|}{(E(\mathfrak{p}_2 + \mathfrak{p}_1) + E(\mathfrak{p}_1))(E(\mathfrak{p}_1) + E(\mathfrak{p}_3))(E(\mathfrak{p}_4) + E(\mathfrak{p}_1))}$$
$$= \int_{\mathbb{R}^3} d\mathfrak{p}_2 |\hat{\chi}(-\mathfrak{p}_2)| f(\mathfrak{p}_2), \quad (49)$$

where we introduce f to be the remaining integrand. We will now estimate f. Substituting $\mathfrak{p}_1 \to \mathfrak{p}_1 + \mathfrak{p}_4$, $\mathfrak{p}_3 \to \mathfrak{p}_3 + \mathfrak{p}_4$ we get

$$f(\mathfrak{p}_2) = \int_{\mathbb{R}^3} d\mathfrak{p}_1 \int_{\mathbb{R}^3} d\mathfrak{p}_3 \int_{\mathbb{R}^3} d\mathfrak{p}_4 |\hat{\varphi}(\mathfrak{p}_2 + \mathfrak{p}_1 - \mathfrak{p}_3)\hat{\varphi}(\mathfrak{p}_3)\hat{\varphi}(\mathfrak{p}_1)|$$
$$\times \frac{|\mathfrak{p}_4 \cdot (\mathfrak{p}_1 + \mathfrak{p}_4) - (E(\mathfrak{p}_4) - 1)(1 + E(\mathfrak{p}_1 + \mathfrak{p}_4))| + |\mathfrak{p}_4 \wedge \mathfrak{p}_1|}{N_-(\mathfrak{p}_1 + \mathfrak{p}_4)N_+(\mathfrak{p}_4)(E(\mathfrak{p}_2 + \mathfrak{p}_1 + \mathfrak{p}_4) + E(\mathfrak{p}_4 + \mathfrak{p}_1))}$$
$$\times \frac{1}{(E(\mathfrak{p}_1 + \mathfrak{p}_4) + E(\mathfrak{p}_3 + \mathfrak{p}_4))(E(\mathfrak{p}_4 + \mathfrak{p}_1) + E(\mathfrak{p}_4))}.$$

Since $E(\mathfrak{p}_1 + \mathfrak{p}_4) = E(\mathfrak{p}_4) + \mu \mathfrak{p}_4 \cdot \mathfrak{p}_1 / E(\mathfrak{p}_4)$ for some $\mu \in [0, 1]$, we see that

$$|\mathfrak{p}_4 \cdot (\mathfrak{p}_1 + \mathfrak{p}_4) - (E(\mathfrak{p}_4) - 1)(E(\mathfrak{p}_1 + \mathfrak{p}_4) + 1)| + |\mathfrak{p}_4 \wedge \mathfrak{p}_1| \leq 4|\mathfrak{p}_1||\mathfrak{p}_4|.$$

Notice that we can bound

$$\int_{\mathbb{R}^3} d\mathfrak{p}_4 \frac{|\mathfrak{p}_4|}{E(\mathfrak{p}_3 + \mathfrak{p}_4)E(\mathfrak{p}_4)^2 N_+(\mathfrak{p}_1 + \mathfrak{p}_4)} \leq \mathrm{c}$$

independently of $\mathfrak{p}_1$ and $\mathfrak{p}_3$. Therefore,

$$f(\mathfrak{p}_2) \leq \mathrm{c} \int_{\mathbb{R}^3} d\mathfrak{p}_1 \int_{\mathbb{R}^3} d\mathfrak{p}_3 |\hat{\varphi}(\mathfrak{p}_2 + \mathfrak{p}_1 - \mathfrak{p}_3)\hat{\varphi}(\mathfrak{p}_3)\hat{\varphi}(\mathfrak{p}_1)||\mathfrak{p}_1|.$$

Since $\hat{\varphi}(\mathfrak{k}) = 4\pi\hat{n}(\mathfrak{k})/\mathfrak{k}^2$, we have that f is bounded and compactly supported since $\hat{n}$ is compactly supported, $\mathfrak{p}_1$ and $\mathfrak{p}_2$ are bounded. Thus,

$$\int_{\mathbb{R}^3} d\mathfrak{p}_2 |\hat{\chi}(-\mathfrak{p}_2)| f(\mathfrak{p}_2) \leq 4\pi \|\widehat{\psi^2}\|_1 \|f\|_\infty \int_{\mathrm{supp} f} |\mathfrak{p}|^{-2} d\mathfrak{p} \leq \mathrm{c}_\varphi(\psi, \psi) \quad (50)$$

for a constant c_φ depending on φ.

Next, we take a peek at the case $a_{\tau_1} = a_{\tau_2} = 1$ and $a_{\tau_3} = a_{\tau_4} = -1$. The corresponding integral over η gives

$$\frac{1}{2\pi} \int_{-\infty}^{\infty} d\eta \frac{1}{(\mathrm{i}E(\mathfrak{p}_1) - \eta)(\mathrm{i}E(\mathfrak{p}_2) - \eta)(-\mathrm{i}E(\mathfrak{p}_3) - \eta)(-\mathrm{i}E(\mathfrak{p}_4) - \eta)}$$
$$= \frac{1}{(E(\mathfrak{p}_2) + E(\mathfrak{p}_3))(E(\mathfrak{p}_2) + E(\mathfrak{p}_4))(E(\mathfrak{p}_1) + E(\mathfrak{p}_4))}$$
$$+ \frac{1}{(E(\mathfrak{p}_2) + E(\mathfrak{p}_3))(E(\mathfrak{p}_1) + E(\mathfrak{p}_3))(E(\mathfrak{p}_1) + E(\mathfrak{p}_4))}.$$

Observe now that the corresponding first factor in (46) can be bounded by

$$c \cdot 4|\mathfrak{p}_2||\mathfrak{p}_3|/N_+(\mathfrak{p}_2)N_-(\mathfrak{p}_3)) \,.$$

Now, we do similar variable transforms as above and arrive at an analog of (50). □

We set $\rho_{1,r}(\mathfrak{x}) := \mathrm{tr}_{\mathbb{C}^4}\,(\alpha Q_1 - \gamma_c)\,(\mathfrak{x},\mathfrak{x})$. Its electric potential $U := |\cdot|^{-1} * \rho_1$ is called the Uehling potential.

Lemma 4. *The Uehling potential is bounded.*

Proof. It will be useful to introduce the function C

$$\begin{aligned} C(y) :=& \frac{1}{2}y^2 \int_0^1 \mathrm{d}x(1-x^2)\log[1+y^2(1-x^2)/4] \\ =& \frac{1}{3}y^2\left[\left(1-\frac{2}{y^2}\right)\sqrt{1+\frac{4}{y^2}}\log\frac{\sqrt{1+4/y^2}+1}{\sqrt{1+4/y^2}-1}+\frac{4}{y^2}-\frac{5}{3}\right] \end{aligned} \tag{51}$$

(Serber [24] and Uehling [29]).

The kernel $\hat{Q}_1(\mathfrak{p},\mathfrak{q})$ of Q_1 in momentum space is

$$\hat{Q}_1(\mathfrak{p},\mathfrak{q}) = (2\pi)^{-5/2}\int_{-\infty}^{\infty}\mathrm{d}\eta\frac{\boldsymbol{\alpha}\cdot\mathfrak{p}+\beta-\mathrm{i}\eta}{\mathfrak{p}^2+1+\eta^2}\hat{\varphi}(\mathfrak{p}-\mathfrak{q})\frac{\boldsymbol{\alpha}\cdot\mathfrak{q}+\beta-\mathrm{i}\eta}{\mathfrak{q}^2+1+\eta^2}\,, \tag{52}$$

which leads to

$$\mathrm{tr}_{\mathbb{C}^4}\hat{Q}_1(\mathfrak{p},\mathfrak{q}) = 2^{-1/2}\pi^{-3/2}\hat{\varphi}(\mathfrak{p}-\mathfrak{q})\frac{\mathfrak{p}\cdot\mathfrak{q}+1-E(\mathfrak{p})E(\mathfrak{q})}{E(\mathfrak{p})E(\mathfrak{q})(E(\mathfrak{p})+E(\mathfrak{q}))} \tag{53}$$

by a straightforward calculation with $E(\mathfrak{p}) = \sqrt{\mathfrak{p}^2+1}$. In configuration space we obtain introducing $\mathfrak{r} = \mathfrak{p} - \mathfrak{k}/2$ and $\mathfrak{q} = \mathfrak{p} + \mathfrak{k}/2$

$$\begin{aligned} \mathrm{tr}_{\mathbb{C}^4}Q_1(\mathfrak{x},\mathfrak{y}) &= (2\pi)^{-3}\int_{\mathbb{R}^3}\mathrm{d}\mathfrak{r}\int_{\mathbb{R}^3}\mathrm{d}\mathfrak{q}e^{\mathrm{i}\mathfrak{r}\cdot\mathfrak{x}}\mathrm{tr}_{\mathbb{C}^4}\hat{Q}_1(\mathfrak{r},\mathfrak{q})e^{-\mathrm{i}\mathfrak{q}\cdot\mathfrak{y}} \\ &= (2\pi)^{-3}\int\mathrm{d}\mathfrak{p}\int\mathrm{d}\mathfrak{k}\mathrm{tr}_{\mathbb{C}^4}\hat{Q}_1(\mathfrak{p}-\mathfrak{k}/2,\mathfrak{p}+\mathfrak{k}/2)e^{\mathrm{i}\mathfrak{p}\cdot(\mathfrak{x}-\mathfrak{y})}e^{-\mathrm{i}\mathfrak{k}\cdot(\mathfrak{x}+\mathfrak{y})/2} \\ &=: \tilde{Q}\left(\mathfrak{x}-\mathfrak{y},\frac{\mathfrak{x}+\mathfrak{y}}{2}\right). \end{aligned} \tag{54}$$

Setting $\boldsymbol{\xi} := \mathfrak{x} - \mathfrak{y}$ we remark: the "limits" $\lim_{\mathfrak{y}\to\mathfrak{x}}\mathrm{tr}_{\mathbb{C}^4}Q_1(\mathfrak{x},\mathfrak{y})$ and $\lim_{\xi\to 0}\tilde{Q}(\xi,\mathfrak{x})$ are formally the same. Define $\hat{\tilde{Q}}(\xi,\cdot)$ to be the Fourier transform of $\tilde{Q}$ with respect to the second variable for fixed $\xi \neq 0$, i.e., formally

$$\begin{aligned} \hat{\tilde{Q}}(\xi,\mathfrak{k}) &= (2\pi)^{-3/2}\int_{\mathbb{R}^3}\mathrm{d}\mathfrak{p}\ \mathrm{tr}_{\mathbb{C}^4}\hat{Q}_1(\mathfrak{p}-\mathfrak{k}/2,\mathfrak{p}+\mathfrak{k}/2)e^{\mathrm{i}\mathfrak{p}\cdot\boldsymbol{\xi}} \\ &= \frac{1}{4\pi^3}\hat{\varphi}(\mathfrak{k})\int_{\mathbb{R}^3}\mathrm{d}\mathfrak{p}\frac{\mathfrak{p}^2-\frac{\mathfrak{k}^2}{4}+1-E(\mathfrak{p}-\frac{\mathfrak{k}}{2})E(\mathfrak{p}+\frac{\mathfrak{k}}{2})}{E(\mathfrak{p}-\frac{\mathfrak{k}}{2})E(\mathfrak{p}+\frac{\mathfrak{k}}{2})\left(E(\mathfrak{p}-\frac{\mathfrak{k}}{2})+E(\mathfrak{p}+\frac{\mathfrak{k}}{2})\right)}e^{\mathrm{i}\mathfrak{p}\cdot\boldsymbol{\xi}}. \end{aligned} \tag{55}$$

We note that the integral (55) is logarithmically divergent at $\boldsymbol{\xi} = 0$ independently of the form of the external potential φ. This shows – as already remarked above – that the limit $\lim_{\mathfrak{y}\to\mathfrak{x}} \mathrm{tr}_{\mathbb{C}^4} Q^\varphi(\mathfrak{x},\mathfrak{y})$ only exists, if φ vanishes.

According to Pauli and Rose [22], (5)–(9), we can separate $\hat{\hat{Q}}$ into two terms

$$\hat{\hat{Q}}(\boldsymbol{\xi},\mathfrak{k}) = F_1(\boldsymbol{\xi},\mathfrak{k}) + \hat{\varphi}(\mathfrak{k})\mathfrak{k}^2 F_0(\boldsymbol{\xi},\mathfrak{k}/|\mathfrak{k}|)\,, \tag{56}$$

with F_0 as defined in (34). F_1 is finite at $\boldsymbol{\xi} = 0$. We have

$$\begin{aligned}\widehat{\rho_{1,r}}(\mathfrak{k}) &= F_1(0,\mathfrak{k})\\ &= \frac{\hat{\varphi}(\mathfrak{k})}{4\pi^3}\int_{\mathbb{R}^3} \frac{\mathfrak{p}^2 - \mathfrak{k}^2/4 + 1 - E(\mathfrak{p}-\frac{\mathfrak{k}}{2})E(\mathfrak{p}+\frac{\mathfrak{k}}{2})}{E(\mathfrak{p}-\frac{\mathfrak{k}}{2})E(\mathfrak{p}+\frac{\mathfrak{k}}{2})\left(E(\mathfrak{p}-\frac{\mathfrak{k}}{2})+E(\mathfrak{p}+\frac{\mathfrak{k}}{2})\right)} + \mathfrak{k}^2\frac{\mathfrak{p}^2\sin^2\theta + 1}{4E(\mathfrak{p})^5}\mathrm{d}\mathfrak{p}\\ &= \frac{1}{4\pi^2}\hat{\varphi}(\mathfrak{k})C(\mathfrak{k})\,,\end{aligned} \tag{57}$$

where C is the function defined in (51). While all summands in the latter formula decreases like $|\mathfrak{p}|^{-3}$ for large $|\mathfrak{p}|$ and therefore the corresponding parts of the integral are logarithmically divergent, the integrand decreases as $|\mathfrak{p}|^{-5}$.

The Uehling potential is obviously the relevant part of $|\cdot|^{-1} * \rho_1$. Recall that the nuclear potential is $\varphi = |\cdot|^{-1} * n$; consequently $\hat{\varphi}(\mathfrak{k}) = 4\pi\hat{n}(\mathfrak{k})/\mathfrak{k}^2$. Thus the Fourier transform of the vacuum polarization potential using (57) gives

$$\hat{U}(\mathfrak{k}) = \frac{\hat{\varphi}(\mathfrak{k})C(\mathfrak{k})}{\pi|\mathfrak{k}|^2} = 4\frac{\hat{n}(\mathfrak{k})C(\mathfrak{k})}{|\mathfrak{k}|^4} \tag{58}$$

which is spherically symmetric and compactly supported under our assumptions on the charge distributions of the nucleus.

Under our general assumptions on the nuclear charge density the Uehling potential U is bounded, continuous, and decreasing exponentially at infinity. Thus, it is plainly bounded. □

Eventually we need to bound the exchange operator X. To formulate the next result we fix the following notation: let $C_{p,q}$ be the optimal constant in the generalized Young inequality, i.e., $\|f*g\|_r \leq C_{p,q}\|f\|_{p,w}\|g\|_q$, $1 < p,q,r < \infty$, $r^{-1}+1 = p^{-1}+q^{-1}$.

Lemma 5. *Let $\psi \in L^3(G)\cap L^2(G)$. Then*

$$\left|\int \mathrm{d}x \int \mathrm{d}y \frac{\overline{\psi(x)}X(x,y)\psi(y)}{|\mathfrak{x}-\mathfrak{y}|}\right| \leq \sqrt{C_{3/2,3/2}\|1/|\cdot|^2\|_{3/2,w}}\|Q^\varphi\|_2\|\psi\|_3^2\,, \tag{59}$$

and for every $\epsilon > 0$ there exists a constant $C_\epsilon > 0$ such that

$$\left|\int \mathrm{d}x \int \mathrm{d}y \frac{\overline{\psi(x)}Q^\varphi(x,y)\psi(y)}{|\mathfrak{x}-\mathfrak{y}|}\right| \leq \epsilon\|\psi\|_3^2 + C_\epsilon\|\psi\|_2^2\,. \tag{60}$$

We note that Lemma 5 implies by Sobolev's inequality that the form $(\psi, X\psi)$ is relatively bounded to $(\psi, D_\varphi\psi)$ with form bound zero.

Proof. We first note that $X(x,y) = Q^\varphi(x,y)/|\mathfrak{x}-\mathfrak{y}|$. Since Q^φ is a Hilbert-Schmidt operator we get using the Schwarz inequality

$$\begin{aligned} L &:= \left| \int \mathrm{d}x \int \mathrm{d}y \frac{\overline{\psi(x)} Q^\varphi(x,y)\psi(y)}{|\mathfrak{x}-\mathfrak{y}|} \right| \\ &\leq \left(\int \mathrm{d}x \int \mathrm{d}y \frac{|\psi(x)|^2 |\psi(y)|^2}{|\mathfrak{x}-\mathfrak{y}|^2} \right)^{1/2} \left(\int \mathrm{d}x \int \mathrm{d}y |Q^\varphi(x,y)|^2 \right)^{1/2} . \end{aligned} \tag{61}$$

The second factor of the right hand side is the Hilbert-Schmidt norm $\|Q^\varphi\|_2$ of Q^φ.

To estimate the first factor we decompose the kernel into two functions $f(\mathfrak{x}) := \chi_{B_R(0)}(\mathfrak{x})/|\mathfrak{x}|^2$ and the rest g, i.e., $1/|\mathfrak{x}|^2 = f(\mathfrak{x}) + g(\mathfrak{x})$.

Thus, using inequality (61) we get

$$L \leq \left[(|\psi|^2 * f, |\psi|^2)^{1/2} + (|\psi|^2 * g, |\psi|^2)^{1/2} \right] \|Q^\varphi\|_2 . \tag{62}$$

The first summand yields by the Hölder and the generalized Young inequality (see, e.g., Reed and Simon II, p. 32)

$$(|\psi|^2 * f, |\psi|^2) \leq C_{3/2,3/2} \|\psi^2\|_{3/2}^2 \|f\|_{3/2,w} , \tag{63}$$

where w indicates the weak-norm.

Picking the radius $R = \infty$, i.e., $g = 0$, yields immediately (59).

To prove (60) we also use (63) but pick the radius $R > 0$ sufficiently small: in this case we need to bound also the second summand containing g; we use again Hölder's inequality now followed by using Young's inequality

$$(|\psi|^2 * g, |\psi|^2) \leq \|\psi^2\|_1^2 \|g\|_\infty . \tag{64}$$

Thus, the first factor on the right side of (33) is bounded by

$$\sqrt{C_{3/2,3/2}} \|\psi^2\|_{3/2} \|f\|_{3/2,w}^{1/2} + \|\psi^2\|_1 \|g\|_\infty^{1/2} .$$

Since $\|f\|_{3/2,w}$ tends to zero as R tends to zero, the claimed inequality follows.

Taken together these Lemmata prove Theorem 1.

5 The Many Electron-Positron Problem

To simplify matters we drop all the renormalization terms of the previous chapter and simply consider the fully normal ordered Hamiltonian $\mathbb{H} = \mathbb{D}_\varphi + \alpha\mathbb{W}$ with the electron space to be specified.

5.1 The One-Particle Density Matrix

For later purposes (suggested by (87)) it will be useful to introduce certain operators γ and Υ in $\mathfrak{H}$ with kernels

$$\gamma_\rho(x,y) := \rho(:\psi^*(x)\psi(y):) \tag{65}$$
$$\Upsilon_\rho(x,y) := \rho(:\psi(y)\psi(x):)\ , \tag{66}$$

where ρ is any given state. The operator γ is called the **one-particle density matrix** (1-pdm) of the state ρ. It has the property

$$-\Lambda_- \leq \gamma \leq \Lambda_+\ , \tag{67}$$
$$D_0\gamma \in \mathfrak{S}_1(\mathfrak{H}), \tag{68}$$

if ρ has finite kinetic energy. We write $\mathfrak{S}_\varphi$ for the set of states fulfilling (67) and (68). The trace $q := -e\,\mathrm{tr}\,\gamma$ is called the charge of γ.

As we will see, these operators will allow us to express one-particle expectations in a non-field theoretic way similar to expressions known from atomic physics.

5.2 Unboundedness for Dilatationally Invariant Electron Subspaces

We will suppose that $\mathfrak{H}_\pm^{\varphi_q}$ are invariant under dilatations and assume vanishing mass, i.e., $m = 0$. Otherwise no specific requirements on $\mathfrak{H}_\pm^{\varphi_q}$ and the associated orthogonal projections $\Lambda_\pm$ are made. Solovej suggested to consider

$$f := \sqrt{1-\epsilon^2}|0\rangle + \epsilon|++--\rangle\ , \tag{69}$$

where $|++--\rangle$ is a two-electron-two-positron state with non-vanishing matrix element $\mathrm{Re}\,\langle 0|\mathbb{W}|++--\rangle \neq 0$. We get

$$(f, \mathbb{H}f) = 2\epsilon \mathrm{Re}\,\langle 0|\mathbb{W}|++--\rangle + O(\epsilon^2) \tag{70}$$

for $\epsilon \to 0$. Thus, choosing the sign of ϵ and picking ϵ close to 0, we can make $(f, \mathbb{H}f) < 0$. Then, dilating the state multiplies the energy by a common positive factor that can be chosen arbitrarily which means that the energy is unbounded from below.

5.3 The Hartree-Fock Approximation

Hartree-Fock States

We consider the set H of states $\rho \in \mathfrak{B}(\mathfrak{F})'$ with the following properties

- For any product of an odd number of creation and annihilation operators $d_1, d_2, \cdots d_{2k-1}$
$$\rho(d_1 \cdots d_{2k-1}) = 0 \tag{71}$$
- For any four creation and annihilation operators a, b, c, d
$$\rho(abcd) = \rho(ab)\rho(cd) - \rho(ac)\rho(bd) + \rho(ad)\rho(bc) \tag{72}$$
- The kinetic energy $\rho(\mathbb{T})$ (with quantization with respect to the free Dirac operator, i.e., $\varphi_q = 0$, is finite.

We call the elements in H Hartree-Fock states or quasi-free states.

An example of Hartree-Fock states are the well known Slater determinants:

Lemma 6. *Let $e_1, \ldots, e_n \in \mathfrak{H}_+ \cap H^1(G)$ be n orthonormal spinors,*
$$s := (n!)^{-1/2} e_1 \wedge \cdots \wedge e_n = a^*(e_1) \cdots a^*(e_n)|0\rangle$$
the corresponding Slater determinant, and ρ_s its state, i.e.,
$$\rho_s(A) := (s, As)\ .$$

Proof. 1. Take $2k-1$ creation and annihilation operators $d_1, \cdots, d_{2k-1}$. Then
$$\rho_s(d_1 \cdots d_{2k-1}) = \langle 0|a(e_n) \cdots a(e_1) d_1 \cdots d_{2k-1} a^*(e_1) \cdots a^*(e_n)|0\rangle$$
is the vacuum expectation of an odd number of creation and annihilation operators and thus vanishing.

2. Let a, b, c, and d be four creation or annihilation operators. We can – modulo ordering – either have two positron creation and two positron annihilation operators, or one of each kind, or two electron creation and two electron creation operators. In all other cases both sides of the claimed inequality are vanishing trivially. The cases when positron operators are present are easy, since they anticommute with electron operators. We therefore concentrate on the case of two electron creators and two electron annihilators and assume $f, g, h, i \in \mathfrak{H}_+$. (We will also assume that the product $d_1, \ldots, d_4$ is already normal ordered, since the other cases differ from this one by only two simple additional terms, a term that is quadratic in the operators a, b, c, d and a constant.) We write $f = \sum_\nu \alpha_\nu e_\nu$, $g = \sum_\nu \beta_\nu e_\nu$,$h = \sum_\nu \gamma_\nu e_\nu$,and $i = \sum_\nu \delta_\nu e_\nu$. Then the claimed inequality is

$$\langle 0|a(e_n) \cdots a(e_1) a^*(f) a^*(g) a(h) a(i) a^*(e_1) \cdots a^*(e_n)|0\rangle \tag{73}$$
$$= \langle 0|a(e_n) \cdots a(e_1) a^*(f) a^*(g) a^*(e_1) \cdots a^*(e_n)|0\rangle \tag{74}$$
$$\times \langle 0|a(e_n) \cdots a(e_1) a(h) a(i) a^*(e_1) \cdots a^*(e_n)|0\rangle \tag{75}$$
$$-\langle 0|a(e_n) \cdots a(e_1) a^*(f) a(h) a^*(e_1) \cdots a^*(e_n)|0\rangle \tag{76}$$
$$\times \langle 0|a(e_n) \cdots a(e_1) a^*(g) a(i) a^*(e_1) \cdots a^*(e_n)|0\rangle \tag{77}$$
$$+\langle 0|a(e_n) \cdots a(e_1) a^*(f) a(i) a^*(e_1) \cdots a^*(e_n)|0\rangle \tag{78}$$
$$\times \langle 0|a(e_n) \cdots a(e_1) a^*(g) a(h) a^*(e_1) \cdots a^*(e_n)|0\rangle\ . \tag{79}$$

Lines (74) and (75) vanish, since the number of creation operators and the number of annihilation operators is not equal. Thus the claim is equivalent to

$$\begin{aligned}
&\langle 0|a(e_n)\cdots a(e_1)a^*(f)a^*(g)a(h)a(i)a^*(e_1)\cdots a^*(e_n)|0\rangle && (80)\\
&= -\langle 0|a(e_n)\cdots a(e_1)a^*(f)a(h)a^*(e_1)\cdots a^*(e_n)|0\rangle && (81)\\
&\quad\times \langle 0|a(e_n)\cdots a(e_1)a^*(g)a(i)a^*(e_1)\cdots a^*(e_n)|0\rangle && (82)\\
&\quad+\langle 0|a(e_n)\cdots a(e_1)a^*(f)a(i)a^*(e_1)\cdots a^*(e_n)|0\rangle && (83)\\
&\quad\times \langle 0|a(e_n)\cdots a(e_1)a^*(g)a(h)a^*(e_1)\cdots a^*(e_n)|0\rangle\,. && (84)
\end{aligned}$$

Inserting the expansion and observing that the summands that contain an index κ, λ, μ, and ν that is bigger than n, vanish we get

$$\begin{aligned}
(73) &= \sum_{\kappa,\lambda,\mu,\nu\in\{1,\ldots n\}} \alpha_\kappa\beta_\lambda\gamma_\mu\delta_\nu \\
&\qquad \langle 0|a(e_n)\cdots a(e_1)a^*(e_\kappa)a^*(e_\lambda)a(e_\mu)a(e_\nu)a^*(e_1)\cdots a^*(e_n)|0\rangle\\
&= \sum_{\kappa,\lambda\in\{1,\ldots n\}} \alpha_\kappa\beta_\lambda(\gamma_\lambda\delta_\kappa-\gamma_\kappa\delta_\lambda)\\
&\qquad \langle 0|a(e_n)\cdots a(e_1)a^*(e_\kappa)a^*(e_\lambda)a(e_\lambda)a(e_\kappa)a^*(e_1)\cdots a^*(e_n)|0\rangle\\
&= \sum_{\kappa,\lambda\in\{1,\ldots n\}} \alpha_\kappa\beta_\lambda(\gamma_\kappa\delta_\lambda-\gamma_\lambda\delta_\kappa)\,.
\end{aligned} \tag{85}$$

On the other hand

$$\begin{aligned}
(81)+\ldots+(84) &= \sum_{\kappa,\lambda\in\{1,\ldots n\}} (-\alpha_\kappa\beta_\lambda\gamma_\kappa\delta_\lambda+\alpha_\kappa\delta_\kappa\beta_\lambda\gamma_\lambda)\\
&\qquad \langle 0|a(e_n)\cdots a(e_1)a^*(e_\kappa)a(\kappa)a^*(e_1)\cdots a^*(e_n)|0\rangle\\
&\qquad \langle 0|a(e_n)\cdots a(e_1)a^*(\lambda)a(\lambda)a^*(e_1)\cdots a^*(e_n)|0\rangle\\
&= \sum_{\kappa,\lambda\in\{1,\ldots n\}} (-\alpha_\kappa\beta_\lambda\gamma_\kappa\delta_\lambda+\alpha_\kappa\delta_\kappa\beta_\lambda\gamma_\lambda)
\end{aligned} \tag{86}$$

which equals (85). □

The interaction energy in a quasi-free state ρ is

$$
\begin{aligned}
\rho(\mathbb{W}) &= \frac{1}{2}\int_{\mathbb{R}^3} \mathrm{d}\mathfrak{x} \int_{\mathbb{R}^3} \mathrm{d}\mathfrak{y} \frac{\rho(:\psi^*(\mathfrak{x})\psi(\mathfrak{x})\psi^*(\mathfrak{y})\psi(\mathfrak{y})):}{|\mathfrak{x}-\mathfrak{y}|} \\
&= W_D + W_P - W_X \\
&:= \frac{1}{2}\int_{\mathbb{R}^3} \mathrm{d}\mathfrak{x} \int_{\mathbb{R}^3} \mathrm{d}\mathfrak{y} \frac{\rho(:\psi^*(\mathfrak{x})\psi(\mathfrak{x}):)\rho(:\psi^*(\mathfrak{y})\psi(\mathfrak{y})):)}{|\mathfrak{x}-\mathfrak{y}|} \\
&\quad + \frac{1}{2}\int_{\mathbb{R}^3} \mathrm{d}\mathfrak{x} \int_{\mathbb{R}^3} \mathrm{d}\mathfrak{y} \frac{\rho(:\psi^*(\mathfrak{x})\psi^*(\mathfrak{y}):)\rho(:\psi(\mathfrak{y})\psi(\mathfrak{x})):)}{|\mathfrak{x}-\mathfrak{y}|} \\
&\quad - \frac{1}{2}\int_{\mathbb{R}^3} \mathrm{d}\mathfrak{x} \int_{\mathbb{R}^3} \mathrm{d}\mathfrak{y} \frac{\rho(:\psi^*(\mathfrak{x})\psi(\mathfrak{y}):)\rho(:\psi^*(\mathfrak{y})\psi(\mathfrak{x})):)}{|\mathfrak{x}-\mathfrak{y}|},
\end{aligned} \tag{87}
$$

where we use that the normal ordering distributes according to the rules of the Hartree-Fock states, a fact due to Hundertmark (see [23]).

Note: all three integrals, W_D, the direct energy, W_P, the pairing energy, and W_X, the exchange energy, are positive.

The Hartree-Fock Functional

Using the above, the energy in Hartree-Fock states ρ is

$$
\rho(\mathbb{H}) = \underbrace{\mathrm{tr}(D_\varphi\gamma_\rho) + \underbrace{\frac{\alpha}{2}\int_G \mathrm{d}x \int_G \mathrm{d}y \frac{\gamma_\rho(x,x)\gamma_\rho(y,y) - |\gamma_\rho(x,y)|^2}{|\mathfrak{x}-\mathfrak{y}|}}_{\alpha Q(\gamma_\rho):=\alpha D(\gamma_\rho,\gamma_\rho)-\alpha X(\gamma_\rho,\gamma_\rho):=}}_{\mathcal{E}(\gamma_\rho):=}
$$

$$
+ \frac{\alpha}{2}\int_G \mathrm{d}x \int_G \mathrm{d}y \frac{\overline{\Upsilon_\rho(x,y)}\Upsilon_\rho(x,y)}{|\mathfrak{x}-\mathfrak{y}|} \tag{88}
$$

We can define $\mathcal{E}$ on $\mathfrak{S}_\varphi$ and call it the **relativistic Hartree-Fock functional.**

Positivity of the Energy

For any Hartree-Fock state ρ there exists a Hartree-Fock state σ such that $\gamma_\rho = \gamma_\sigma$ and $\Upsilon_\sigma = 0$. Moreover, for any $\gamma \in \mathfrak{S}_\varphi$, there exists a Hartree Fock state ρ such that $\gamma_\rho = \gamma$ and $\Upsilon_\rho = 0$. Thus, the minimization of Hartree-Fock states is equivalent to the minimization of $\rho(\mathbb{H})$ over Hartree-Fock states.

Since $-\Lambda_- \leq \gamma \leq \Lambda_+$ is equivalent to $0 \leq \Gamma := \Lambda_- + \gamma \leq 1$ we have

$$
\Gamma^2 \leq \Gamma \tag{89}
$$

Introducing $\Gamma_{++} := \Lambda_+\Gamma\Lambda_+$, $\Gamma_{+-} := \Lambda_+\Gamma\Lambda_-$, $\Gamma_{-+} := \Lambda_-\Gamma\Lambda_+$, and $\Gamma_{--} := \Lambda_-\Gamma\Lambda_-$ (and analogously for other operators) Inequality (89) implies

$$
\gamma_{++}^2 + \gamma_{+-}\gamma_{-+} = \Gamma_{++}^2 + \Gamma_{+-}\Gamma_{-+} \leq \Gamma_{++} = \gamma_{++} \tag{90}
$$

$$
\gamma_{--}^2 + 2\gamma_{--} + \gamma_{-+}\gamma_{+-} = \Gamma_{--}^2 + \Gamma_{-+}\Gamma_{+-} \leq \Gamma_{--} = \gamma_{--} + \Lambda_- \tag{91}
$$

or equivalently

$$\gamma_{++}^2 + \gamma_{+-}\gamma_{-+} \le \gamma_{++} \tag{92}$$
$$\gamma_{--}^2 + \gamma_{-+}\gamma_{+-} \le -\gamma_{--} \,. \tag{93}$$

With the inequalities (92) and (93), picking $\varphi_q = \varphi$ and the fact that the direct part is non-negative we have

$$E(\gamma) \ge \mathrm{tr}(D_\varphi \gamma) - \alpha X(\gamma,\gamma) \ge \mathrm{tr}(|D_\varphi|\gamma^2) - \alpha X(\gamma,\gamma) \,. \tag{94}$$

$$\begin{aligned} X(\gamma,\gamma) &= \frac{1}{2}\int_G \mathrm{d}x \int_G \mathrm{d}y |\gamma(x,y)|^2 |\mathfrak{x}-\mathfrak{y}|^{-1} \\ &\le \frac{\pi}{4}\int_G \mathrm{d}y (\gamma(\cdot,y), |\mathfrak{p}|\gamma(\cdot,y)) \le \frac{\pi}{4}\mathrm{tr}(|D_0|\gamma^2) \,. \end{aligned} \tag{95}$$

This shows in the case $\varphi_q = \varphi = 0$ that the critical fine structure constant is at least $\pi/4$. In fact this value is sharp. One can prove

Theorem 2 (Bach et al. [3], Hundertmark et al. [15]). *Assume $\varphi_q = \varphi = 0$. Then $\forall_{\gamma \in \mathfrak{S}_\varphi}\ \mathcal{E}(\gamma) \ge 0$ if and only if $0 \le \alpha \le 4/\pi$.*

This result – with a constant that is not sharp – was heuristically derived by Chaix and Iracane [6] and Chaix et al. [7]. The above proof that the critical constant is at least $4/\pi$ which is due to Bach et al. [3]; the reverse direction is a variational calculation due to Hundertmark et al. [15] which partly follows an idea of Evans et al. [10].

Inequalities (94) and (95) can be also useful for showing positivity for general $\varphi = \varphi_q$. However, in this case, one needs to compare $|D_0|$ and $|D_\varphi|$. This can in fact be done in some physically important cases. We will treat here the case $\varphi = eZ/|\cdot|$ with a robust proof giving a positivity result for small Z.

We need an inequality of the type $|D_0| \le c|D_Z|$, where $D_Z = D\varphi$ for some nonnegative c. This is implied by $|D_0|^2 \le c^2|D_Z|^2$, since taking roots is operator monotone. This is equivalent to $\|D_0\psi\| \le c\|D_Z\psi\|$ for all $\psi \in H^1(G)$. Therefore, we have

$$\|D_Z\psi\| \ge \|D_0\psi\| - \alpha Z\|\psi|\cdot|^{-1}\| \ge (1-2\alpha Z)\|D_0\psi\| \,. \tag{96}$$

Thus the desired inequality holds, if $\alpha Z \le 1/2$. In fact one can show

Theorem 3 (Brummelhuis et al. [5]). *Assume $\varphi = \varphi_q = \sqrt{\alpha}Z/|\cdot|$. If $\alpha \le (4/\pi)(1-g^2)^{1/2}(\sqrt{4g^2+9}-4g)/3$. Then $\mathcal{E}(\gamma) \ge 0$ for all $\gamma \in \mathfrak{S}_\varphi$.*

In fact, this has been further generalized. We now set D_g for the operator $-\mathrm{i}\boldsymbol{\alpha}\cdot\nabla + m\beta - g/|\cdot|$:

Theorem 4 (Morozov [20]). *For all $a, b \in (3/2, 3/2)$ and for all $m \geq 0$ we have*

$$D_a^2 \geq \nu_{a,b} D2_b \, ,$$

where $\nu_{a,b}$ is the maximal value of ν satisfying the two inequalities

$$\nu + \frac{C_{a,b}^2 (a - \nu b)^2}{(C_{a,b}^2 - \nu) C_{a,a-\nu b}^2} \leq 1 \text{ and } 0 \leq \nu \leq C_{a,b}^2$$

with

$$C_{a,b} := \frac{1}{3 - 4b^2} \left(-4|b - a| + \sqrt{16(b-a)^2 + (3 - 4b^2)(3 - 4a^2)} \right) .$$

We would like to mention at this case, that the choice of the Furry picture, i.e., picking the quantization potential φ_q to be the external potential φ, is not arbitrary. In fact this choice gives the most stable ground state energy, a choice suggested by Mittleman [19]. We prove here a variant of a Theorem of Bach et al. [3].

It is also possible to include the case of several nuclei and of a magnetic field (Bach et al. [2]). In this case $\mathfrak{A}_q := \mathfrak{A}$ and

$$\varphi(\mathfrak{x}) := \sum_{\kappa=1}^{K} \frac{e Z_\kappa}{|\mathfrak{x} - \mathfrak{R}_\kappa|} , \tag{97}$$

where $Z_1, \ldots, Z_K$ are the atomic numbers of the involved nuclei and $\mathfrak{R}_1, \ldots, \mathfrak{R}_\kappa$ are their positions. A convenient quantization potential φ_q is given by through the help of a Voronoi decomposition and includes also the magnetic vector potential:

$$\varphi_q(\mathfrak{x}) := - \sum_{\kappa=1}^{K} \frac{e Z_\kappa \chi_{\Upsilon_\kappa}(\mathfrak{x})}{|\mathfrak{x} - \mathfrak{R}_\kappa|} . \tag{98}$$

Here $\Upsilon_\kappa := \{\mathfrak{x} \in \mathbb{R}^3 : |\mathfrak{x} - \mathfrak{R}_\kappa| \leq |\mathfrak{x} - \mathfrak{R}_k|, \forall k = 1, \ldots, K\}$ denotes the κ-th Voronoi cell and χ_M is the characteristic function of the set M, and $Z := \max\{Z_1, \ldots Z_K\}$. The result, also called stability of matter for this model, is

Theorem 5. *Pick $\mathfrak{H}_+ := [\chi_{[0,\infty)}(D_{\varphi_q, \mathfrak{A}})](\mathfrak{H})$ as electron subspace. Let $L_{1/2,3}$ be the constant in the Lieb-Thirring inequality*[2] *for moments of order $1/2$. If $\epsilon \in (0,1)$, $\alpha \in [0, 4/\pi]$ and $Z \in [0, \infty)$ are such that*

$$1 - \epsilon - \pi^2 \alpha^2 / 16 - 4(1/\epsilon - 1) \alpha^2 Z^2 > 0 \, ,$$

and

$$\frac{26296 \pi L_{1/2,3} (1/\epsilon - 1)^2}{105 (1 - \epsilon - \pi^2 \alpha^2/16 - 4(1/\epsilon - 1)\alpha^2 Z^2)^{3/2}} \alpha^3 Z^2 \leq 1 \, ,$$

---

[2] See Appendix A.1.

$\mathfrak{R}_1, \ldots, \mathfrak{R}_K \in \mathbb{R}^3$ *pairwise different, and* $|\nabla \times \mathfrak{A}| \in L^2(\mathbb{R}^3)$ *then*

$$\mathcal{E}(\gamma) + \sum_{1 \leq \kappa < \lambda \leq K} \frac{\alpha Z_\kappa Z_\lambda}{|\mathfrak{R}_\kappa - \mathfrak{R}_\lambda|} + \frac{1}{8\pi} \int_{\mathbb{R}^3} |\nabla \times \mathfrak{A}(\mathfrak{x})|^2 d\mathfrak{x} \geq 0$$

for all $\gamma \in \mathfrak{S}_{\varphi_q, \mathfrak{A}}$.

Note also that ϵ is a free parameter that we can use to optimize the value of α and Z. Numerical evaluation gives for the physical value $\alpha \approx 1/137.0359895$ stability up to $\alpha Z \approx 0.489576$, i.e., $Z \approx 67.089649$.

The Optimal Quantization

We define $\mathcal{S}$ to be the set of all orthogonal projections Λ such that Λ and $\Lambda^\perp$ leaves the domain of D_φ invariant. We consider

$$E_\Lambda := \inf_{-\Lambda \leq \gamma \leq \Lambda^\perp} \mathcal{E}(\gamma) \, . \tag{99}$$

Theorem 6 (Bach et al. [3]). *Consider* D_φ*, where the electric potential is generated by an electric charge of finite self-interaction. For any* Λ *in* $\mathcal{S}$ *that is not equal to* $\Lambda^\varphi := \chi_{(-\infty,0)}(D_\varphi)$ *we get*

$$E(\Lambda) < E(\Lambda^\varphi) = 0 \, . \tag{100}$$

Note, that this result can be strengthened in many ways (see, e.g., the original result [3]).

The Selfconsistent Vacuum Polarization in the Mean Field Picture

Recently Hainzl, Lewin, and Séré [11, 12] have treated the vacuum polarization of the Electron-Positron field in the mean-field approximation. Firstly, we will review their result partly using our language. Secondly, we would like to show that the self-consistency requirement of Hainzl et al. is in fact nothing but a different formulation of the minmax requirement of Mittleman [19] when the vacuum polarization term is included.

The Minimization of the Energy in the Free Picture

In order to treat the vacuum polarization the model is regularized by introducing a form factor for the nucleus and an ultra-violet cutoff for the electron-positron field at momentum $|\mathfrak{p}| \leq a$. In particular one requires that the Coulomb norm $D(n, n)$ of the charge distribution n of the nucleus is finite. We use – as before – the following notation: Γ is the unrenormalized density matrix and γ is the charge density matrix with respect to the free quantization. Equipped with the above regularization we can go out and think of minimizing $\mathcal{E}(\Gamma - \Lambda_0)$, since we have using (94) and (95)

$$\begin{aligned}\mathcal{E}(\gamma) &= \operatorname{tr}(|D_0||\gamma|) - \alpha X(\gamma,\gamma) - \alpha D(n,\rho_\gamma) + D(\rho_\gamma,\rho_\gamma) \\ &\geq \alpha\left(D(\rho_\gamma - n, \rho_\gamma - n) - D(n,n)\right) \geq -\alpha D(n,n)\,, \quad (101)\end{aligned}$$

i.e., the energy is bounded from below. With this starting point, however, it cannot be expected that the minimizer γ is a trace class operator. In particular, the energy will be negative implying that $\gamma = 0$ is not the minimizer. Hainzl et al. show that the minimizer Γ fulfills the equation

$$\Gamma = \chi_{(-\infty,0)}(D_\varphi + \alpha W_\gamma))\,, \tag{102}$$

where $W_\gamma \psi = \rho_\gamma * |\cdot|^{-1}\psi - X_\gamma \psi$. Here X_γ is the exchange operator of the density matrix γ which has the integral kernel $X_\gamma(x,y) := \gamma(x,y)/|\mathfrak{x}-\mathfrak{y}|$. This $\Gamma = \Lambda_0 + \gamma$ should then be viewed as the density of the polarized vacuum. Of course – to connect with the usual normalization – the energy should be shifted such that the energy for the minimizing Γ is vanishing.

Equation (102) can indeed be solved as Hainzl et al. show. In [12] the equation is solved for small fine structure constant α and momentum cutoff a by a fixed point argument. In [11] the equation is solved variationally for all values of α and a.

The Optimal Quantization of the Electron-Positron Field in Hartree-Fock Approximation: Implementing the Minimax Procedure of Mittleman [19]

As described in Sect. 4.1 the vacuum polarization is mathematically stemming from the one-particle operators that occur in the difference of the non-normal ordered and normal ordered selfinteraction of the electron-positron field with the subtraction of the free quantities, i.e., the operators in the last line of Equation (32). Thus, the corresponding Hartree-Fock functional is the functional $\mathcal{E}$ of (88) plus the trace of the one-particle vacuum polarization potential, i.e.,

$$\mathcal{E}_{vpq:}(\gamma) := \mathcal{E}(\gamma) + \alpha \operatorname{tr}(P_{\varphi_q}\gamma)\,, \tag{103}$$

where

$$P_0 := |\cdot|^{-1} * \operatorname{tr}_{\mathbb{C}^4}(\Lambda_-^{\varphi_q} - \Lambda_-^0) - X_{\Lambda_-^{\varphi_q} - \Lambda_-^0}\,. \tag{104}$$

As remarked earlier, the operator P_0 is not well defined as it stands, since $\Lambda_-^{\varphi_q} - \Lambda_-^0)$ is infinite on the diagonal.

If we disregard this problem for the moment and follow the suggestion of Mittleman [19] to search for the projection Λ_- that gives the most stable quantization. By using the optimality result, Theorem 6 of Bach, Barbaroux, Helffer, Siedentop the maximal ground state energy is obtained when the

the quantization is taken with respect to the negative spectral subspace of the one-particle terms of the Hamiltonian. Using this fact the optimal Λ_- is determined by

$$\Lambda_- = \chi_{(-\infty,0)}\left(D_\varphi + \alpha|\cdot|^{-1} * \mathrm{tr}_{\mathbb{C}^4}(\Lambda_- - \Lambda_-^0) - \alpha X_{\Lambda_- - \Lambda_-^0}\right) . \tag{105}$$

This (102) has been also obtained by Hainzl, Lewin, and Séré [11, 12] for minimizing the total energy including the Dirac sea.

In order to make the expressions meaningful, we need to manipulate these terms once more. There are several ways of doing this. Here we will discuss two ways: (i) An ultra-violet cutoff, i.e., straight forward ad hoc regularization of the problem and (ii) charge renormalization, i.e., adding a counting term on the diagonal following the physical principle W3 of Weisskopf.

Ultra-Violet Cutoff: In this case we can immediately use the results by Hainzl et al. [11, 12]. They show that the equation has a (unique) solution by two methods: a variational method provided the cutoff is small enough, and a variational method. In particular the solution of this equation is not only the minimizer of the energy in the ultra-violet cutoff free picture; it also maximizes the energy in the sense of Mittleman and the corresponding Hartree-Fock ground state energy is zero.

Charge Renormalization: Instead of introducing an ad hoc ultra-violet cutoff, we can follow the principle of Weisskopf and apply W3. The treatment of the vacuum polarization as treated in Sect. 4 shows, that the charge renormalization introduced there makes the vacuum potential finite as well. Using again the optimality result (Theorem 6), (36) suggests the following choice of the quantization

$$\Lambda_-^{\varphi_q} = \chi_{(-\infty,0)}\left(D_\varphi + \alpha|\cdot|^{-1} * \rho - \alpha X\right) , \tag{106}$$

where, as before, $\rho(\mathfrak{x}) = \mathrm{tr}_{\mathbb{C}^4}(\Lambda_-^{\varphi_q} - \Lambda_-^0 - \gamma_c)(\mathfrak{x},\mathfrak{x})$. Note that this deviates from the one in the previous subsection by the extra regularizing term on the direct part. Hainzl et al. obtain this (in cut-off version) from the non-charge regularized equation by redefining the coupling constants (apart from a sign in C ([12, Formula (21)]).

A Appendix

A.1 The Lieb-Thirring Inequality

($d = 3$, $\gamma = 1/2$) Given a positive constant μ, a real vector field $\mathbf{A}$ with square integrable gradients, and a real valued function V in $L^2(\mathbb{R}^3)$, we have for $V_+ := (|V| + V)/2$

$$\mathrm{tr}\left\{\left[(-\mathrm{i}\mu\nabla - \mathbf{A})^2 - V\right]_-^{1/2}\right\} \leq \frac{L_{1/2,3}}{\mu^3}\int_{\mathbb{R}^3} V_+^2$$

(see Lieb and Thirring [18] for the case $\mathbf{A} = 0$ and Avron, Herbst, and Simon [1] for the general case).

References

1. J. Avron, I. Herbst, and B. Simon. Schrödinger operators with magnetic fields. I. general interactions. *Duke Math. J.*, 45(4):847–833, December 1978.
2. Volker Bach, Jean-Marie Barbaroux, Bernard Helffer, and Heinz Siedentop. Stability of matter for the Hartree-Fock functional of the relativistic electron-positron field. *Doc. Math.*, 3:353–364 (electronic), 1998.
3. Volker Bach, Jean-Marie Barbaroux, Bernard Helffer, and Heinz Siedentop. On the stability of the relativistic electron-positron field. *Comm. Math. Phys.*, 201:445–460, 1999.
4. G.E. Brown and D.G. Ravenhall. On the interaction of two electrons. *Proc. Roy. Soc. London Ser. A.*, 208:552–559, 1951.
5. Raymond Brummelhuis, Norbert Röhrl, and Heinz Siedentop. Stability of the relativistic electron-positron field of atoms in Hartree-Fock approximation: Heavy elements. *Doc. Math., J. DMV*, 6:1–8, 2001.
6. P. Chaix and D. Iracane. From quantum electrodynamics to mean-field theory: I. The Bogoliubov-Dirac-Fock formalism. *J. Phys. B.*, 22(23):3791–3814, December 1989.
7. P. Chaix, D. Iracane, and P.L. Lions. From quantum electrodynamics to mean-field theory: II. Variational stability of the vacuum of quantum electrodynamics in the mean-field approximation. *J. Phys. B.*, 22(23):3815–3828, December 1989.
8. P.-A.-M. Dirac. Théorie du positron. In Cockcroft, J. Chadwick, F. Joliot, J. Joliot, N. Bohr, G. Gamov, P.A.M. Dirac, and W. Heisenberg, editors, *Structure et propriétés des noyaux atomiques. Rapports et discussions du septieme conseil de physique tenu à Bruxelles du 22 au 29 octobre 1933 sous les auspices de l'institut international de physique Solvay. Publies par la commission administrative de l'institut.*, pp. 203–212. Paris: Gauthier-Villars. XXV, 353 S., 1934.
9. P.A.M. Dirac. Discussion of the infinite distribution of electrons in the theory of the positron. *Proc. Camb. Philos. Soc.*, 30:150–163, 1934.
10. William Desmond Evans, Peter Perry, and Heinz Siedentop. The spectrum of relativistic one-electron atoms according to Bethe and Salpeter. *Comm. Math. Phys.*, 178(3):733–746, July 1996.
11. Christian Hainzl, Mathieu Lewin, and Éric Séré. Self-consistent solution for the polarized vacuum in a no-photon qed model. *lanl.arXiv.org*, http://xxx.lanl.gov/pdf/physics/0404047, 2004.
12. Christian Hainzl, Mathieu Lewin, and Éric Séré. Existence of a stable polarized vacuum in the Bogoliubov-Dirac-Fock approximation. *Commun. Math. Phys.*, 257(3), 2005.
13. Christian Hainzl and Heinz Siedentop. Non-perturbative mass and charge renormalization in relativistic no-photon quantum electrodynamics. *Comm. Math. Phys.*, 243(2):241–260, 2003.
14. W. Heisenberg. Bemerkungen zur Diracschen Theorie des Positrons. *Z. Phys.*, 90:209–231, 1934.
15. Dirk Hundertmark, Norbert Röhrl, and Heinz Siedentop. The sharp bound on the stability of the relativistic electron-positron field in Hartree-Fock approximation. *Comm. Math. Phys.*, 211(3):629–642, May 2000.

16. Tosio Kato. *Perturbation Theory for Linear Operators*, volume 132 of *Grundlehren der mathematischen Wissenschaften*. Springer-Verlag, Berlin, 1 edition, 1966.
17. Elliott H. Lieb, Heinz Siedentop, and Jan Philip Solovej. Stability and instability of relativistic electrons in classical electromagnetic fields. *J. Statist. Phys.*, 89(1-2):37–59, 1997. Dedicated to Bernard Jancovici.
18. Elliott H. Lieb and Walter E. Thirring. Inequalities for the moments of the eigenvalues of the Schrödinger Hamiltonian and their relation to Sobolev inequalities. In Elliott H. Lieb, Barry Simon, and Arthur S. Wightman, editors, *Studies in Mathematical Physics: Essays in Honor of Valentine Bargmann*. Princeton University Press, Princeton, 1976.
19. Marvin H. Mittleman. Theory of relativistic effects on atoms: Configuration-space Hamiltonian. *Phys. Rev. A*, 24(3):1167–1175, September 1981.
20. Sergey Morozov. Extension of a minimax principle for Coulomb-Dirac operators. Master's thesis, Mathematisches Institut, Ludwig-Maximilians-Universität, Theresienstr. 39, 80333 München, Germany, August 2004.
21. G. Nenciu. Self-adjointness and invariance of the essential spectrum for Dirac operators defined as quadratic forms. *Comm. Math. Phys.*, 48:235–247, 1976.
22. W. Pauli and M.E. Rose. Remarks on the Polarization Effects in the Positron Theory. *Phys. Rev., II. Ser.*, 49:462–465, 1936.
23. Norbert Röhrl. *Stabilität und Instabilität des relativistischen Elektronen-Positronen-Felds in Hartree-Fock-Näherung*. Typoskript-Edition. Hieronymus, München, 1 edition, December 2000.
24. Robert Serber. Linear modifications in the Maxwell field equations. *Phys. Rev., II. Ser.*, 48:49–54, 1935.
25. Barry Simon. *Functional Integration and Quantum Physics*. Academic Press Inc. [Harcourt Brace Jovanovich Publishers], New York, 1979.
26. J. Sucher. Foundations of the relativistic theory of many-electron atoms. *Phys. Rev. A*, 22(2):348–362, August 1980.
27. J. Sucher. Foundations of the relativistic theory of many-electron bound states. *International Journal of Quantum Chemistry*, 25:3–21, 1984.
28. J. Sucher. Relativistic many-electron Hamiltonians. *Phys. Scripta*, 36:271–281, 1987.
29. E.A. Uehling. Polarization effects in the positron theory. *Phys. Rev., II. Ser.*, 48:55–63, 1935.
30. V. Weisskopf. Über die Elektrodynamik des Vakuums auf Grund der Quantentheorie des Elektrons. *Math.-Fys. Medd., Danske Vid. Selsk.*, 16(6):1–39, 1936.

Index

Lecture Notes in Physics

For information about earlier volumes
please contact your bookseller or Springer
LNP Online archive: springerlink.com

Vol.648: S. G. Karshenboim, E. Peik (Eds.), Astrophysics, Clocks and Fundamental Constants

Vol.649: M. Paris, J. Rehacek (Eds.), Quantum State Estimation

Vol.650: E. Ben-Naim, H. Frauenfelder, Z. Toroczkai (Eds.), Complex Networks

Vol.651: J. S. Al-Khalili, E. Roeckl (Eds.), The Euroschool Lectures of Physics with Exotic Beams, Vol.I

Vol.652: J. Arias, M. Lozano (Eds.), Exotic Nuclear Physics

Vol.653: E. Papantonoupoulos (Ed.), The Physics of the Early Universe

Vol.654: G. Cassinelli, A. Levrero, E. de Vito, P. J. Lahti (Eds.), Theory and Appplication to the Galileo Group

Vol.655: M. Shillor, M. Sofonea, J. J. Telega, Models and Analysis of Quasistatic Contact

Vol.656: K. Scherer, H. Fichtner, B. Heber, U. Mall (Eds.), Space Weather

Vol.657: J. Gemmer, M. Michel, G. Mahler (Eds.), Quantum Thermodynamics

Vol.658: K. Busch, A. Powell, C. Röthig, G. Schön, J. Weissmüller (Eds.), Functional Nanostructures

Vol.659: E. Bick, F. D. Steffen (Eds.), Topology and Geometry in Physics

Vol.660: A. N. Gorban, I. V. Karlin, Invariant Manifolds for Physical and Chemical Kinetics

Vol.661: N. Akhmediev, A. Ankiewicz (Eds.) Dissipative Solitons

Vol.662: U. Carow-Watamura, Y. Maeda, S. Watamura (Eds.), Quantum Field Theory and Noncommutative Geometry

Vol.663: A. Kalloniatis, D. Leinweber, A. Williams (Eds.), Lattice Hadron Physics

Vol.664: R. Wielebinski, R. Beck (Eds.), Cosmic Magnetic Fields

Vol.665: V. Martinez (Ed.), Data Analysis in Cosmology

Vol.666: D. Britz, Digital Simulation in Electrochemistry

Vol.667: W. D. Heiss (Ed.), Quantum Dots: a Doorway to Nanoscale Physics

Vol.668: H. Ocampo, S. Paycha, A. Vargas (Eds.), Geometric and Topological Methods for Quantum Field Theory

Vol.669: G. Amelino-Camelia, J. Kowalski-Glikman (Eds.), Planck Scale Effects in Astrophysics and Cosmology

Vol.670: A. Dinklage, G. Marx, T. Klinger, L. Schweikhard (Eds.), Plasma Physics

Vol.671: J.-R. Chazottes, B. Fernandez (Eds.), Dynamics of Coupled Map Lattices and of Related Spatially Extended Systems

Vol.672: R. Kh. Zeytounian, Topics in Hyposonic Flow Theory

Vol.673: C. Bona, C. Palenzula-Luque, Elements of Numerical Relativity

Vol.674: A. G. Hunt, Percolation Theory for Flow in Porous Media

Vol.675: M. Kröger, Models for Polymeric and Anisotropic Liquids

Vol.676: I. Galanakis, P. H. Dederichs (Eds.), Half-metallic Alloys

Vol.677: A. Loiseau, P. Launois, P. Petit, S. Roche, J.-P. Salvetat (Eds.), Understanding Carbon Nanotubes

Vol.678: M. Donath, W. Nolting (Eds.), Local-Moment Ferromagnets

Vol.679: A. Das, B. K. Chakrabarti (Eds.), Quantum Annealing and Related Optimization Methods

Vol.680: G. Cuniberti, G. Fagas, K. Richter (Eds.), Introducing Molecular Electronics

Vol.681: A. Llor, Statistical Hydrodynamic Models for Developed Mixing Instability Flows

Vol.682: J. Souchay (Ed.), Dynamics of Extended Celestial Bodies and Rings

Vol.683: R. Dvorak, F. Freistetter, J. Kurths (Eds.), Chaos and Stability in Planetary Systems

Vol.684: J. Dolinšek, M. Vilfan, S. Žumer (Eds.), Novel NMR and EPR Techniques

Vol.685: C. Klein, O. Richter, Ernst Equation and Riemann Surfaces

Vol.686: A. D. Yaghjian, Relativistic Dynamics of a Charged Sphere

Vol.687: J. W. LaBelle, R. A. Treumann (Eds.), Geospace Electromagnetic Waves and Radiation

Vol.688: M. C. Miguel, J. M. Rubi (Eds.), Jamming, Yielding, and Irreversible Deformation in Condensed Matter

Vol.689: W. Pötz, J. Fabian, U. Hohenester (Eds.), Quantum Coherence

Vol.690: J. Asch, A. Joye (Eds.), Mathematical Physics of Quantum Mechanics

Vol.691: S. S. Abdullaev, Construction of Mappings for Hamiltonian Systems and Their Applications

Vol.692: J. Frauendiener, D. J. W. Giulini, V. Perlick (Eds.), Analytical and Numerical Approaches to Mathematical Relativity

Vol.694: H. Schwoerer, J. Magill, B. Beleites (Eds.), Lasers and Nuclei

Vol.695: J. Dereziński, H. Siedentop (Eds.), Large Coulomb Systems